Monographs in Computer Science

Editors

David Gries
Fred B. Schneider

Monographs in Computer Science

Abadi and Cardelli, **A Theory of Objects**

Benosman and Kang [editors], **Panoramic Vision: Sensors, Theory, and Applications**

Broy and Stølen, **Specification and Development of Interactive Systems: FOCUS on Streams, Interfaces, and Refinement**

Brzozowski and Seger, **Asynchronous Circuits**

Burgin, **Super-Recursive Algorithms**

Cantone, Omodeo, and Policriti, **Set Theory for Computing: From Decision Procedures to Declarative Programming with Sets**

Castillo, Gutiérrez, and Hadi, **Expert Systems and Probabilistic Network Models**

Downey and Fellows, **Parameterized Complexity**

Feijen and van Gasteren, **On a Method of Multiprogramming**

Herbert and Spärck Jones [editors], **Computer Systems: Theory, Technology, and Applications**

Leiss, **Language Equations**

Levin, Heydon, and Mann, **Software Configuration Management with VESTA**

McIver and Morgan [editors], **Programming Methodology**

McIver and Morgan [editors), **Abstraction, Refinement and Proof for Probabilistic Systems**

Misra, **A Discipline of Multiprogramming: Programming Theory for Distributed Applications**

Nielson [editor], **ML with Concurrency**

Paton [editor], **Active Rules in Database Systems**

Selig, **Geometrical Methods in Robotics**

Selig, **Geometric Fundamentals of Robotics, Second Edition**

Shasha and Zhu, **High Performance Discovery in Time Series: Techniques and Case Studies**

Tonella and Potrich, **Reverse Engineering of Object Oriented Code**

J.M. Selig

Geometric Fundamentals of Robotics

Second Edition

J.M. Selig
London South Bank University
Faculty of Business, Computing and Information Management
London, SE1 0AA
U.K.
seligjm@lsbu.ac.uk

Series Editors:
David Gries
Cornell University
Department of Computer Science
Ithaca, NY 14853
U.S.A.

Fred B. Schneider
Cornell University
Department of Computer Science
Ithaca, NY 14853
U.S.A.

Mathematics Subject Classification (2000): 70B15, 70E60, 53A17, 22E99

ISBN 978-1-4419-1929-8   e-ISBN 978-0-387-27274-0

Printed on acid-free paper.

©2005 Springer Science+Business Media Inc.
Softcover reprint of the hardcover 2nd edition 2005
Based on *Geometrical Methods in Robotics,* Springer New York ©1996.

9 8 7 6 5 4 3 2 1

springeronline.com

To Kathy

Preface

This book is an extended and corrected version of an earlier work, "Geometrical Methods in Robotics" published by Springer-Verlag in 1996. I am extremely glad of the opportunity to publish this work which contains many corrections and additions. The extra material, two new chapters and several new sections, reflects some of the advances in the field over the past few years as well as some material that was missed in the original work.

As before this book aims to introduce Lie groups and allied algebraic and geometric concepts to a robotics audience. I hope that the power and elegance of these methods as they apply to problems in robotics is still clear. By now the pioneering work of Ball is well known. However, the work of Study and his colleagues is not so widely appreciated, at least not in the English speaking world. This book is also an attempt to bring at least some of their work to the attention of a wider audience.

In the first four chapters, a careful exposition of the theory of Lie groups and their Lie algebras is given. All examples used to illustrate these ideas, except for the simplest ones, are taken from robotics. So, unlike most standard texts on Lie groups, emphasis is placed on a group that is not semi-simple—the group of proper Euclidean motions in three dimensions. In particular, the continuous subgroups of this group are found, and the elements of its Lie algebra are identified with the surfaces of the lower Reuleaux pairs. These surfaces were first identified by Reuleaux in the latter half of the 19th century. They allow us to associate a Lie algebra element to every basic mechanical joint. The motions allowed by the joint are then just the one-parameter subgroups generated by the

Lie algebra element. A detailed study of the exponential map and its derivative is given for the rotation and rigid body motion groups.

Chapter 5 looks at some geometrical problems that are basic to robotics and the theory of mechanisms. Having developed in the previous chapter the description of robot kinematics using exponentials of Lie algebra elements, these ideas are used to generalise and simplify some standard results in kinematics. The chapter looks at the kinematics of 3-joint wrists and 3-joint regional manipulators.

Some of the classical theory of ruled surfaces and line complexes is introduced in Chapter 6. This material also benefits from the Lie algebra point of view. For robotics, the most important ruled surfaces are the cylindrical hyperboloid and the cylindroid. A full description of these surfaces is given.

In Chapter 7, the theory of group representations is introduced. Once again, the emphasis is on the group of proper Euclidean motions. Many representations of this group are used in robotics. A benefit of this is that it allows a concise statement and proof of the 'Principle of Transference', a result that, until recently, had the status of a 'folk theorem' in the mechanism theory community.

Ball's theory of screws underlies much of the work in this book. Ball's treatise was written at the turn of the twentieth century, just before Lie's and Cartan's work on continuous groups. The infinitesimal screws of Ball can now be seen as elements of the Lie algebra of the group of proper Euclidean motions. In Chapter 8, on screw systems, the linear subspaces of this Lie algebra are explored. The Gibson–Hunt classification of these systems is derived using a group theoretic approach.

Clifford algebra is introduced in Chapter 9. Again, attention is quickly specialised to the case of the Clifford algebra for the group of proper Euclidean motions. This is something of an esoteric case in the standard mathematical literature, since it is the Clifford algebra of a degenerate bilinear form. This algebra is a very efficient vehicle for carrying out computation both in the group and in some of its geometrical representations. Moreover, it allows us to define the Study quadric, an algebraic variety that contains the elements of the group of proper Euclidean motions.

Chapter 10 explores this Clifford algebra in more detail. It is shown how points, lines and planes can be represented in this algebra, and how geometric operations can be modelled by algebraic operations in the algebra. The results are used to look at the kinematics of six-joint industrial robots and prove an important theorem concerning designs of robots that have solvable inverse kinematics.

The Study quadric is more fully explored in Chapter 11, where its subspaces and quotients are examined in some depth. The intersection theory of the variety is introduced and used to solve some simple enumerative problems like the number of postures of the general 6-R robot.

Chapters 12, 13 and 14 cover the statics and dynamics of robots. The dual space to the Lie algebra is identified with the space of wrenches, that is, force-torque vectors. This facilitates a simple description of some standard problems in robotics, in particular, the problem of gripping solid objects. The group theory helps to isolate the surfaces that cannot be completely immobilised without friction. They turn out to be exactly the surfaces of the lower Reuleaux pairs.

In order to deal with the dynamics of robots, the inertia properties of rigid bodies must be studied. In standard dynamics texts, the motion of the centre of mass and the rotation about the centre of mass are treated separately. For robots, it is more convenient to use a six-dimensional notation, which does not separate the rotational and translational motion. This leads to a six-by-six inertia matrix for a rigid body and also allows a modern exposition of some ideas due to Ball, namely conjugate screws and principal screws of inertia. The standard theory of robot dynamics is presented in two ways, first as a simple Newtonian-style approach, and then using Lagrangian dynamics. The Lagrangian approach leads to a simple study of small oscillations of the end-effector of a robot and reintroduces what Ball termed harmonic screws. The neat formalism used means that the equations of motion for a simple robot can be studied quite easily. This advantage is used to look at the design of robots with a view to simplifying their dynamics. Several approaches to this problem are considered.

The dynamics of robots with end-effector constraints and the dynamics of robots with star structures is also investigated. This allows the description of the dynamics of parallel manipulators and some simple examples of these are presented.

In Chapter 15 some deeper applications of differential geometry are explored. Three applications are studied: the mobility of overconstrained mechanisms, the control of robots along geodesic paths, and hybrid control.

The original book was never intended as an encyclopedic account of "robot geometry", but over the last few years this field has expanded so much that it is no longer even feasible to catalogue the omissions. The criterion for selecting material for this book is still a reliance on the methods outlined in the first few chapters of the book, essentially elementary differential geometry.

However, one omission that I would like to mention is the field of robot vision. A central problem in robot vision is to find the rigid motion undergone by the camera using information derived from the images. There are many other interesting geometric problems in this area, see Kanatani [61] for example. I feel that this area is so large and with very specific problems that it deserves separate treatment.

I would like to thank the many people who pointed out errors in the original book, in particular Charles Wampler, Andreas Ruf and Ross McAree. I met Pertti Lounesto shortly before his untimely death in 2002. Naturally he found an error in the chapter on Clifford algebra in the original book, but this is almost a source of pride for me. His plans to apply his considerable knowledge and skill to mathematical problems in robotics were tragically cut short.

It is also with sadness that I report that Ken Hunt and Joe Duffy both passed away in 2002. Both made substantial contributions to the fields of robotics and kinematics and both will be greatly missed.

London 2003

J.M. Selig

seligjm@lsbu.ac.uk

Contents

Preface   **vii**

1 Introduction   **1**
  1.1 Theoretical Robotics?   1
  1.2 Robots and Mechanisms   2
  1.3 Algebraic Geometry   4
  1.4 Differential Geometry   7

2 Lie Groups   **11**
  2.1 Definitions and Examples   12
  2.2 More Examples — Matrix Groups   15
    2.2.1 The Orthogonal Group $O(n)$ 15
    2.2.2 The Special Orthogonal Group $SO(n)$ 16
    2.2.3 The Symplectic Group $Sp(2n, \mathbb{R})$ 17
    2.2.4 The Unitary Group $U(n)$ 18
    2.2.5 The Special Unitary Group $SU(n)$ 18
  2.3 Homomorphisms   18
  2.4 Actions and Products   21
  2.5 The Proper Euclidean Group   23
    2.5.1 Isometries   23
    2.5.2 Chasles's Theorem   25
    2.5.3 Coordinate Frames   27

3 Subgroups   **31**
  3.1 The Homomorphism Theorems 31
  3.2 Quotients and Normal Subgroups 34
  3.3 Group Actions Again 36
  3.4 Matrix Normal Forms 37
  3.5 Subgroups of $SE(3)$ 41
  3.6 Reuleaux's Lower Pairs 44
  3.7 Robot Kinematics 46

4 Lie Algebra   **51**
  4.1 Tangent Vectors 51
  4.2 The Adjoint Representation 54
  4.3 Commutators 57
  4.4 The Exponential Mapping 61
      4.4.1 The Exponential of Rotation Matrices 63
      4.4.2 The Exponential in the Standard Representation of $SE(3)$ 66
      4.4.3 The Exponential in the Adjoint Representation of $SE(3)$ 68
  4.5 Robot Jacobians and Derivatives 71
      4.5.1 The Jacobian of a Robot 71
      4.5.2 Derivatives in Lie Groups 73
      4.5.3 Angular Velocity 75
      4.5.4 The Velocity Screw 76
  4.6 Subalgebras, Homomorphisms and Ideals 77
  4.7 The Killing Form 80
  4.8 The Campbell–Baker–Hausdorff Formula 81

5 A Little Kinematics   **85**
  5.1 Inverse Kinematics for 3-R Wrists 85
  5.2 Inverse Kinematics for 3-R Robots 89
      5.2.1 Solution Procedure 89
      5.2.2 An Example 92
      5.2.3 Singularities 94
  5.3 Kinematics of Planar Motion 98
      5.3.1 The Euler–Savaray Equation 101
      5.3.2 The Inflection Circle 103
      5.3.3 Ball's Point 104
      5.3.4 The Cubic of Stationary Curvature 105
      5.3.5 The Burmester Points 106
  5.4 The Planar 4-Bar 108

6 Line Geometry   **113**
  6.1 Lines in Three Dimensions 113
  6.2 Plücker Coordinates 115
  6.3 The Klein Quadric 117
  6.4 The Action of the Euclidean Group 119

6.5 Ruled Surfaces 123
    6.5.1 The Regulus 124
    6.5.2 The Cylindroid 126
    6.5.3 Curvature Axes 128
6.6 Line Complexes 130
6.7 Inverse Robot Jacobians 133
6.8 Grassmannians 135

7 Representation Theory   **139**
7.1 Definitions 139
7.2 Combining Representations 142
7.3 Representations of $SO(3)$ 148
7.4 $SO(3)$ Plethyism 151
7.5 Representations of $SE(3)$ 153
7.6 The Principle of Transference 158

8 Screw Systems   **163**
8.1 Generalities 163
8.2 2-systems 167
    8.2.1 The Case $\mathbb{R}^2$ 169
    8.2.2 The Case $SO(2) \times \mathbb{R}$ 169
    8.2.3 The Case $SO(3)$ 170
    8.2.4 The Case $H_p \ltimes \mathbb{R}^2$ 170
    8.2.5 The Case $SE(2)$ 171
    8.2.6 The Case $SE(2) \times \mathbb{R}$ 171
    8.2.7 The Case $SE(3)$ 172
8.3 3-systems 175
    8.3.1 The Case $\mathbb{R}^3$ 176
    8.3.2 The Case $SO(3)$ 176
    8.3.3 The Case $SE(2)$ 176
    8.3.4 The Case $H_p \ltimes \mathbb{R}^2$ 177
    8.3.5 The Case $SE(2) \times \mathbb{R}$ 177
    8.3.6 The Case $SE(3)$ 177
8.4 Identification of Screw Systems 183
    8.4.1 1-systems and 5-systems 183
    8.4.2 2-systems 184
    8.4.3 4-systems 188
    8.4.4 3-systems 189
8.5 Operations on Screw Systems 193

9 Clifford Algebra   **197**
9.1 Geometric Algebra 199
9.2 Clifford Algebra for the Euclidean Group 206
9.3 Dual Quaternions 210
9.4 Geometry of Ruled Surfaces 214

10 A Little More Kinematics   221
  10.1 Clifford Algebra of Points, Lines and Planes 221
      10.1.1 Planes 221
      10.1.2 Points 222
      10.1.3 Lines 223
  10.2 Euclidean Geometry 224
      10.2.1 Incidence 224
      10.2.2 Meets 225
      10.2.3 Joins—The Shuffle product 226
      10.2.4 Perpendicularity—The Contraction 228
  10.3 Pieper's Theorem 231
      10.3.1 Robot Kinematics 231
      10.3.2 The T^3 Robot 234
      10.3.3 The PUMA 238

11 The Study Quadric   241
  11.1 Study's Soma 241
  11.2 Linear Subspaces 245
      11.2.1 Lines 245
      11.2.2 3-planes 246
      11.2.3 Intersections of 3-planes 248
      11.2.4 Quadric Grassmannians 250
  11.3 Partial Flags and Projections 252
  11.4 Some Quadric Subspaces 255
  11.5 Intersection Theory 256
      11.5.1 Postures for General 6-R Robots 262
      11.5.2 Conformations of the 6–3 Stewart Platform 264
      11.5.3 The Tripod Wrist 266
      11.5.4 The 6-6 Stewart Platform 267

12 Statics   271
  12.1 Co-Screws 271
  12.2 Forces, Torques and Wrenches 272
  12.3 Wrist Force Sensor 274
  12.4 Wrench at the End-Effector 276
  12.5 Gripping 278
  12.6 Friction 283

13 Dynamics   287
  13.1 Momentum and Inertia 287
  13.2 Robot Equations of Motion 292
      13.2.1 Equations for a Single Body 292
      13.2.2 Serial Robots 293
      13.2.3 Change in Payload 296
  13.3 Recursive Formulation 296

13.4 Lagrangian Dynamics of Robots 300
    13.4.1 Euler–Lagrange Equations 301
    13.4.2 Derivatives of the Generalised Inertia Matrix 303
    13.4.3 Small Oscillations 304
13.5 Hamiltonian Dynamics of Robots 306
13.6 Simplification of the Equations of Motion 309
    13.6.1 Decoupling by Design 309
    13.6.2 Ignorable Coordinates 312
    13.6.3 Decoupling by Coordinate Transformation 316

14 Constrained Dynamics   **321**
14.1 Trees and Stars 321
    14.1.1 Dynamics of Tree and Star Structures 323
    14.1.2 Link Velocities and Accelerations 324
    14.1.3 Recursive Dynamics for Trees and Stars 325
14.2 Serial Robots with End-Effector Constraints 327
    14.2.1 Holonomic Constraints 327
    14.2.2 Constrained Dynamics of a Rigid Body 330
    14.2.3 Constrained Serial Robots 331
14.3 Constrained Trees and Stars 333
    14.3.1 Systems of Freedom 333
    14.3.2 Parallel Mechanisms 334
14.4 Dynamics of Planar 4-Bars 336
14.5 Biped Walking 340
14.6 The Stewart Platform 343

15 Differential Geometry   **349**
15.1 Metrics, Connections and Geodesics 349
15.2 Mobility of Overconstrained Mechanisms 355
15.3 Controlling Robots Along Helical Trajectories 360
15.4 Hybrid Control 363
    15.4.1 What is Hybrid Control? 363
    15.4.2 Constraints 364
    15.4.3 Projection Operators 365
    15.4.4 The Second Fundamental Form 369

References   **373**

Index   **383**

Geometric Fundamentals
of Robotics

1
Introduction

1.1 Theoretical Robotics?

In May 2000 there was a meeting at the National Science Foundation in Arlington Virginia on "The Interplay between Mathematics and Robotics". Many leading experts in the U.S. discussed the importance of mathematics in robotics and also the role that robotic problems could play in the development of mathematics. The experts gave a broad overview of the problems they saw as important and worth studying. Their list was long and touched on many branches of mathematics and many areas in robotics.

Robotics is a practical discipline. It grew out of engineers' ability to build very sophisticated machines that combine computer control with electro-mechanical actuators and sensors. Any theory in the subject must take account of what is practically possible with real machines. Nevertheless, there is clearly a place for a theoretical side to the subject.

Of course, by definition theory is always useless, otherwise it wouldn't be theory! But surely all disciplines recognise the need for sound theoretical underpinnings. The question really is whether the theoretical underpinnings of robotics are distinct or just a part of the general theory used in the disciplines that make up robotics. One cannot sensibly separate say, a theory of robot mechanisms from the general theory of mechanisms and linkages. However, there is something special about robotics and that is the central importance of the group of rigid body motions $SE(3)$. That is not to say that theory not involving this group is not robotics nor that other disciplines can't profitably use this group. Its just that I see this as a major theme running through much of robotics: The links of a robot are not really rigid, but to a first approximation they are. The motions allowed by the joints of the robot are rigid body motions. The payload carried by the robot's end-effector is more often than not a rigid body. Standard analysis of the kinematics, dynamics and control of these robots all reflect this rigid body approach.

In essence, this book is concerned with the geometry of the group of rigid body motions $SE(3)$, and its applications robotics. Before embarking on this we look at a little history and background.

1.2   Robots and Mechanisms

Although modern industrial robots are only a few decades old, their antecedents are mechanisms and linkages, which have centuries of history. Indeed, examples like cranes and other lifting devices are among the oldest machines used by humans.

Such machines consist of rigid links connected by joints. The inspiration for these devices may well have come from animal skeletons. The slow but steady development and innovation of such mechanisms has continued throughout history and in all parts of the world. However, the industrial revolution in Europe in the late eighteenth century created a huge demand for these devices, a demand that was met in excess.

One of the most pressing problems was how to turn the reciprocating motion generated by piston engines into rotary motion. Many people, including Watt, proposed approximate straight-line linkages. However, it was not until 1864 that a French naval officer named Peaucellier invented an eight-bar mechanism that does the job exactly. By that time, machining methods had become more accurate and lubrication technology had improved, so the practical importance of the discovery had disappeared. However, mathematicians were intrigued by the machine since the design was easily adapted to mechanically invert curves. Hence, it could be used to study 'inversion geometry'. It also begged the question, Just which curves can be traced by linkages?

In 1876, Kempe, the London solicitor and amateur mathematician, proved that all algebraic curves can be traced by mechanisms. Later, Koenigs proved a similar theorem for curves in space. At around this time, many mathematicians were interested in the theory of mechanisms—Chebychev, Schonflies and Darboux, to name but a few. By far the most intensively studied machine was and probably remains the 4-bar mechanism. The device is ubiquitous in mechanical engineering, as it is an extraordinary design element. It has been used for everything from door hinges to the tilting mechanism for high speed trains. Even so, there still remain some unanswered questions concerning the geometrical capabilities of this mechanism.

Mathematicians were also interested in more general problems. Clifford developed geometric algebras, modelled on Hamilton's quaternions.

Around the turn of the twentieth century, Ball developed screw theory, which dealt with infinitesimal rigid body motions and was mainly used to look at problems concerning statics and dynamics.

A little later, Study looked at the geometry of the set of all finite rigid body motions. For this work he invented dual quaternions, which could represent translations as well as rotations.

After the first world war, mathematicians seem to have turned away from the study of mechanisms. There have been notable exceptions; see, for example, Thurston and Weeks [120]. However, it would be true to say that the subject remains a backwater, away from the main development of modern mathematics.

Mechanical engineers also neglected mechanisms during this period. Although the machines remained of great practical importance, academic engineers were more interested in other areas. The notable exception here was the Moscow school, led by Artobolevskii. In particular, Dimentberg was using screw theory in 1948 to analyse closed kinematic chains.

In the 1950s, Freudenstein, at Columbia University, began a revival of the subject in the west. More or less simultaneously Hunt and Phillips, at Monash and Sydney respectively, championed Ball's screw theory for analysing spatial mechanisms.

Perhaps the main difference between robots and mechanisms is that mechanisms are usually designed for particular functions and hence usually have only a few, normally just one, degree of freedom. Robots are supposed to be general purpose machines with, consequently, many degrees of freedom. The geometrical analyses of mechanisms and robots are to all intents and purposes the same. Geometry here refers to both kinematics and dynamics. Kinematics studies the possible movements the machine can make, irrespective of any forces or inertias, while dynamics looks at how the machine will move, taking forces and inertias into account.

In 1954, Devol patented what he called a "programmed articulated transfer device". This machine derived from telechirs used in the nuclear industry for the remote handling of radioactive materials and computer (numerically) controlled machine tools. Engleberger, a Columbia University student, realised that Devol's machine was essentially a robot. In 1956, he bought the patent and set up Unimation. The first industrial robot was installed by General Motors in 1961. In 1968, Kawasaki bought a licence from Unimation to manufacture robots in Japan. In 1978, Unimation introduced the PUMA, an acronym for 'programmable universal machine for assembly'. This robot was the result of a study for General Motors on automating assembly tasks. In 1979, SCARA (selective compliance assembly robot arm) was introduced in Japan for assembly of printed circuit boards. The mechanisms community was well placed to contribute to the study and design of the new machines.

In the 1980s, the subject of robotics attracted a lot of funding and hence the attention of other researchers. In particular, electrical engineers became interested in the problem of controlling robots. Computer scientists saw robots as vehicles for testing their ideas about artificial intelligence. The recession of the late 1980s meant a decline in the funding for robotics. However, there remained a substantial interest in the subject in many disciplines.

The 1990s saw the robotics industry moving back into profit. However, there do not seem to have been any major innovations in the robots sold to the manufacturing industry. The academic robotics community grew very large, supporting several major international conferences each year. The large size of the community reflects a large diversity of interests. So it is difficult to discern any particularly strong themes in robotics research. However, there was a lot of interest in robot surgery and, after the success of the sojourner mars rover, in mobile robots.

As mentioned at the beginning of this introduction, the theme of 'mathematical robotics' has reached some sort of maturity. There are several researchers around the world interested in these problems and one or two books in the area. However, at the time of writing I don't think it could be called a community or a fully fledged subject area.

In the following couple of sections, we briefly review some modern geometry. We cannot expect to do justice to these large subjects in a couple of short sections, and so the reader is advised to consult the texts cited at the end of each section. This material will be more or less assumed in the rest of the book, so these sections should be viewed as defining our terminology and notation.

1.3 Algebraic Geometry

The subject of algebraic geometry has a long history. It goes back to the work of Descartes, who introduced coordinates and described curves by equations. Even Newton studied cubic curves.

In the nineteenth century, algebraic geometry was more or less synonymous with the study of invariants. The invariants in question were quantities that remain unchanged after linear changes of coordinates—"collineations" in the old terminology. Examples are discriminants like the $B^2 - AC$ of a quadratic function. Hilbert solved most of the problems of classical invariant theory in a ground breaking series of papers in the early 1890s. To do this, he introduced many new algebraic methods. Despite the objections of some of his contemporaries, these ideas are now the bedrock of the subject, and they paved the way for much of the rest of twentieth century mathematics. In particular, we consider the ring of polynomial functions in several variables. The solutions to systems of equations correspond to an ideal in the ring, that is, subsets of polynomials closed under the operations of addition and multiplication by arbitrary polynomials in the ring. Geometrically, the ideal representing the set of common zeros of a system of polynomials consists of all polynomials that also vanish on the set. Of course, the polynomials of the ideal may also vanish elsewhere. There followed a long period when methods of commutative algebra were applied to geometrical problems. This resulted in many notable successes; in particular, sound definitions for the dimension and degree of these algebraic sets were developed.

In more modern times, the subject has undergone yet another revision of its foundations. In the 1960s, Grothendieck introduced schemes, which generalised the notion of algebraic sets. The language of schemes is extremely powerful and very general, and there is no doubt that it will eventually find applications in robotics as well as in many other fields.

However, in this work we will take a fairly traditional view of the subject. This is done to make the work easily accessible and to stay close to the practical applications we want to study.

As hinted above, the basic object of study is the set of solutions to an algebraic equation or system of algebraic equations. We will call such a set an algebraic set or a **variety**. Note, however, that many authors reserve the word "variety" for an irreducible algebraic set. An algebraic set is irreducible if it cannot be decomposed into the union of two or more algebraic sets. The irreducible sets of a variety will be referred to as its components.

If we consider the variables in the polynomials as coordinates of an affine space, then the solution set forms an **affine variety**. Usually, we can make better progress if the polynomials are homogeneous, that is, all terms having the same total degree. In this case, we may think of the variables as homogeneous coordinates in some projective space. A projective space has homogeneous coordinates $(x_0 : x_1 : \cdots : x_n)$. If the coordinates can be complex, we write the space as $\mathbb{PC}^n$ when the coordinate must be real numbers, then we denote the space $\mathbb{PR}^n$. Unlike affine coordinates, it is only the ratios of the coordinates that are important. The point referred to by a set of homogeneous coordinates is unaffected if we multiply all the coordinates by a non-zero constant. Another way to think of projective space $\mathbb{PC}^n$ is as the set of lines through the origin in $\mathbb{C}^{n+1}$. This is closer to the origins of the subject, which began with the study of perspective in drawings.

Notice that the homogeneous polynomials change when the homogeneous coordinates undergo multiplication by a constant. But the set of zeros of a homogeneous polynomial is independent of such a transformation. Hence, such a set defines a subset of a projective space. Thus, we may define a **quasi-projective variety** as the set of zeros for a system of homogeneous polynomials.

The variety defined by the zeros of a single homogeneous equation forms a space with one less dimension than the projective space we are working in. Such a space is often referred to as a hypersurface.

If the homogeneous equations are all linear, that is, of degree 1, then we can use the results of linear algebra to describe the situation. A single linear equation determines a hyperplane with dimension $n-1$. In general, a collection of j hyperplanes will intersect in a linear space of dimension $n-j$. In particular, n hyperplanes will normally intersect in a single point. Notice that a linear space of dimension k in $\mathbb{PC}^n$ corresponds to a plane through the origin in $\mathbb{C}^{n+1}$ of dimension $k+1$. Hence, a linear subspace of dimension k is isomorphic to the projective space $\mathbb{PC}^k$.

Notice that it is usually simpler to use the codimension of the space rather than the dimension. A subspace of dimension $n - j$ in an ambient space of dimension n is said to have codimension j. Hence, hypersurfaces have codimension 1. The intersection of j hypersurfaces normally has codimension j; when this is the case, we say that the subspace is a **complete intersection**.

A variety comprising the zeros of a degree-2 equation is called a **quadric**. The equation of a quadric can be written neatly in matrix form as

$$\mathbf{x}^T M \mathbf{x} = 0$$

where M is a symmetric matrix of coefficients and $\mathbf{x}$ is the vector of coordinates.

The hyperplane tangent to a hypersurface is essentially given by the partial derivatives of the equation of the hypersurface. This is because the gradient of the left-hand side of the equation gives a vector tangent to the surface. Hence, the tangent to a quadric at a point $\mathbf{p}$ is given by the hyperplane with equation

$$\mathbf{p}^T M \mathbf{x} = 0.$$

This is an example of a more general construction for quadrics. If $\mathbf{p}$ is an arbitrary point in the projective space, then the hyperplane defined by $\mathbf{p}^T M \mathbf{x} = 0$ is called the polar plane to $\mathbf{p}$. We can also find the polar plane to a linear subspace by intersecting the polar hyperplanes of its points. If the original linear space has dimension k, then its polar plane will be a linear space of dimension $n - k - 1$.

A variety with codimension greater than 1 can be thought of as the intersection of a number of hypersurfaces, locally at least. So, to find the plane tangent to a point on a variety of smaller dimension, we simply take the intersection of the tangent hyperplanes of the hypersurfaces in which it lies.

A singularity on a variety is a point where the tangent space, as defined above, is larger than the dimension of the variety. In particular, for a quadric, we have that $\mathbf{p}$ is a singular point if and only if

$$\mathbf{p}^T M = \mathbf{0}.$$

Hence, the quadric is non-singular unless

$$\det(M) = 0.$$

The type of the singularity here depends on the rank of the coefficient matrix M. In the extreme case that the rank is 1, the quadric degenerates to a repeated hyperplane. When the rank is 2, the quadric is a pair of hyperplanes, the singular points lying in the intersection of the planes.

The degree of a hypersurface is simply the degree of the equation that defines it. More generally, the degree of an algebraic variety of dimension j is defined as the number of intersections of the variety with a general linear subspace with codimension j, that is, with complimentary dimension. The number of intersections must be counted with the correct multiplicity. A crucial result here is Bézout's theorem. Roughly speaking, this says that the intersection of a variety of degree d_1 with a variety of degree d_2 is a variety of degree $d_1 d_2$. In particular, the degree of a complete intersection will be the product of the degrees of the hypersurfaces that define it.

A common technique in the subject is linear projection. In the simplest case, we project from a point in $\mathbb{PC}^n$ to a hyperplane. Suppose the point, which we call the centre of the projection, is given by $\mathbf{c} = (c_0 : c_1 : \cdots : c_n)$. Let the hyperplane be the solution of the single linear equation

$$L(\mathbf{x}) = a_0 x_0 + a_1 x_1 + \cdots + a_n x_n = 0.$$

To find the image of a point $\mathbf{x}$ in $\mathbb{PC}^n$, we draw a line from $\mathbf{c}$ to $\mathbf{x}$. The image of $\mathbf{x}$ under the projection is the point where the line meets the hyperplane. In coordinates, this is given by

$$pr(\mathbf{x}) = L(\mathbf{c})\mathbf{x} - L(\mathbf{x})\mathbf{c}.$$

Notice that the mapping is undefined for the point $\mathbf{c}$ itself. So, technically, the projection is not a well-defined mapping at all. However, it is an example of a rational mapping; such mappings are only required to be well defined on the complement of an algebraic set.

The concept of a projection can be extended to centres that are linear subspaces. If the centre has dimension j, then we may project onto a linear subspace of dimension $n - j - 1$, so long as this subspace does not meet the centre. The image of an arbitrary point $\mathbf{x}$ is given by the unique intersection of the subspace with the $j + 1$-plane spanned by the centre and $\mathbf{x}$. Once again, this is a rational map, with the centre as exceptional set, that is, the set with no image.

There are now a couple of very good introductions to the subject, Reid [92] and Harris [44]. A more extensive review of the subject may found in Griffiths and Harris [42] and Hartshorne [45]. The older classic Semple and Roth [109] is still worth reading, not least for its wealth of examples.

1.4 Differential Geometry

Newton certainly applied his calculus to the geometry of curves. Euler also did some early work in this area, but it was probably the discoveries of Gauss on the curvature of surfaces that founded the subject of differential geometry. After Riemann, the subject was a mature discipline.

Many results were intrinsic; that is, they did not depend on how the surface was situated in space. This led to the definition of **manifolds**. These are spaces that look locally like Euclidean space $\mathbb{R}^n$, so that it is possible to do calculus on them. Globally, however, these spaces may not be isomorphic to $\mathbb{R}^n$. A manifold consists of a finite number of coordinate patches, each patch isomorphic to an open set in $\mathbb{R}^n$. These patches provide local coordinate systems. Where the patches overlap, the two coordinates systems are related to each other by a smooth change of coordinates. The classic example is the sphere, which can be specified by a pair of patches, one covering the northern hemisphere and the other the southern hemisphere. The overlap between these patches will be an open set around the equator.

The maps between manifolds that we are interested in are differentiable maps. These can be defined on the patches so long as they agree on the overlaps.

Tangent vectors to manifolds can be defined in several equivalent ways. One possibility would be to define them on the coordinate patches. However, a more geometrical way to do it is to think about curves in the manifold. Notice that a smooth curve in a manifold is an example of a differentiable map from the manifold $\mathbb{R}$ to the manifold under consideration. We may consider two curves through a point to be equivalent if they have the same first derivative at the point. The equivalence class represents a tangent vector at the point. Notice that we have a space of tangent vectors at each point in the manifold. We denote the tangent space to a manifold M at a point x by TM_x. Each of these vector spaces is isomorphic to the others. However, there are many isomorphisms between them, and there is no reason to choose one over the others. So, without further information, we cannot compare tangent vectors at one point to those at another. We may choose a tangent vector at each point. If we do this smoothly, then the result is a vector field on the manifold. Tangent vector fields can also be viewed as differential operators, and so differential geometry is important for the theory of partial differential equations.

A differentiable map between two manifolds induces a map between the tangent spaces. Suppose p is a differentiable map from a manifold M to a manifold N; then by considering smooth curves and their equivalence classes we obtain a linear map, written p_* or dp from TM_x to $TN_{p(x)}$. This induced map is the Jacobian of the map.

The dual of a vector space V is the set of linear functionals on the vector space V^*, that is, functions $\mathbf{f}$ from V to $\mathbb{R}$ which obey

$$\mathbf{f}(a\mathbf{u} + b\mathbf{v}) = a\mathbf{f}(\mathbf{u}) + b\mathbf{f}(\mathbf{v})$$

for any $\mathbf{u}, \mathbf{v} \in V$ and $a, b \in \mathbb{R}$. If we take the dual vector space to the space of tangent vectors at some point on the manifold TM_x, we obtain the space of cotangent vectors, T^*M_x. A smooth field of cotangent vectors is known as a differential form. Differential forms were introduced by Elie Cartan in the 1920s. They have many useful properties and are by now an indispensable tool in differential geometry and hence the theory of ordinary and partial differential equations. Surely it is only a matter of time before differential forms appear on undergraduate syllabi.

Two theorems are basic to modern differential geometry, the first of which is the inverse function theorem. Consider a smooth map p from a manifold M to a manifold N. Suppose that the map sends a point $a \in M$ to a point $b = p(a) \in N$. Further, suppose that the Jacobian of the map p_* has non-zero determinant at this point; this of course requires that the dimensions of the two manifolds are the same. In such a case, the inverse function theorem tells us that there is a unique smooth function q, defined in a neighbourhood of b, such that $q(b) = a$ and the composite $p \circ q$ is the identity function on the neighbourhood of b. Effectively, the function q is the inverse of p. However, the theorem only guarantees its existence in a neighbourhood of b; it says nothing about the extent of that neighbourhood. Another way of expressing this theorem is to say that the equation $p(x) = y$ can be solved for x so long as y is near b. For a proof, see [13, Chap. II sect. 6] or [88, Chap. 19].

The second important theorem is the implicit function theorem. Here, we look at a smooth map of the form $f : M \times N \longrightarrow W$ near the point $(a, b) \in M \times N$, where $f(a, b) = c \in W$. We may think of the manifold N as a set of parameters. Fixing these parameters defines a new map $f_b : M \longrightarrow W$ given by

$$f_b(x) = f(x, b).$$

Assuming that this map has a non-singular Jacobian requires that M and W have the same dimension. With these hypotheses, the implicit function theorem guarantees that there is a unique smooth map h from a neighbourhood of b to M such that $h(b) = a$ and $f(h(y), y) = c$. Notice that this means that if we are given an equation $f(x, y) = c$, then we can solve for x in terms of y, provided, of course, we are near a known solution (a, b). This theorem can be proved by applying the inverse function theorem to the map $F : M \times N \longrightarrow W \times N$ given by $F(x, y) = (f(x, y), y)$. The inverse of this is the map $F^{-1} : W \times N \longrightarrow M \times N$ given by $F^{-1}(w, y) = (H(w, y), y)$. The mapping we want is then given by $h(y) = H(c, y)$. For a formal proof, see [88, Chap. 19].

A consequence of the implicit function theorem is that non-singular algebraic varieties are manifolds.

The final definitions here concern maps of one manifold into another. A smooth map from a manifold M to a manifold N is called an **immersion** if at all points of M the Jacobian of the map is an injection. An **embedding** is a map from M to N where the image of M in N is homeomorphic to M itself. Notice that an embedding is necessarily also an immersion, but an immersion need not be an embedding. This is because a point in the image of an immersion may have come from several points in the domain. Using these ideas, it is possible to study the extrinsic geometry of one manifold embedded in another.

There are several excellent texts on the subject. O'Neill [81] is a nice introduction which keeps to concrete examples of curves and surfaces in $\mathbb{R}^3$. Schutz [99] takes a physicist's view, concentrating on applications in relativity and gauge theory. Berger and Gostiaux [8] give a modern introduction to the subject. Bishop and Crittenden [11] and Auslander [5] take an abstract approach to the subject but are quite readable. The classic reference work is Kobayashi and Nomizu [65].

2
Lie Groups

The concept of a group was introduced into mathematics by Cayley in the 1860s, generalising the older notion of "substitutions". The theory of substitutions studied the symmetries of algebraic equations generated by permutations of their roots. The theory was already highly developed; in particular Galois had developed a method to determine whether an algebraic equation can be solved by radicals. Although the work was done before 1832, it was not until 1843 that it gained a wide audience when it was popularised by Liouville.

In the late 1800s, many workers tried to extend these ideas to differential equations. Most notably, Lie and Killing developed the idea of "continuous groups". One of their first examples was the group of isometries of three-dimensional space. This could also be called the group of proper rigid motions in $\mathbb{R}^3$. This group is perhaps the most important one for robotics, and it is this that we will concentrate on in these notes. It was first extensively studied by Clifford [20] and Study [118]. Cartan developed a classification of semi-simple groups, and it is this that is the focus of most textbooks. However, the group of proper rigid motions is not semi-simple and hence is ignored by most modern textbooks. For a more detailed history of the subject of Lie groups, see Coleman [22] and Hawkins [47].

2.1   Definitions and Examples

Formally, a Lie Group is defined to be a smooth manifold G with a distinguished point or element e, together with two continuous functions

$$mult : G \times G \longrightarrow G,$$
$$inv : G \longrightarrow G.$$

The binary operation $mult$ is sometimes called the group operation. The image of an element under the map inv is the inverse of that element. These functions are required to satisfy the usual group axioms— $mult$ must be associative, inv must be a bijection, and e must be the identity element. Usually, for group elements g, g_1, g_2, $g_3 \in G$, we abbreviate the functions as

$$mult(g_1, g_2) = g_1 g_2 \qquad \text{and} \qquad inv(g) = g^{-1}.$$

The group axioms can then be written as

$$\begin{aligned} eg = ge = g & \qquad \text{(identity)}, \\ g^{-1}g = gg^{-1} = e & \qquad \text{(inverses)}, \\ g_1(g_2 g_3) = (g_1 g_2)g_3 = g_1 g_2 g_3 & \ \text{(associativity)}. \end{aligned}$$

For a Lie group, we further require that the functions $mult$ and inv must be differentiable mappings.

Our first example is $\mathbb{R}^n$. This is a **commutative group**, that is, a group in which the multiplication is a commutative binary operation. The identity element is the zero vector $e = (0, 0, \ldots, 0)^T$. The group operation $mult$ is given by vector addition: given two group elements

$$\mathbf{u} = (u_1, u_2, \ldots, u_n)^T, \qquad \mathbf{v} = (v_1, v_2, \ldots, v_n)^T,$$

their sum is simply

$$mult(\mathbf{u}, \mathbf{v}) = \mathbf{u} + \mathbf{v} = (u_1 + v_1, u_2 + v_2, \ldots, u_n + v_n)^T.$$

The inverse of a group element is given by multiplying it by -1:

$$inv(\mathbf{u}) = -\mathbf{u} = (-u_1, -u_2, \ldots, -u_n)^T.$$

These operations are manifestly continuous, and it is an easy matter to check that they satisfy the group axioms. The manifold of the group is simply the vector space $\mathbb{R}^n$.

For our second example, let us take the group of unit modulus complex numbers. Any element of the group has the form $z = \cos\theta + i\sin\theta$, where the real number θ lies between zero and 2π, $(0 \le \theta < 2\pi)$. The group operation is complex multiplication:

$$\begin{aligned} z_1 z_2 &= (\cos\theta_1 + i\sin\theta_1)(\cos\theta_2 + i\sin\theta_2) \\ &= (\cos\theta_1\cos\theta_2 - \sin\theta_1\sin\theta_2) + i(\sin\theta_1\cos\theta_2 + \cos\theta_1\sin\theta_2) \\ &= \cos(\theta_1 + \theta_2) + i\sin(\theta_1 + \theta_2). \end{aligned}$$

Again, this is continuous since addition and multiplication are continuous.

The inverse of an element is given by its complex conjugate, and hence the identity element is the complex number 1. Again, this is a commutative group since complex multiplication commutes. The manifold of the group is more interesting, though. As the parameter θ increases from 0 to 2π in $z = \cos\theta + i\sin\theta$, we obtain every possible group element just once, but at $\theta = 2\pi$ we return to our starting value $z = 1$. Hence, the group manifold has the topology of a circle. Alternatively, we could picture the group elements as making up the unit circle in the complex plane; see Figure 2.1.

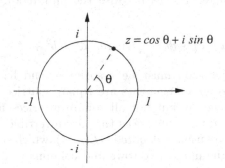

FIGURE 2.1. The Group of Unit Modulus Complex Numbers

Our next example is the unit modulus quaternions. Hamilton's quaternions are numbers of the form

$$q = a + bi + cj + dk.$$

The numbers a, b, c and d are real, while i, j and k are the quaternionic units. The set of quaternions forms an associative algebra. Addition is componentwise:

$$q_1 + q_2 = (a_1 + b_1 i + c_1 j + d_1 k) + (a_2 + b_2 i + c_2 j + d_2 k)$$
$$= (a_1 + a_2) + (b_1 + b_2)i + (c_1 + c_2)j + (d_1 + d_2)k.$$

Multiplication is defined by the famous relations

$$i^2 = j^2 = k^2 = -1, \qquad ijk = -1.$$

So, for example, $ij = k$, $jk = i$ and so forth. The linearity of the space means that we can extend these definitions to arbitrary linear combinations of the quaternionic units. For two arbitrary quaternions, we have

$$q_1 q_2 = (a_1 + b_1 i + c_1 j + d_1 k)(a_2 + b_2 i + c_2 j + d_2 k)$$
$$= (a_1 a_2 - b_1 b_2 - c_1 c_2 - d_1 d_2) + (a_1 b_2 + a_2 b_1 + c_1 d_2 - c_2 d_1)i +$$
$$(a_1 c_2 + a_2 c_1 - b_1 d_2 + b_2 d_1)j + (a_1 d_2 + a_2 d_1 + b_1 c_2 - b_2 c_1)k.$$

Notice that this multiplication is not commutative, for instance $ij = -ji$. The quaternionic conjugate is given by

$$q^* = (a + bi + cj + dk)^* = a - bi - cj - dk.$$

Hence, the modulus of any quaternion is given by the square root of

$$q^*q = a^2 + b^2 + c^2 + d^2.$$

We obtain a Lie group by restricting our attention to quaternions for which $q^*q = 1$. The group operation is the quaternionic multiplication described above, while the inverse of such a quaternion is simply given by its quaternion conjugate. The identity element is the quaternion 1. The manifold underlying this group can be identified with the unit sphere in $\mathbb{R}^4$, that is, a three-dimensional sphere. This is because the unit modulus quaternions are given by

$$q = a + bi + cj + dk \qquad \text{where} \quad a^2 + b^2 + c^2 + d^2 = 1.$$

Since our group operation must be associative and have inverses, we are led to consider groups whose elements are square matrices. Certainly, matrix multiplication is associative, but not all square matrices have inverses. If we restrict our attention to $n \times n$ matrices that are invertible, then we obtain Lie groups. These groups are usually denoted $GL(n, \mathbb{R})$, which stands for the general linear group, with n the number of rows and columns in the matrices. The $\mathbb{R}$ refers to the field of scalars; that is, the matrix entries are real numbers. We use this notation to distinguish this group from the group of complex non-singular matrices $GL(n, \mathbb{C})$. However, we will usually abbreviate $GL(n, \mathbb{R})$ to $GL(n)$. Note that $GL(1)$ is the commutative group of real numbers excluding zero; the group operation is multiplication. When $n \geq 2$, $GL(n)$ is not commutative, since matrix multiplication is not in general commutative. Matrix multiplication is continuous though, since it is simply a sequence of additions and multiplications. To specify an $n \times n$ matrix A, we must give its n^2 entries (a_{ij}). Hence, the group manifold of $GL(n)$ lies in $\mathbb{R}^{n^2}$, where the coordinates of $\mathbb{R}^{n^2}$ are given by the n^2 entries (a_{ij}). It cannot be all of $\mathbb{R}^{n^2}$, however, because we must exclude the matrices with zero determinant. These are solutions of the equation $\det(A) = 0$, which is a polynomial equation in the coordinates (a_{ij}). So the singular matrices form an affine algebraic variety in $\mathbb{R}^{n^2}$. We conclude that the manifold of $GL(n)$ is an open set in $\mathbb{R}^{n^2}$ and has dimension n^2.

If we multiply two matrices with unit determinant, the result is another matrix with determinant 1. This is because in general we have

$$\det(AB) = \det(A)\det(B).$$

Taking all the $n \times n$ matrices with determinant 1, we get more examples of Lie groups. These groups are usually called special linear groups, written $SL(n)$, or $SL(n, \mathbb{R})$ to make the ground field explicit. Each matrix in the group $SL(n)$ satisfies the degree n polynomial equation $\det(A) = 1$. That is, elements of the group lie on a non-singular algebraic variety in $\mathbb{R}^{n^2}$. This variety is the group manifold, and its dimension is $n^2 - 1$. For example, the group manifold of $SL(2)$ is the 3-dimensional quadric given by the equation

$$a_{11}a_{22} - a_{12}a_{21} = 1$$

in $\mathbb{R}^4$.

2.2   More Examples — Matrix Groups

Several other matrix groups have been defined. Collectively, these matrix groups are also known as the classical groups; see Weyl [126]. These groups can be thought of as the symmetry groups for particular kinds of metric spaces. As with finite groups, we think of the group elements as operators that act on the space but leave some feature or aspect of the space invariant. For example, the unit modulus complex numbers described above can be thought of as anticlockwise rotations of the complex plane. These rotations leave circles centred at the origin invariant. More significantly, the Hermitian form $z_1^* z_2$ is invariant with respect to the action. In fact, this group is $U(1)$; see below.

2.2.1   The Orthogonal Group $O(n)$

This is the group that preserves the positive definite bilinear form

$$\mathbf{x} \cdot \mathbf{y} = \mathbf{x}^T \mathbf{y} = (x_1 y_1 + x_2 y_2 + \cdots + x_n y_n) =$$

$$(x_1, x_2, \cdots, x_n) \begin{pmatrix} 1 & 0 & \cdots & 0 \\ 0 & 1 & \cdots & 0 \\ \vdots & \vdots & \ddots & \vdots \\ 0 & 0 & \cdots & 1 \end{pmatrix} \begin{pmatrix} y_1 \\ y_2 \\ \vdots \\ y_n \end{pmatrix}.$$

This is the standard scalar product between n-dimensional vectors. Group elements will be $n \times n$ matrices, M say. The effect of such an element on the vectors is given by

$$\mathbf{x}' = M\mathbf{x}.$$

The scalar product of two vectors after transformation must be the same as before the transformation, so that

$$\mathbf{x}' \cdot \mathbf{y}' = \mathbf{x}^T M^T M \mathbf{y} = \mathbf{x}^T \mathbf{y}.$$

Hence, matrices of the orthogonal group must satisfy

$$M^T M = I_n,$$

where I_n is the $n \times n$ identity matrix. This last equation is now an alternative definition of the group. The identity element in the group is I_n itself and the group operation is matrix multiplication again. Notice that for two orthogonal matrices M_1 and M_2 we have

$$(M_1 M_2)^T M_1 M_2 = M_2^T M_1^T M_1 M_2 = M_2^T I_n M_2 = I_n.$$

That is, the product of two orthogonal matrices is again orthogonal.

Now since $\det(M^T) = \det(M)$, by the rule for the determinant of a product we have that $\det(M)^2 = 1$ and hence $\det(M) = \pm 1$. The manifold for this group consists of two disconnected components. One part consists of those matrices that have $\det(M) = +1$; these include the identity element I_n and can be thought of as rotations about the origin. The other part consists of those orthogonal matrices with $\det(M) = -1$; these are usually thought of as reflections. For example, in two dimensions, $O(2)$ consists of matrices of the form

$$\begin{pmatrix} \cos\theta & -\sin\theta \\ \sin\theta & \cos\theta \end{pmatrix} \quad \text{and} \quad \begin{pmatrix} \cos\theta & \sin\theta \\ \sin\theta & -\cos\theta \end{pmatrix}.$$

The first of these correspond to anticlockwise rotation by θ about the origin, while the second are reflections in a line through the origin at an angle of $\theta/2$ from the first axis, that is, the line $x_2 \cos(\theta/2) - x_1 \sin(\theta/2) = 0$.

2.2.2   The Special Orthogonal Group $SO(n)$

This is simply the group of determinant 1 orthogonal matrices. So, by the remarks above, it is the group of rotations about the origin in n-dimensional space.

For our purposes, it is $SO(2)$ and $SO(3)$ that will be most important, since these are the rigid body rotations about a fixed centre in two and three dimensions. The group $SO(2)$ consists of matrices of the form

$$\begin{pmatrix} \cos\theta & -\sin\theta \\ \sin\theta & \cos\theta \end{pmatrix}.$$

The group manifold is a circle once more . The matrices of $SO(3)$ cannot be parameterised so neatly. This is because the underlying manifold does not admit a global coordinate system. In fact, the group manifold of $SO(3)$ is homeomorphic to the 3-dimensional real projective plane $\mathbb{PR}^3$. If $(w : x : y : z)$ are homogeneous coordinates for $\mathbb{PR}^3$, then a possible homeomorphism is given by

$$(w : x : y : z) \longmapsto$$

$$\frac{1}{\Delta} \begin{pmatrix} w^2 + x^2 - y^2 - z^2 & 2(xy - wz) & 2(xz + wy) \\ 2(xy + wz) & w^2 - x^2 + y^2 - z^2 & 2(yz - wx) \\ 2(xz - wy) & 2(yz + wx) & w^2 - x^2 - y^2 + z^2 \end{pmatrix},$$

where $\Delta = w^2 + x^2 + y^2 + z^2$. It is straightforward but rather tedious to check that this indeed gives an orthogonal matrix with unit determinant. To show that this is indeed a homeomorphism, we need to give the inverse mapping. This is a little more tricky since there is no global coordinate system. An $SO(3)$ matrix is given by

$$R = \begin{pmatrix} r_{11} & r_{12} & r_{13} \\ r_{21} & r_{22} & r_{23} \\ r_{31} & r_{32} & r_{33} \end{pmatrix}.$$

Consider the following mapping to $\mathbb{PR}^3$:

$$R \longmapsto \left(1 + r_{11} + r_{22} + r_{33} : \frac{1}{2}(r_{32} - r_{23}) : \frac{1}{2}(r_{13} - r_{31}) : \frac{1}{2}(r_{21} - r_{21}) \right).$$

This map is an inverse to the original one—don't forget that the homogeneous coordinates of projective space are only defined up to an overall non-zero factor. The trouble is that the map is not defined on all of $SO(3)$ and it is not onto: the plane $w = 0$ is not in the image. We can patch together the full inverse from this and similar maps that cover other parts of $\mathbb{PR}^3$. We use the above map in the region where $1 + r_{11} + r_{22} + r_{33} \neq 0$; then we use

$$R \longmapsto \left(\frac{1}{2}(r_{32} - r_{23}) : 1 + r_{11} - r_{22} - r_{33} : \frac{1}{2}(r_{12} + r_{21}) : \frac{1}{2}(r_{13} + r_{31}) \right),$$

where $1 + r_{11} - r_{22} - r_{33} \neq 0$ and

$$R \longmapsto \left(\frac{1}{2}(r_{13} - r_{31}) : \frac{1}{2}(r_{12} + r_{21}) : 1 - r_{11} + r_{22} - r_{33} : \frac{1}{2}(r_{23} + r_{32}) \right),$$

where $1 - r_{11} + r_{22} - r_{33} \neq 0$ and finally

$$R \longmapsto \left(\frac{1}{2}(r_{21} - r_{12}) : \frac{1}{2}(r_{13} + r_{31}) : \frac{1}{2}(r_{23} + r_{32}) : 1 - r_{11} - r_{22} + r_{33} \right)$$

in the region where $1 - r_{11} - r_{22} + r_{33} \neq 0$.

These four maps agree on the regions where they overlap and together cover all of $\mathbb{PR}^3$. Moreover, they invert the original map from $\mathbb{PR}^3$ to $SO(3)$. Hence, the two spaces are homeomorphic.

2.2.3   The Symplectic Group $Sp(2n, \mathbb{R})$

This is the group of symmetries that preserves the bilinear anti-symmetric form

$$\mathbf{q}_1 \cdot \mathbf{p}_2 - \mathbf{q}_2 \cdot \mathbf{p}_1 = (\mathbf{q}_1^T, \mathbf{p}_1^T) \begin{pmatrix} 0 & I_n \\ -I_n & 0 \end{pmatrix} \begin{pmatrix} \mathbf{q}_2 \\ \mathbf{p}_2 \end{pmatrix}.$$

Here, $\mathbf{q}_i$ and $\mathbf{p}_i$ are n-dimensional vectors. These are the groups of canonical transformations of Hamiltonian mechanics. However, they were first studied because $Sp(4, \mathbb{R})$ turns out to be the symmetry group of line complexes, that is, linear systems of lines in three dimensions; see Weyl [126]. Note that we use the notation $Sp(2n, \mathbb{R})$ to avoid confusion with $Sp(n)$, which are symmetry groups associated with quaternionic spaces.

2.2.4   The Unitary Group $U(n)$

If we allow complex scalars, we get more groups. The unitary groups preserve the Hermitian form

$$\mathbf{z}^\dagger \mathbf{w} = (z_1^*, z_2^*, \ldots, z_n^*) \begin{pmatrix} w_1 \\ w_2 \\ \vdots \\ w_n \end{pmatrix}.$$

The "$\dagger$" here represents the Hermitian conjugate obtained by transposition and complex conjugation. Hence, the matrices of the unitary group satisfy

$$M^\dagger M = I_n.$$

Using an argument similar to the one we used above for orthogonal matrices, we see that matrices satisfying this equation do form a group.

From the equation for the determinant of a product of two matrices, we see that the determinant of a unitary matrix must be a unit modulus complex number:

$$1 = \det(M^\dagger M) = \det(M^\dagger)\det(M) = \det(M)^* \det(M).$$

2.2.5   The Special Unitary Group $SU(n)$

The special unitary group consists of unitary matrices with unit determinant. For example, the group $SU(2)$ consists of matrices of the form

$$\begin{pmatrix} a+ib & c+id \\ -c+id & a-ib \end{pmatrix},$$

where the real parameters a, b, c and d satisfy

$$a^2 + b^2 + c^2 + d^2 = 1.$$

Hence, we may identify the elements of $SU(2)$ with the points of a three-dimensional sphere in $\mathbb{R}^4$.

In fact, we can even think of $GL(n)$ and $SL(n)$ as groups of symmetries of a bilinear form. To do this, we must use scalars from the ring $\mathbb{R} \times \mathbb{R}$; see Porteous [88, Chap. 11].

2.3   Homomorphisms

Now that we have some examples of groups, we can discuss mappings between groups. As with finite groups, we require maps between groups to preserve the group structure. So a map $f : G \longrightarrow H$ between two groups must satisfy

$$f(g_1 g_2) = f(g_1) f(g_2) \qquad \text{for all} \quad g_1, g_2 \in G.$$

For Lie groups, the map must also be a differentiable map. When all these conditions are satisfied, we call the map a Lie group homomorphism or, for brevity, just a homomorphism.

The simplest examples of homomorphisms between Lie groups are probably inclusions. For example, we could map the unit modulus complex numbers to the unit modulus quaternions:

$$x + yi \longmapsto x + yi + 0j + 0k.$$

In a similar fashion, we have the following inclusions for our matrix groups:

$$SL(n) \longrightarrow GL(n), \qquad SO(n) \longrightarrow O(n), \qquad SU(n) \longrightarrow U(n).$$

In fact, we can include any of our matrix groups in the appropriate general linear group:

$$O(n) \longrightarrow GL(n), \qquad Sp(2n, \mathbb{R}) \longrightarrow GL(2n).$$

We can also include each matrix group in a group of higher dimension. For example, we have an inclusion $SO(n) \longrightarrow SO(n+1)$ given, in partitioned matrix form, by

$$M \longmapsto \begin{pmatrix} M & 0 \\ 0 & 1 \end{pmatrix}, \qquad \text{for all} \quad M \in SO(n).$$

Inclusions are injective, that is, 1-to-1 mappings. As an example of a surjective homomorphism that is an onto mapping, consider the following map $\pi : SU(2) \longrightarrow SO(3)$:

$$\pi : \begin{pmatrix} a + ib & c + id \\ -c + id & a - ib \end{pmatrix} \longmapsto$$

$$\begin{pmatrix} a^2 + b^2 - c^2 - d^2 & 2(bc - ad) & 2(ac + bd) \\ 2(bc + ad) & a^2 - b^2 + c^2 - d^2 & 2(cd - ab) \\ 2(bd - ac) & 2(cd + ab) & a^2 - b^2 - c^2 + d^2 \end{pmatrix}.$$

Remember that $a^2 + b^2 + c^2 + d^2 = 1$ here.

Notice that if M is in $SU(2)$, then both M and $-M$ are taken to the same element of $SO(3)$ by the homomorphism. In particular

$$\pi : \begin{pmatrix} 1 & 0 \\ 0 & 1 \end{pmatrix} \longmapsto \begin{pmatrix} 1 & 0 & 0 \\ 0 & 1 & 0 \\ 0 & 0 & 1 \end{pmatrix} \quad \text{and} \quad \pi : \begin{pmatrix} -1 & 0 \\ 0 & -1 \end{pmatrix} \longmapsto \begin{pmatrix} 1 & 0 & 0 \\ 0 & 1 & 0 \\ 0 & 0 & 1 \end{pmatrix}.$$

Because the pre-image of each element of $SO(3)$ consists of two elements from $SU(2)$, this map is called a double covering. This does not mean that the group $SU(2)$ is simply two copies of $SO(3)$. Indeed we have already seen that the group manifold of $SU(2)$ is a 3-sphere: a connected manifold. The homomorphism is a differentiable map between the group manifolds. Topologically this mapping from the 3-sphere to the 3-dimensional real projective space is an example of a non-trivial fibre bundle. It is not possible to find a continuous function $s : SO(3) \longrightarrow SU(2)$ that is a left inverse of π. That is, no continuous function s exists that satisfies $\pi \circ s = id_{SO(3)}$. In other words, it is not possible to choose one element of the pre-image of π for every point of $SO(3)$ in a continuous fashion. The choice can be made locally, over a small patch, but not globally over all of $SO(3)$. The non-existence of such a map is a consequence of the fact that the 3-sphere is not homeomorphic to two copies of 3-dimensional real projective space.

A simpler example of a non-trivial double covering is given by mapping the edge of a Möbius band to its centre circle; see Figure 2.2.

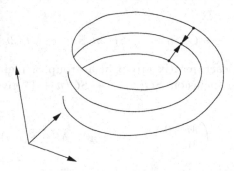

FIGURE 2.2. The Edge of a Möbius Band Double Covers Its Centre Circle

Finally, we look at some isomorphisms between the low-dimensional matrix groups. Isomorphisms are bijective, that is, they are surjective and injective homomorphisms. The isomorphisms we present do not generalise to higher dimensions; they are sometimes called accidental isomorphisms.

The unit modulus complex numbers $U(1)$ are isomorphic to the group $SO(2)$. The isomorphism is given by

$$\cos\theta + i\sin\theta \longmapsto \begin{pmatrix} \cos\theta & -\sin\theta \\ \sin\theta & \cos\theta \end{pmatrix}.$$

The unit modulus quaternions are isomorphic to $SU(2)$:

$$a + bi + cj + dk \longmapsto \begin{pmatrix} a+ib & c+id \\ -c+id & a-ib \end{pmatrix}.$$

Hence, the unit modulus quaternions also double cover $SO(3)$.

The symplectic group $Sp(2,\mathbb{R})$ is isomorphic to $SL(2)$. The isomorphism is trivial; the point is that the symplectic group satisfies

$$\begin{pmatrix} a & c \\ b & d \end{pmatrix} \begin{pmatrix} 0 & 1 \\ -1 & 0 \end{pmatrix} \begin{pmatrix} a & b \\ c & d \end{pmatrix} = \begin{pmatrix} 0 & 1 \\ -1 & 0 \end{pmatrix}.$$

Simplification yields the single independent equation $ad - bc = 1$; that is, the matrix must have unit determinant.

2.4   Actions and Products

We may combine groups to obtain new groups. The simplest combination is the direct product. Given two groups G and H, elements of their direct product, written $G \times H$, are pairs of elements (g, h), where $g \in G$ and $h \in H$. The group operation is simply

$$(g_1, h_1)(g_2, h_2) = (g_1 g_2, h_1 h_2).$$

That is, we multiply the elements from the different groups separately. In a moment, we will see a different way to combine groups, but first we discuss the concept of group actions.

Many of the groups we have defined above have been described as groups of symmetry operations on certain spaces. The time has come to give a formal definition of what is meant by this. Consider a group G and a manifold X. We say that G acts on X if there is a differentiable map

$$a : G \times X \longrightarrow X$$

that satisfies

$$a(e, x) = x \qquad \text{for all} \quad x \in X \tag{A1}$$

and

$$a(g_1, a(g_2, x)) = a(g_1 g_2, x) \qquad \text{for all} \quad x \in X \quad \text{and all} \quad g_1, g_2 \in G. \tag{A2}$$

Often, when no confusion can arise we simply write the action $a(g, x)$ as $g(x)$; so our second axiom would read $g_1(g_2(x)) = g_1 g_2(x)$. Notice that the axioms imply that the map $g(x)$ is a differentiable isomorphism (diffeomorphism) for each $g \in G$, since the inverse map will be given by $g^{-1}(x)$.

We have already seen examples of group actions. First, if we take X to be the group manifold of G itself, then G acts on itself by left multiplication: $l(g_1, g) = g_1 g$. We also have a right action of G on itself, given by $r(g_1, g) = g g_1^{-1}$. We use the inverse of g_1 here so that axiom (A2) is satisfied. Note also that, on any space X, we always have the trivial group action of any group G: $g(x) = x$ for all $g \in G$ and all $x \in X$; not a very interesting construction but sometimes a useful one.

Suppose X is a vector space, $\mathbb{R}^n$ say. The matrix groups now act on this space by matrix multiplication:

$$a(M, \mathbf{v}) = M\mathbf{v}$$

where $M \in GL(n)$ for example and $\mathbf{v} \in \mathbb{R}^n$. Linear actions of groups on vector spaces are called representations, that is, actions which satisfy

$$g(h_1 + h_2) = g(h_1) + g(h_2) \qquad \text{for all} \quad g \in G \quad \text{and all} \quad h_1, h_2 \in H. \tag{A3}$$

Much of group theory involves finding representations of groups. Equivalently, we could think of this problem as finding homomorphisms from the group to $GL(n)$, the value of n here giving the dimension of the representation.

We will also be interested in group actions on other spaces. For example, the space of all possible lines in $\mathbb{R}^3$ is a non-singular quadric in $\mathbb{PR}^5$, the Klein quadric. The group of rigid body motions acts on this space; lines are moved into lines. See Chapter 6.

Now suppose we have a group G and a commutative group H together with a linear action of G on H. That is, a map $G \times H \longrightarrow H$ given by $g(h)$ satisfying all three axioms given above. The **semi-direct product** of G and H, written $G \ltimes H$, has the same elements as the direct product, that is, pairs of the form (g, h), where $g \in G$ and $h \in H$. However, the product of two elements is defined as

$$(g_1, h_1)(g_2, h_2) = (g_1 g_2, h_1 + g_1(h_2)).$$

Since the second group is commutative, we write the group operation here as "+". The identity element in such a group is (e, e), and the inverse of an element (g, h) is given by

$$(g, h)^{-1} = (g^{-1}, -g^{-1}(h)).$$

Finally, to show that this is a group, we must show that the product is associative:

$$(g_1, h_1)\Big((g_2, h_2)(g_3, h_3)\Big) = (g_1, h_1)(g_2 g_3, h_2 + g_2(h_3))$$

$$= (g_1 g_2 g_3, h_1 + g_1(h_2 + g_2(h_3)))$$

and

$$\Big((g_1, h_1)(g_2, h_2)\Big)(g_3, h_3) = (g_1 g_2, h_1 + g_1(h_2))(g_3, h_3)$$

$$= (g_1 g_2 g_3, h_1 + g_1(h_2) + g_1 g_2(h_3))).$$

Equality of these two expressions is guaranteed by (A3).

Finally, in this section we introduce two common actions of matrix groups. Consider the space of all $n \times n$ matrices; call this space $M(n)$. This space is in fact a vector space; we may add square matrices and multiply by scalars, and it is not hard to check that the axioms for a vector space are satisfied. We may take as basis elements for this space the matrices M_{ij} in which the element in the ij-th position is 1 and all others are zero. Hence, we can see that this space has dimension n^2.

Now consider the matrix group $GL(n)$; an action of this group on $M(n)$ is given by

$$S(g, M) = gMg^{-1}.$$

A transformation with this shape is called a **similarity**. This is an action since it satisfies both of our axioms, (A1): $S(e, M) = eMe = M$ and (A2):

$$S(g_1, S(g_2, M)) = S(g_1, g_2 M g_2^{-1}) = g_1 g_2 M g_2^{-1} g_1^{-1}$$

$$= (g_1 g_2) M (g_1 g_2)^{-1} = S(g_1 g_2, M).$$

In fact, the action also satisfies (A3):

$$S(g, M_1 + M_2) = g(M_1 + M_2)g^{-1} = gM_1g^{-1} + gM_2g^{-1} = S(g, M_1) + S(g, M_2)$$

and so is a representation of the group. Notice that we can use the same representation for any subgroup of $GL(n)$ by restricting the maps to the subgroup.

The second common action is given by

$$C(g, M) = gMg^T.$$

A transformation with this shape is usually called a **congruence**. Notice that if the original matrix M is symmetric, that is, if it satisfies $M^T = M$, then the transformed matrix will also be symmetric. Hence, it is common to restrict the action to the vector space of $n \times n$ symmetric matrices. This space has dimension $n(n + 1)/2$ and will be denoted $M^S(n)$. Using arguments analogous to the ones above, we can easily show that this action satisfies our three axioms and thus gives another representation of $GL(n)$.

Note that the orthogonal matrices are defined as the set of matrices that preserve the identity matrix under congruence transformations. To be precise, in Section 2.2.1 we defined the orthogonal group as the set of matrices satisfying $g^T g = I_n$ rather than $gg^T = I_n$. However, for symmetric matrices it doesn't matter whether we define the congruence as gMg^T or $g^T Mg$.

2.5   The Proper Euclidean Group

At long last we are in a position to describe the group of rigid body transformations. This is the group of transformations of the vector space $\mathbb{R}^n$ that preserves the Euclidean metric.

2.5.1   Isometries

Transformations that preserve a metric are known as isometries, hence we could also call this group the isometry group of $\mathbb{R}^n$. Now if $\mathbf{u}^T = (u_1, u_2, \ldots, u_n)$ and $\mathbf{v}^T = (v_1, v_2, \ldots, v_n)$ are two vectors that transform to $\mathbf{u}'$ and $\mathbf{v}'$, respectively, then the transformation is a rigid body motion provided

$$|\mathbf{u} - \mathbf{v}|^2 = |\mathbf{u}' - \mathbf{v}'|^2.$$

There are two possibilities: either we could transform vectors by adding a constant vector $\mathbf{t}$ say, to each, or we could multiply the vectors by an orthogonal matrix. The second of these possibilities is allowed because we can write

$$|\mathbf{u} - \mathbf{v}|^2 = \mathbf{u} \cdot \mathbf{u} - 2\mathbf{u} \cdot \mathbf{v} + \mathbf{v} \cdot \mathbf{v}$$

and orthogonal matrices preserve the scalar product. More generally, we may perform a combination of both possibilities, so that the effect of a general rigid transformation on an arbitrary vector will be given by the matrix equation

$$\mathbf{v}' = M\mathbf{v} + \mathbf{t}$$

where M is an orthogonal matrix and $\mathbf{t}$ is a constant vector. Hence, these transformations can be written as pairs $(M, \mathbf{t})$, with $M \in O(n)$ and $\mathbf{t} \in \mathbb{R}^n$. Note that it is a straightforward matter of linear algebra to show that different pairs give rise to different transformations.

To find the product of two such pairs, we look at the effect of two successive transformations on a single vector:

$$\mathbf{v}' = M_1\mathbf{v} + \mathbf{t}_1 \quad \text{and then} \quad \mathbf{v}'' = M_2\mathbf{v}' + \mathbf{t}_2 = M_2 M_1 \mathbf{v} + M_2 \mathbf{t}_1 + \mathbf{t}_2.$$

The product of two of these transformations is thus

$$(M_2, \mathbf{t}_2)(M_1, \mathbf{t}_1) = (M_2 M_1, M_2 \mathbf{t}_1 + \mathbf{t}_2).$$

The group of rigid body motions in $\mathbb{R}^n$ is thus the semi-direct product of the orthogonal group with $\mathbb{R}^n$ itself. We will denote it $E(n)$ for the Euclidean group:

$$E(n) = O(n) \ltimes \mathbb{R}^n.$$

From the arguments above, it can be seen that group elements of the form $(I_n, \mathbf{t})$ correspond to translations in $\mathbb{R}^n$; each vector in the space is translated by an amount $\mathbf{t}$. We have also seen above that the group $O(n)$ has two disconnected components. Orthogonal matrices with determinant 1 correspond to rotations about the origin in $\mathbb{R}^n$. They also form the group $SO(n)$. Orthogonal matrices with determinant -1 correspond to reflections. Since no physical machine is capable of effecting a reflection on rigid bodies, we will confine our attentions to the group of proper rigid body transformations, which we will write $SE(n)$:

$$SE(n) = SO(n) \ltimes \mathbb{R}^n.$$

Luckily, we do not have to work with the pairs defined above. There is a convenient $(n + 1)$-dimensional representation of $SE(n)$, that is, an injective homomorphism $SE(n) \longrightarrow GL(n + 1)$ given by

$$(R, \mathbf{t}) \longmapsto \begin{pmatrix} R & \mathbf{t} \\ 0 & 1 \end{pmatrix},$$

where we have used a partitioned form for the matrix on the right. This is a homomorphism, since multiplying these matrices exactly replicates the product of the pairs

$$\begin{pmatrix} R_2 & \mathbf{t}_2 \\ 0 & 1 \end{pmatrix} \begin{pmatrix} R_1 & \mathbf{t}_1 \\ 0 & 1 \end{pmatrix} = \begin{pmatrix} R_2 R_1 & R_2 \mathbf{t}_1 + \mathbf{t}_2 \\ 0 & 1 \end{pmatrix}.$$

The inverse of such a matrix is conveniently given by

$$\begin{pmatrix} R & \mathbf{t} \\ 0 & 1 \end{pmatrix}^{-1} = \begin{pmatrix} R^T & -R^T\mathbf{t} \\ 0 & 1 \end{pmatrix}$$

since R is an orthogonal matrix.

In the case we are most interested in, $SE(3)$, we use 4×4 matrices. The representation is sometimes called the homogeneous representation because of its connection with projective transformations; see Section 3.3.

The action of these rotation-translation pairs on vectors in $\mathbb{R}^n$ is also reproduced by this representation. Consider the affine subspace of $\mathbb{R}^{n+1}$ consisting of vectors of the form

$$\tilde{\mathbf{v}} = \begin{pmatrix} v_1 \\ v_2 \\ \vdots \\ v_n \\ 1 \end{pmatrix}.$$

This space is isomorphic to $\mathbb{R}^n$; we could write elements of the space in partitioned form as $\tilde{\mathbf{v}}^T = (\mathbf{v}^T, 1)$. The action of the matrix representation on these points is simply

$$\begin{pmatrix} R & \mathbf{t} \\ 0 & 1 \end{pmatrix} \begin{pmatrix} \mathbf{v} \\ 1 \end{pmatrix} = \begin{pmatrix} R\mathbf{v} + \mathbf{t} \\ 1 \end{pmatrix}$$

which precisely models the action of the pairs on vectors in $\mathbb{R}^n$. For rigid body motions in two dimensions, we can draw a picture; see Figure 2.3. Although the group of matrices of the given form acts on all vectors in $\mathbb{R}^3$, those lying on the plane $z = 1$ stay on that plane. Furthermore, the Euclidean metric on the plane is preserved by the group.

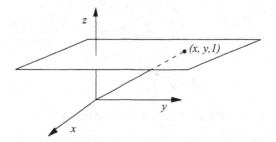

FIGURE 2.3. A Plane in 3-Space

2.5.2   Chasles's Theorem

In this section a statement and proof of Chasles's theorem is given. This is one of the earliest results in this field; see the discussion in Hunt [54, p. 49]. The theorem states that all proper rigid body motions in 3-dimensional space, with the exception of pure translations, are equivalent to a screw motion, that is, a rotation about a line together with a translation along the line. If the line passes through the origin, we can write the screw motion as

$$\begin{pmatrix} R(\theta, \hat{\mathbf{v}}) & \frac{\theta}{2\pi}p\hat{\mathbf{v}} \\ 0 & 1 \end{pmatrix}.$$

Here, $R(\theta, \hat{\mathbf{v}})$ represents a 3×3 rotation matrix about an axis in the direction of the unit vector $\hat{\mathbf{v}}$ and through an angle of θ. The number p is called the pitch of the screw, it is the distance moved along the axis for a complete turn about the axis. More generally, the axis of the screw motion may not pass through the origin. If $\mathbf{u}$ is the position vector of some point on the axis, then we can obtain the matrix representing such a screw motion by translating the screw axis back to the origin, performing the screw motion about the line through the origin, and finally translating the line back to its original position; see Figure 2.4. In terms of matrices we have

$$\begin{pmatrix} I_3 & \mathbf{u} \\ 0 & 1 \end{pmatrix} \begin{pmatrix} R(\theta, \hat{\mathbf{v}}) & \frac{\theta}{2\pi}p\hat{\mathbf{v}} \\ 0 & 1 \end{pmatrix} \begin{pmatrix} I_3 & -\mathbf{u} \\ 0 & 1 \end{pmatrix} = \begin{pmatrix} R & \frac{\theta}{2\pi}p\hat{\mathbf{v}} + (I_3 - R)\mathbf{u} \\ 0 & 1 \end{pmatrix},$$

where we have suppressed the explicit dependence of R on θ and $\hat{\mathbf{v}}$ on the right. Notice that this amounts to a similarity transformation of the original matrix. In group theory, the operation that sends a group element x to gxg^{-1} is called a conjugation, where g is another group element.

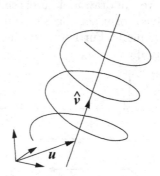

FIGURE 2.4. A Screw Motion

Now, to show that any proper rigid motion can be thought of as a screw motion, we must show that an arbitrary matrix of the form

$$\begin{pmatrix} R & \mathbf{t} \\ 0 & 1 \end{pmatrix}$$

can be written in the above form. Assuming that we can find the angle of rotation and axis of the 3×3 rotation matrix (a simple way of doing this can

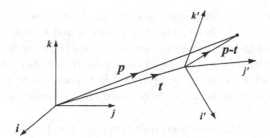

FIGURE 2.5. A vector referred to two coordinate frames

be found in Section 4.4), this amounts to solving the following system of linear equations for p and $\mathbf{u}$:

$$\frac{\theta}{2\pi}p\hat{\mathbf{v}} + (I_3 - R)\mathbf{u} = \mathbf{t}.$$

The matrix $(I_3 - R)$ is singular since R has 1 as an eigenvalue. This eigenvalue corresponds to the unit eigenvector $\hat{\mathbf{v}}$; a rotation leaves its axis fixed. We choose p so that the linear equations are consistent. So $(I_3 - R)\hat{\mathbf{v}} = \mathbf{0}$. Moreover, since R^T is simply a rotation in the opposite sense, we have $(I_3 - R^T)\hat{\mathbf{v}} = \mathbf{0}$. This allows us to conclude that $\hat{\mathbf{v}}^T(I_3 - R) = \mathbf{0}^T$, so taking the scalar product of our original equations with $\hat{\mathbf{v}}$ gives

$$\hat{\mathbf{v}} \cdot (I_3 - R)\mathbf{u} = 0 = \hat{\mathbf{v}} \cdot \mathbf{t} - \frac{\theta}{2\pi}p.$$

The pitch of the transformation is thus given by

$$p = \frac{2\pi}{\theta}\hat{\mathbf{v}} \cdot \mathbf{t}.$$

Since the equations are consistent, they can be solved. We get a line of solutions of the form

$$\mathbf{u} = \mathbf{u}_0 + \lambda\hat{\mathbf{v}}$$

where $\mathbf{u}_0$ is any particular solution and λ is an arbitrary constant. This constructive proof depends on the matrix $(I_3 - R)$ having rank 2, which is true so long as $R \neq I_3$. In the case $R = I_3$, corresponding to pure translations, this method fails.

2.5.3 Coordinate Frames

So far we have taken an active view of rigid body motions. That is, we assume a fixed coordinate frame and then the matrix of a rigid body motion is thought of as moving the points around. This viewpoint can be extended to describe the position and orientation of a rigid body. If we agree on a standard 'home' position for the body, any subsequent position and orientation of the body is described by the rigid motion that moves the home position of the body to its current position and orientation.

There is an alterative viewpoint however. In the passive viewpoint a coordinate frame is fixed in the body, now the position and orientation of the body is given by the coordinate transform that expresses the coordinates of points in the current frame in terms of those in the home frame. For historical reasons, this passive viewpoint seems to be preferred in robotics even though the active viewpoint is often simpler because there is only one coordinate system to worry about.

These two viewpoints are related quite simply. The transformation given by the passive view is the inverse of the active transformation. To see this assume that the new frame has origin at a point with position vector $\mathbf{t}$ with respect to the original frame. Now an arbitrary point $\mathbf{p}$ in the original frame has position vector $\mathbf{p}' = \mathbf{p} - \mathbf{t}$ with respect to the new frame; see Figure 2.5. In general the orientation of the new frame will be different to that of the original, assume it is given by a rotation R. That is, the basis vectors of the new frame $\mathbf{i}'$, $\mathbf{j}'$ and $\mathbf{k}'$, are given by

$$\mathbf{i}' = R\mathbf{i}, \quad \mathbf{j}' = R\mathbf{j}, \quad \mathbf{k}' = R\mathbf{k}.$$

If we write the position vector of the point as

$$\mathbf{p} = x\mathbf{i} + y\mathbf{j} + z\mathbf{k}$$

in the original frame, then in the new frame the point will have coordinates

$$\mathbf{p}' = x'\mathbf{i}' + y'\mathbf{j}' + z'\mathbf{k}',$$

where $x' = \mathbf{p}' \cdot \mathbf{i}' = (\mathbf{p} - \mathbf{t})^T R\mathbf{i}$ and so forth. Hence the new coordinates can be written in terms of the old as

$$\begin{pmatrix} x' \\ y' \\ z' \end{pmatrix} = R^T \begin{pmatrix} x \\ y \\ z \end{pmatrix} - R^T \mathbf{t}.$$

This can be summarised as,

$$\begin{pmatrix} \mathbf{p}' \\ 1 \end{pmatrix} = \begin{pmatrix} R^T & -R^T\mathbf{t} \\ 0 & 1 \end{pmatrix} \begin{pmatrix} \mathbf{p} \\ 1 \end{pmatrix} = \begin{pmatrix} R & \mathbf{t} \\ 0 & 1 \end{pmatrix}^{-1} \begin{pmatrix} \mathbf{p} \\ 1 \end{pmatrix}.$$

Suppose now that X' is a matrix representing a rigid motion referred to the dashed coordinate frame. How can we transform this matrix to the original, undashed coordinate? Assume that X' sends a point $\mathbf{p}_1'$ to $\mathbf{p}_2'$ so that

$$X' \begin{pmatrix} \mathbf{p}_1' \\ 1 \end{pmatrix} = \begin{pmatrix} \mathbf{p}_2' \\ 1 \end{pmatrix}.$$

In the undashed coordinates the corresponding points are $\mathbf{p}_1$ and $\mathbf{p}_2$; using the result above we can write

$$\begin{pmatrix} R & \mathbf{t} \\ 0 & 1 \end{pmatrix} X' \begin{pmatrix} R & \mathbf{t} \\ 0 & 1 \end{pmatrix}^{-1} \begin{pmatrix} \mathbf{p}_1 \\ 1 \end{pmatrix} = \begin{pmatrix} \mathbf{p}_2 \\ 1 \end{pmatrix}.$$

Hence we can conclude that in the undashed coordinates the matrix of the transformation is given by

$$X = \begin{pmatrix} R & t \\ 0 & 1 \end{pmatrix} X' \begin{pmatrix} R & t \\ 0 & 1 \end{pmatrix}^{-1}.$$

Hint: So can compute that in the attained coordinates the masses of the components ... each ...

$$\sum_{k} \binom{n}{k} W \left(\frac{n}{k-1} \right) W \left(\frac{k}{n} \right)$$

3
Subgroups

3.1 The Homomorphism Theorems

For any group, a **subgroup** is a subset of elements of the original group that is closed under the group operation. That is, the product of any two elements of the subgroup is again an element of the subgroup. For Lie groups, we have the concept of a Lie subgroup. In addition to the closure requirement, the subgroup must also be a submanifold of the group manifold of the original group. It is quite possible to have subgroups of Lie groups that are not Lie subgroups. However, when we talk about the subgroups of a Lie group we will always mean a Lie subgroup. So, for consistency, the group manifold of a discrete group will be thought of as a zero-dimensional manifold. For example, the trivial group has just a single element, the identity element. We will write this group as $0 = \{e\}$; notice that 0 is a subgroup of every group.

The last chapter provides us with many examples—any matrix group is a subgroup of $GL(n)$ for the appropriate n. We also have that $SO(n)$ is a subgroup of $O(n)$ and $SU(n)$ is contained in $U(n)$.

As with discrete groups, we have some homomorphism theorems. First of all, the image of a homomorphism is always a subgroup of the codomain. Also, the kernel of a homomorphism, that is, the set of elements that map to the identity in the codomain, is always a subgroup of the domain. To prove these statements, consider a homomorphism h, between two groups G and Q, $h : G \longrightarrow Q$. The fact that this is a homomorphism means that for any pair of elements $g_1, g_2 \in G$ we have

$$h(g_1)h(g_2) = h(g_1 g_2).$$

So the product of two elements in the image of the homomorphism is also in the image. Hence, the image is closed under the group product and is thus a subgroup. For the second of the theorems, consider two arbitrary elements of the kernel, k_1 and k_2. By definition, we have

$$h(k_1) = e \qquad \text{and} \qquad h(k_2) = e,$$

so for the product of these two we have

$$h(k_1 k_2) = h(k_1)h(k_2) = ee = e,$$

which demonstrates that the product of two elements of the kernel is also in the kernel. Hence, the kernel of a homomorphism is a subgroup. The fact that these subgroups are also Lie subgroups is a consequence of the fact that Lie group homomorphisms must be differentiable mappings.

If the kernel of a homomorphism is trivial, that is, the trivial group, then we can infer that the homomorphism is injective. To see this, assume the contrary; suppose g_1 and g_2 have the same image under the homomorphism $h(g_1) = h(g_2)$. So

$$e = h(g_1)h(g_2)^{-1} = h(g_1)h(g_2^{-1}) = h(g_1 g_2^{-1}),$$

showing that the group element $g_1 g_2^{-1}$ must be in the kernel, contradicting our hypothesis.

As an application of some of these ideas, we will look at conjugations. Let G be a Lie group and g an element of G; then there is a homomorphism from G to G given by

$$h_g : G \longrightarrow G,$$
$$h_g(x) = gxg^{-1}$$

for all $x \in G$. This is certainly a homomorphism since for any two elements of G we have

$$h_g(x_1)h_g(x_2) = gx_1 g^{-1} gx_2 g^{-1} = gx_1 x_2 g^{-1} = h_g(x_1 x_2).$$

Such a homomorphism is called a **conjugation**. We already used such a mapping in the last chapter when we discussed screw motions in $\mathbb{R}^3$.

Notice that group elements that are fixed by a conjugation are precisely those that commute with g:

$$h_g(x) = x \quad \Leftrightarrow \quad gxg^{-1} = x \quad \Leftrightarrow \quad gx = xg.$$

A homomorphism given by a conjugation is always an isomorphism. We can show this in two stages; first, conjugations are injective since the kernel of a conjugation is trivial:

$$h_g(x) = e \quad \Leftrightarrow \quad gxg^{-1} = e \quad \Leftrightarrow \quad x = g^{-1}eg \quad \Leftrightarrow \quad x = e.$$

Second, conjugations are surjective since for any $y \in G$ we can always solve $h_g(x) = y$, that is, find a pre-image for y. The solution is simply given by $y = g^{-1}xg$. In fact, we can see that the inverse isomorphism to a conjugation $h_g(x) = gxg^{-1}$ is the conjugation by the inverse element to g, $h_{g^{-1}}(x) = g^{-1}xg$.

In the last chapter, we defined the group $O(n)$ to be the matrix group that preserves the bilinear form I_n under congruence. What about other bilinear forms? Suppose we considered a bilinear form given by an $n \times n$ symmetric matrix Q. This would determine a metric on $\mathbb{R}^n$ given by

$$Q(\mathbf{u}, \mathbf{v}) = \mathbf{u}^T Q \mathbf{v},$$

where $\mathbf{u}$ and $\mathbf{v}$ are arbitrary vectors in $\mathbb{R}^n$. A linear transformation of $\mathbb{R}^n$ that preserves this metric would have to satisfy

$$Q(\mathbf{u}, \mathbf{v}) = Q(M\mathbf{u}, M\mathbf{v}),$$

which implies

$$\mathbf{u}^T Q \mathbf{v} = \mathbf{u}^T M^T Q M \mathbf{v}$$

or, since the vectors are arbitrary, simply

$$M^T Q M = Q.$$

As before, we get a group of matrices that satisfy this equation. However, if Q is positive definite, then by Sylvester's law of inertia, see Cohn [21, sect. 8.2], we can always find a non-singular matrix P that transforms the identity matrix into Q by a congruence:

$$P^T I_n P = Q \qquad \text{or} \qquad I_n = (P^{-1})^T Q P^{-1}.$$

If we conjugate $GL(n)$ by this element P, then the resulting isomorphism maps our new matrix group to $O(n)$. That is, $h_P(M) = PMP^{-1}$ is in $O(n)$:

$$
\begin{aligned}
(PMP^{-1})^T (PMP^{-1}) &= (P^{-1})^T M^T P^T PMP^{-1}, \\
&= (P^{-1})^T M^T Q M P^{-1}, \\
&= (P^{-1})^T Q P^{-1}, \\
&= I_n.
\end{aligned}
$$

So the group of matrices that preserve Q is isomorphic to $O(n)$, which we have already met.

On the other hand, suppose Q was not positive definite, only non-degenerate. In general, we can find a congruence that will turn Q into a diagonal matrix with p 1s and q -1s:

$$P^T Q P = \begin{pmatrix} I_p & 0 \\ 0 & -I_q \end{pmatrix}.$$

Here, $p + q = n$ and (p, q) is the **index** of Q, that is, the number of positive and negative eigenvalues that Q has. Again, it is possible to show that groups of matrices that preserve a metric with a given index are isomorphic. So we need only consider the group of matrices that preserve the diagonal matrix given above:

$$M^T \begin{pmatrix} I_p & 0 \\ 0 & -I_q \end{pmatrix} M = \begin{pmatrix} I_p & 0 \\ 0 & -I_q \end{pmatrix}.$$

This gives us a new sequence of Lie groups, which we will call $O(p,q)$ and $SO(p,q)$ for their unit determinant subgroups. These groups have several applications; for instance, $SO(3,1)$ is the Lorentz group of special relativity.

We have a similar result for the symplectic groups $Sp(2n, \mathbb{R})$. Recall that these groups were defined as the groups of matrices that preserve the anti-symmetric matrix

$$E_{2n} = \begin{pmatrix} 0 & I_n \\ -I_n & 0 \end{pmatrix},$$

under a congruence. Now, any non-degenerate $2n \times 2n$ anti-symmetric matrix can be transformed to E_{2n} by a suitable congruence; see Cohn [21, sect. 8.5] for example. Hence, the group of matrices that preserve any non-degenerate $2n \times 2n$ anti-symmetric matrix will be isomorphic to $Sp(2n, \mathbb{R})$.

3.2   Quotients and Normal Subgroups

Each subgroup H of a group G gives an equivalence relation on G. Under this relation, two elements of G are equivalent if there is an element of the subgroup H that links them:

$$g_1 \equiv g_2 \Leftrightarrow g_1 = hg_2, \quad \text{for some } h \in H.$$

This is an equivalence relation since it is reflexive, symmetric and transitive:

reflexivity $g \equiv g \Leftrightarrow g = eg,$

symmetry $g_1 \equiv g_2 \Leftrightarrow g_1 = hg_2 \Leftrightarrow g_2 = h^{-1}g_1 \Leftrightarrow g_2 \equiv g_1,$

transitivity $g_1 \equiv g_2 \quad \& \quad g_2 \equiv g_3 \Leftrightarrow g_1 = h_1g_2 \quad \& \quad g_2 = h_2g_3$
                 $\Leftrightarrow g_1 = (h_1h_2)g_3 \Leftrightarrow g_1 \equiv g_3.$

The equivalence classes are usually called **cosets** by group theorists. The space of cosets, or equivalence classes, for such a relation is called the **quotient** of G by H, written G/H. We can denote a point in G/H by $[g]$, signifying the equivalence class of g. So if $h \in H$, then $[hg] = [g]$. The projection $G \longrightarrow G/H$ given by $g \longmapsto [g]$ is a differentiable mapping. So the quotient space is a manifold, but not necessarily a Lie group. Such a manifold is also called a **coset space** or a **homogeneous space**.

As an example, consider the $U(1)$ subgroup of $SU(2)$ given by elements of the form

$$\begin{pmatrix} \cos\theta + i\sin\theta & 0 \\ 0 & \cos\theta - i\sin\theta \end{pmatrix}.$$

The quotient here is the one-dimensional complex projective space $\mathbb{PC}^1$. This manifold is known to be isomorphic to a two-dimensional sphere, the so-called Riemann 2-sphere. Using homogeneous coordinates for the projective space, the projection from $SU(2)$ onto the quotient $\mathbb{PC}^1$ is given by

$$\begin{pmatrix} w & z \\ -z^* & w^* \end{pmatrix} \longmapsto (w : z).$$

Recall that homogeneous coordinates $(w : z)$ and $(\lambda w : \lambda z)$ refer to the same point so long as λ is a non-zero complex number.

When is the quotient space actually a Lie group? The answer to this question is, when the subgroup is a normal subgroup. A **normal subgroup** N is one that is fixed by any conjugation. This is often written $gNg^{-1} = N$. This is shorthand for

$$N \text{ is normal} \quad \Leftrightarrow \quad gng^{-1} \in N, \quad \text{for all } g \in G \text{ and all } n \in N.$$

Notice that this does not mean that we must have $gng^{-1} = n$, only that $gng^{-1} \in N$.

Now if we take the quotient of a group G by a normal subgroup N, then multiplication of the equivalence classes is well defined:

$$[g_1][g_2] = [g_1 g_2].$$

Multiplying any two representatives from the equivalence classes on the left will give a result that is always in the equivalence class on the right of the equation above:

$$(n_1 g_1)(n_2 g_2) = n_1 (g_1 n_2 g_1^{-1}) g_1 g_2 = (n_1 n_3) g_1 g_2.$$

Here $n_1 g_1$ is some element in $[g_1]$; likewise $n_2 g_2 \in [g_2]$ and $n_3 = g_1 n_2 g_1^{-1}$. This must be an element of N if N is normal. Notice also that the projection from G to G/N is a homomorphism in this case.

Consider the example $U(n)/SU(n)$. Here, the unit determinant matrices form a normal subgroup since for any $P \in U(n)$ and any $M \in SU(n)$ we have

$$\det(PMP^{-1}) = \det(P)\det(M)\det(P^{-1}) = \det(M)$$

That is, a conjugation can only turn unit determinant matrices into other unit determinant matrices. By the results above, the quotient should have the structure of a Lie group. It is not too hard to see that this must be the group $U(1)$:

$$U(n)/SU(n) = U(1),$$

where the mapping from $U(n)$ to the quotient is given by the determinant function $P \longmapsto \det(P)$.

3.3 Group Actions Again

We return to the subject of group actions. Recall that an action of a group G on a space X is given by a differentiable map $G \times X \longrightarrow X$ that satisfies

$$e(x) = x \qquad \text{for all} \quad x \in X \tag{A1}$$

and

$$g_1(g_2(x)) = g_1 g_2(x) \qquad \text{for all} \quad x \in X \quad \text{and all} \quad g_1, g_2 \in G. \tag{A2}$$

To each point x in the space X there is a subgroup of G. The **isotropy group** of a point is the subgroup of elements that leave the point fixed:

$$\mathcal{I}_x = \{g \in G \mid g(x) = x\}.$$

To see that this is a subgroup, observe that by (A1), e is in $\mathcal{I}_x$. The inverse of any element in the set is again in the set since $g^{-1}(x) = g^{-1}(g(x)) = e(x) = x$ by (A1) and (A2). Also, by (A2) the product of any two elements of $\mathcal{I}_x$ is also in the set since $g_1 g_2(x) = g_1(g_2(x)) = g_1(x) = x$.

The subspace of points reachable from some point by the action is called the **orbit** through that point:

$$\mathcal{O}_x = \{y \in X \mid g(x) = y \quad \text{for some} \quad g \in G\}.$$

If the orbit through any point is the whole of the space X, then we say that the group acts **transitively** on the space or that the group action is a **transitive action**.

Suppose H is a subgroup of G; then the quotient space G/H carries an action of G. This action is given by

$$g([g_1]) = [gg_1] \qquad \text{for all} \quad g, g_1 \in G.$$

At any point in G/H, the isotropy group for the action is just H since $h([g_1]) = [hg_1] = [g_1]$ for all $h \in H$.

For example, consider the symmetries of the projective plane $\mathbb{PR}^n$. A point in $\mathbb{PR}^n$ is given in homogeneous coordinates as $\mathbf{x} = (x_0 : x_1 : \cdots : x_n)$. The group $GL(n+1)$ acts on this space if we regard the homogeneous coordinates as coordinates of an $n+1$-dimensional vector space $\mathbf{x} = (x_0, x_1, \ldots, x_n)^T$ and $\mathbf{x}' = M\mathbf{x}$ with M a non-singular matrix. Now if the matrix is a diagonal matrix all of whose entries are $\lambda \neq 0$, that is $D(\lambda) = \lambda I_{n+1}$, then $\mathbf{x}' = D(\lambda)\mathbf{x} = \mathbf{x}$ since the homogeneous coordinates $(x_0 : x_1 : \cdots : x_n)$ and $(\lambda x_0 : \lambda x_1 : \cdots : \lambda x_n)$ represent the same point in projective space. These diagonal matrices are isomorphic to the group $GL(1)$ the multiplicative group of non-zero real numbers. For each point in projective space, these diagonal matrices are the isotropy group of the action of $GL(n+1)$; moreover, the group is normal in

$GL(n+1)$ since these diagonal matrices commute with all other matrices. Hence, we have a new sequence of groups given by the quotient

$$PGL(n) = GL(n+1)/GL(1).$$

These are the **projective groups**; they are the symmetry groups of the projective spaces. It is not difficult to show that $PGL(n)$ acts transitively on $\mathbb{PR}^n$ with trivial isotropy group. The group $PGL(2)$ is the group of projective transformations of 2-dimensional space much used in computer graphics. Notice that elements of these groups are not matrices but equivalence classes of matrices.

3.4   Matrix Normal Forms

In this section, we collect some well-known results on normal, or canonical, forms for matrices. The general situation is the following: we consider the space of matrices as a vector space; see section 2.4. These vector spaces carry representations of matrix groups. The aim is to find a matrix in each orbit of the group to represent the orbit. In this way, any matrix can be transformed to one of the normal forms by the action of some group element. Every matrix in an orbit can be transformed to the normal form representing that orbit. Only the simplest of results will be proved. The point of repeating these well-known theorems is that they are usually dispersed through the linear algebra literature and are often confused with each other. In particular, it is easy to confuse results that apply over the complex numbers with ones that apply only to matrices with real entries. Usually, the results are simpler when we consider complex matrices. Note that we have already made use of several of the results.

The first theorem concerns matrices transforming under similarity. Let $M(n)$ be the vector space of complex $n \times n$ matrices; this space has complex dimension n^2. The group $GL(n, \mathbb{C})$ acts on this space by similarity: $\tilde{M} = gMg^{-1}$, where $g \in GL(n, \mathbb{C})$ and $M, \tilde{M} \in M(n)$. This type of transformation occurs in the theory of systems of linear first order differential equations, where the transformation represents a linear transformation of the dependent variables. There are two commonly used normal forms in this case. This reflects the fact that the choice of normal forms is far from unique. Which normal form is used in any particular situation is an aesthetic decision. However, a 'nice' normal form will usually simplify calculation. The first possible normal form in this case is rational normal form. Any square matrix is similar to a block diagonal matrix

$$\begin{pmatrix} B_1 & 0 & \cdots & 0 \\ 0 & B_2 & \cdots & 0 \\ \vdots & \vdots & \ddots & \vdots \\ 0 & 0 & \cdots & B_r \end{pmatrix},$$

where each block has the form

$$B_i = \begin{pmatrix} 0 & 1 & 0 & \cdots & 0 \\ 0 & 0 & 1 & \cdots & 0 \\ \vdots & \vdots & \vdots & \ddots & \vdots \\ 0 & 0 & 0 & \cdots & 1 \\ a_0 & a_1 & a_2 & \cdots & a_p \end{pmatrix}.$$

This form is often used in control theory. Alternatively, we can use Jordan normal form. Any square matrix is also similar to a block diagonal matrix of the form

$$\begin{pmatrix} A_1 & 0 & \cdots & 0 \\ 0 & A_2 & \cdots & 0 \\ \vdots & \vdots & \ddots & \vdots \\ 0 & 0 & \cdots & A_s \end{pmatrix},$$

where the Jordan blocks have the form

$$A_i = \begin{pmatrix} \lambda_i & 1 & 0 & \cdots & 0 & 0 \\ 0 & \lambda_i & 1 & \cdots & 0 & 0 \\ \vdots & \vdots & \vdots & \ddots & \vdots & \vdots \\ 0 & 0 & 0 & \cdots & \lambda_i & 1 \\ 0 & 0 & 0 & \cdots & 0 & \lambda_i \end{pmatrix}.$$

The λs are the eigenvalues of the matrix. Hence, if the eigenvalues of a matrix are all distinct, the Jordan normal form of the matrix will be diagonal. A convenient way of describing a Jordan matrix is given by the Segre symbols. The Segre symbol of a matrix lists the sizes of the Jordan blocks associated with each eigenvalue. Thus, the Segre symbol of the Jordan matrix

$$\begin{pmatrix} \lambda_1 & 0 & 0 & 0 \\ 0 & \lambda_2 & 1 & 0 \\ 0 & 0 & \lambda_2 & 0 \\ 0 & 0 & 0 & \lambda_2 \end{pmatrix}$$

would be $[(1), (21)]$, since the first eigenvalue has just one Jordan block of size 1 while the second eigenvalue has two blocks, one 2×2 block and the other of size 1; see Cohn [21, sect. 11.4]. For real matrices transforming under $GL(n)$, there is a real Jordan form that is slightly different from the above, see Shilov [111, sect. 6.6].

Next, we look at symmetric matrices; $M^S(n)$, under orthogonal congruence. That is, we have an action of the orthogonal group $O(n)$ given by $M' = g^T M g$, with $g \in O(n)$ and $M, \tilde{M} \in M^S(n)$. Notice that we have switched back to real matrices here. For elements of the orthogonal group, similarity and congruence are the same since $g^T = g^{-1}$ in this case. Now, any symmetric matrix is congruent to a diagonal one, so in this case the normal form is given by $D(\lambda_1, \ldots, \lambda_n)$, that is, a diagonal matrix whose entries are the eigenvalues of M. In other words, the λs are the roots of the equation

$$\det(M - \lambda I_n) = 0.$$

The above can be shown by using the Gramm–Schmidt process to construct a basis for $\mathbb{R}^n$ that is orthogonal with respect to the quadratic form defined by M; see [21, sect. 8.2].

If we allow the group elements to come from the larger group $GL(n)$, then we can rescale and make all the eigenvalues $+1$, -1 or 0. That is, once we have diagonalised the matrix, we can transform with a diagonal matrix whose entries are $1/\sqrt{\lambda_i}$ for positive eigenvalues, $1/\sqrt{-\lambda_i}$ for negative eigenvalues, and 1 for zero eigenvalues. This gives us Sylvester's law of inertia: Any symmetric matrix is congruent to a diagonal matrix with entries $+1$, -1 or 0. Another way of specifying the normal form of such a matrix is to give the rank and semi-index. The **rank** is the number of non-zero eigenvalues and the **semi-index** is the number of positive eigenvalues minus the number of negative ones. These numbers can be thought of as functions with domain $M^S(n)$. We have seen that they are constants on the orbits and hence are invariant under the action of the group. Expanding the field of scalars back to the complex numbers, that is, looking at matrices with complex entries under the action of the group $GL(n, \mathbb{C})$, allows us to rescale with square roots of negative numbers. So here the normal form has only 1s or 0s on the leading diagonal, and the only invariant is the rank. Congruences of symmetric matrices are used in the study of quadric varieties and quadratic forms.

A normal form for anti-symmetric matrices under congruence is given by matrices of the form

$$\begin{pmatrix} 0 & I_r & 0 \\ -I_r & 0 & 0 \\ 0 & 0 & 0 \end{pmatrix}.$$

The rank is the only invariant; it is always even, $2r$ in the above matrix. This can be shown using a process similar to the Gramm–Schmidt one. In this case, we take vectors in $\mathbb{R}^n$ two at a time and normalise so that $\mathbf{v}^T M \mathbf{u} = -\mathbf{u}^T M \mathbf{v} = 1$. In this way, we can construct a basis in which the alternating form represented by M transforms to the given normal form; see Cohn [21, sect. 8.5].

The next result concerns pairs of symmetric matrices. The group $GL(n)$ acts by congruence on each matrix: $(\tilde{M}_1, \tilde{M}_2') = (g^T M_1 g, g^T M_2 g)$. We only consider the special case when M_1 is positive definite. In this case M_1 can be transformed to the identity matrix. Using an orthogonal transformation will not now affect the first matrix, so we may use such a transformation to diagonalise the second matrix. Hence, the normal form is $(I_n, D(\lambda_1, \ldots, \lambda_n))$. The λs are roots of the equation $\det(I_n - \lambda \tilde{M}_2) = 0$ where $\tilde{M}_2 = g^T M_2 g$ and $I_n = g^T M_1 g$. So the eigenvalue equation can be written

$$\det(g^T) \det(M_1 - \lambda M_2) \det(g) = 0$$

and then finally, since g is a non-singular matrix

$$\det(M_1 - \lambda M_2) = 0.$$

This result is used in dynamics when the kinetic and potential energies are given by quadratic forms.

The last result here relates to pencils of symmetric matrices. It is used for classifying pencils of quadrics, that is 2-dimensional linear systems of quadrics. In a linear system, we may take linear combinations of the matrices. This is useful because if we are studying the set of zeros common to a pair of quadratic equations, then the solutions will also satisfy any linear combination of the equations. The group is bigger now: $GL(n, \mathbb{C}) \times GL(2, \mathbb{C})$. The action of $GL(n, \mathbb{C})$ is as before, but now we have an action of $GL(2, \mathbb{C})$ given by

$$(\tilde{M}_1, \tilde{M}_2) = (aM_1 + bM_2, cM_1 + dM_2) \qquad \text{where} \qquad \begin{pmatrix} a & b \\ c & d \end{pmatrix} \in GL(2, \mathbb{C}).$$

We refer to the action of $GL(n, \mathbb{C})$ as a change of coordinates (in $\mathbb{C}^n$), while the action of $GL(2, \mathbb{C})$ is a change of basis in the pencil. Now we restrict our attention to non-singular, or regular, pencils. Such a pencil has some elements that are non-singular; hence, without loss of generality, we may assume that M_2 is non-singular since we can always change the basis of the pencil to achieve this result. Thus, we can form the matrix $M_2^{-1}M_1$. This matrix is no longer symmetric. A change of coordinates affects $M_2^{-1}M_1$ by a similarity transformation. Hence, by the remarks above, such a transformation cannot change the eigenvalues or Segre symbol of the matrix. Let us look at the action of $GL(2, \mathbb{C})$ on the eigenvalues and eigenvector of this matrix; they will satisfy

$$M_2^{-1}M_1\mathbf{v} = \lambda\mathbf{v},$$
$$M_1\mathbf{v} = \lambda M_2\mathbf{v}.$$

In terms of the new basis, we have

$$M_1 = \frac{d}{(ad-bc)}\tilde{M}_1 - \frac{b}{(ad-bc)}\tilde{M}_2, \qquad M_2 = -\frac{c}{(ad-bc)}\tilde{M}_1 + \frac{a}{(ad-bc)}\tilde{M}_2.$$

Substituting this into the eigenvector equation, we get

$$\left(\frac{d}{(ad-bc)}\tilde{M}_1 - \frac{b}{(ad-bc)}\tilde{M}_2\right)\mathbf{v} = \lambda\left(-\frac{c}{(ad-bc)}\tilde{M}_1 + \frac{a}{(ad-bc)}\tilde{M}_2\right)\mathbf{v},$$

$$\tilde{M}_1\mathbf{v} = \left(\frac{\lambda a + b}{\lambda c + d}\right)\tilde{M}_2\mathbf{v}.$$

That is, the eigenvector is unchanged by the change of basis, but the eigenvalue undergoes a projective transformation. That is, think of λ as the point $(\lambda : 1)$ in $\mathbb{PC}^1$. A projective transformation is then given by an element of the group $PGL(1, \mathbb{C})$. With a little more effort, we can show that the Jordan blocks of $M_2^{-1}M_1$ are unchanged by a basis change in the pencil. Hence, we can completely specify a regular pencil by giving the Segre symbol of $M_2^{-1}M_1$ and a list of its eigenvalues, remembering that if we subject the eigenvalues to a projective transformation we obtain the same pencil. In the case where all the eigenvalues are distinct, it is easy to see that the normal form is given by $(I_n, D(\lambda_1, \ldots, \lambda_n))$; see Harris [44, Lecture 22]. From the above, we could work out the normal forms for all the regular pencils, but rather than take up space here, refer to Hodge and Pedoe [52, Book IV Chap. XIII part 11]. Hodge and Pedoe also treat the case of singular pencils. Once again, there will be a finer classification for real pencils.

Now any three distinct points in $\mathbb{PC}^1$ are projectively equivalent to the three points $(1 : 0)$, $(1 : 1)$, $(0 : 1)$. So in the case $n = 3$, the regular pencils are classified completely by their Segre symbol. This is the classically well-known case of pencils of conics in $\mathbb{PC}^2$.

In concluding this section, we remark that there are many other classification problems, where a solution to such a problem means a list of all orbits of a group acting on some space. Chapter 8 below is a detailed look at one such problem. However, many classification problems remain unsolved. For example, at present there is no general classification for nets of quadrics, that is 3-dimensional linear systems of quadrics.

3.5   Subgroups of $SE(3)$

As a general problem, finding all the subgroups of a Lie group is unsolved. But in particular cases it can be done. In the case of $SE(3)$ it is possible because the group is a semi-direct product of two low-dimensional groups. The discrete subgroups of $SE(3)$ are important because they are the crystallographic groups, that is, the possible symmetry groups for atoms in crystals. These are computed in most elementary texts on crystallography.

Although the higher-dimensional subgroups of $SE(3)$ are tabulated in several places, Hervé [49] for example, no derivation seems to exist in the kinematics or mathematics literature. The derivation is sometimes given as an elementary exercise in mathematics texts.

Certainly, $\mathbb{R}^3$ is a subgroup of $SE(3)$. Using the four-dimensional representation of the group, elements of the subgroup are given by matrices of the form

$$\begin{pmatrix} I_3 & \mathbf{t} \\ 0 & 1 \end{pmatrix}.$$

This subgroup is a normal subgroup since conjugations of elements of the subgroup give

$$\begin{pmatrix} R & \mathbf{u} \\ 0 & 1 \end{pmatrix} \begin{pmatrix} I_3 & \mathbf{t} \\ 0 & 1 \end{pmatrix} \begin{pmatrix} R^T & -R^T\mathbf{u} \\ 0 & 1 \end{pmatrix} = \begin{pmatrix} I_3 & R\mathbf{t} \\ 0 & 1 \end{pmatrix}.$$

The quotient by this normal subgroup is $SE(3)/\mathbb{R}^3 = SO(3)$.

Now any subgroup of $SE(3)$ will restrict to a subgroup of $\mathbb{R}^3$. That is, if G is a subgroup of $SE(3)$, then the intersection $A = \mathbb{R}^3 \cap G$ will be a subgroup of $\mathbb{R}^3$. The possible subgroups of $\mathbb{R}^3$ are simple to find:

$$A = \mathbb{R}^3, \quad \mathbb{R}^2, \quad \mathbb{R}, \quad p\mathbb{Z}, \quad p\mathbb{Z} \times \mathbb{R}, \quad p\mathbb{Z} \times \mathbb{R}^2, \quad p\mathbb{Z} \times q\mathbb{Z},$$
$$p\mathbb{Z} \times q\mathbb{Z} \times \mathbb{R}, \quad p\mathbb{Z} \times q\mathbb{Z} \times r\mathbb{Z}, \quad 0.$$

where p, q and r are real numbers and $h\mathbb{Z}$ is the additive group with elements

$$\{\ldots, -2h, -h, 0, h, 2h, 3h, \ldots\}.$$

The projection from $SE(3)$ to its quotient $SO(3)$ is a homomorphism, so any subgroup G of $SE(3)$ projects to a subgroup, H, of $SO(3)$. Since $SO(3)$ is rather small, only three-dimensional, it is possible to find all its subgroups by inspection. The only possibilities are

$$H = 0, \quad SO(2), \quad \text{or} \quad SO(3).$$

We have listed here only the connected subgroups, since we are only interested in connected subgroups of $SE(3)$. The disconnected subgroups of $SO(3)$ consist of the discrete subgroups, called point groups in crystallography, and $O(2)$. The point groups are the symmetry groups of regular polyhedra. We can think of the $O(2)$ subgroup as the group of rotations about a fixed axis together with a rotation of π about an axis perpendicular to the original.

We have reduced the problem to finding pairs of subgroups A and H that can combine to form a subgroup of $SE(3)$. We can simplify things a little further by observing that if G is a subgroup of $SE(3)$, then its restriction to $\mathbb{R}^3$ will be a normal subgroup of G; in other words, $G/A = H$. This means that we only need to check that A is preserved by conjugation with elements of H. In general, we have

$$\begin{pmatrix} R & \mathbf{0} \\ 0 & 1 \end{pmatrix} \begin{pmatrix} I_3 & \mathbf{t} \\ 0 & 1 \end{pmatrix} \begin{pmatrix} R^T & \mathbf{0} \\ 0 & 1 \end{pmatrix} = \begin{pmatrix} I_3 & R\mathbf{t} \\ 0 & 1 \end{pmatrix}.$$

So we are looking for a group of vectors $\mathbf{t}$ that are invariant with respect to a subgroup of the rotations.

For example, if $H = SO(3)$, all possible rotations, then only $A = \mathbb{R}^3$ or $A = 0$ are possible. The subgroup $A = \mathbb{R}^2$, for instance, is not possible here, since there will always be an element in $H = SO(3)$ that rotates vectors out of any plane $\mathbb{R}^2$.

When $H = SO(2)$ it is possible to have $A = \mathbb{R}^3, \mathbb{R}^2, \mathbb{R}$, or 0. However, when $A = \mathbb{R}^2$ the plane of the vectors in A must be the same as the plane of rotations determined by H. In this case, we get $G = SO(2) \ltimes \mathbb{R}^2 = SE(2)$, the group of rigid body motions in the plane. For $A = \mathbb{R}$ the vectors in this line must be normal to the plane of the rotations determined by H, since vectors

TABLE 3.1. The Connected Subgroups of $SE(3)$

Dimension	Subgroups		
6	$SO(3) \ltimes \mathbb{R}^3 = SE(3)$		
5	–		
4	$SO(2) \ltimes \mathbb{R}^3 = SE(2) \times \mathbb{R}$		
3	$SO(2) \ltimes \mathbb{R}^2 = SE(2)$	$SO(3)$ $\mathbb{R}^3$ $H_p \ltimes \mathbb{R}^2$	
2	$SO(2) \times \mathbb{R}$	$\mathbb{R}^2$	
1	$SO(2)$	$\mathbb{R}$	H_p

normal to the plane are parallel to the axis of rotations in that plane. The group obtained here is a group of cylindric motions, rotations about a line together with translations along the line. If the line is the x-axis, then a typical element will look like

$$\begin{pmatrix} 1 & 0 & 0 & x \\ 0 & \cos\theta & -\sin\theta & 0 \\ 0 & \sin\theta & \cos\theta & 0 \\ 0 & 0 & 0 & 1 \end{pmatrix}.$$

For $H = SO(2)$ we have two more possibilities: $A = p\mathbb{Z}$ and $A = p\mathbb{Z} \times \mathbb{R}^2$. The first of these gives the screw motions we saw at the end of the previous chapter:

$$\begin{pmatrix} 1 & 0 & 0 & p\theta/2\pi \\ 0 & \cos\theta & -\sin\theta & 0 \\ 0 & \sin\theta & \cos\theta & 0 \\ 0 & 0 & 0 & 1 \end{pmatrix}.$$

Although the group of elements with this shape is isomorphic to $\mathbb{R}$, it is not conjugate to $\mathbb{R}$ in $SE(3)$. That is, no conjugation in $SE(3)$ will turn such a group into the subgroup $\mathbb{R}$. Hence, to distinguish them, we will denote these groups H_p. The second possibility thus gives rise to the subgroup $H_p \ltimes \mathbb{R}^2$.

When $H = 0$ the only connected possibilities are $A = \mathbb{R}^3, \mathbb{R}^2, \mathbb{R}, 0$. The results are summarised in Table 3.1.

Notice that there are many versions of each subgroup in $SE(3)$; for example, for every different plane $\mathbb{R}^2$ in 3-space $\mathbb{R}^3$ there is a subgroup $SE(2)$ of transformations that preserve that plane. However, the different copies of this subgroup are all conjugate. That is, there is an element of $SE(3)$ that will transform one subgroup into another. This is most easily seen by observing that there is always a rigid body transformation that will map a given plane onto another plane. The list above gives a classification of subgroups of $SE(3)$ up to conjugacy. Observe that for the screw motion subgroups H_p, different real values of p give different conjugacy classes of subgroups. The number p corresponds to the pitch of the screw motion, the distance advanced for a complete turn. Negative values of p give left-handed screw motions.

The distinction between the subgroups and their conjugacy class is important; if we want to look at the intersections of the subgroups, then we need to know about the subgroup itself not just its conjugacy class. For example, suppose we want to know the intersection of a pair of subgroups both conjugate to $SO(3)$. Assume one is the set of rotations about a point $\mathbf{p}$ and the other consists of the rotations about $\mathbf{q}$. It is not difficult to see that if $\mathbf{p} \neq \mathbf{q}$, then the rotations common to both subgroups will be the subgroup of rotations about the line joining $\mathbf{p}$ and $\mathbf{q}$. Clearly this subgroup is conjugate to $SO(2)$, the group of rotations about an axis. However, if $\mathbf{p} = \mathbf{q}$, then the intersection is just the $SO(3)$ of rotations about the common point $\mathbf{p} = \mathbf{q}$. A detailed discussion of the intersections of subgroups of $SE(3)$ is given by Hervé in [49].

3.6   Reuleaux's Lower Pairs

In 1875, Franz Reuleaux defined what he called lower pairs; see [93]. These were pairs of surfaces, identical in form, but one solid and one hollow. The pair of surfaces fit together but can move relative to each other while remaining in surface contact. Reuleaux listed six possible surfaces with this property: the sphere, the plane, the cylinder, any surface of revolution, any surface of translation, and any surface with a helicoidal symmetry. Waldron [122] showed that these are the only possibilities; his methods involved solving a linear partial differential equation; see section 12.5 later. The presentation of this result given below, follows Selig and Rooney [101]. The reason for Reuleaux's interest was that he thought that these pairs would make simple joints. For example, the sphere gives a ball and socket joint. In order to articulate two rigid bodies, we attach one of the surfaces to one of the bodies and fix the mating surface of the pair to the other body. A surface of revolution gives a pair of surfaces that can rotate relative to each other; this would be a hinge joint or what mechanical engineers call a **revolute joint**. A surface of translation yields a **prismatic joint**. A helicoidal pair of surfaces can be found on any nut and matching bolt; see Figure 3.1. Reuleaux also considered higher pairs, which were more complicated, involving cams, belts, and so forth. We will not consider these here, so for us a Reuleaux pair will be understood to mean a lower pair.

In the present setting, we recognise a Reuleaux pair as a surface in $\mathbb{R}^3$ that is invariant under the action of a subgroup of $SE(3)$. To find all possible Reuleaux pairs, we could investigate the orbits of all the possible subgroups. However, if we are systematic we can make things a little simpler. First, we look at the one-dimensional subgroups: $SO(2)$, $\mathbb{R}$ and the screw motions H_p. In each case, the orbit through a general point of $\mathbb{R}^3$ is one-dimensional, a curve. Exceptionally, the orbit through a point may be zero-dimensional, that is, fixed; this occurs if the point is on the axis of rotation. For these subgroups, we can build an

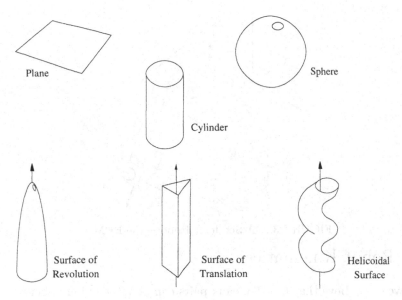

FIGURE 3.1. The Surfaces of the Six Reuleaux Lower Pairs

invariant surface as follows: Take a plane curve, which must be normal to the translation direction for $\mathbb{R}$ and must contain the rotation axis or screw axis for $SO(2)$ or H_p. Now produce a surface by 'extruding' the surface with the subgroup—think of taking the orbit through each point on the curve. In this way $\mathbb{R}$ will produce surfaces of translation, $SO(2)$ surfaces of revolution, and H_p helicoidal surfaces. It is not difficult to see that any surface invariant under the action of one of these subgroups can be formed in this way.

To investigate the surfaces invariant under higher-dimensional subgroups, notice that the higher-dimensional subgroups will contain several of the one-dimensional subgroups. So a surface invariant under a larger subgroup will have to be a combination of two or more surfaces of translation, rotation, or helicoidal surfaces. The only possibilities are:

- The cylinder: a surface of translation and of rotation about the same axis. The cylinder is also a helicoidal surface of any pitch about the cylinder's axis. The maximal symmetry group for the cylinder is $SO(2) \times \mathbb{R}$.

- The sphere: a surface of rotation about any axis through its centre. Here the maximal symmetry group is $SO(3)$.

- The plane: a surface of translation about any line parallel to a tangent and a surface of rotation about any normal. The maximal symmetry group of the plane is $SO(2) \ltimes \mathbb{R}^2 = SE(2)$.

It is straightforward to check that the orbits in $\mathbb{R}^3$ of the other subgroups, $SE(3)$, $SE(2) \times \mathbb{R}$, $\mathbb{R}^3$, and $H_p \ltimes \mathbb{R}^2$ are three-dimensional and so do not give Reuleaux pairs.

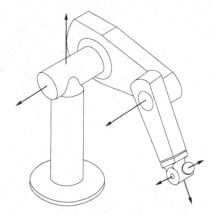

FIGURE 3.2. A Six-Joint Robot—the PUMA

3.7 Robot Kinematics

We have seen above that the Reuleaux pairs can be used as joints between the rigid members of a mechanism. For robot arms, it is usual to have six members connected in series by six one-parameter joints. The joints are often revolute joints but can also be prismatic and more rarely even helical. One-parameter joints are chosen because they are easy to drive—a simple motor for a revolute joint or a hydraulic ram for a prismatic joint. The number of joints is six so that the end-effector, or gripper, has six degrees of freedom: three positional and three orientational. This, of course, corresponds to the dimensionality of the group of rigid body transformations $SE(3)$. Redundant robots, with more than six joints, have been built for special purposes. The end-effector still has six degrees of freedom, but now the machine has more flexibility but at the cost of a harder control problem.

Consider an ordinary six-joint robot; see Figure 3.2 for example. Suppose we know the joint variables (angles or lengths) for each joint. How can we work out the position and orientation of the end-effector? This problem is called the **forward kinematic problem** for the robot. The solution is straightforward. First, we choose a 'home' configuration for the robot. In this position, all the joint variables will be taken as zero. The final position and orientation of the end-effector will be specified by giving the rigid transformation that takes the end-effector from its home position and orientation to its final configuration. Let us call this transformation $K(\boldsymbol{\theta})$, where $\boldsymbol{\theta} = (\theta_1, \theta_2, \theta_3, \theta_4, \theta_5, \theta_6)^T$ shows the dependence on the joint variables. The first joint, θ_1, is the one nearest the base, the next joint along the arm has variable θ_2, and so on, until the last; the one nearest the end-effector which has joint variable θ_6.

From the above, we know that the rigid body transformation given by a one-parameter joint has the form

$$X_p(\theta) = \begin{pmatrix} 1 & 0 & 0 & p\theta/2\pi \\ 0 & \cos\theta & -\sin\theta & 0 \\ 0 & \sin\theta & \cos\theta & 0 \\ 0 & 0 & 0 & 1 \end{pmatrix}$$

for a helical joint or a revolute joint when $p = 0$, and for a prismatic joint we have

$$X_t(\theta) = \begin{pmatrix} 1 & 0 & 0 & \theta \\ 0 & 1 & 0 & 0 \\ 0 & 0 & 1 & 0 \\ 0 & 0 & 0 & 1 \end{pmatrix}.$$

But these are for motions about particular axes. We can find the matrices for different axes by conjugating the matrices above. For example, the matrices representing rotations about a line parallel to the x-axis but through the point $\mathbf{x} = (0, 1, 0)^T$ are given by

$$\begin{pmatrix} I_3 & \mathbf{x} \\ 0 & 1 \end{pmatrix} X_0(\theta) \begin{pmatrix} I_3 & -\mathbf{x} \\ 0 & 1 \end{pmatrix} = \begin{pmatrix} 1 & 0 & 0 & 0 \\ 0 & \cos\theta & -\sin\theta & 1 - \cos\theta \\ 0 & \sin\theta & \cos\theta & \sin\theta \\ 0 & 0 & 0 & 1 \end{pmatrix}.$$

We must find these matrices for the home position of each of the robot's six joints. Call the matrices obtained in this way $A_1(\theta_1)$, $A_2(\theta_2)$, ..., $A_6(\theta_6)$.

Now to get the end-effector from home to its final configuration, we can perform the following sequence of moves: first move the final joint from 0 to the final value θ_6; this does not affect the positions of the joints lower down the arm. Next move the fifth joint from 0 to the final value θ_5; again none of the lower joints are affected by this. Continue moving the joints one by one down the arm until at last we move the first joint from 0 to the final value θ_1. The solution to the problem is thus seen to be

$$K(\boldsymbol{\theta}) = A_1(\theta_1)A_2(\theta_2)A_3(\theta_3)A_4(\theta_4)A_5(\theta_5)A_6(\theta_6).$$

It is also common to use tool-frame coordinates, and to give the forward kinematics in terms of this frame. The **tool-frame** is a coordinate system fixed to the robot's end-effector. In this formulation a passive, coordinate transformation, is sought. This is the transformation that converts coordinates in a frame fixed in the base link of the robot to coordinates in the tool frame. Suppose that, in the robot's home position the active transformation from the base link frame to the tool frame is given by the matrix B. The overall transformation between the base link frame and the tool frame in an arbitrary configuration of the robot will be given by the transformation, B to the tool frame in the home position, followed by the transformation of the end-effector itself. That is $K(\boldsymbol{\theta})B$. This is still an active transformation, to convert it to a coordinate transformation, as we saw above in section 2.5.3, we simply invert the matrix. Hence, the kinematics in terms of tool-frame coordinates is given by

$$\left(K(\boldsymbol{\theta})B\right)^{-1} = B^{-1}A_6^{-1}(\theta_6)A_5^{-1}(\theta_5)A_4^{-1}(\theta_4)A_3^{-1}(\theta_3)A_2^{-1}(\theta_2)A_1^{-1}(\theta_1)$$

Since the matrices $A_i(\theta_i)$ form one-parameter subgroups it is clear that $A_i^{-1}(\theta_i) = A_i(-\theta_i)$.

In either case we can regard the joint variables $\theta_1, \theta_2, \ldots, \theta_6$ as the coordinates of a topological space, usually called **joint space**. Joint space is in fact a smooth manifold. For example, if all the joints are revolute, then joint space is a 6-torus, the Cartesian product of six circles. The forward kinematics gives us a mapping from joint space to $SE(3)$:

$$\rho : (\theta_1, \theta_2, \theta_3, \theta_4, \theta_5, \theta_6) \longmapsto A_1(\theta_1)A_2(\theta_2)A_3(\theta_3)A_4(\theta_4)A_5(\theta_5)A_6(\theta_6).$$

In tool-frame coordinates we get a very slightly different mapping. In both cases the mapping is differentiable since in coordinates it is given by differentiable functions, sums and products of sines, and cosines and so forth. Finding right inverses for this mapping, that is to say, finding possible sets of joint parameters given the final position and orientation of the end-effector, is the problem of **inverse kinematics**, a considerably harder problem. In Chapters 5 and 10 below we study some practically important special cases of the inverse kinematics problem. In general however, there is no closed-form solution, but of course there are good numerical approaches to the problem.

Finally here, notice that if we can solve the inverse kinematics of a robot in the active viewpoint then it is simple to solve it in the tool-frame view. In the tool-frame approach we would be given a coordinate transformation K_*, and asked to find the joint angles corresponding to this move. This can be done by converting to an active transformation in the base-frame $K(\boldsymbol{\theta}) = (BK_*)^{-1}$, and then using the solution in this view. Still another version of the inverse kinematics problem specifies the move to be preformed by the robot in terms of an active transformation but relative to the current tool-frame. For example, suppose that the x-axis of the tool-frame is aligned with an axis in the robot's gripper; now we might want to command the robot to move a little way along this axis or perhaps turn about this axis. To find the joint angles in this case we need to remember the robot's current position of course; suppose this is given by the active transformation K_c of the robot's end-effector from its home position to its current position. Further suppose that the required move is given by K_Δ in the tool-frame. So the final position of the end-effector is given by multiplying K_c by K_Δ, but before we can do this we must transform K_Δ into the base frame coordinates. This is done with a conjugation, $(K_cB)K_\Delta(K_cB)^{-1}$. So the overall active transformation of the end-effector from its home position is given in the base frame by,

$$K(\boldsymbol{\theta}) = (K_cB)K_\Delta(K_cB)^{-1}K_c = K_cBK_\Delta B^{-1}.$$

Clearly there are several other ways that the inverse kinematics problem can be posed, but in all cases we can transform the required move into a form suitable for solution in the active viewpoint using the base frame.

4
Lie Algebra

In 1900 R.S. Ball first published his treatise on "the Theory of Screws" [6]. The finite screws he describes are clearly rigid body motions. Ball also describes instantaneous rigid body motions as 'twists'; these clearly correspond to elements of the Lie algebra of the group $SE(3)$. Ball's instantaneous screws are elements of the projective Lie algebra of the group $SE(3)$, that is rays through the origin in the Lie algebra. Although roughly contemporary with the work of Lie and Killing, Ball's work had a rather different focus from the emerging theory of Lie groups and algebras. We hope to show the connections here. We begin by looking at Lie algebras in general.

4.1 Tangent Vectors

There are several ways of describing Lie algebras. Initially, they were thought of as 'infinitesimal' group elements, that is, group elements very near the identity. Later, this became the tangent space to the identity element. There are now several possible, but equivalent, ways of defining tangent vectors to a manifold. We will use what is referred to as the 'geometer's definition' by Bröcker and Jänich [15], although a somewhat simplified version. Consider a smooth path through the identity in a group G, that is, a smooth mapping $\gamma : \mathbb{R} \longrightarrow G$ such that $\gamma(0) = e$. Now we introduce an equivalence relation on these paths: two paths are considered equivalent if their first derivatives at 0 are the same. A tangent vector is an equivalence class for this relation. The space of equivalence classes can be shown to be a vector space. Essentially, this involves looking at the derivatives in some local coordinate system around e. Note that in the differential geometry literature the above definition usually refers to 1-jets of paths.

As an example, let's look at the tangent space to the identity in $O(n)$ and $SO(n)$. In Section 2.2.1, we saw that the group $O(n)$ consists of two disconnected pieces and that one of these pieces, the one that contains the identity, is just $SO(n)$. So $O(n)$ and $SO(n)$ have the same tangent space at the identity and hence the same Lie algebra. In the standard matrix representation of $SO(n)$ a path through the identity is given by a matrix-valued function, $\gamma(t) = M(t)$, where $M(0) = I_n$ and $M(t)^T M(t) = I_n$. Differentiating this last relation, we get

$$\frac{d}{dt} M(t)^T M(t) + M(t)^T \frac{d}{dt} M(t) = 0.$$

When $t = 0$ we get

$$\dot{M}(0)^T + \dot{M}(0) = 0.$$

Hence, the tangent space to the identity consists of anti-symmetric matrices. This gives a simple way to find the dimension of the groups, since the dimension of a manifold is the same as the dimension of a tangent space. The dimensions of the groups $O(n)$ and $SO(n)$ are thus the dimensions of the vector spaces of $n \times n$ anti-symmetric matrices, which is simply $\frac{1}{2}n(n-1)$.

We will adopt the convention that the Lie algebra of a group will be denoted by the same name as the group but in lower case, so the Lie algebra of $SO(n)$ will be written $so(n)$ and the Lie algebra of $SE(n)$ will be $se(n)$.

We can perform similar calculations for $u(n)$ and $su(n)$. Unitary matrices satisfy the relation $U^\dagger U = I_n$. Differentiating a path in the group and restricting to the identity, we obtain

$$\dot{U}(0)^\dagger + \dot{U}(0) = 0.$$

Hence, the Lie algebra of the unitary groups consists of Hermitian matrices. The special unitary groups satisfy the additional requirement that the determinant of the matrices must be 1; this translates to the requirement that Lie algebra elements must be traceless. Thus, the Lie algebra $su(2)$ consists of matrices of the form

$$\begin{pmatrix} ai & b+ci \\ -b+ci & -ai \end{pmatrix},$$

where a, b and c are real numbers. The Lie algebra $su(2)$ is therefore a 3-dimensional vector space. In general, the dimension of $u(n)$ is n^2 and the dimension of $su(n)$ is $n^2 - 1$.

For our next example, we look at the velocities of rigid bodies. Think of a smooth path in the group $SE(3)$. This could be thought of as a parameterised sequence of rigid transformations. Applying the sequence to some rigid body, we obtain a smooth movement of the body through space. If the parameter is taken to be time, then the derivative will give a velocity. Velocities of rigid bodies are therefore, essentially, elements of the Lie algebra $se(3)$. Suppose

$$\gamma : t \longmapsto \begin{pmatrix} R(t) & \mathbf{x}(t) \\ 0 & 1 \end{pmatrix}$$

is a curve in the group $SE(3)$, parameterised by time t. The derivative at the identity element will thus be of the form

$$S = \begin{pmatrix} \Omega & \mathbf{v} \\ 0 & 0 \end{pmatrix},$$

with Ω a skew 3×3 matrix and $\mathbf{v}$ a three-component vector. These Lie algebra elements are essentially Ball's screws, or rather his twists. To get a clearer picture of all this, consider what happens to a single point $\mathbf{p} = (x, y, z)^T$ on the rigid body. Its position at any time is given by

$$\begin{pmatrix} \mathbf{p}(t) \\ 1 \end{pmatrix} = \begin{pmatrix} R(t) & \mathbf{x}(t) \\ 0 & 1 \end{pmatrix} \begin{pmatrix} \mathbf{p}(0) \\ 1 \end{pmatrix}.$$

So its velocity at $t = 0$, is given by

$$\begin{pmatrix} \dot{x}(0) \\ \dot{y}(0) \\ \dot{z}(0) \end{pmatrix} = \Omega \begin{pmatrix} x(0) \\ y(0) \\ z(0) \end{pmatrix} + \begin{pmatrix} v_x \\ v_y \\ v_z \end{pmatrix}.$$

We know that generally rigid transformations are screw motions, so generally we have $\mathbf{x} = (I - R)\mathbf{u} + \theta p \hat{\mathbf{v}}/2\pi$, and thus the velocity of the point is

$$\dot{\mathbf{p}} = \Omega(\mathbf{p} - \mathbf{u}) + \dot{\theta}\frac{p}{2\pi}\hat{\mathbf{v}}.$$

Since Ω is a 3×3 anti-symmetric matrix, we could write it as

$$\Omega = \begin{pmatrix} 0 & -\omega_z & \omega_y \\ \omega_z & 0 & -\omega_x \\ -\omega_y & \omega_x & 0 \end{pmatrix}$$

so that for any vector $\mathbf{x}$

$$\Omega \mathbf{x} = \boldsymbol{\omega} \times \mathbf{x},$$

where $\boldsymbol{\omega} = (\omega_x, \omega_y, \omega_z)^T$ and $\times$ represents the vector product. Now we can identify $\boldsymbol{\omega}$ as the instantaneous angular velocity of the motion and $\dot{\theta}\frac{p}{2\pi}\hat{\mathbf{v}}$ as the instantaneous velocity of a point on the axis of the motion. The vector $(\mathbf{p} - \mathbf{u})$ is the position vector of the point $\mathbf{p}$ relative to a point on the axis of the motion. It will often be convenient below to write these Lie algebra elements as six component vectors:

$$\mathbf{s} = \begin{pmatrix} \boldsymbol{\omega} \\ \mathbf{v} \end{pmatrix} = \begin{pmatrix} \omega_x \\ \omega_y \\ \omega_z \\ v_x \\ v_y \\ v_z \end{pmatrix},$$

where $\mathbf{v} = \mathbf{u} \times \boldsymbol{\omega} + \dot{\theta}\frac{p}{2\pi}\hat{\mathbf{v}}$ is a characteristic linear velocity of the motion. With these definitions, we see that the velocity of a point $\mathbf{p}$ on a rigid body moving with instantaneous screw $\mathbf{s}$ is given by

$$\dot{\mathbf{p}} = \boldsymbol{\omega} \times \mathbf{p} + \mathbf{v}.$$

Finally here a few words about 3×3 anti-symmetric matrices. These turn up frequently in robotics because they form the Lie algebra of the rotation group $SO(3)$. Above we have indicated how we can transform 3-dimensional vectors into 3×3 anti-symmetric matrices, however there doesn't seem to be a standard notation for this. Some author use the following,

$$[\boldsymbol{\omega}\times] = \begin{pmatrix} 0 & -\omega_z & \omega_y \\ \omega_z & 0 & -\omega_x \\ -\omega_y & \omega_x & 0 \end{pmatrix}.$$

In relativity it is common to introduce the alternating tensor $\mathcal{E}_{ijk}$, defined by

$$\mathcal{E}_{ijk} = \begin{cases} 1, & \text{if } ijk \text{ is an even permutation of 123,} \\ -1, & \text{if } ijk \text{ is an odd permutaion of 123,} \\ 0, & \text{otherwise.} \end{cases}$$

Now if the 3-dimensional vector has components $\boldsymbol{\omega}^T = (\omega_1, \omega_2, \omega_3)$, then we can write the corresponding anti-symmetric matrix as

$$(\Omega)_{ij} = \mathcal{E}_{ijk}\omega_k,$$

where summation on k is intended. Yet another approach might be to use the properties of the adjoint representation of $SO(3)$, see the following section. Then we could write

$$\Omega = \text{ad}(\boldsymbol{\omega}).$$

However, we will often be working with the group $SE(3)$ and hence we would have to show clearly that we meant the adjoint representation of $SO(3)$ and not $SE(3)$. In this work the anti-symmetric matrix corresponding to a vector $\boldsymbol{\omega}$ will simply be denoted by capitalising the name of the vector and recording that $\Omega\mathbf{x} = \boldsymbol{\omega} \times \mathbf{x}$ for any 3-vector $\mathbf{x}$.

4.2   The Adjoint Representation

Consider a lie group G and the conjugation by an element $g \in G$. This gives a smooth mapping from the manifold of G back to itself. Under this mapping, the identity element is fixed, since $geg^{-1} = e$. The differential of this map, that is its Jacobian, maps the tangent space at the identity to itself. Moreover, if we consider all Jacobians of all group elements we get a representation of the group acting on its Lie algebra. Recall that a representation is a linear action of the group; see Section 2.4. A simple path in the group is given in a matrix representation by

$$\gamma : t \longmapsto I + tX + t^2 Q(t)$$

where X is the Lie algebra element and $Q(t)$ the remainder that ensures that the image of the path stays in the group. Now, if we conjugate by g and then differentiate and set $t = 0$, we get gXg^{-1}. So the action is given by

$$\mathrm{Ad}(g)X = gXg^{-1}, \qquad \text{for all} \quad g \in G.$$

We have seen in the previous chapter how conjugations give group actions; what is new here is that the action is linear. For any two scalars α and β, we have

$$\mathrm{Ad}(g)(\alpha X_1 + \beta X_2) = g(\alpha X_1 + \beta X_2)g^{-1} = \alpha g X_1 g^{-1} + \beta g X_2 g^{-1}$$
$$= \alpha \, \mathrm{Ad}(g)X_1 + \beta \, \mathrm{Ad}(g)X_2.$$

This representation of the group on its Lie algebra is called the **adjoint representation** of the group.

As usual, our examples will be $so(3)$ and $se(3)$. For the rotations, we must calculate the product

$$\Omega' = R\Omega R^T,$$

where Ω is a skew matrix

$$\Omega = \begin{pmatrix} 0 & -\omega_z & \omega_y \\ \omega_z & 0 & -\omega_x \\ -\omega_y & \omega_x & 0 \end{pmatrix}.$$

To facilitate the computation, we will write the rotation matrix as partitioned into three orthogonal vectors:

$$R = \begin{pmatrix} \mathbf{r}_1^T \\ \hline \mathbf{r}_2^T \\ \hline \mathbf{r}_3^T \end{pmatrix}.$$

The advantage of this is that, because of the relation $R^T R = I_3$, the vectors must be mutually orthogonal unit vectors:

$$\mathbf{r_i} \cdot \mathbf{r_j} = \begin{cases} 1, & \text{if } i = j, \\ 0, & \text{if } i \neq j. \end{cases}$$

Moreover, the relation $\det(R) = 1$ means that the triple product $\mathbf{r}_1 \cdot (\mathbf{r}_2 \times \mathbf{r}_3) = 1$. From these relations, we infer

$$\mathbf{r}_1 \times \mathbf{r}_2 = \mathbf{r}_3, \qquad \mathbf{r}_2 \times \mathbf{r}_3 = \mathbf{r}_1, \qquad \mathbf{r}_3 \times \mathbf{r}_1 = \mathbf{r}_2.$$

Lastly, recall that $\Omega \mathbf{v} = \boldsymbol{\omega} \times \mathbf{v}$ for any vector $\mathbf{v}$ and where $\boldsymbol{\omega} = (\omega_x, \omega_y, \omega_z)^T$. Armed with this knowledge and a little vector algebra, we have

$$R\Omega R^T = R\Big(\boldsymbol{\omega} \times \mathbf{r}_1 \,|\, \boldsymbol{\omega} \times \mathbf{r}_2 \,|\, \boldsymbol{\omega} \times \mathbf{r}_3 \Big)$$

$$= \begin{pmatrix} 0 & \mathbf{r}_1 \cdot (\boldsymbol{\omega} \times \mathbf{r}_2) & \mathbf{r}_1 \cdot (\boldsymbol{\omega} \times \mathbf{r}_3) \\ \mathbf{r}_2 \cdot (\boldsymbol{\omega} \times \mathbf{r}_1) & 0 & \mathbf{r}_2 \cdot (\boldsymbol{\omega} \times \mathbf{r}_3) \\ \mathbf{r}_3 \cdot (\boldsymbol{\omega} \times \mathbf{r}_1) & \mathbf{r}_3 \cdot (\boldsymbol{\omega} \times \mathbf{r}_2) & 0 \end{pmatrix}.$$

Rearranging cyclically, we get

$$R\Omega R^T = \begin{pmatrix} 0 & \boldsymbol{\omega} \cdot (\mathbf{r}_2 \times \mathbf{r}_1) & \boldsymbol{\omega} \cdot (\mathbf{r}_3 \times \mathbf{r}_1) \\ \boldsymbol{\omega} \cdot (\mathbf{r}_1 \times \mathbf{r}_2) & 0 & \boldsymbol{\omega} \cdot (\mathbf{r}_3 \times \mathbf{r}_2) \\ \boldsymbol{\omega} \cdot (\mathbf{r}_1 \times \mathbf{r}_3) & \boldsymbol{\omega} \cdot (\mathbf{r}_2 \times \mathbf{r}_3) & 0 \end{pmatrix}$$

$$= \begin{pmatrix} 0 & -\mathbf{r}_3 \cdot \boldsymbol{\omega} & \mathbf{r}_2 \cdot \boldsymbol{\omega} \\ \mathbf{r}_3 \cdot \boldsymbol{\omega} & 0 & -\mathbf{r}_1 \cdot \boldsymbol{\omega} \\ -\mathbf{r}_2 \cdot \boldsymbol{\omega} & \mathbf{r}_1 \cdot \boldsymbol{\omega} & 0 \end{pmatrix}.$$

Now, we can see that in terms of the vector $\boldsymbol{\omega}$ the adjoint representation is given by $R\boldsymbol{\omega}$. That is, the adjoint representation of $SO(3)$ is the same as its defining representation on $\mathbb{R}^3$. Again, this is an accidental property of three dimensions, which does not generalise.

The adjoint action of $SE(3)$ on its Lie algebra is now fairly simple to compute. For a typical Lie algebra element in 4×4 matrix form, we have

$$\begin{pmatrix} \Omega' & \mathbf{v}' \\ 0 & 0 \end{pmatrix} = \begin{pmatrix} R & \mathbf{t} \\ 0 & 1 \end{pmatrix} \begin{pmatrix} \Omega & \mathbf{v} \\ 0 & 0 \end{pmatrix} \begin{pmatrix} R^T & -R^T \mathbf{t} \\ 0 & 1 \end{pmatrix} = \begin{pmatrix} R\Omega R^T & R\mathbf{v} - R\Omega R^T \mathbf{t} \\ 0 & 0 \end{pmatrix}.$$

We saw above that $R\Omega R^T$ is equivalent to $R\boldsymbol{\omega}$. So $R\Omega R^T \mathbf{t} = (R\boldsymbol{\omega}) \times \mathbf{t}$. Now we, write

$$T = \begin{pmatrix} 0 & -t_z & t_y \\ t_z & 0 & -t_x \\ -t_y & t_x & 0 \end{pmatrix}$$

so that $T\mathbf{x} = \mathbf{t} \times \mathbf{x}$ for any vector $\mathbf{x}$. The term $-R\Omega R^T \mathbf{t}$ can now be written as $TR\boldsymbol{\omega}$. In the six-component vector form of the Lie algebra, the representation has the form

$$\begin{pmatrix} \boldsymbol{\omega}' \\ \mathbf{v}' \end{pmatrix} = \begin{pmatrix} R & 0 \\ TR & R \end{pmatrix} \begin{pmatrix} \boldsymbol{\omega} \\ \mathbf{v} \end{pmatrix}.$$

In other words, a rotation by R followed by a translation $\mathbf{t}$ is represented by the 6×6 matrix

$$\begin{pmatrix} R & 0 \\ TR & R \end{pmatrix}.$$

It is straightforward to show that these matrices do form a representation of $SE(3)$. The product of two of them gives a third:

$$\begin{pmatrix} R_1 & 0 \\ T_1 R_1 & R_1 \end{pmatrix} \begin{pmatrix} R_2 & 0 \\ T_2 R_2 & R_2 \end{pmatrix} = \begin{pmatrix} R_1 R_2 & 0 \\ (T_1 + R_1 T_2 R_1^T) R_1 R_2 & R_1 R_2 \end{pmatrix}.$$

The inverse of such a matrix is given by

$$\begin{pmatrix} R & 0 \\ TR & R \end{pmatrix}^{-1} = \begin{pmatrix} R^T & 0 \\ -R^T T & R^T \end{pmatrix}.$$

To have the inverse in exactly the same form as the original matrices, that is, with the bottom left element in the shape TR, we could always write $-R^T T = -(R^T TR)R^T$.

4.3 Commutators

Now let us look at the differential of the adjoint representation. Suppose that g is a matrix in a group, close to the identity, so it can be approximated by $g \approx I + tX + \theta(t^2)$, where X is a Lie algebra element. The inverse of g is then approximately $g^{-1} \approx I - tX + \theta(t^2)$. Conjugation of a general element Y in the Lie Algebra gives

$$(I + tX)Y(I - tX) = Y + t(XY - YX) + \theta(t^2).$$

Differentiating and setting $t = 0$ gives $XY - YX$. The commutator of two Lie algebra elements is another Lie algebra element. Hence, we get a binary operation defined on the Lie algebra of a group. This operation is usually written $[X, Y] = XY - YX$ and is called the **Lie bracket**, or commutator, of two elements. The existence of this binary operation is why these spaces are called algebras. Note that the multiplication of two matrices representing Lie algebra elements does not in general produce another Lie algebra element.

From the definition, we can see that the Lie bracket is a linear operation:

$$[aX_1 + bX_2, Y] = (aX_1 + bX_2)Y - Y(aX_1 + bX_2)$$
$$= a(X_1 Y - Y X_1) + b(X_2 Y - Y X_2) = a[X_1, Y] + b[X_2, Y]$$

and also

$$[X, aY_1 + bY_2] = X(aY_1 + bY_2) - (aY_1 + bY_2)X$$
$$= a(XY_1 - Y_1 X) + b(XY_2 - Y_2 X) = a[X, Y_1] + b[X, Y_2].$$

These relations are extremely useful since they mean that we only have to find the commutators for pairs of basis elements. All other elements of the Lie algebra are linear combinations of the basis elements, and hence commutators for them can be found from the commutators of the basis elements. If we have a basis $\{X_1, X_2, \ldots, X_n\}$ for the algebra, then the Lie bracket of any pair of basis elements will be a linear combination of basis elements:

$$[X_i, X_j] = C_{ij}^1 X_1 + C_{ij}^2 X_2 + \cdots + C_{ij}^n X_n.$$

The constants C_{ij}^k are called the **structure constants** of the algebra. Giving the structure constants for a Lie algebra specifies the algebra completely.

The Lie bracket is anti-symmetric, $[X_1, X_2] = -[X_2, X_1]$, so for any element X we have $[X, X] = 0$. This also means that the structure constants obey $C_{ji}^k = -C_{ij}^k$. This further reduces the number of commutators, or structure constants, we have to give to specify the Lie algebra.

The commutator is not associative; it does, however, obey the Jacobi identity:

$$[X_1, [X_2, X_3]] + [X_2, [X_3, X_1]] + [X_3, [X_1, X_2]] = 0.$$

Essentially, this relation says that the Lie bracket is a derivation, since rearranging the Jacobi identity gives the Leibnitz rule for the derivation of a product:

$$[X_1, [X_2, X_3]] = [[X_1, X_2], X_3] + [X_2, [X_1, X_3]].$$

Another way to look at the above is to think of it as defining a representation of the Lie algebra on itself. Again, we call this representation the **adjoint representation** but write it in lower case:

$$\mathrm{ad}(X)Y = XY - YX = [X, Y].$$

This might be a little confusing. The difficulty arises because we are thinking about vector spaces of matrices. Elements of the space are vectors, since they can be added and multiplied by scalars. However, the elements of the space are also matrices. For example, the space of anti-symmetric, traceless matrices form a Lie algebra. So, consider a Lie algebra of $m \times m$ matrices. The dimension of the algebra as a vector space is not dependent on the dimensions of the constituent matrices; suppose the dimension of the algebra is n. Now, $\mathrm{ad}(X)$ is another matrix which acts on the vector space, hence it will be given by an $n \times n$ matrix. Given a basis for the Lie algebra as above, the kj-th element of the matrix $\mathrm{ad}(X_i)$ will be the structure constant C_{ij}^k.

Our first example is $so(3)$. This Lie algebra consists of 3×3 anti-symmetric matrices. As a basis for this 3-dimensional vector space, let us take

$$X = \begin{pmatrix} 0 & 0 & 0 \\ 0 & 0 & -1 \\ 0 & 1 & 0 \end{pmatrix}, \qquad Y = \begin{pmatrix} 0 & 0 & 1 \\ 0 & 0 & 0 \\ -1 & 0 & 0 \end{pmatrix}, \qquad Z = \begin{pmatrix} 0 & -1 & 0 \\ 1 & 0 & 0 \\ 0 & 0 & 0 \end{pmatrix}.$$

The commutators are now easy to work out:

$$[X, Y] = \begin{pmatrix} 0 & -1 & 0 \\ 1 & 0 & 0 \\ 0 & 0 & 0 \end{pmatrix} = Z,$$

$$[Y, Z] = \begin{pmatrix} 0 & 0 & 0 \\ 0 & 0 & -1 \\ 0 & 1 & 0 \end{pmatrix} = X,$$

$$[Z, X] = \begin{pmatrix} 0 & 0 & 1 \\ 0 & 0 & 0 \\ -1 & 0 & 0 \end{pmatrix} = Y.$$

In this case, we see that $\mathrm{ad}(X) = X$, $\mathrm{ad}(Y) = Y$, and $\mathrm{ad}(Z) = Z$. This is rather special and not a general result.

Next we look at $su(2)$. This Lie algebra consists of traceless Hermitian 2×2 matrices; see Section 4.1. As a basis for the algebra, we can take

$$\sigma_x = \frac{1}{2} \begin{pmatrix} i & 0 \\ 0 & -i \end{pmatrix}, \qquad \sigma_y = \frac{1}{2} \begin{pmatrix} 0 & 1 \\ -1 & 0 \end{pmatrix}, \qquad \sigma_z = \frac{1}{2} \begin{pmatrix} 0 & i \\ i & 0 \end{pmatrix}.$$

These are the Pauli spin matrices of quantum physics. Computing the commutators of these matrices gives the following results:

$$[\sigma_x, \sigma_y] = \frac{1}{2} \begin{pmatrix} 0 & i \\ i & 0 \end{pmatrix} = \sigma_z,$$

$$[\sigma_y, \sigma_z] = \frac{1}{2} \begin{pmatrix} i & 0 \\ 0 & -i \end{pmatrix} = \sigma_x,$$

$$[\sigma_z, \sigma_x] = \frac{1}{2} \begin{pmatrix} 0 & 1 \\ -1 & 0 \end{pmatrix} = \sigma_y.$$

In this case, we see that the adjoint representation is given by

$$\mathrm{ad}(\sigma_x) = X, \quad \mathrm{ad}(\sigma_y) = Y, \quad \mathrm{ad}(\sigma_z) = Z.$$

with X, Y, and Z the 3×3 matrices given above.

Notice that the last two examples are isomorphic to each other. Moreover, both of these algebras are exactly the algebra of 3-dimensional vectors under the action of the vector product and also the Lie algebra to the group of unit quaternions.

This is rather useful for us, but it is accidental. There is no generalisation of this to four or more dimensions; for example $so(4)$, the Lie algebra of $SO(4)$, is six-dimensional. Writing the basis vectors X_{12}, X_{13}, X_{14}, X_{23}, X_{24} and X_{34}, the commutation relation can be summarised neatly as

$$\begin{aligned}
[X_{ij}, X_{ik}] &= X_{jk}, \; 1 \le i < j < k \le 4, \\
[X_{ij}, X_{jk}] &= -X_{ik}, \; 1 \le i < j < k \le 4, \\
[X_{ij}, X_{kj}] &= X_{ik}, \; 1 \le i < k < j \le 4. \\
[X_{ij}, X_{kl}] &= 0, \quad \text{if no indices match.}
\end{aligned}$$

For the six-dimensional Lie algebra of the Euclidean group $SE(3)$, we can use the following generators:

$$S(\boldsymbol{\omega}_i) = \begin{pmatrix} 0 & 0 & 0 & 0 \\ 0 & 0 & -1 & 0 \\ 0 & 1 & 0 & 0 \\ 0 & 0 & 0 & 0 \end{pmatrix}, \quad S(\mathbf{v}_i) = \begin{pmatrix} 0 & 0 & 0 & 1 \\ 0 & 0 & 0 & 0 \\ 0 & 0 & 0 & 0 \\ 0 & 0 & 0 & 0 \end{pmatrix},$$

$$S(\boldsymbol{\omega}_j) = \begin{pmatrix} 0 & 0 & 1 & 0 \\ 0 & 0 & 0 & 0 \\ -1 & 0 & 0 & 0 \\ 0 & 0 & 0 & 0 \end{pmatrix}, \quad S(\mathbf{v}_j) = \begin{pmatrix} 0 & 0 & 0 & 0 \\ 0 & 0 & 0 & 1 \\ 0 & 0 & 0 & 0 \\ 0 & 0 & 0 & 0 \end{pmatrix},$$

$$S(\boldsymbol{\omega}_k) = \begin{pmatrix} 0 & -1 & 0 & 0 \\ 1 & 0 & 0 & 0 \\ 0 & 0 & 0 & 0 \\ 0 & 0 & 0 & 0 \end{pmatrix}, \quad S(\mathbf{v}_k) = \begin{pmatrix} 0 & 0 & 0 & 0 \\ 0 & 0 & 0 & 0 \\ 0 & 0 & 0 & 1 \\ 0 & 0 & 0 & 0 \end{pmatrix}.$$

The commutation relation for these generators, from which the commutators of any elements of the algebra can be found, are then easily computed. The results are

$$\begin{array}{lll} [\boldsymbol{\omega}_i, \boldsymbol{\omega}_j] = \boldsymbol{\omega}_k, & [\boldsymbol{\omega}_j, \boldsymbol{\omega}_k] = \boldsymbol{\omega}_i, & [\boldsymbol{\omega}_k, \boldsymbol{\omega}_i] = \boldsymbol{\omega}_j, \\ [\boldsymbol{\omega}_i, \mathbf{v}_j] = \mathbf{v}_k, & [\boldsymbol{\omega}_j, \mathbf{v}_k] = \mathbf{v}_i, & [\boldsymbol{\omega}_k, \mathbf{v}_i] = \mathbf{v}_j, \\ [\boldsymbol{\omega}_i, \mathbf{v}_k] = -\mathbf{v}_j, & [\boldsymbol{\omega}_j, \mathbf{v}_i] = -\mathbf{v}_k, & [\boldsymbol{\omega}_k, \mathbf{v}_j] = -\mathbf{v}_i, \\ [\boldsymbol{\omega}_i, \mathbf{v}_i] = \mathbf{0}, & [\boldsymbol{\omega}_j, \mathbf{v}_j] = \mathbf{0}, & [\boldsymbol{\omega}_k, \mathbf{v}_k] = \mathbf{0}, \\ [\mathbf{v}_i, \mathbf{v}_j] = \mathbf{0}, & [\mathbf{v}_j, \mathbf{v}_k] = \mathbf{0}, & [\mathbf{v}_k, \mathbf{v}_i] = \mathbf{0}. \end{array}$$

Notice that the $\boldsymbol{\omega}$s behave like a copy of $so(3)$ while the $\mathbf{v}$s all commute with each other. This reflects the fact that $SO(3)$ and $\mathbb{R}^3$ are subgroups of $SE(3)$. The adjoint representation can now be calculated. We use the following column vectors as a basis for the six-dimensional vector space:

$$\boldsymbol{\omega}_i = \begin{pmatrix} 1 \\ 0 \\ 0 \\ 0 \\ 0 \\ 0 \end{pmatrix}, \quad \boldsymbol{\omega}_j = \begin{pmatrix} 0 \\ 1 \\ 0 \\ 0 \\ 0 \\ 0 \end{pmatrix}, \quad \boldsymbol{\omega}_k = \begin{pmatrix} 0 \\ 0 \\ 1 \\ 0 \\ 0 \\ 0 \end{pmatrix},$$

$$\mathbf{v}_i = \begin{pmatrix} 0 \\ 0 \\ 0 \\ 1 \\ 0 \\ 0 \end{pmatrix}, \quad \mathbf{v}_j = \begin{pmatrix} 0 \\ 0 \\ 0 \\ 0 \\ 1 \\ 0 \end{pmatrix}, \quad \mathbf{v}_k = \begin{pmatrix} 0 \\ 0 \\ 0 \\ 0 \\ 0 \\ 1 \end{pmatrix}.$$

Now, bearing in mind that $\mathrm{ad}(X)Y = [X, Y]$, the adjoint representation of this Lie algebra is given by

$$\mathrm{ad}(\boldsymbol{\omega}_i) = \begin{pmatrix} X & 0 \\ 0 & X \end{pmatrix}, \quad \mathrm{ad}(\mathbf{v}_i) = \begin{pmatrix} 0 & 0 \\ X & 0 \end{pmatrix},$$

$$\mathrm{ad}(\boldsymbol{\omega}_j) = \begin{pmatrix} Y & 0 \\ 0 & Y \end{pmatrix}, \quad \mathrm{ad}(\mathbf{v}_j) = \begin{pmatrix} 0 & 0 \\ Y & 0 \end{pmatrix},$$

$$\mathrm{ad}(\boldsymbol{\omega}_k) = \begin{pmatrix} Z & 0 \\ 0 & Z \end{pmatrix}, \quad \mathrm{ad}(\mathbf{v}_k) = \begin{pmatrix} 0 & 0 \\ Z & 0 \end{pmatrix},$$

where X, Y and Z are the 3×3 matrices we met above in connection with the adjoint representation of $SO(3)$.

4.4   The Exponential Mapping

Another way of looking at Lie algebra elements is as left-invariant vector fields on the group. So far, we have only mentioned vectors at the identity; for a vector field, we need a tangent at every group element. Multiplication by a group element on the left gives an isomorphism of the group manifold $g : G \longrightarrow G$, where $g(g_1) = gg_1$. This action of the group on its underlying manifold induces an action on the vector fields of the manifold. The left-invariant vector fields are fixed by this group action. Now any left-invariant vector field restricts to a tangent vector at the identity, a Lie algebra element. However, given a tangent vector at the identity, we can produce a left-invariant vector field. All we do is to left translate the original vector to every point on the manifold. If X is a matrix representing a tangent vector at the identity, then the tangent vector at the point g of the group will be given by gX. Hence, there is a one-to-one correspondence between tangent vectors at the identity and left-invariant vector fields.

The integral curves of these left-invariant vector fields play an important role in what follows. Integral curves of a vector field are curves that are tangent to the field at each point. For a left-invariant vector field, such a curve would satisfy the differential equation

$$\frac{d\gamma}{dt} = \gamma X.$$

This equation has an analytic solution. The solution that passes through the identity element is

$$\gamma(t) = e^{tX}.$$

The exponential of a matrix X can be expanded into a power series:

$$e^X = 1 + X + X^2/2 + \cdots + X^n/n! + \cdots.$$

For exponentials of matrices, we have the relation

$$e^X e^Y = e^{X+Y} \qquad \text{if and only if} \qquad [X, Y] = 0.$$

That is, we may only add the exponents in a product of exponentials if the exponents commute. Certainly, the elements $t_1 X$ and $t_2 X$ commute. This means that the elements of the group of the form e^{tX} form a subgroup:

$$e^{t_1 X} e^{t_2 X} = e^{(t_1 + t_2)X} \qquad \text{and} \qquad e^{tX} e^{-tX} = I.$$

These are the one-dimensional or one-parameter subgroups of the group. Each Lie algebra element generates a one-parameter subgroup in this way.

The exponential function can also be regarded as giving a mapping from the Lie algebra to the group. This mapping is neither injective nor surjective in general. However, near the identity it is a homeomorphism. That is, there is a neighbourhood of 0 in the Lie algebra that maps homeomorphically to a neighbourhood of the identity in the group. In this neighbourhood, there is an inverse mapping, usually called the logarithm, given by the well-known Mercator series

$$\log(g) = (g - I) - \frac{1}{2}(g - I)^2 + \frac{1}{3}(g - I)^3 - \frac{1}{4}(g - I)^4 + \cdots.$$

When g is too far from the identity, this series fails to converge; see for example Curtis [23, Chap.4].

The determinant of a matrix exponential is given by the exponential of the trace of the matrix:

$$\det(e^X) = e^{\text{Tr}(X)}.$$

The trace of a matrix, $\text{Tr}()$ is the sum of its diagonal entries. If the eigenvalues of the matrix X are distinct, then this relation is simple to prove by diagonalising the matrix. However, the relationship is generally true. The relation shows that the exponential of a matrix has unit determinant if and only if the matrix is traceless. This is why the Lie algebras $so(n)$, $su(n)$ and $sl(n)$ consist of traceless matrices.

For some Lie algebras, we can be more specific about the exponential map. For example, consider $su(2)$. Recall from Section 4.1 that a typical Lie algebra element $\mathbf{m}$ is represented in the adjoint representation by a matrix of the form

$$\text{ad}(\mathbf{m}) = M = \begin{pmatrix} ai & b + ci \\ -b + ci & -ai \end{pmatrix}.$$

A straightforward calculation reveals that $M^2 = -(a^2 + b^2 + c^2)I_2$. Now, if we put $M = \sqrt{a^2 + b^2 + c^2} X$ we will have that $X^2 = -I_2$. So let us identify $\det(M) = a^2 + b^2 + c^2$ with a parameter t^2 then we can substitute this into the definition of the exponential to obtain

$$e^{tX} = (1 - \frac{t^2}{2!} + \frac{t^4}{4!} \cdots)I_2 + (t - \frac{t^3}{3!} + \frac{t^5}{5!} \cdots)X = \cos(t)I_2 + \sin(t)X.$$

This can also be written in terms of the original Lie algebra element M,

$$e^M = \cos(t)I_2 + \frac{1}{t}\sin(t)M.$$

Remember we can find t from the determinant of M.

The fact that this relation is linear means that the logarithm is simple to find, if U is an element of $su(2)$, then

$$\log(U) = t\csc(t)U - t\cot(t)I_2.$$

Here, t can be found from the fact that $\mathrm{Tr}(U) = 2\cos(t)$. The logarithm is not globally defined on the group, as mentioned above. In fact it is clear that the formula given above fails when $t = \pm\pi$ radians.

4.4.1   The Exponential of Rotation Matrices

Next we look at the 3×3 matrices representing the algebra $so(3)$. These matrices are anti-symmetric, and a direct calculation shows that a 3×3 anti-symmetric matrix

$$\Omega = \begin{pmatrix} 0 & -\omega_z & \omega_y \\ \omega_z & 0 & -\omega_x \\ -\omega_y & \omega_x & 0 \end{pmatrix},$$

satisfies a cubic equation

$$\Omega^3 + \theta^2\Omega = 0,$$

where $\theta^2 = \omega_x^2 + \omega_y^2 + \omega_z^2$. Rather than follow the method used to find the exponential in $su(2)$ as above, a more direct approach will be developed. This involves writing the anti-symmetric matrix as a sum of mutually annihilating idempotents.

Consider the three matrices

$$P_0 = \frac{1}{\theta^2}(\Omega + i\theta I_3)(\Omega - i\theta I_3) = \frac{1}{\theta^2}\Omega^2 + I_3,$$

$$P_+ = \frac{-1}{2\theta^2}\Omega(\Omega - i\theta I_3) = \frac{-1}{2\theta^2}\Omega^2 + \frac{i}{2\theta}\Omega,$$

$$P_- = \frac{-1}{2\theta^2}\Omega(\Omega + i\theta I_3) = \frac{-1}{2\theta^2}\Omega^2 - \frac{i}{2\theta}\Omega.$$

It is easy to see that these matrices annihilate each other since, for example,

$$P_0 P_+ = \frac{-1}{2\theta^4}\Omega(\Omega + i\theta I_3)(\Omega - i\theta I_3)^2 = \frac{-1}{2\theta^4}(\Omega^3 + \theta^2\Omega)(\Omega - i\theta I_3) = 0$$

using the cubic equation satisfied by Ω. In general we have that, $P_0 P_+ = 0$, $P_0 P_- = 0$ and $P_+ P_- = 0$.

These matrices can be found by expanding the reciprocal of the cubic into partial fractions, see [113]. One consequence of this is that the sum of the matrices is the identity matrix

$$P_0 + P_+ + P_- = I_3.$$

This can also be checked by direct computation. The fact that these matrices are idempotents is now easily proved, for instance,

$$P_0 = I_3 P_0 = (P_0 + P_+ + P_-)P_0 = P_0^2$$

and in general, $P_0^2 = P_0$, $P_+^2 = P_+$ and $P_-^2 = P_-$.

The final property we need is that a linear combination of the idempotents gives us back Ω,

$$\Omega = i\theta P_- - i\theta P_+.$$

The point of these manipulations is that if we raise Ω to some power then because the 'P' matrices are mutually annihilating there are no cross terms. Moreover, since the 'P's are idempotent, only their coefficients are effected by the power

$$\Omega^n = (-i\theta)^n P_+ + (i\theta)^n P_-.$$

Hence the exponential of the matrix Ω can be found as

$$e^\Omega = P_0 + e^{-i\theta} P_+ + e^{i\theta} P_-.$$

Now we can replace the idempotents by their definitions in terms of Ω to get

$$e^\Omega = I_3 + \frac{i}{2\theta}\left(e^{-i\theta} - e^{i\theta}\right)\Omega - \frac{1}{2\theta^2}\left(e^{i\theta} + e^{-i\theta} - 2\right)\Omega^2.$$

Finally, replacing the complex exponential by trigonometic functions we have

$$e^\Omega = I_3 + \frac{1}{\theta}\sin\theta\,\Omega + \frac{1}{\theta^2}(1 - \cos\theta)\Omega^2.$$

This relation is usually known as the Rodrigues formula, and written as $e^{\theta\mathbf{v}}\mathbf{u} = \mathbf{u} + \sin\theta\mathbf{v}\times\mathbf{u} + (1 - \cos\theta)\mathbf{v}\times(\mathbf{v}\times\mathbf{u})$.

The inverse function, the logarithm, is not hard to find. Suppose that we are given an arbitrary 3×3 special orthogonal matrix, that is, an element of $SO(3)$, R say. We can find the angle θ and the anti-symmetric matrix Ω as follows. Notice that $\mathrm{Tr}(I_3) = 3$, $\mathrm{Tr}(\Omega) = 0$ and $\mathrm{Tr}(\Omega^2) = -2\theta^2$. Comparing R with the exponential of a Lie algebra element, we have

$$R = e^\Omega = I_3 + \frac{1}{\theta}\sin\theta\,\Omega + \frac{1}{\theta^2}(1 - \cos\theta)\Omega^2.$$

so the trace of R gives

$$\mathrm{Tr}(R) = \mathrm{Tr}(I_3) + \frac{1}{\theta}\sin\theta\,\mathrm{Tr}(\Omega) + \frac{1}{\theta^2}(1 - \cos\theta)\mathrm{Tr}(\Omega^2) = 1 + 2\cos\theta.$$

To find the anti-symmetric matrix Ω observe that since the matrix Ω is anti-symmetric, its square Ω^2 must be symmetric like I_3. Hence, if we compute $R - R^T$ we will obtain

$$R - R^T = \frac{2}{\theta} \sin \theta \, \Omega.$$

In other words, the logarithm is given by

$$\log(R) = \frac{\theta}{2 \sin \theta} (R - R^T).$$

The method fails when $\theta = \pm\pi$, since $\sin \pi = 0$.

This formula does not generalise to $SE(3)$ very easily, so a slightly different formula will be derived. We seek a formula of the form

$$\Omega = aI_3 + bR + cR^2$$

where a, b and c are to be determined. Only powers of R up to 2 are needed since R satisfies a cubic equation. This follows from the Cayley–Hamilton theorem, but it is not hard to find the cubic relation

$$R^3 - \text{Tr}(R)R^2 + \text{Tr}(R)R - I_3 = 0.$$

In terms of the idempotents we have

$$I_3 = P_0 + P_+ + P_-,$$
$$R = P_0 + e^{-i\theta} P_+ + e^{i\theta} P_-,$$
$$R^2 = P_0 + e^{-2i\theta} P_+ + e^{2i\theta} P_-$$

and also

$$\Omega = -i\theta P_+ + i\theta P_-.$$

Substituting these relations into the equations above and comparing coefficients of P_0, P_+ and P_-, we obtain three linear equations in the unknowns a, b and c:

$$\begin{aligned}
a + b \quad\;\; + c \quad\;\;\; &= \quad 0, \\
a + be^{-i\theta} + ce^{-2i\theta} &= -i\theta, \\
a + be^{i\theta} \;\; + ce^{2i\theta} &= \quad i\theta.
\end{aligned}$$

These linear equations are exactly the same equations that would be derived in a Lagrange interpolation problem. So there is a standard solution to the problem. In other words, the problem is now equivalent to finding the polynomial $p(x) = a + bx + cx^2$, which passes through the three points $(x_1, y_1) = (1, 0)$, $(x_2, y_2) = (e^{-i\theta}, -i\theta)$ and $(x_3, y_3) = (e^{i\theta}, i\theta)$. The standard solution is

$$p(x) = \frac{(x - x_2)(x - x_3)}{(x_1 - x_2)(x_1 - x_3)} y_1 + \frac{(x - x_3)(x - x_1)}{(x_2 - x_3)(x_2 - x_1)} y_2 + \frac{(x - x_1)(x - x_2)}{(x_3 - x_1)(x_3 - x_1)} y_3.$$

After some manipulation this leads to the results

$$a = \frac{-\theta}{2\sin\theta}(1 + 2\cos\theta), \quad b = \frac{\theta}{\sin\theta}(1 + \cos\theta), \quad c = \frac{-\theta}{2\sin\theta}.$$

That is,

$$\log(R) = \frac{-\theta}{2\sin\theta}\left((1 + 2\cos\theta)I_3 - 2(1 + \cos\theta)R + R^2\right).$$

4.4.2   The Exponential in the Standard Representation of $SE(3)$

For $se(3)$, a general matrix from the standard 4×4 representation has the form

$$S = \begin{pmatrix} \Omega & \mathbf{u} \\ 0 & 0 \end{pmatrix}$$

where Ω is a 3×3 anti-symmetric matrix and $\mathbf{u}$ is an arbitrary 3-dimensional vector; see Section 4.3. When Ω is non-zero, these matrices satisfy

$$S^4 + \theta^2 S^2 = 0,$$

where $\theta^2 = \omega_x^2 + \omega_y^2 + \omega_z^2$ as above.

If Ω is zero, we have $S^2 = 0$, so for a pure translation we have

$$\exp\begin{pmatrix} 0 & \mathbf{u} \\ 0 & 0 \end{pmatrix} = \begin{pmatrix} I_3 & \mathbf{u} \\ 0 & 1 \end{pmatrix}.$$

In general the polynomial satisfied by the matrices has a repeated factor,

$$S^4 + \theta^2 S^2 = S^2(S + i\theta I_4)(S - i\theta I_4).$$

As before we can find three mutually annihilating idempotents, which will be labelled, P_0, P_+ and P_- again. But to recover the matrix S we need a nilpotent N_0. This nilpotent satisfies $N_0^2 = 0$, $N_0 P_+ = N_0 P_- = 0$ and $N_0 P_0 = N_0$.

These matrices are given by

$$P_0 = \frac{1}{\theta^2}(S^2 + \theta^2 I_4), \qquad N_0 = \frac{1}{\theta^2}S(S^2 + \theta^2 I_4),$$

$$P_+ = \frac{1}{2i\theta^3}S^2(S - i\theta I_4),$$

$$P_- = \frac{-1}{2i\theta^3}S^2(S + i\theta I_4).$$

Again, the relations given above can be checked by direct computation. The linear relation for the matrix S is now

$$S = N_0 - i\theta P_+ + i\theta P_-.$$

As usual, the sum of the idempotents gives the identity matrix, so exponentiating we get

$$e^S = P_0 + N_0 + e^{-i\theta}P_+ + e^{i\theta}P_-.$$

This is particularly simple since P_0 does not appear in the linear relation for S. Finally, substituting back the Ss we have the result

$$e^S = I_4 + S + \frac{1}{\theta^2}(1 - \cos\theta)S^2 + \frac{1}{\theta^3}(\theta - \sin\theta)S^3.$$

Notice that if the Lie algebra element has pitch zero, that is if it represents a pure rotation, then it will satisfy the polynomial equation $S^3 = -\theta^2 S$. Hence it is easy to see that in this case the exponential is given by

$$e^S = I_4 + \frac{1}{\theta}\sin\theta S + \frac{1}{\theta^2}(1 - \cos\theta)S^2.$$

For pure translations the exponential is even simpler, since in this case the corresponding Lie algebra element satisfies $S^2 = 0$ and hence $e^S = I_4 + S$ for translations.

To find the logarithm let us write a 4×4 matrix in this representation of $SE(3)$ as

$$G = e^S = \begin{pmatrix} R & \mathbf{t} \\ 0 & 1 \end{pmatrix}.$$

As usual R is a 3×3 rotation matrix and $\mathbf{t}$ is a translation vector. Now we seek a relation of the form,

$$S = aI_4 + bG + cG^2 + dG^3.$$

The equations for the powers of G in terms of the idempotents and nilpotent are

$$\begin{aligned}
I_4 &= P_0 + P_+ + P_-, \\
G &= P_0 + N_0 + e^{-i\theta}P_+ + e^{i\theta}P_-, \\
G^2 &= P_0 + 2N_0 + e^{-2i\theta}P_+ + e^{2i\theta}P_-, \\
G^3 &= P_0 + 3N_0 + e^{-3i\theta}P_+ + e^{3i\theta}P_-
\end{aligned}$$

and

$$S = N_0 - i\theta P_+ + i\theta P_-.$$

This leads to the system of linear equations

$$\begin{aligned}
a + b \quad + c \quad\quad + d \quad\quad &= \quad 0, \\
b \quad + 2c \quad\quad + 3d \quad\quad &= \quad 1, \\
a + be^{-i\theta} + ce^{-2i\theta} + de^{-3i\theta} &= -i\theta, \\
a + be^{i\theta} + ce^{2i\theta} + de^{3i\theta} &= \quad i\theta.
\end{aligned}$$

The presence of the nilpotent element turns this problem into a Hermite interpolation problem. That is, we can find the unknown co-efficients by finding the polynomial that passes through the same three points as in the previous section but also matches the derivative at the point (x_1, y_1). Again there is a reasonably simple formula for the interpolating polynomial and from this the coefficients can be computed

$$a = \frac{1}{8} \csc^3 \frac{\theta}{2} \sec \frac{\theta}{2} (\theta \cos 2\theta - \sin \theta)$$

$$b = -\frac{1}{8} \csc^3 \frac{\theta}{2} \sec \frac{\theta}{2} (\theta \cos \theta + 2\theta \cos 2\theta - \sin \theta - \sin 2\theta)$$

$$c = \frac{1}{8} \csc^3 \frac{\theta}{2} \sec \frac{\theta}{2} (2\theta \cos \theta + \theta \cos 2\theta - \sin \theta - \sin 2\theta)$$

$$d = -\frac{1}{8} \csc^3 \frac{\theta}{2} \sec \frac{\theta}{2} (\theta \cos \theta - \sin \theta).$$

So the logarithm is given by

$$\log(G) = \frac{1}{8} \csc^3 \frac{\theta}{2} \sec \frac{\theta}{2} \Big((\theta \cos 2\theta - \sin \theta) I_4 - (\theta \cos \theta + 2\theta \cos 2\theta - \sin \theta$$
$$- \sin 2\theta) G + (2\theta \cos \theta + \theta \cos 2\theta - \sin \theta - \sin 2\theta) G^2 - (\theta \cos \theta - \sin \theta) G^3 \Big).$$

Here θ can be found from the relation $\mathrm{Tr}(G) = 2(1 + \cos \theta)$ and as usual the relation is only valid when $-\pi < \theta < \pi$.

4.4.3   The Exponential in the Adjoint Representation of $SE(3)$

In the adjoint representation a Lie algebra element can be written as a matrix of the form

$$Z = \mathrm{ad}(\mathbf{s}) = \begin{pmatrix} \Omega & 0 \\ V & \Omega \end{pmatrix}.$$

The matrix Ω here is as before, a 3×3 anti-symmetric matrix, but now V is another 3×3 anti-symmetric matrix. In general we would expect a 6×6 matrix to satisfy a degree-6 polynomial equation; these matrices however, satisfy a degree-5 equation

$$Z^5 + 2\theta^2 Z^3 + \theta^4 Z = 0,$$

where once again $\theta^2 = \omega_x^2 + \omega_y^2 + \omega_z^2$. The polynomial relation can be verified easily using the following relation between pairs of 3×3 anti-symmetric matrices,

$$\Omega^2 V \Omega^2 + \theta^2 (\Omega V + V \Omega) = -\theta^4 V.$$

In turn this relation can be verified by using the connection between these matrices and vector products, $V\mathbf{x} = \mathbf{v} \times \mathbf{x}$ for any vector $\mathbf{x}$, and then using the familiar formula for vector triple products.

The degree-5 equation above factorises as

$$Z^5 + 2\theta^2 Z^3 + \theta^4 Z = Z(Z + i\theta I_6)^2 (Z - i\theta I_6)^2.$$

So this time we have two repeated factors and hence we seek three idempotents as before but two nilpotents, one associated with each repeated factor:

$$P_0 = \frac{1}{\theta^4}(Z + i\theta I_6)^2 (Z - i\theta I_6)^2$$
$$= (Z^4 + 2\theta^2 Z^2 + \theta^4 I_6)/\theta^4,$$

$$P_+ = \frac{-1}{4\theta^4} Z(Z - i\theta I_6)^2 (2Z + 3i\theta I_6)$$
$$= -(2Z^4 - i\theta Z^3 + 4\theta^2 Z^2 - 3i\theta^3 Z)/4\theta^4,$$

$$N_+ = \frac{1}{4\theta^4} Z(Z + i\theta I_6)(Z - i\theta I_6)^2$$
$$= (Z^4 - i\theta Z^3 + \theta^2 Z^2 - i\theta^3 Z)/4\theta^4,$$

$$P_- = \frac{-1}{4\theta^4} Z(Z + i\theta I_6)^2 (2Z - 3i\theta I_6)$$
$$= -(2Z^4 + i\theta Z^3 + 4\theta^2 Z^2 + 3i\theta^3 Z)/4\theta^4,$$

$$N_- = \frac{1}{4\theta^4} Z(Z + i\theta I_6)^2 (Z - i\theta I_6)$$
$$= (Z^4 + i\theta Z^3 + \theta^2 Z^2 + I_6\theta^3 Z)/4\theta^4.$$

By inspection we have that

$$Z = -i\theta P_+ - i\theta N_+ + i\theta P_- + i\theta N_-.$$

Notice that with matrices P and N such that $P^2 = P$, $N^2 = 0$ and $PN = N$, the kth power of their sum is simply, $(P+N)^k = P+kN$. Hence the exponential of a matrix Z from the adjoint representation of $se(3)$ can be written as

$$e^Z = P_0 + e^{-i\theta} P_+ - i\theta e^{-i\theta} N_+ + e^{i\theta} P_- + i\theta e^{i\theta} N_-.$$

Expanding the idempotents and nilpotents in terms of powers of Z we finally obtain the result

$$e^Z = I_6 + \frac{1}{2\theta}(3\sin\theta - \theta\cos\theta)Z + \frac{1}{2\theta^2}(4 - 4\cos\theta - \theta\sin\theta)Z^2 +$$
$$\frac{1}{2\theta^3}(\sin\theta - \theta\cos\theta)Z^3 + \frac{1}{2\theta^4}(2 - 2\cos\theta - \theta\sin\theta)Z^4.$$

To find an expression for the logarithm of a matrix from the adjoint representation of $SE(3)$, suppose that a general element of the representation is given as

$$H = e^Z = \begin{pmatrix} R & 0 \\ TR & R \end{pmatrix}.$$

Here, T is a 3×3 anti-symmetric matrix corresponding to the translation vector. The relation we seek here is then of the form

$$Z = \alpha I_6 + \beta H + \gamma H^2 + \delta H^3 + \epsilon H^4 + \zeta H^5.$$

In terms of the idempotents and nilpotents the powers of H are

$$
\begin{aligned}
I_6 &= P_0 + P_+ + P_-, \\
H &= P_0 + e^{-i\theta} P_+ - i\theta e^{-i\theta} N_+ + e^{i\theta} P_- + i\theta e^{i\theta} N_-, \\
H^2 &= P_0 + e^{-2i\theta} P_+ - 2i\theta e^{-2i\theta} N_+ + e^{2i\theta} P_- + 2i\theta e^{2i\theta} N_-, \\
H^3 &= P_0 + e^{-3i\theta} P_+ - 3i\theta e^{-3i\theta} N_+ + e^{3i\theta} P_- + 3i\theta e^{3i\theta} N_-, \\
H^4 &= P_0 + e^{-4i\theta} P_+ - 4i\theta e^{-4i\theta} N_+ + e^{4i\theta} P_- + 4i\theta e^{4i\theta} N_-, \\
H^5 &= P_0 + e^{-5i\theta} P_+ - 5i\theta e^{-5i\theta} N_+ + e^{5i\theta} P_- + 5i\theta e^{5i\theta} N_-.
\end{aligned}
$$

Substituting these results into the equation above and comparing with

$$Z = -i\theta P_+ - i\theta N_+ + i\theta P_- + i\theta N_-,$$

leads to six linear equations for the unknown coefficients. Once again the problem is effectively a Hermite interpolation problem, this time with derivative constraints at the points x_2 and x_3. After some computation, best done with a computer algebra system such as Maple or *Mathematica*, the results are

$$\alpha = -\frac{1}{8\sin^3\theta}(2\theta + 4\theta\cos\theta - \theta\cos 3\theta - \sin 2\theta - \sin 3\theta),$$

$$\beta = \frac{1}{8\sin^3\theta}(8\theta + 8\theta\cos\theta + 2\theta\cos 2\theta -$$
$$2\sin\theta - 3\sin 2\theta - 2\sin 3\theta - \sin 4\theta),$$

$$\gamma = -\frac{1}{8\sin^3\theta}(8\theta + 11\theta\cos\theta + 4\theta\cos 2\theta + \theta\cos 3\theta -$$
$$3\sin\theta - 4\sin 2\theta - 3\sin 3\theta - \sin 4\theta),$$

$$\delta = \frac{1}{8\sin^3\theta}(4\theta + 8\theta\cos\theta + 2\theta\cos 2\theta -$$
$$2\sin\theta - 3\sin 2\theta - 2\sin 3\theta),$$

$$\epsilon = -\frac{1}{8\sin^3\theta}(2\theta + \theta\cos\theta - \sin\theta - \sin 2\theta),$$

$$\zeta = 0.$$

This gives the rather lengthy quartic polynomial

$$\log(H) = \frac{-1}{8\sin^3\theta}\Big((2\theta + 4\theta\cos\theta - \theta\cos 3\theta - \sin 2\theta - \sin 3\theta)I_6$$
$$- (8\theta + 8\theta\cos\theta + 2\theta\cos 2\theta - 2\sin\theta - 3\sin 2\theta - 2\sin 3\theta - \sin 4\theta)H$$

$$+ (8\theta + 11\theta \cos\theta + 4\theta \cos 2\theta + \theta \cos 3\theta - 3\sin\theta - 4\sin 2\theta - 3\sin 3\theta - \sin 4\theta)H^2$$
$$- (4\theta + 8\theta \cos\theta + 2\theta \cos 2\theta - 2\sin\theta - 3\sin 2\theta - 2\sin 3\theta)H^3$$
$$+ (2\theta + \theta \cos\theta - \sin\theta - \sin 2\theta)H^4\Big).$$

The angle θ can be found from the relation $\mathrm{Tr}(H) = 2\,\mathrm{Tr}(R) = 2 + 4\cos\theta$.

4.5   Robot Jacobians and Derivatives

4.5.1   The Jacobian of a Robot

In the previous chapter, we saw that the possible rigid motions about a one-degree of freedom joint comprise a one-parameter subgroup of the Euclidean group. Above we have just seen that these subgroups can be generated from Lie algebra elements using the exponential mapping. So the A-matrices of robot kinematics can be written as exponentials, $A(\theta) = e^{\theta \mathbf{s}}$. The $\mathbf{s}$ here is the Lie algebra element corresponding to the particular joint; hence, we will call it the **joint screw**. Actually, the result of the exponential map will be a group element here. To get a matrix representing the group element we would exponentiate a matrix representation of the Lie algebra. With six joints connected serially as in a robot arm, we can write the kinematic matrix as

$$K(\boldsymbol{\theta}) = e^{\theta_1 \mathbf{s}_1} e^{\theta_2 \mathbf{s}_2} e^{\theta_3 \mathbf{s}_3} e^{\theta_4 \mathbf{s}_4} e^{\theta_5 \mathbf{s}_5} e^{\theta_6 \mathbf{s}_6}.$$

Note that here the $\mathbf{s}_i$s are the joint screws for the joints in their home position. This 'product-of-exponentials' formula was first published by Brockett in [16]. One of the advantages of this formalism is that it allows us to find derivatives very easily. For example, suppose we wanted to know the velocity of the robot's final link. The above equation is completely general; it does not depend on any particular representation of the Lie algebra, but here we must use the 4×4 representation since we want to talk about points in $\mathbb{R}^3$. We will write S_i for the 4×4 matrix representing s_i. Now let $\mathbf{p} = (x, y, z)^T$ be a point on the end-effector when the robot is in its home position. The velocity of this point is given by

$$\begin{pmatrix} \dot{\mathbf{p}} \\ 0 \end{pmatrix} = \frac{d}{dt} K(\boldsymbol{\theta}) \begin{pmatrix} \mathbf{p} \\ 1 \end{pmatrix}.$$

The joint variables θ_i will in general be functions of time, so the time derivative of the kinematic matrix will be given by

$$\frac{d}{dt} K(\boldsymbol{\theta}) = \dot{\theta}_1 \frac{\partial K}{\partial \theta_1} + \dot{\theta}_2 \frac{\partial K}{\partial \theta_2} + \dot{\theta}_3 \frac{\partial K}{\partial \theta_3} + \dot{\theta}_4 \frac{\partial K}{\partial \theta_4} + \dot{\theta}_5 \frac{\partial K}{\partial \theta_5} + \dot{\theta}_6 \frac{\partial K}{\partial \theta_6}.$$

In the home position $\theta_1 = \theta_2 = \theta_3 = \theta_4 = \theta_5 = \theta_6 = 0$, and so the partial derivatives of the kinematic matrix are

$$\frac{\partial K}{\partial \theta_i} = S_i, \qquad i = 1, 2, \ldots, 6.$$

The velocity of the point is thus

$$\begin{pmatrix} \dot{\mathbf{p}} \\ 0 \end{pmatrix} = \left(\dot{\theta}_1 S_1 + \dot{\theta}_2 S_2 + \dot{\theta}_3 S_3 + \dot{\theta}_4 S_4 + \dot{\theta}_5 S_5 + \dot{\theta}_6 S_6 \right) \begin{pmatrix} \mathbf{p} \\ 1 \end{pmatrix}.$$

This equation has been derived for the home position of the robot. However, there is nothing special about the home position. We could have chosen any position to be home just by choosing to have the joint variables zero at that position. So the above equation is generally true, but when using it we must not forget that now the S_is refer to the joints in their current position.

We can use the computations above to work out the Jacobian of a robot. The Jacobian of a differentiable mapping is the linear map that is induced on the tangent spaces. In some coordinate system, it is given by the matrix of partial derivatives; see for example O'Neill [81, sect. 1.7]. In robotics, the robot Jacobian is usually the Jacobian of the map from joint space to $SE(3)$ given by the forward kinematics, so that

$$\begin{pmatrix} \boldsymbol{\omega} \\ \mathbf{v} \end{pmatrix} = J \dot{\boldsymbol{\theta}},$$

where $\dot{\boldsymbol{\theta}}^T = (\dot{\theta}_1, \dot{\theta}_2, \dot{\theta}_3, \dot{\theta}_4, \dot{\theta}_5, \dot{\theta}_6)$ is the vector of joint velocities, or joint rates. The velocity screw of the robot's end-effector is found by a calculation similar to the one above:

$$\begin{pmatrix} \boldsymbol{\omega} \\ \mathbf{v} \end{pmatrix} = \left(\dot{\theta}_1 \mathbf{s}_1 + \dot{\theta}_2 \mathbf{s}_2 + \dot{\theta}_3 \mathbf{s}_3 + \dot{\theta}_4 \mathbf{s}_4 + \dot{\theta}_5 \mathbf{s}_5 + \dot{\theta}_6 \mathbf{s}_6 \right).$$

Hence, the robot Jacobian can then be written as

$$J = \left(\begin{array}{c|c|c|c|c|c} \mathbf{s}_1 & \mathbf{s}_2 & \mathbf{s}_3 & \mathbf{s}_4 & \mathbf{s}_5 & \mathbf{s}_6 \end{array} \right).$$

That is, the columns of the Jacobian are simply the current joint screws. Robot singularities occur when this matrix has zero determinant. Another way of expressing this is to say that the joint screws are linearly dependent at singularities. Most current robot control systems are sensitive to these singularities, and hence it is of practical importance to know about them.

We can also find the partial derivatives of the joint screws with respect to changes in the joint variables. Suppose $\mathbf{s}_i$ is the home position of the ith joint screw, and let's write $\mathbf{s}_i(\boldsymbol{\theta})$ for the current position of the joint. Then we can write

$$\mathbf{s}_1(\boldsymbol{\theta}) = \mathbf{s}_1, \quad \mathbf{s}_2(\boldsymbol{\theta}) = \mathrm{Ad}(e^{\theta_1 \mathbf{s}_1}) \mathbf{s}_2, \quad \mathbf{s}_3(\boldsymbol{\theta}) = \mathrm{Ad}(e^{\theta_1 \mathbf{s}_1} e^{\theta_2 \mathbf{s}_2}) \mathbf{s}_3, \ldots$$
$$\ldots, \mathbf{s}_6(\boldsymbol{\theta}) = \mathrm{Ad}(e^{\theta_1 \mathbf{s}_1} \cdots e^{\theta_5 \mathbf{s}_5}) \mathbf{s}_6.$$

Now $\mathrm{Ad}(e^{\mathbf{s}}) = e^{\mathrm{ad}(\mathbf{s})}$ and $\mathrm{ad}(\mathbf{s}_i) \mathbf{s}_j = [\mathbf{s}_i, \mathbf{s}_j]$, so that the derivatives at the home position are given by

$$\frac{\partial \mathbf{s}_j}{\partial \theta_i} = \begin{cases} [\mathbf{s}_i, \mathbf{s}_j] & \text{if } i < j, \\ 0 & \text{if } i \geq j. \end{cases}$$

Once again, we have done the computations for the home position, but the results extend to the general case. These results will be important when we come to look at the dynamics of robots later. However we can see immediately from the above that the determinant of the Jacobian matrix will always be independent of both θ_1 and θ_6.

4.5.2 Derivatives in Lie Groups

So far we have just considered differentiating along one-parameter subgroups, but what about more general paths in the group? Suppose we want to consider a path in a Lie group as an exponential,

$$g(t) = e^{X(t)}.$$

What is the derivative of this?

The problem is that $dX/dt = \dot{X}$ does not necessarily commute with X so differentiating the series expansion of the exponential we get

$$\frac{d}{dt}e^X = \dot{X} + \frac{1}{2!}(\dot{X}X + X\dot{X}) + \frac{1}{3!}(\dot{X}X^2 + X\dot{X}X + X^2\dot{X}) + \cdots$$

$$\cdots + \frac{1}{(k+1)!}(\dot{X}X^k + X\dot{X}X^{k-1} + \cdots + X^k\dot{X}) + \cdots.$$

Hausdorff showed that

$$\left(\frac{d}{dt}e^X\right)e^{-X} = \dot{X} + \frac{1}{2!}[X, \dot{X}] + \frac{1}{3!}[X, [X, \dot{X}]] + \frac{1}{4!}[X, [X, [X, \dot{X}]]] + \cdots;$$

see [46]. The right-hand side of this equation will be abbreviated to X_d. Notice that this matrix is composed of sums of iterated commutators and hence X_d is an element of the Lie algebra. The Hausdorff formula above implies that

$$\frac{d}{dt}e^X = X_d e^X.$$

This can be shown by induction as in [60] or, as is done here, by the following argument from [17]. Consider the expression $e^{\mu X(t)}$ where μ is an independent variable; now differentiate the expression with respect to μ and t in both possible orders,

$$\frac{\partial^2}{\partial t \partial \mu}e^{\mu X(t)} = \frac{\partial}{\partial t}\left(e^{\mu X(t)}X(t)\right) = X_d^\mu e^{\mu X(t)}X(t) + e^{\mu X(t)}\dot{X}(t)$$

and

$$\frac{\partial^2}{\partial \mu \partial t}e^{\mu X(t)} = \frac{\partial}{\partial \mu}\left(X_d^\mu e^{\mu X(t)}\right) = \left(\frac{\partial}{\partial \mu}X_d^\mu\right)e^{\mu X(t)} + X_d^\mu e^{\mu X(t)}X(t),$$

where
$$X_d^\mu = \frac{\partial}{\partial t}\left(e^{\mu X(t)}\right)e^{-\mu X(t)}.$$

Assuming that the derivatives in either order are equal we have
$$\left(\frac{\partial}{\partial \mu}X_d^\mu\right) = e^{\mu X(t)}\dot{X}(t)e^{-\mu X(t)}.$$

Clearly we want a relation for X_d^μ for $\mu = 1$; this can be found by integration. Before doing this we can write the formula in the adjoint representation of the group; here a Lie algebra element X will be written as a vector $\mathbf{x}$, so
$$\left(\frac{\partial}{\partial \mu}\mathbf{x}_d^\mu\right) = \mathrm{Ad}(e^{\mu X(t)})\dot{\mathbf{x}} = e^{\mathrm{ad}(\mu \mathbf{x})}\dot{\mathbf{x}}.$$

Now we can integrate with respect to μ to get the formula
$$\mathbf{x}_d = \mathbf{x}_d^1 = \int_0^1 e^{\mathrm{ad}(\mu \mathbf{x})}\, d\mu \quad \dot{\mathbf{x}}$$

Notice that $\mathbf{x}_d^0 = 0$ since the group element is a constant when $\mu = 0$. Finally, the Hausdorff formula given above can be recovered by integrating the series for the exponential term-by-term.

In the adjoint representation of the group we have the neat expansion
$$\mathbf{x}_d = \sum_{k=0}^{\infty} \frac{1}{(k+1)!}\, \mathrm{ad}^k(\mathbf{x})\dot{\mathbf{x}}$$

where $\mathrm{ad}(\ )$ denotes the adjoint representation of the Lie algebra as usual.

In the rotation group $SO(3)$, this means that $\mathbf{x}_d$ corresponds to the angular velocity of the motion. In $SE(3)$, the group of rigid body motions, the corresponding vector is the velocity screw of the motion.

The relations above have been computed using a fixed coordinate system, however it is common when studying the dynamics of a single rigid body to use a coordinate system fixed in the body. Suppose that the active transformation from the fixed frame of reference to the body-fixed frame is given by $e^{X(t)}$. In these coordinates the velocity is given by $X_b = e^{-X}X_d e^X$ and so we have the relation
$$X_b = e^{-X}\left(\frac{d}{dt}e^X\right)$$

for the velocity in the body-fixed frame. This leads to the following Hausdorff formula for the body-fixed velocity,
$$\mathbf{x}_b = \sum_{k=0}^{\infty} \frac{(-1)^k}{(k+1)!}\, \mathrm{ad}^k(\mathbf{x})\dot{\mathbf{x}}.$$

In the remainder of this section explicit formulas for the fixed-frame velocities in the rotation group and the group of rigid motions will be found. For robot dynamics it turns out that it is simpler to use velocities referred to a single inertial frame of reference, see Chapter 13 below.

4.5.3   Angular Velocity

In the rotation group the adjoint representation is the standard 3×3 representation by anti-symmetric matrices. So the results found in Section 4.4.1 above can be used directly to evaluate the infinite sum in the Hausdorff formula

$$\omega_d = \sum_{k=0}^{\infty} \frac{1}{(k+1)!} \Omega^k \dot{\omega},$$

where the notation has been changed to reflect the fact that the rotation group is under consideration here.

Using the mutually annihilating idempotents found in Section 4.4.1, we have that

$$\sum_{k=0}^{\infty} \frac{1}{(k+1)!} \Omega^k = I_3 + \sum_{k=1}^{\infty} \frac{(i\theta)^k}{(k+1)!} P_- + \sum_{k=1}^{\infty} \frac{(-i\theta)^k}{(k+1)!} P_+ .$$

Using the relation $P_0 + P_+ + P_- = I_3$, this can be rewritten using summations starting at $k = 0$,

$$\sum_{k=0}^{\infty} \frac{1}{(k+1)!} \Omega^k = P_0 + \sum_{k=0}^{\infty} \frac{(i\theta)^k}{(k+1)!} P_- + \sum_{k=0}^{\infty} \frac{(-i\theta)^k}{(k+1)!} P_+ .$$

This is easily summed to give

$$\sum_{k=0}^{\infty} \frac{1}{(k+1)!} \Omega^k = P_0 + \frac{1}{i\theta} \left(e^{i\theta} - 1 \right) P_- - \frac{1}{i\theta} \left(e^{-i\theta} - 1 \right) P_+ .$$

Substituting for the 'Ps' gives the result

$$\sum_{k=0}^{\infty} \frac{1}{(k+1)!} \Omega^k = I_3 + \frac{1}{\theta^2} (1 - \cos \theta) \Omega + \frac{1}{\theta^3} (\theta - \sin \theta) \Omega^2 .$$

That is,

$$\omega_d = \left(I_3 + \frac{1}{\theta^2} (1 - \cos \theta) \Omega + \frac{1}{\theta^3} (\theta - \sin \theta) \Omega^2 \right) \dot{\omega}.$$

In many situations it would be more useful to have $\dot{\omega}$ as the subject of the equation, for example, for numerical simulations. Such an inversion is possible in general, see [60] for example. The result is another infinite series, however it is not necessary to follow this route since the above formula can be inverted more directly using the idempotents. Recall that $P_0 + P_+ + P_- = I_3$ so that

$$(a P_0 + b P_+ + c P_-)(\frac{1}{a} P_0 + \frac{1}{b} P_+ + \frac{1}{c} P_-) = I_3 .$$

That is, we can invert an expression in the 'Ps' by inverting the coefficients of each idempotent. So we can write

$$\left(\sum_{k=0}^{\infty} \frac{1}{(k+1)!} \Omega^k \right)^{-1} = P_0 + \frac{i\theta}{(e^{i\theta} - 1)} P_- - \frac{i\theta}{(e^{-i\theta} - 1)} P_+ .$$

Substituting for the idempotents gives the formula

$$\dot{\omega} = \left(I_3 - \frac{1}{2}\Omega + \left(\frac{1}{\theta^2} + \frac{\sin\theta}{2\theta(\cos\theta - 1)} \right)\Omega^2 \right)\omega_d,$$

or perhaps more neatly using half angles,

$$\dot{\omega} = \left(I_3 - \frac{1}{2}\Omega + \left(\frac{1}{\theta^2} - \frac{1}{2\theta}\cot\frac{\theta}{2} \right)\Omega^2 \right)\omega_d.$$

An equivalent equation to the above appears in [19].

4.5.4   The Velocity Screw

Next, the computations are repeated for the adjoint representation of the group of rigid body motions. Here we have

$$\sum_{k=0}^{\infty} \frac{1}{(k+1)!} Z^k = P_0 + \sum_{k=0}^{\infty} \frac{(-i\theta)^k}{(k+1)!} P_+ + \sum_{k=0}^{\infty} \frac{k(-i\theta)^k}{(k+1)!} N_+$$

$$+ \sum_{k=0}^{\infty} \frac{(i\theta)^k}{(k+1)!} P_- + \sum_{k=0}^{\infty} \frac{k(i\theta)^k}{(k+1)!} N_-.$$

Evaluating the infinite sums gives

$$\sum_{k=0}^{\infty} \frac{1}{(k+1)!} Z^k = P_0 + \frac{1}{-i\theta}(e^{-i\theta} - 1)P_+ + \frac{1}{-i\theta}((-i\theta - 1)e^{-i\theta} + 1)N_+$$

$$+ \frac{1}{i\theta}(e^{i\theta} - 1)P_- + \frac{1}{i\theta}((i\theta - 1)e^{i\theta} + 1)N_-.$$

Substituting for the idemponents and nilpotents gives

$$\sum_{k=0}^{\infty} \frac{1}{(k+1)!} Z^k = I_6 + \frac{1}{2\theta^2}(4 - \theta\sin\theta - 4\cos\theta)Z + \frac{1}{2\theta^3}(4\theta - 5\sin\theta + \theta\cos\theta)Z^2$$

$$+ \frac{1}{2\theta^4}(2 - \theta\sin\theta - 2\cos\theta)Z^3 + \frac{1}{2\theta^5}(2\theta - 3\sin\theta + \theta\cos\theta)Z^4.$$

Again this can be inverted using the relation

$$(a_0 P_0 + a_+ P_+ + b_+ N_+ + a_- P_- + b_- N_-)(\frac{1}{a_0}P_0 + \frac{1}{a_+}P_+ - \frac{b_+}{a_+^2}N_+ + \frac{1}{a_-}P_- - \frac{b_-}{a_-^2}N_-)$$

$$= (P_0 + P_+ + P_-) = I_6.$$

The computations are a little more than can be comfortably done by hand, but are readily computed using a computer algebra system such as Maple or *Mathematica*. Let us write the elements of the Lie algebra as six component vectors $\mathbf{s}$, that is $Z = \mathrm{ad}(\mathbf{s})$. So if we let $\mathbf{s}_d$ be the Lie algebra element satisfying $\frac{d}{dt}e^Z = \mathrm{ad}(\mathbf{s}_d)e^Z$, then we get the result

$$\dot{\mathbf{s}} = \left(I_6 - \frac{1}{2}Z + \left(\frac{2}{\theta^2} + \frac{\theta + 3\sin\theta}{4\theta(\cos\theta - 1)} \right)Z^2 + \left(\frac{1}{\theta^4} + \frac{\theta + \sin\theta}{4\theta^3(\cos\theta - 1)} \right)Z^4 \right) \mathbf{s}_d.$$

Notice the absence of a term in Z^3 in the above. This formula appears in [17].

4.6 Subalgebras, Homomorphisms and Ideals

Given a Lie group, we have seen how we can find its Lie algebra. On the other hand, if we are given a Lie algebra, that is a vector space with an anti-symmetric product that satisfies the Jacobi identities, then is there a group for which this is the Lie algebra? The answer is yes, since we can construct the group by exponentiating the Lie algebra. However, the correspondence between Lie groups and Lie algebras is not one-to-one. Recall that $SU(2)$, $O(3)$ and $SO(3)$ are different groups, but they all have the same Lie algebra. Now $O(3)$ is a disconnected group; it has two separate pieces. The piece that contains the identity is isomorphic to $SO(3)$, and hence these two groups have the same Lie algebra. If we compare $SU(2)$ with $SO(3)$, then we must remember that $SU(2)$ double covers $SO(3)$; see Section 2.3. Another way to think of this covering is to think of the covering map as a quotient of $SU(2)$ by the discrete subgroup

$$\mathbb{Z}_2 = \left\{ \begin{pmatrix} 1 & 0 \\ 0 & 1 \end{pmatrix}, \begin{pmatrix} -1 & 0 \\ 0 & -1 \end{pmatrix} \right\}.$$

One of the consequences of this is that a path in $SU(2)$ from I_2 to $-I_2$ becomes a closed loop when projected to $SO(3)$, starting and ending at I_3. This loop cannot be continuously deformed to a point; however, any closed loop in $SU(2)$ can be shrunk to a point. In topology, a space with the property that every closed loop can be shrunk to a point in a continuous manner is said to be a **simply connected** space. The group $SU(2)$ is simply connected but $SO(3)$ is not. In elementary texts, non-simply connected spaces are often depicted as spaces with 'holes' or voids in them. Here we see an example of a space without such a void but that is still non-simply connected. Now it is true that there is a one-to-one correspondence between Lie algebras and connected, simply connected Lie groups. Moreover, every connected but non-simply connected Lie group is the quotient of a simply connected one by a discrete group. The consequence of all this is that much of the study of Lie groups can be reduced to the study of their Lie algebras, which is usually somewhat easier. This means that we will need to know the correspondence between the Lie group concepts we are interested in and those for Lie algebras.

Before we look at the above in more detail, we note that this is the origin of the "soup plate trick"; see [9]. Imagine holding a soup plate in the palm of your hand. Now rotate the plate through a complete turn without spilling the soup! In the rotation group $SO(3)$ the move represents a closed loop. However, you will have noticed that if you began with your elbow pointing towards the floor, at the end of the move your elbow will be pointing upwards. Although the plate has returned to its original position, your arm is left twisted. This twist can be removed by turning the plate through a further complete turn, a total of 4π radians. This demonstrates that although the turn of 2π radians is a closed loop it is not a contractible loop. The double loop, that is, tracing the loop twice, is contractible. This has implications for robot arms, though the precise nature of the implications is still unclear. Suppose we try to reproduce the trick with a six-joint industrial robot (this is good fun in practice!). The forward kinematics gives a continuous mapping from the joint space into the group $SE(3)$; in general, this mapping is many-to-one. The different pre-images of a point in the group are different postures of the arm, that is, different configurations of the arm but with the same position and orientation of the end-effector. If we rotate about a single joint, then we return to exactly the same position of the arm; hence we get a contractible loop. A move that follows a non-contractible loop will cause the arm to change posture. Usually, such a loop will pass through a singular position of the robot, a position where the determinant of the Jacobian is zero. However, it is not clear whether this is always the case. In general, very little is known about the topology of the forward kinematic mapping.

A **subalgebra** of a Lie algebra is a vector subspace of the Lie algebra that is closed under the Lie bracket operation. Clearly, the subalgebras of a given algebra correspond to subgroups in the group, not discrete subgroups, however, but only those with one or more dimensions.

The Jacobian of a homomorphism, that is, its differential, is a mapping between tangent vectors and hence gives a mapping between the Lie algebras. Remember, a homomorphism must map the identity of the first group to the identity of the second. Now, the fact that the homomorphism preserves the group structure means that the Lie algebra structure is preserved by the Jacobian. That is, commutators must commute with the Jacobian map.

An isomorphism, that is, an invertible homomorphism between groups, will induce an isomorphism between their Lie algebras. By the remarks above on simply connected groups, we see that the converse is not true; an isomorphism between Lie algebras does not mean that the corresponding Lie groups are isomorphic, only that their simply connected covers are isomorphic.

In Section 2.3 we saw that the group of unit quaternions is isomorphic to $SU(2)$. Thus, their corresponding Lie algebras should be isomorphic too. The unit quaternions satisfy $q^*q = 1$; hence, the Lie algebra of this group satisfies $\dot{q}^* + \dot{q} = 0$. That is, the Lie algebra of the group is the set of pure quaternions. A possible isomorphism is given by

$$i \longmapsto 2\sigma_x, \qquad j \longmapsto 2\sigma_y, \qquad k \longmapsto 2\sigma_z.$$

Hence, the Lie algebra structure on the pure quaternions is given by quaternion multiplication:

$$\left[\frac{i}{2},\frac{j}{2}\right] = \frac{k}{2} = \frac{ij}{2}, \qquad \left[\frac{j}{2},\frac{k}{2}\right] = \frac{i}{2} = \frac{jk}{2}, \qquad \left[\frac{k}{2},\frac{i}{2}\right] = \frac{j}{2} = \frac{ki}{2}.$$

An **ideal** in an algebra is a subset of the algebra that is invariant under multiplication by any element of the algebra. So for a Lie algebra $\mathcal{L}$, the elements of an ideal $\mathcal{I}$ satisfy

$$Z \in \mathcal{I} \implies [X, Z] \in \mathcal{I} \quad \text{for all} \quad X \in \mathcal{L}.$$

Ideals are always subalgebras since they must be closed under the Lie bracket operation. In fact, ideals correspond to normal subgroups. To see this, suppose that N is a normal subgroup of a group G, and suppose that the corresponding Lie algebras are $\mathcal{N}$ and $\mathcal{G}$. Now since N is a normal subgroup, for any element $n \in N$ we have $gng^{-1} \in N$ for any group element g. When g and n are near the identity element, we can take logarithms, and on the Lie algebra we get

$$gX_n g^{-1} = \mathrm{Ad}(g)X_n \in \mathcal{N}$$

where $\exp(X_n) = n$ is in the subalgebra $\mathcal{N}$. Taking differentials, as in Section 4.3, we get

$$X_g X_n - X_n X_g = [X_g, X_n] \in \mathcal{N}.$$

So, as claimed, the Lie algebra of a normal subgroup is an ideal.

As with a normal subgroup, we can take the quotient by an ideal. That is, we can define an equivalence relation on a Lie algebra. Two elements are considered equivalent if they differ by an element of the ideal. The set of equivalence classes under this relation is again a Lie algebra. The Lie bracket is defined by

$$\left[[X_1],[X_2]\right] = \left[[X_1, X_2]\right].$$

This relation is well defined on the equivalence classes since if we use another representative of the equivalence class the result is unchanged. If X_n is in the ideal, then $[X_1 + X_n] = [X_1]$, and hence

$$\left[[X_1 + X_n, X_2]\right] = \left[[X_1, X_2] + [X_n, X_2]\right] = \left[[X_1, X_2] + X_m\right] = \left[[X_1, X_2]\right].$$

where X_m is also in the ideal. As usual, we write the quotient as $\mathcal{G}/\mathcal{N}$.

A Lie algebra that has no ideals, apart from the trivial ideal 0 and the whole of the algebra, is called simple. If the only ideals of the algebra are non-commutative, then we say that the algebra is semi-simple. With these definitions, the algebra $so(3)$ is semi-simple. However, $se(3)$ is not. This is easily seen since $se(3)$ has a commutative ideal generated by the elements $\mathbf{v}_i$, $\mathbf{v}_j$ and $\mathbf{v}_k$; see Section 4.3.

Finally, notice that if we have a normal subgroup N of a group G, then the adjoint representation restricted to the Lie algebra of N is still a linear representation of G. This is because the Lie algebra of N is invariant under the action of the group.

4.7 The Killing Form

Every Lie algebra has an invariant, symmetric bilinear form. This form is called the Killing form and is defined as follows. Suppose X and Y are elements of a Lie algebra $\mathcal{G}$. The adjoint representation of the algebra is determined by the relation $\mathrm{ad}(X)Y = [X, Y]$. This is a matrix representation of the algebra, so $\mathrm{ad}(X)$ is a matrix with the same dimensions as the algebra. The Killing form is given by

$$< X, Y >= \mathrm{Tr}(\mathrm{ad}(X)\,\mathrm{ad}(Y)),$$

where $\mathrm{Tr}()$ is the trace of the matrix. Since for any square matrices we have $\mathrm{Tr}(AB) = \mathrm{Tr}(BA)$, the Killing form is seen to be symmetric. The bilinearity comes from the fact that the trace of a matrix is a linear operation: $\mathrm{Tr}((A + B)C) = \mathrm{Tr}(AC + BC) = \mathrm{Tr}(AC) + \mathrm{Tr}(BC)$.

As examples, we look at $so(3)$ first. We write the generators of the algebra as

$$\mathbf{i} = \begin{pmatrix} 1 \\ 0 \\ 0 \end{pmatrix}, \qquad \mathbf{j} = \begin{pmatrix} 0 \\ 1 \\ 0 \end{pmatrix}, \qquad \mathbf{k} = \begin{pmatrix} 0 \\ 0 \\ 1 \end{pmatrix}.$$

The commutation relations in Section 4.3 then give us the adjoint representation as

$$\mathrm{ad}(\mathbf{i}) = \begin{pmatrix} 0 & 0 & 0 \\ 0 & 0 & -1 \\ 0 & 1 & 0 \end{pmatrix}, \quad \mathrm{ad}(\mathbf{j}) = \begin{pmatrix} 0 & 0 & 1 \\ 0 & 0 & 0 \\ -1 & 0 & 0 \end{pmatrix}, \quad \mathrm{ad}(\mathbf{k}) = \begin{pmatrix} 0 & -1 & 0 \\ 1 & 0 & 0 \\ 0 & 0 & 0 \end{pmatrix}.$$

On these basis elements, the Killing form is thus

$$\langle \mathbf{i}, \mathbf{i} \rangle = -2, \quad \langle \mathbf{i}, \mathbf{j} \rangle = 0, \quad \langle \mathbf{i}, \mathbf{k} \rangle = 0,$$
$$\langle \mathbf{j}, \mathbf{j} \rangle = -2, \quad \langle \mathbf{j}, \mathbf{k} \rangle = 0, \quad \langle \mathbf{k}, \mathbf{k} \rangle = -2.$$

For a general pair of vectors in $so(3)$, we can write the Killing form as

$$\langle x_1\mathbf{i} + y_1\mathbf{j} + z_1\mathbf{k}, x_2\mathbf{i} + y_2\mathbf{j} + z_2\mathbf{k} \rangle = (x_1, y_1, z_1) \begin{pmatrix} -2 & 0 & 0 \\ 0 & -2 & 0 \\ 0 & 0 & -2 \end{pmatrix} \begin{pmatrix} x_2 \\ y_2 \\ z_2 \end{pmatrix}.$$

This is just a multiple of the usual scalar product on $\mathbb{R}^3$.

Similar calculations for $se(3)$ give the Killing form for this algebra. If we write two arbitrary elements of $se(3)$ as six component vectors, $\mathbf{s}_1^T = (\boldsymbol{\omega}_1^T, \mathbf{v}_1^T)$ and $\mathbf{s}_2^T = (\boldsymbol{\omega}_2^T, \mathbf{v}_2^T)$, the Killing form is given by

$$\langle \mathbf{s}_1, \mathbf{s}_2 \rangle = (\boldsymbol{\omega}_1^T, \mathbf{v}_1^T) \begin{pmatrix} -2I_3 & 0 \\ 0 & 0 \end{pmatrix} \begin{pmatrix} \boldsymbol{\omega}_2 \\ \mathbf{v}_2 \end{pmatrix} = -2\boldsymbol{\omega}_1 \cdot \boldsymbol{\omega}_2.$$

This form is degenerate.

4.8   The Campbell–Baker–Hausdorff Formula

As we have already seen, if we multiply matrix exponentials together we cannot simply add the exponents. However, it is possible to say a little more about this problem. In general, for two Lie algebra elements X and Y, we have

$$e^X e^Y = e^{f(X,Y)}.$$

An important point here is that the function $f = \log(e^X e^Y)$ is not just differentiable but actually analytic. We can write down the first few terms in the Taylor expansion by comparing

$$e^X e^Y = \left(I + X + \frac{X^2}{2} + \frac{X^3}{6} + \cdots\right)\left(I + Y + \frac{Y^2}{2} + \frac{Y^3}{6} + \cdots\right)$$

with the expansion for $e^{f(X,Y)}$. The first few terms are

$$f(X,Y) = X + Y + \frac{1}{2}[X,Y] + \frac{1}{12}([X,[X,Y]] + [Y,[Y,X]]) + \cdots.$$

This is one form of the Campbell–Baker–Hausdorff theorem. An important part of the theorem states that the higher terms are also given by elements of the Lie algebra, that is, in terms of bracket expressions like $[X,[X,Y]]$, rather than powers of X and Y. So if $X_1, X_2 \in so(3)$, a three-dimensional algebra, then we can write

$$e^{\theta_1 X_1} e^{\theta_2 X_2} = e^{\phi(aX_1 + bX_2 + c[X_1,X_2])},$$

since the higher degree bracket expressions must lie in the vector space spanned by X_1, X_2 and $[X_1, X_2]$. (If X_1 and X_2 are dependent, the problem is trivial.)

We can find formulas for a, b, c and the angle ϕ using the formula for the exponential mapping found above. In fact the relation $f = \log(e^X e^Y)$ is a relation on the Lie algebra, so we can use any faithful representation of the Lie algebra $so(3) = su(2)$ here. If we work with $su(2)$ our computations will be shorter, so let us represent a rotation by θ about a unit vector $\mathbf{v}$ as a complex 2×2 matrix,

$$M = \frac{\theta_1}{2}\begin{pmatrix} iv_x & v_y + iv_z \\ -v_y + iv_z & -iv_x \end{pmatrix}.$$

Recall from Section 4.3 that the basis elements in this algebra contain a factor $1/2$; this is why the half-angle appears here. Multiplying the exponentials gives

$$e^{\frac{\theta_1}{2}X_1} e^{\frac{\theta_2}{2}X_2} = (\cos\frac{\theta_1}{2}I_2 + \sin\frac{\theta_1}{2}X_1)(\cos\frac{\theta_2}{2}I_2 + \sin\frac{\theta_2}{2}X_2)$$

$$= \cos\frac{\theta_1}{2}\cos\frac{\theta_2}{2}I_2 + \sin\frac{\theta_1}{2}\cos\frac{\theta_2}{2}X_1 + \cos\frac{\theta_1}{2}\sin\frac{\theta_2}{2}X_2 + \sin\frac{\theta_1}{2}\sin\frac{\theta_2}{2}X_1X_2.$$

Suppose we write the mapping from a 3-vector to the 2×2 traceless hermitian matrices as

$$R(\mathbf{v}) = \begin{pmatrix} iv_x & v_y + iv_z \\ -v_y + iv_z & -iv_x \end{pmatrix}.$$

Then a simple computation confirms the relation

$$R(\mathbf{x}_1)R(\mathbf{x}_2) = -(\mathbf{x}_1 \cdot \mathbf{x}_2)I_2 + R(\mathbf{x}_1 \times \mathbf{x}_2).$$

From this we can see that

$$X_1 X_2 = -(\mathbf{x}_1 \cdot \mathbf{x}_2) + \frac{1}{2}[X_1, X_2].$$

Hence, the product of exponentials can be written

$$e^{\frac{\theta_1}{2}X_1}e^{\frac{\theta_2}{2}X_2} = \left(\cos\frac{\theta_1}{2}\cos\frac{\theta_2}{2} - \sin\frac{\theta_1}{2}\sin\frac{\theta_2}{2}(\mathbf{x}_1 \cdot \mathbf{x}_2)\right)I_2$$
$$+ \sin\frac{\theta_1}{2}\cos\frac{\theta_2}{2}X_1 + \cos\frac{\theta_1}{2}\sin\frac{\theta_2}{2}X_2 + \frac{1}{2}\sin\frac{\theta_1}{2}\sin\frac{\theta_2}{2}[X_1, X_2].$$

This can be compared with the exponential of the target form

$$e^{\frac{\theta_1}{2}X_1}e^{\frac{\theta_2}{2}X_2} = e^{\frac{\phi}{2}(aX_1+bX_2+\frac{c}{2}[X_1,X_2])}.$$

Notice the factor $\frac{c}{2}[X_1, X_2]$; this is because Lie algebra elements are $\frac{1}{2}X_1$ and $\frac{1}{2}X_2$, so their commutator is $\frac{1}{4}[X_1, X_2]$. Comparing the two expansions gives

$$\cos\frac{\phi}{2} = \cos\frac{\theta_1}{2}\cos\frac{\theta_2}{2} - \sin\frac{\theta_1}{2}\sin\frac{\theta_2}{2}(\mathbf{x}_1 \cdot \mathbf{x}_2),$$
$$a\sin\frac{\phi}{2} = \sin\frac{\theta_1}{2}\cos\frac{\theta_2}{2},$$
$$b\sin\frac{\phi}{2} = \cos\frac{\theta_1}{2}\sin\frac{\theta_2}{2},$$
$$c\sin\frac{\phi}{2} = \sin\frac{\theta_1}{2}\sin\frac{\theta_2}{2}.$$

We must be a little careful about changing the representation of the Lie algebra; for example, in the adjoint representation of $so(3)$ the element $\frac{\theta}{2}R(\mathbf{v})$ would be represented by the 3×3 anti-symmetric matrix $\theta\,\mathrm{ad}(\mathbf{v})$. Now,

$$[R(\mathbf{x}_1),\, R(\mathbf{x}_2)] = 2R(\mathbf{x}_1 \times \mathbf{x}_2)$$

but in the adjoint representation,

$$[\mathrm{ad}(\mathbf{x}_1),\, \mathrm{ad}(\mathbf{x}_2)] = \mathrm{ad}(\mathbf{x}_1 \times \mathbf{x}_2).$$

That is, there is no half factor now. So the corresponding formula in the adjoint representation

$$e^{\theta_1\,\mathrm{ad}(\mathbf{x}_1)}e^{\theta_2\,\mathrm{ad}(\mathbf{x}_2)} = e^{\phi(a\,\mathrm{ad}\,\mathbf{x}_1+b\,\mathrm{ad}(\mathbf{x}_2)+c[\mathrm{ad}(\mathbf{x}_1),\mathrm{ad}(\mathbf{x}_2)])},$$

with the constants a, b and c as above.

Given three rotations it should be possible to find analytic functions a, b, c and ϕ such that

$$e^{\theta_1 \operatorname{ad}(\mathbf{x}_1)} e^{\theta_2 \operatorname{ad}(\mathbf{x}_2)} e^{\theta_3 \operatorname{ad}(\mathbf{x}_3)} = e^{\phi(a \operatorname{ad}\mathbf{x}_1 + b \operatorname{ad}(\mathbf{x}_2) + c \operatorname{ad}(\mathbf{x}_3))}.$$

This would require a knowledge of the structure constants, that is the coefficients C_{ij}^k such that $\mathbf{x}_i \times \mathbf{x}_j = C_{ij}^1 \mathbf{x}_1 + C_{ij}^2 \mathbf{x}_2 + C_{ij}^3 \mathbf{x}_3$.

Other forms of the Campbell–Baker–Hausdorff theorem can be found in almost any textbook on Lie groups and Lie algebras; for example, see Gilmore [38], Postnikov [90] and Miller [76]. From the product of exponentials representation of robot kinematics that we met in Section 4.5, we might reasonably expect that the Campbell–Baker–Hausdorff theorem would be central to a study of the kinematics of robots. Unfortunately, this does not seem to be the case. In the next chapter, we study several more specific examples, and in each case we resort to less general methods. The problem seems to be one of complexity; for more than two exponentials, the Campbell–Baker–Hausdorff relations become very large and difficult to work with.

5

A Little Kinematics

5.1 Inverse Kinematics for 3-R Wrists

In Section 3.7, it was stated that the general problem of inverse kinematics for robots is rather difficult. However, for the 3-R wrist it is possible to give a general solution. This is because we are only dealing with three joints, and since the joints are all revolute, intersecting at a single point, we can reduce everything to the rotation group $SO(3)$. So consider a three-joint wrist, as in Figure 5.1, with the joint axes aligned along the unit vectors $\mathbf{v}_1$, $\mathbf{v}_2$ and $\mathbf{v}_3$. These are the home positions of the axes, and we will assume that these vectors are linearly independent. If the home position of the wrist has linearly dependent joint axes, then we can always move it a little and use a non-singular home position.

The forward kinematics are given by the product of three exponentials:

$$e^{\theta_1 ad(\mathbf{v}_1)} e^{\theta_2 ad(\mathbf{v}_2)} e^{\theta_3 ad(\mathbf{v}_3)} = R.$$

For the inverse kinematic problem, we assume that the rotation matrix R is given. It is the orientation of the end-effector that we want to achieve. What we are seeking are the joint angles θ_1, θ_2 and θ_3 that will yield this objective. In Section 4.4, we saw an expansion for the exponential in the algebra $so(3)$, so that for two three-dimensional vectors $\mathbf{v}$ and $\mathbf{u}$ we have

$$e^{\theta\mathbf{v}}\mathbf{u} = \mathbf{u} + \sin\theta(\mathbf{v} \times \mathbf{u}) + (1 - \cos\theta)\mathbf{v} \times (\mathbf{v} \times \mathbf{u}).$$

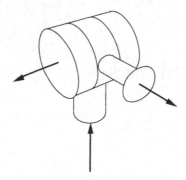

FIGURE 5.1. A 3-R Wrist

Now, since $\mathbf{v}_1$ and $\mathbf{v}_3$ are eigenvectors of the first and third exponentials, respectively, we can form the following equation from the forward kinematics:

$$\mathbf{v}_1^T R \mathbf{v}_3 = \mathbf{v}_1 \cdot \mathbf{v}_3 + \sin\theta_2 \mathbf{v}_1 \cdot (\mathbf{v}_2 \times \mathbf{v}_3) + (1 - \cos\theta_2)\mathbf{v}_1 \cdot (\mathbf{v}_2 \times (\mathbf{v}_2 \times \mathbf{v}_3)).$$

Notice that this equation is linear in the sine and cosine of the second angle and independent of the others. Let us write it as

$$a = \sin\theta_2 - b\cos\theta_2$$

with

$$a = \frac{\mathbf{v}_1^T R \mathbf{v}_3 - (\mathbf{v}_1 \cdot \mathbf{v}_2)(\mathbf{v}_2 \cdot \mathbf{v}_3)}{\mathbf{v}_1 \cdot (\mathbf{v}_2 \times \mathbf{v}_3)}, \qquad b = \frac{(\mathbf{v}_1 \cdot \mathbf{v}_2)(\mathbf{v}_2 \cdot \mathbf{v}_3) - (\mathbf{v}_1 \cdot \mathbf{v}_3)}{\mathbf{v}_1 \cdot (\mathbf{v}_2 \times \mathbf{v}_3)}.$$

Substituting into the trigonometric identity $\cos^2\theta_2 + \sin^2\theta_2 = 1$, gives the quadratic equation

$$(b^2 + 1)\cos^2\theta_2 + 2ab\cos\theta_2 + (a^2 - 1) = 0.$$

In general, we will get two solutions for $\cos\theta_2$ and $\sin\theta_2$. After a little algebra, we get

$$\cos\theta_2 = \frac{-ab}{b^2 + 1} \pm \frac{\sqrt{1 - a^2 + b^2}}{b^2 + 1}, \qquad \sin\theta_2 = \frac{a}{b^2 + 1} \pm \frac{b\sqrt{1 - a^2 + b^2}}{b^2 + 1}.$$

We will return to discuss the discriminant of this problem in a moment, but first we find expressions for the two other angles. For either solution for θ_2 we can set

$$\mathbf{x} = e^{-\theta_2 ad(\mathbf{v}_2)}\mathbf{v}_1, \qquad \text{and} \quad \mathbf{y} = e^{\theta_2 ad(\mathbf{v}_2)}\mathbf{v}_3.$$

We obtain

$$\mathbf{y}^T R \mathbf{v}_3 = \mathbf{y}^T e^{\theta_1 ad(\mathbf{v}_1)}\mathbf{y} = (\mathbf{y} \cdot \mathbf{v}_1)^2 + (1 - (\mathbf{y} \cdot \mathbf{v}_1)^2)\cos\theta_1,$$

$$(\mathbf{y} \times \mathbf{v}_1)^T R \mathbf{v}_3 = (\mathbf{y} \times \mathbf{v}_1)^T e^{\theta_1 ad(\mathbf{v}_1)} \mathbf{y} = (\mathbf{y} \times \mathbf{v}_1)^2 \sin \theta_1,$$

$$\mathbf{v}_1^T R \mathbf{x} = \mathbf{x}^T e^{\theta_3 ad(\mathbf{v}_3)} \mathbf{x} = (\mathbf{x} \cdot \mathbf{v}_3)^2 + (1 - (\mathbf{x} \cdot \mathbf{v}_3)^2) \cos \theta_3,$$

$$\mathbf{v}_1^T R(\mathbf{x} \times \mathbf{v}_3) = \mathbf{x}^T e^{\theta_3 ad(\mathbf{v}_3)}(\mathbf{x} \times \mathbf{v}_3) = (\mathbf{x} \times \mathbf{v}_3)^2 \sin \theta_3.$$

The above results are very general, but we can use these formulas for a particular wrist. For example, the "roll-pitch-yaw" wrist has a home position where $\mathbf{v}_1 = \mathbf{k}$, $\mathbf{v}_2 = \mathbf{j}$ and $\mathbf{v}_3 = \mathbf{i}$. If we write the required rotation as

$$R = \begin{pmatrix} r_{11} & r_{12} & r_{13} \\ r_{21} & r_{22} & r_{23} \\ r_{31} & r_{32} & r_{33} \end{pmatrix},$$

then we can simply substitute into the above formulas to find the inverse kinematic relations of the wrist:

$$a = r_{13} \qquad \text{and} \qquad b = 0.$$

Hence

$$\cos \theta_2 = \pm \sqrt{1 - r_{13}^2} \qquad \text{and} \qquad \sin \theta_2 = r_{13}.$$

The two intermediate vectors $\mathbf{x}$ and $\mathbf{y}$ are given by

$$\mathbf{x} = \begin{pmatrix} -\sin \theta_2 \\ 0 \\ \cos \theta_2 \end{pmatrix} \qquad \text{and} \qquad \mathbf{y} = \begin{pmatrix} \cos \theta_2 \\ 0 \\ -\sin \theta_2 \end{pmatrix}.$$

The two other angles are then given by

$$\cos \theta_1 = \left(r_{11} \cos \theta_2 - r_{31} \sin \theta_2 - \sin^2 \theta_2 \right) / \cos^2 \theta_2,$$

$$\sin \theta_1 = -r_{21} / \cos \theta_2,$$

$$\cos \theta_3 = \left(r_{33} \cos \theta_2 - r_{31} \sin \theta_2 - \sin^2 \theta_2 \right) / \cos^2 \theta_2,$$

$$\sin \theta_3 = -r_{32} / \cos^2 \theta_2.$$

We return now to the discriminant $\Delta = (1 - a^2 + b^2)$ for the general 3-R wrist. This quantity appears in the solution given above for the second joint angle θ_2. Clearly, if this quantity is negative, then the rotation R is not reachable by the wrist. However, for the joint angle to be real we also need the cosine and sine of the angle to have modulus less than 1. Fortunately this doesn't place any extra restrictions on the problem. This can be seen by solving the linear equation for $\sin \theta_2$ and $\cos \theta_2$ in terms of tan-half-angles, that is substituting, $\sin \theta_2 = 2t/(1 + t^2)$ and $\cos \theta_2 = (1 - t^2)/(1 + t^2)$, where $t = \tan(\theta_2/2)$. The result is a quadratic equation in t with exactly the same discriminant, and since the tangent t can take any real value the sign of the discriminant alone determines whether the solutions are real. Note that this method was not used above to solve for the angle because it can never give a result $\theta_2 = \pm\pi$, whereas by solving for the sine and cosine separately any angle can be found.

To make progress in studying this discriminant we will choose coordinates for the vectors $\mathbf{v}_1$, $\mathbf{v}_2$ and $\mathbf{v}_3$. Remember, these are the positions on the joint axes of the wrist in its home position. Hence, we choose any system of coordinate consistent with the design parameters of the wrist. There are only two parameters to specify the design of this type of wrist, the angles between the first and second and between the second and third joint axes. We will label these parameters ϕ_{12} and ϕ_{23} respectively. Now we can choose the coordinates so that the joint axes are

$$\mathbf{v}_1 = \begin{pmatrix} \cos\phi_{12} \\ -\sin\phi_{12} \\ 0 \end{pmatrix}, \quad \mathbf{v}_2 = \begin{pmatrix} 1 \\ 0 \\ 0 \end{pmatrix}, \quad \mathbf{v}_3 = \begin{pmatrix} \cos\phi_{23} \\ \sin\phi_{23}\cos\theta \\ \sin\phi_{23}\sin\theta \end{pmatrix},$$

where the angle θ is the angle between the plane formed by the first and second joint and the plane determined by the second and third joint. The scalar and vector products in the definitions of the constants a and b can now be evaluated. Substituting into the discriminant inequality, $\Delta \geq 0$ gives

$$\sin^2\phi_{12}\sin^2\phi_{23} - (\mathbf{v}_1^T R\mathbf{v}_3 - \cos\phi_{12}\cos\phi_{23})^2 \geq 0.$$

This gives us a relation between the design parameters and the reachable rotations R. The maximum and minimum values of $\mathbf{v}_1^T R\mathbf{v}_3$ are 1 and -1 respectively, since the joint vectors are unit vectors. So we can ask, what do the design parameters have to be in order that the wrist can reach all possible rotations? To answer this we can use the two extreme values to get a pair of inequalities,

$$-(\cos\phi_{12} - \cos\phi_{23})^2 \geq 0 \quad \text{and} \quad -(\cos\phi_{12} + \cos\phi_{23})^2 \geq 0,$$

where the trigonometric identity $\sin^2\phi = 1-\cos^2\phi$ has been used to simplify the expressions. The left-hand sides of both of these inequalities is clearly negative semi-definite, hence the only way to satisfy them is to have both left-hand sides vanish.

So to reach all possible rotations, the wrist must have design parameters, $\phi_{12} = \pm\frac{\pi}{2}$ and $\phi_{23} = \pm\frac{\pi}{2}$. It is not difficult to see that if the design parameters are different from this, then there will be rotations that cannot be reached. Imagine a wrist with very small angles between its successive joints; clearly such a device would only be able to wobble a little about its first joint. These observations are originally due to Paul and Stevenson [84].

The Jacobian is given by the matrix

$$J(\theta_1, \theta_2, \theta_3) = \begin{pmatrix} \mathbf{v}_1 & | & e^{\theta_1 ad(\mathbf{v}_1)}\mathbf{v}_2 & | & e^{\theta_1 ad(\mathbf{v}_1)}e^{\theta_2 ad(\mathbf{v}_2)}\mathbf{v}_3 \end{pmatrix},$$

$$= \begin{pmatrix} \mathbf{v}_1 & | & e^{\theta_1 ad(\mathbf{v}_1)}\mathbf{v}_2 & | & e^{\theta_1 ad(\mathbf{v}_1)}\mathbf{y} \end{pmatrix},$$

where as above $\mathbf{y} = e^{\theta_2 ad(\mathbf{v}_2)}\mathbf{v}_3$. The determinant of the Jacobian is thus the scalar triple product

$$\det(J) = (\mathbf{v}_1 \times \mathbf{v}_2) \cdot \mathbf{y}.$$

After a little computation, we obtain the relation

$$\det(J) = \Big(\cos\theta_2 \mathbf{v}_1 + \sin\theta_2(\mathbf{v}_1 \times \mathbf{v}_2)\Big) \cdot (\mathbf{v}_2 \times \mathbf{v}_3).$$

Recall the constant b introduced above, which can be written $b = (\mathbf{v}_1 \times \mathbf{v}_2) \cdot (\mathbf{v}_2 \times \mathbf{v}_3)/\mathbf{v}_1 \cdot (\mathbf{v}_2 \times \mathbf{v}_3)$. So we can write

$$\frac{\det(J)}{\mathbf{v}_1 \cdot (\mathbf{v}_2 \times \mathbf{v}_3)} = \cos\theta_2 + b\sin\theta_2.$$

Using the results we found above for the sine and cosine of the second joint angle this becomes

$$\frac{\det(J)}{\mathbf{v}_1 \cdot (\mathbf{v}_2 \times \mathbf{v}_3)} = \pm\sqrt{1 - a^2 + b^2}.$$

The two solutions to the inverse kinematic problem correspond to the two different postures, or poses, of the wrist, that is, the two different positions for the joints giving the same overall rotation matrix. These different solutions are distinguished by the different signs of the square root of the discriminant, but this is just the sign of the determinant of the Jacobian; the factor $\mathbf{v}_1 \cdot (\mathbf{v}_2 \times \mathbf{v}_3)$ is just a constant. This means that if $\det(J)$ is positive the wrist is in one posture, and if it is negative we are in the other posture. Put another way, it is possible to tell which of the two postures the wrist is in by computing the sign of $\det(J)$.

One consequence of all this is that the wrist cannot change posture without encountering a singularity. A singularity for a robot is a configuration where the Jacobian drops rank. This implies a linear dependence among the joints and hence, at these positions, the robot has fewer degrees-of-freedom than the number of joints would imply.

For any robot, a continuous change in the joint angles will produce a continuous change in $\det(J)$. However, changing from a position with $\det(J) > 0$ to one where $\det(J) < 0$ we must somewhere encounter a position with $\det(J) = 0$.

5.2   Inverse Kinematics for 3-R Robots

Next we look at the inverse kinematics for a manipulator with three arbitrary revolute joints. Such a manipulator is sometimes called a regional manipulator.

5.2.1   Solution Procedure

Assume that a point on the last link has home position $\mathbf{p}_0$. If we want this point to move to the position $\mathbf{p}$, to what must the joint angles be set? The equations we have to solve are

$$e^{\theta_1 S_1} e^{\theta_2 S_2} e^{\theta_3 S_3} \begin{pmatrix} \mathbf{p_0} \\ 1 \end{pmatrix} = \begin{pmatrix} \mathbf{p} \\ 1 \end{pmatrix},$$

where

$$S_i = \begin{pmatrix} ad(\mathbf{v}_i) & -ad(\mathbf{v}_i)\mathbf{r}_i \\ 0 & 0 \end{pmatrix}$$

is the Lie algebra element representing the ith joint, $\mathbf{v}_i$ the direction of the joint axis, and $\mathbf{r}_i$ a point on the axis; see Section 4.4.2. The fact that the joints are all revolute joints means that the exponential has a simple form:

$$e^{\theta_i S_i} = I_4 + \sin \theta_i S_i + (1 - \cos \theta_i) S_i^2.$$

We have assumed that $\mathbf{v}_i$ are unit vectors so that the joint angles are included explicitly in the exponentials. The effect of such a rotation on a vector in space is given by

$$\begin{pmatrix} \mathbf{p}' \\ 1 \end{pmatrix} = e^{\theta_i S_i} \begin{pmatrix} \mathbf{p} \\ 1 \end{pmatrix},$$

with

$$\mathbf{p}' = \mathbf{p} + \sin \theta_i \mathbf{v}_i \times (\mathbf{p} - \mathbf{r}_i) + (1 - \cos \theta_i) \mathbf{v}_i \times (\mathbf{v}_i \times (\mathbf{p} - \mathbf{r}_i)).$$

Notice here that the quantities $\mathbf{p}' \cdot \mathbf{v}_i$ and $(\mathbf{p}' - \mathbf{r}_i)^2$ are independent of the joint angle. Writing $\mathbf{p}' \cdot \mathbf{v}_i = \mathbf{p} \cdot \mathbf{v}_i$ expresses the fact that as the joint rotates, the point $\mathbf{p}'$ lies in a plane perpendicular to the joint axis. The equation $(\mathbf{p}' - \mathbf{r}_i)^2 = (\mathbf{p} - \mathbf{r}_i)^2$ likewise reflects the fact that the point $\mathbf{p}'$ always lies a fixed distance from the joint axis.

Now, let us write

$$\begin{pmatrix} \mathbf{a} \\ 1 \end{pmatrix} = e^{\theta_3 S_3} \begin{pmatrix} \mathbf{p_0} \\ 1 \end{pmatrix} \qquad \text{and} \qquad \begin{pmatrix} \mathbf{b} \\ 1 \end{pmatrix} = e^{-\theta_1 S_1} \begin{pmatrix} \mathbf{p} \\ 1 \end{pmatrix}.$$

The kinematic equations at the beginning of this section can then be written

$$e^{\theta_2 S_2} \begin{pmatrix} \mathbf{a} \\ 1 \end{pmatrix} = \begin{pmatrix} \mathbf{b} \\ 1 \end{pmatrix}.$$

The points $\mathbf{a}$ and $\mathbf{b}$ lie on circles parameterised by the first and last joint angles:

$$\mathbf{a} = \mathbf{p_0} + \sin \theta_3 \mathbf{v}_3 \times (\mathbf{p_0} - \mathbf{r}_3) + (1 - \cos \theta_3) \mathbf{v}_3 \times (\mathbf{v}_3 \times (\mathbf{p_0} - \mathbf{r}_3)),$$
$$\mathbf{b} = \mathbf{p} - \sin \theta_1 \mathbf{v}_1 \times (\mathbf{p} - \mathbf{r}_1) + (1 - \cos \theta_1) \mathbf{v}_1 \times (\mathbf{v}_1 \times (\mathbf{p} - \mathbf{r}_1)).$$

These two points must lie in a plane perpendicular to the second joint axis, so we have $\mathbf{a} \cdot \mathbf{v}_2 = \mathbf{b} \cdot \mathbf{v}_2$. This gives us a linear equation in the unknowns, $\cos \theta_1$, $\sin \theta_1$, $\cos \theta_3$ and $\sin \theta_3$.

Next, we can set equal the distance of the points $\mathbf{a}$ and $\mathbf{b}$ from the axis of the second joint: $(\mathbf{a} - \mathbf{r}_2)^2 = (\mathbf{b} - \mathbf{r}_2)^2$. At first sight this look like an equation of degree 2 in the variables, but notice that both the terms $\mathbf{a} - \mathbf{r}_2$ and $\mathbf{b} - \mathbf{r}_2$ have the general form

$$\mathbf{c} - \mathbf{r}_2 = \mathbf{w} + \sin\theta \mathbf{v} \times \mathbf{q} + (1 - \cos\theta)\mathbf{v} \times (\mathbf{v} \times \mathbf{q}).$$

The second and third terms here are clearly orthogonal, so when the square of such an expression is taken, these cross terms vanish. Further, since $\mathbf{v}$ is a unit vector, the squares of the second and third terms will be the same. This means that the only terms in the square of $\mathbf{c} - \mathbf{r}_2$ that are quadratic in $\cos\theta$ and $\sin\theta$ will be

$$(\mathbf{c} - \mathbf{r}_2)^2 = (\cos^2\theta + \sin^2\theta)(\mathbf{q} \cdot \mathbf{q} - (\mathbf{v} \cdot \mathbf{q})^2) + \cdots.$$

Hence, using the trigonometric identity $\cos^2\theta + \sin^2\theta = 1$, all the quadratic terms can be eliminated and the equation $(\mathbf{a} - \mathbf{r}_2)^2 = (\mathbf{b} - \mathbf{r}_2)^2$ is then linear in the variables $\cos\theta_1$, $\sin\theta_1$, $\cos\theta_3$ and $\sin\theta_3$.

This gives us two linear equations in the variables; we get another pair of quadratic equations from the trigonometric identities

$$\cos^2\theta_1 + \sin^2\theta_1 = 1 \quad \text{and} \quad \cos^2\theta_3 + \sin^2\theta_3 = 1.$$

We can use the two linear equations to eliminate two of the variables, say $\cos\theta_3$ and $\sin\theta_3$, This leaves us with two degree-2 equations in two variables; a pair of conics. Hence, the inverse kinematics can be reduced to finding the intersection of a pair of conics, that is, the base points of a pencil of conics; see Section 3.4. This problem can be solved by radicals in several ways. First, if we make the tan-half-angle substitutions,

$$\cos\theta = \frac{1 - t^2}{1 + t^2} \quad \text{and} \quad \sin\theta = \frac{2t}{1 + t^2}.$$

This will automatically satisfy one of the conics, the circle given by the trigonometric identity, substituting into the other conic will give a quartic equation. Alternatively we could choose different conics in the pencil. In particular, a general pencil of conics will contain three degenerate conics. A degenerate conic is a pair of lines; we can find the degenerate conics in a pencil by solving the cubic equation $\det(Q_0 + \lambda Q_1) = 0$, where Q_0 and Q_1 are the 3×3 symmetric matrices representing the two conics. Now if we know a degenerate conic in the pencil, we can find the solutions by intersecting the lines with one of the conics, that is solving two quadratic equations. Given two degenerate conics in the pencil, the solutions are given by intersecting pairs of lines, one from each degenerate conic, this involves solving linear equations only.

In general we will get four solutions, of course some of these solutions may be repeated or complex, (see Section 5.2.3 later). A solution here will assign values to $\cos\theta_1$ and $\sin\theta_1$ and hence we will get a unique value for the joint angle θ_1 (between 0 and 2π) for each solution. Each of the four solutions will give a unique solution for θ_3 via the linear equations for $\cos\theta_3$ and $\sin\theta_3$. This allows us to find the vectors $\mathbf{a}$ and $\mathbf{b}$; we can then use the equation

$$e^{\theta_2 S_2}\begin{pmatrix}\mathbf{a}\\1\end{pmatrix}=\begin{pmatrix}\mathbf{b}\\1\end{pmatrix}$$

to solve for the second joint angle θ_2. Expanding the exponentials gives the vector form of the equation

$$\mathbf{b}=\mathbf{a}+\sin\theta_2\mathbf{v}_2\times(\mathbf{a}-\mathbf{r}_2)+(1-\cos\theta_2)\mathbf{v}_2\times(\mathbf{v}_2\times(\mathbf{a}-\mathbf{r}_2)).$$

Notice that this is linear in the sine and cosine of θ_2; moreover, as we observed above, the vectors, $\mathbf{v}_2\times(\mathbf{a}-\mathbf{r}_2)$ and $\mathbf{v}_2\times(\mathbf{v}_2\times(\mathbf{a}-\mathbf{r}_2))$ are orthogonal. So we can write the solution symbolically as

$$\cos\theta_2=1-\frac{(\mathbf{b}-\mathbf{a})\cdot(\mathbf{v}_2\times(\mathbf{v}_2\times(\mathbf{a}-\mathbf{r}_2)))}{|\mathbf{v}_2\times(\mathbf{v}_2\times(\mathbf{a}-\mathbf{r}_2))|^2},$$

$$\sin\theta_2=\frac{(\mathbf{b}-\mathbf{a})\cdot(\mathbf{v}_2\times(\mathbf{a}-\mathbf{r}_2))}{|\mathbf{v}_2\times(\mathbf{a}-\mathbf{r}_2)|^2}.$$

Hence, we get four solutions for these joint angles and thus, in general, such a robot has four postures. However, it is clear that there are points with fewer postures. This is easily seen from the representation of the problem as a pencil of conics. The conics in the pencil can intersect in 4, 3, 2, 1 or no real points. We will look at this in more detail in a moment.

These results and those of the previous section can be combined to say something about the inverse kinematics of six-joint robots. In the special case where either the first three or final three-joint axes are coincident, it is possible to derive the inverse kinematic relations. Such a manipulator is sometimes called an **elbow manipulator**. The arrangement is common in anthropomorphic designs of robot arms, where the last three joints comprise a spherical wrist. For such a robot, the inverse kinematics can be found by separating the problem into finding the first three joint angles from the position of the wrist centre, the point where the axes of the wrist meet, and then finding the last three-joint axes from the orientation of the end-effector and the rotation due to the first three joints. This result is usually attributed to Pieper [87]; see Section 10.3 below. Notice that this means that such a robot will generally have eight postures, four due to the different solutions for the first three joints, and for each of these two possible wrist poses.

5.2.2   An Example

Now let us turn to a specific example and see how the solutions may be derived in practice. We can specify the example in question by giving the home position of the joints and the point $\mathbf{p}_0$, so let us say that

$$\begin{array}{lll}\mathbf{v}_1=\mathbf{k}, & \mathbf{v}_2=\mathbf{i}, & \mathbf{v}_3=\mathbf{i},\\\mathbf{r}_1=\mathbf{0}, & \mathbf{r}_2=l_1\mathbf{j}, & \mathbf{r}_3=(l_1+l_2)\mathbf{j},\end{array}$$

and $\mathbf{p}_0 = (l_1 + l_2 + l_3)\mathbf{j}$, where, as usual, $\mathbf{i}$, $\mathbf{j}$ and $\mathbf{k}$ are the unit vectors in the x, y and z directions. So, the point on the end-effector will be written $\mathbf{p} = x\mathbf{i} + y\mathbf{j} + z\mathbf{k}$. This mechanism is commonly used for finger joints in robot hands. Moreover, if $l_1 = 0$, then the arrangement is the same as for the first three joints of the PUMA robot. The two circles are parameterised as

$$\mathbf{a} = (l_1 + l_2 + l_3 \cos\theta_3)\mathbf{j} + l_3 \sin\theta_3\mathbf{k},$$
$$\mathbf{b} = (x\cos\theta_1 + y\sin\theta_1)\mathbf{i} + (-x\sin\theta_1 + y\cos\theta_1)\mathbf{j} + z\mathbf{k}.$$

The two linear equations, derived from $\mathbf{a}\cdot\mathbf{v}_2 = \mathbf{b}\cdot\mathbf{v}_2$ and $(\mathbf{a} - \mathbf{r}_2)^2 = (\mathbf{b} - \mathbf{r}_2)^2$, become

$$x\cos\theta_1 + y\sin\theta_1 = 0$$

and

$$(l_1 + l_2 + l_3)^2 - 2l_1(l_1 + l_2 + l_3\cos\theta_3)$$
$$= (x^2 + y^2 + z^2) - 2(l_1 + l_2)y + 2l_2(-x\sin\theta_1 + y\cos\theta_1).$$

We may abbreviate these equations as

$$x\cos\theta_1 + y\sin\theta_1 = 0,$$
$$-x\sin\theta_1 + y\cos\theta_1 = A + B\cos\theta_3,$$

where $A = \{(l_1 + l_2 + l_2)^2 - (x^2 + y^2 + z^2) + 2(l_1 + l_2)y - 2l_1^2 - 2l_1 l_2\}/2l_2$ and $B = -l_1 l_3/l_2$. It seems to make more sense here to eliminate θ_1 rather than θ_3 In this simple situation, the easiest way is probably to square and add the equations. The result is a quadratic:

$$B^2 \cos^2\theta_3 + 2AB\cos\theta_3 + A^2 - x^2 - y^2 = 0.$$

This, together with the trigonometric identity $\cos^2\theta_3 + \sin^2\theta_3 = 1$, forms the pair of conics. However, in this simple situation the solutions are easily found, since one of the equations is independent of the variable $\sin\theta_3$; hence, we may solve for $\cos\theta_3$ using the formula for quadratic equations:

$$\cos\theta_3 = \frac{A}{B} \pm \sqrt{x^2 + y^2}.$$

For each of these two solutions, we have two solutions for the sine of θ_3, that is, $\sin\theta_3 = \pm\sqrt{1 - \cos^2\theta_3}$.

Having found the four solutions for $\cos\theta_3$ and $\sin\theta_3$, we then find the corresponding values for the other angles. Notice that the linear equations above can be rearranged into the form

$$\cos\theta_1 = \frac{y}{x^2 + y^2}(A + B\cos\theta_3),$$
$$\sin\theta_1 = \frac{-x}{x^2 + y^2}(A + B\cos\theta_3).$$

Finally, to find θ_2, we use the relation

$$e^{\theta_2 S_2} \begin{pmatrix} \mathbf{a} \\ 1 \end{pmatrix} = \begin{pmatrix} \mathbf{b} \\ 1 \end{pmatrix}$$

to show that

$$\cos\theta_2 = \frac{(\mathbf{a} - \mathbf{r}_2) \cdot (\mathbf{b} - \mathbf{r}_2) + ((\mathbf{a} - \mathbf{r}_2) \cdot \mathbf{v}_2)^2}{|\mathbf{v}_2 \times (\mathbf{a} - \mathbf{r}_2)|^2},$$

$$\sin\theta_2 = \frac{\mathbf{b} \cdot \mathbf{v}_2 \times (\mathbf{a} - \mathbf{r}_2)}{|\mathbf{v}_2 \times (\mathbf{a} - \mathbf{r}_2)|}.$$

5.2.3 Singularities

Here we return to the general 3-R manipulator and, as promised, we look at how the number of postures can change. This will depend on the position of the point $\mathbf{p}$ on the end-effector and the design parameters of the machine.

For any particular robot its workspace will be divided into regions with different numbers of postures. These regions will be separated by surfaces corresponding to points in joint space where the Jacobian of the manipulator is singular. These surfaces will be referred to as **bifurcation surfaces**. The bifurcation surfaces are also closely connected with the pencil of conics view of the kinematics outlined in Section 5.2.1 above. To see the connection we can factor the forward kinematic mapping through $\mathbb{R}^2 \times \mathbb{R}^3$. The first map is

$$\rho_1 : (\theta_1, \theta_2, \theta_3) \mapsto (\cos\theta_3, \sin\theta_3, p_x, p_y, p_z),$$

where p_x, p_y and p_z are the coordinates of the point on the robot's end-effector. The second map is simply a projection onto the last three coordinates. Clearly the composition of these two maps is just the forward kinematic map from joint space to workspace. The first map is an immersion, at least in the case that the first two joint axes are not coplanar. To see this consider the Jacobian of the map,

$$J_1 = \begin{pmatrix} 0 & 0 & -\sin\theta_3 \\ 0 & 0 & \cos\theta_3 \\ \partial p_x/\partial\theta_1 & \partial p_x/\partial\theta_2 & \partial p_x/\partial\theta_3 \\ \partial p_y/\partial\theta_1 & \partial p_y/\partial\theta_2 & \partial p_y/\partial\theta_3 \\ \partial p_z/\partial\theta_1 & \partial p_z/\partial\theta_2 & \partial p_z/\partial\theta_3 \end{pmatrix}.$$

The rank of this matrix will be 3 so long as the first two columns are not linearly dependent. In a moment we will see that the columns of the Jacobian are given by the velocity of the point $\mathbf{p}$. If the first two joint axes are not coplanar, then the above Jacobian will always have full rank and hence the map is a smooth immersion. The projection map has Jacobian

$$J_2 = \begin{pmatrix} 0 & 0 & 1 & 0 & 0 \\ 0 & 0 & 0 & 1 & 0 \\ 0 & 0 & 0 & 0 & 1 \end{pmatrix},$$

which is clearly always full rank. Now the Jacobian of the composite map, and hence the Jacobian of the forward kinematic of the robot is given by the matrix product

$$J = J_2 J_1.$$

So the only way that the Jacobian of the forward kinematics can be singular is if the image of J_1 intersects the null-space of J_2. This means that at a singularity at least one tangent to the corresponding point in joint space must be in the direction of the fibre of the projection. Now consider an arbitrary point $\mathbf{p}$ in the robot's work space. The fibre sitting above $\mathbf{p}$ is a plane, and the intersection of this plane with the image of ρ_1 is simply the four base-points of the pencil of conics. So we can conclude that, at a singularity, two or more of the base-points in the pencil must coalesce.

Next we turn to the Jacobian itself. As in Section 4.5, we can write the columns of the Jacobian very simply. The derivative of the forward kinematic map gives

$$\begin{pmatrix} \dot{\mathbf{p}} \\ 0 \end{pmatrix} = \left(\dot{\theta}_1 S_1 + \dot{\theta}_2 e^{\theta_1 S_1} S_2 e^{-\theta_1 S_1} + \dot{\theta}_3 e^{\theta_1 S_1} e^{\theta_2 S_2} S_3 e^{-\theta_2 S_2} e^{-\theta_1 S_1} \right) \begin{pmatrix} \mathbf{p} \\ 1 \end{pmatrix}.$$

Comparing this with $\dot{\mathbf{p}} = J\dot{\theta}$, the relation for the Jacobian matrix, the columns of J are

$$J = \left(\mathbf{v}_1 \times (\mathbf{p} - \mathbf{r}_1) | \mathbf{v}_2 \times (\mathbf{p} - \mathbf{r}_2) | \mathbf{v}_3 \times (\mathbf{p} - \mathbf{r}_3) \right).$$

Here we assume that $\mathbf{v}_i$ and $\mathbf{r}_i$ are the current values of these vectors, not the values at the home position as before. Here the ith column gives the velocity of the point $\mathbf{p}$ due to a unit joint velocity at the ith joint, with the other joints remaining stationary. Notice that the direction of this velocity is normal to the plane containing the joint axis and the point $\mathbf{p}$. For the three joints we get three such planes. When the Jacobian is singular, its columns, the normals to the planes will be linearly dependant. This implies that the planes will meet in a line and since $\mathbf{p}$ always lies on the three planes, the line common to the three planes will pass through $\mathbf{p}$. In each of the three planes the common line will meet the corresponding joint axis, or exceptionally will be parallel to it. Hence, we have a geometric condition for the robot to be singular: If the robot is in a singular configuration then there will be a line that meets all the joint axes and passes through $\mathbf{p}$. Conversely, if there is such a line, we may reverse the argument and conclude that the robot will be singular. The significance of this common line is that, to first order, the robot will not be able to move the point $\mathbf{p}$ along this line. Moreover, the line will be normal to the bifurcation surface, that is the surface in workspace traced out by the robot's end-point $\mathbf{p}$ in singular configurations.

The bifurcation surfaces can have singularities themselves. Suppose that it is possible for the robot's end-point to lie on one of the joint axes. If this is the third joint axes, then for any design of 3-R robot this is always true or never possible, depending on the design parameters. So assume $\mathbf{p}$ can lie on the first or second joint. Now there will be two lines through **p** that meet all the joint axes. To see this, take a plane containing **p** on joint 1 say, and also containing one of the other joint axes, 2 or 3. If the plane contains joint 2, then there will generally be a point of intersection between the plane and the final joint. The line joining **p** and this intersection point will meet the other lines; it meets joint 1 at **p** and is coplanar with joint 2. Clearly, if we exchange joints 2 and 3 in this argument we get another line through **p** meeting all the joint axes. So such a point is always a singular point and moreover the singular surface will have two normals at such a point; actually we can always rotate the end of the robot about joint 1 or 2 and hence we get two circles of normal lines. In other words, in general these points will be conical double points on the bifurcation surface.

Next suppose that it is possible for the robot to assume a position where joints 1 and three are coplanar. When **p** lies in this plane the robot will certainly be singular since the line joining **p** and the point where the second joint meets the plane will necessarily meet the first and third joint axes. To see that such a configuration can be a singularity of the bifurcation surface we look at the gradient to the surface; we can differentiate in an arbitrary direction **q**,

$$\nabla_{\mathbf{q}} \det(J) = \det \left(\mathbf{v}_1 \times \mathbf{q} | \mathbf{v}_2 \times (\mathbf{p} - \mathbf{r}_2) | \mathbf{v}_3 \times (\mathbf{p} - \mathbf{r}_3) \right)$$
$$+ \det \left(\mathbf{v}_1 \times (\mathbf{p} - \mathbf{r}_1) | \mathbf{v}_2 \times \mathbf{q} | \mathbf{v}_3 \times (\mathbf{p} - \mathbf{r}_3) \right)$$
$$+ \det \left(\mathbf{v}_1 \times (\mathbf{p} - \mathbf{r}_1) | \mathbf{v}_2 \times (\mathbf{p} - \mathbf{r}_2) | \mathbf{v}_3 \times \mathbf{q} \right).$$

The determinants in the above relation can be expanded as scalar triple products,

$$\nabla_{\mathbf{q}} \det(J) = (\mathbf{v}_1 \times \mathbf{q}) \cdot \left((\mathbf{v}_2 \times (\mathbf{p} - \mathbf{r}_2)) \times (\mathbf{v}_3 \times (\mathbf{p} - \mathbf{r}_3)) \right)$$
$$+ (\mathbf{v}_2 \times \mathbf{q}) \cdot \left((\mathbf{v}_3 \times (\mathbf{p} - \mathbf{r}_3)) \times (\mathbf{v}_1 \times (\mathbf{p} - \mathbf{r}_1)) \right)$$
$$+ (\mathbf{v}_3 \times \mathbf{q}) \cdot \left((\mathbf{v}_1 \times (\mathbf{p} - \mathbf{r}_1)) \times (\mathbf{v}_2 \times (\mathbf{p} - \mathbf{r}_2)) \right).$$

Now, in the case under consideration the first and third joint axes are coplanar; assume that they meet at a point **r** (the case where the axes are parallel is simpler so will be ignored). So we may set $\mathbf{r}_1 = \mathbf{r}_3 = \mathbf{r}$. The fact that the point, **p** lies in the same plane as the first and third joint axes can be expressed by the equation $(\mathbf{p} - \mathbf{r}_1) \cdot (\mathbf{v}_1 \times \mathbf{v}_3) = 0$. Substituting these relations into the expression for the derivative and simplifying gives the result

$$\nabla_{\mathbf{q}} \det(J) = - \left(\mathbf{q} \cdot (\mathbf{v}_1 \times \mathbf{v}_3) \right) \left((\mathbf{p} - \mathbf{r}) \cdot (\mathbf{v}_2 \times (\mathbf{p} - \mathbf{r}_2)) \right).$$

Notice that this means that if the direction of **q** lies in the plane determined by the first and third joints, the gradient vanishes. So if we restrict the bifurcation surface to this plane, we obtain a curve and such a configuration can be a singular point on the curve, usually a cusp. Keeping everything except the first joint angle fixed, we can sweep this configuration about the first joint axis and hence produce a cuspidal edge in the bifurcation surface.

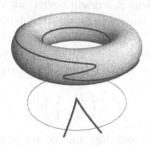

FIGURE 5.2. Circling a Cusp

These cusps on the bifurcation surface can have an interesting effect on the kinematics of the robot. This is best illustrated with a diagram; see Figure 5.2.

The robot can change posture without passing through a singularity by circling a cusp. For the purpose of illustration the workspace of the robot has been restricted to a plane. The path of the robot's end-point, $\mathbf{p}$, is shown as a circle in the workspace. Above each point on this circle is another circle representing one of the joint angles, θ_1 in this case. Hence, moving around the circle in the workspace generates a torus. Above each point on the circle in the workspace are points in the torus, four or two, corresponding to the postures of the robot. The postures of the robot form a curve in the θ_1-torus. This curve has two disjoint components in this case, one component lies on the inner diameter of the torus, the other is more interesting as it folds back on itself. When the circle in the workspace meets the bifurcation curve, the curve of postures in the torus has a tangent parallel to the projection to the workspace. That is, two postures coalesce at these points and there are only three postures corresponding to these points. If the robot is in one of these coalescing postures it cannot pass through the bifurcation curve. However, by reversing direction at these points it is possible for the robot to move into another posture, that is changing posture by passing through a singularity. On the other hand, if $\mathbf{p}$ is at a point between the arms of the cusp, in the region where there are 4 postures, then the robot can move from one posture to different one, passing over the bifurcation curve, around the back of the torus and back to the same point in the workspace but in the other posture.

This phenomenon was first discovered by Burdick [18], the explanation above was heavily influenced by Smith [112]. In the literature these robots have come to be known as cuspidal robots and have been studied in more detail by Wenger [125] and others.

5.3   Kinematics of Planar Motion

In this section, we will look a little more closely at the group of rigid motions in the plane, $SE(2)$. In particular, we will derive some of the classical results of planar kinematics but using some of the mathematical machinery we have just developed.

So far, we have not said very much about the group of rigid motions in the plane; we will remedy this at once. A general 2-D rigid motion can be written as a 3×3 matrix

$$M = \begin{pmatrix} R & \mathbf{t} \\ 0 & 1 \end{pmatrix}.$$

The matrix R here is a 2×2 rotation matrix, an element of the group $SO(2)$, and hence has the general form

$$R = \begin{pmatrix} \cos\theta & -\sin\theta \\ \sin\theta & \cos\theta \end{pmatrix}.$$

Apart from pure translations, for which $R = I_2$, each of these motions is a rotation about some point in the plane. If the centre of rotation is the point with position vector $\mathbf{c} = (c_x, c_y)^T$, then a rotation about this point is given by the conjugation

$$M = \begin{pmatrix} I_2 & \mathbf{c} \\ 0 & 1 \end{pmatrix} \begin{pmatrix} R & 0 \\ 0 & 1 \end{pmatrix} \begin{pmatrix} I_2 & -\mathbf{c} \\ 0 & 1 \end{pmatrix} = \begin{pmatrix} R & (I_2 - R)\mathbf{c} \\ 0 & 1 \end{pmatrix}.$$

Hence, a rotation about the point $\mathbf{c}$ has the general form

$$M = \begin{pmatrix} \cos\theta & -\sin\theta & c_x - c_x\cos\theta + c_y\sin\theta \\ \sin\theta & \cos\theta & c_y - c_x\sin\theta - c_y\cos\theta \\ 0 & 0 & 1 \end{pmatrix}.$$

The assertion above about every motion being a rotation about some point amounts to the fact that we can always find the centre of rotation by solving the 2×2 matrix equation $(I_2 - R)\mathbf{c} = \mathbf{t}$ for $\mathbf{c}$. In fact, $(I_2 - R)$ is usually invertible, and the solution is given by

$$\mathbf{c} = \frac{1}{\Delta}(I_2 - R^T)\mathbf{t}.$$

where $\Delta = 2 - 2\cos\theta$. Clearly, the method fails for $\theta = 0$, that is, for pure translations.

To find the Lie algebra element corresponding to such a rotation, we differentiate with respect to θ and then set $\theta = 0$ to obtain

$$C = \begin{pmatrix} 0 & -1 & c_y \\ 1 & 0 & -c_x \\ 0 & 0 & 0 \end{pmatrix} = \begin{pmatrix} -E_2 & E_2\mathbf{c} \\ 0 & 0 \end{pmatrix}.$$

This partitioned form will be useful later but for the moment we just note that the 2×2 matrix E_2 is given by

$$E_2 = \begin{pmatrix} 0 & 1 \\ -1 & 0 \end{pmatrix}.$$

A general element of the Lie algebra can be written as θC. The matrix C satisfies the cubic equation

$$C^3 + C = 0.$$

This is essentially the same equation satisfied by elements of the adjoint representation of $so(3)$, see Section 4.4.1. Hence, we may simply reuse the formulas found in sections 4.4 and 4.5 for the rotations. For example the exponential of a general Lie algebra element is given by

$$M = e^{\theta C} = I_3 + \sin \theta C + (1 - \cos \theta) C^2.$$

The 3×3 representation of the $se(2)$ is also the adjoint representation. If we write a general element of the Lie algebra as

$$L = \begin{pmatrix} -\theta E_2 & E_2 \mathbf{a} \\ 0 & 0 \end{pmatrix},$$

then we could write this as a 3-vector in the form

$$\mathbf{l} = \begin{pmatrix} a_x \\ a_y \\ \theta \end{pmatrix}.$$

Now we have that $\mathrm{ad}(\mathbf{l}) = L$, and hence we can use the formula for the derivative in $SO(3)$. Let $L_d = (de^L/dt)L^{-1}$; then

$$\mathbf{l}_d = \left(I_3 + \frac{1}{\theta^2}(1 - \cos \theta)L + \frac{1}{\theta^3}(\theta - \sin \theta)L^2 \right)\dot{\mathbf{l}},$$

where $\mathrm{ad}(\mathbf{l}_d) = L_d$.

The similarity between formulas in $SO(3)$ and $SE(2)$ breaks down when we look at the Campbell–Baker–Hausdorff formula. As for $so(3)$, the Lie algebra $se(2)$ is only three-dimensional, so we can find an explicit Campbell–Baker–Hausdorff formula. However, the commutation relations of the two algebras are different; for planar motions we get

$$e^{\theta_1 C_1} e^{\theta_2 C_2} = e^{\phi(\alpha C_1 + \beta C_2 + \gamma [C_1, C_2])}$$

with

$$\phi = \theta_1 + \theta_2,$$
$$\alpha = \tan \frac{\theta_1}{2} \Big/ \left(\tan \frac{\theta_1}{2} + \tan \frac{\theta_2}{2} \right),$$

$$\beta = \tan\frac{\theta_2}{2} \left/ \left(\tan\frac{\theta_1}{2} + \tan\frac{\theta_2}{2} \right), \right.$$

$$\gamma = \tan\frac{\theta_1}{2} \tan\frac{\theta_2}{2} \left/ \left(\tan\frac{\theta_1}{2} + \tan\frac{\theta_2}{2} \right). \right.$$

Compare these results with those at the end of Section 4.8 for $so(3)$.

Now, we consider the effect of a rotation about a fixed point on other points in the plane. Let $\mathbf{p} = (x, y)^T$ be a general point in the plane. Rotating about the point $\mathbf{c}$ takes $\mathbf{p}$ to the point $\mathbf{p}(t)$ given by

$$\begin{pmatrix} \mathbf{p}(t) \\ 1 \end{pmatrix} = e^{\theta(t)C} \begin{pmatrix} \mathbf{p} \\ 1 \end{pmatrix}.$$

Notice that here we do not assume that the rotation is uniform, so θ is just some smooth function of the parameter t. For convenience, however, we will assume that $\theta(0) = 0$. We can always achieve this by simply choosing to measure the parameter t from this point. If we differentiate the above relation, we get

$$\begin{pmatrix} \dot{\mathbf{p}}(t) \\ 0 \end{pmatrix} = \dot{\theta}(t)C e^{\theta(t)C} \begin{pmatrix} \mathbf{p} \\ 1 \end{pmatrix}.$$

Writing ω for the angular velocity $\omega = \dot{\theta}(0)$ and setting $t = 0$ in the rest of the equation, we obtain

$$\dot{\mathbf{p}}(0) = \omega \begin{pmatrix} c_y - y \\ -c_x + x \end{pmatrix} = -\omega E_2(\mathbf{p} - \mathbf{c}).$$

This gives the linear velocity of the point $\mathbf{p}$. For the sake of brevity, we will write $\dot{\mathbf{p}}$ for $\dot{\mathbf{p}}(0)$ in what follows. Now, the effect of the matrix E_2 is a rotation through $-\pi/2$ radians; hence, if we take the scalar product of both sides of the above equation with $(\mathbf{p} - \mathbf{c})$ we can eliminate the angular velocity ω.

$$\dot{\mathbf{p}} \cdot (\mathbf{p} - \mathbf{c}) = 0.$$

We see that the linear velocity is normal to the position vector of $\mathbf{p}$ relative to the centre of rotation $\mathbf{c}$.

Differentiating a second time gives

$$\begin{pmatrix} \ddot{\mathbf{p}}(t) \\ 0 \end{pmatrix} = \ddot{\theta}(t)C e^{\theta(t)C} \begin{pmatrix} \mathbf{p} \\ 1 \end{pmatrix} + \dot{\theta}^2(t)C^2 e^{\theta(t)C} \begin{pmatrix} \mathbf{p} \\ 1 \end{pmatrix}.$$

Setting $t = 0$ again gives

$$\ddot{\mathbf{p}} = -\dot{\omega} E_2(\mathbf{p} - \mathbf{c}) - \omega^2(\mathbf{p} - \mathbf{c}).$$

Again, we can remove the explicit dependence on the angular velocity and its derivative by observing that the acceleration of the point satisfies the vector equation

$$\ddot{\mathbf{p}} \cdot (\mathbf{p} - \mathbf{c}) + \dot{\mathbf{p}} \cdot \dot{\mathbf{p}} = 0.$$

The third derivative is given by

$$\begin{pmatrix} \mathbf{p}^{(3)}(t) \\ 0 \end{pmatrix} = \theta^{(3)}(t) C e^{\theta(t)C} \begin{pmatrix} \mathbf{p} \\ 1 \end{pmatrix} + 3\ddot{\theta}\dot{\theta} C^2 e^{\theta(t)C} \begin{pmatrix} \mathbf{p} \\ 1 \end{pmatrix} + \dot{\theta}^3(t) C^3 e^{\theta(t)C} \begin{pmatrix} \mathbf{p} \\ 1 \end{pmatrix}.$$

When $t = 0$ we get

$$\mathbf{p}^{(3)} = -\ddot{\omega} E_2 (\mathbf{p} - \mathbf{c}) - 3\dot{\omega}\omega (\mathbf{p} - \mathbf{c}) + \omega^3 E_2 (\mathbf{p} - \mathbf{c}).$$

Once more we can remove the explicit dependence on the angular velocity to get the equation

$$\mathbf{p}^{(3)} \cdot (\mathbf{p} - \mathbf{c}) + 3\ddot{\mathbf{p}} \cdot \dot{\mathbf{p}} = 0.$$

A similar calculation for the fourth derivative gives

$$\mathbf{p}^{(4)} \cdot (\mathbf{p} - \mathbf{c}) + 4\mathbf{p}^{(3)} \cdot \dot{\mathbf{p}} + 3\ddot{\mathbf{p}} \cdot \ddot{\mathbf{p}} = 0.$$

To summarise, we have derived the vector equations

$$\dot{\mathbf{p}} \cdot (\mathbf{p} - \mathbf{c}) = 0, \tag{C1}$$
$$\ddot{\mathbf{p}} \cdot (\mathbf{p} - \mathbf{c}) + \dot{\mathbf{p}} \cdot \dot{\mathbf{p}} = 0, \tag{C2}$$
$$\mathbf{p}^{(3)} \cdot (\mathbf{p} - \mathbf{c}) + 3\ddot{\mathbf{p}} \cdot \dot{\mathbf{p}} = 0, \tag{C3}$$
$$\mathbf{p}^{(4)} \cdot (\mathbf{p} - \mathbf{c}) + 4\mathbf{p}^{(3)} \cdot \dot{\mathbf{p}} + 3\ddot{\mathbf{p}} \cdot \ddot{\mathbf{p}} = 0. \tag{C4}$$

These equations relate the velocity and higher derivatives of a point on a circular path to the position of the point and the position of the centre. They can also be derived very neatly by differentiating the 'distance squared' $(\mathbf{p} - \mathbf{c})^2$; see Porteous [89, Chap. 3].

We may use the first two of the equations above to find the instantaneous centre of curvature of a point on a general curve. Recall that the curvature circle to a point on a curve is the unique circle that has three-point contact with the curve at the point. This means that the curvature circle meets the curve and the velocity and that acceleration vectors agree at the intersection. So, if we know the velocity and acceleration of some point on the curve, the first two equations above give a pair of linear equations in the two unknown coordinates of the centre of curvature, c_x and c_y.

5.3.1 The Euler–Savaray Equation

Now consider a lamina moving in the plane; see Figure 5.3. We may describe the motion of the lamina by the sequence of rigid motions it undergoes, $M(t) = e^{L(t)}$, that is a path in the group. The matrix $L(t)$ determines a path in the Lie algebra of the group. As we saw above, almost all Lie algebra elements correspond to points in the original plane:

$$L(t) = \theta(t) \begin{pmatrix} -E_2 & E_2 \mathbf{r}(t) \\ 0 & 0 \end{pmatrix}.$$

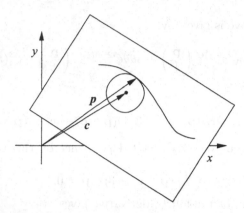

FIGURE 5.3. A Curve Generated by a Point on a Moving Lamina

As usual, we will assume that $\theta(0) = 0$ and write $\omega = \dot{\theta}(0)$ and $\dot{\omega} = \ddot{\theta}(0)$. The curve $t \longmapsto \mathbf{r}(t)$ is called the **fixed centrode** in the kinematics literature, see [54, §3.3] for example.

The idea here is to relate the fixed centrode to the curvature of curves traced out by points on the moving lamina. So let $\mathbf{p}$ be a point on the moving lamina. Its position at time t is given by

$$\begin{pmatrix} \mathbf{p}(t) \\ 1 \end{pmatrix} = e^{L(t)} \begin{pmatrix} \mathbf{p} \\ 1 \end{pmatrix}.$$

To find the centre of curvature of this path at $\mathbf{p}$, we need the first and second derivatives.

$$\begin{pmatrix} \dot{\mathbf{p}}(t) \\ 0 \end{pmatrix} = L_d(t) e^{L(t)} \begin{pmatrix} \mathbf{p} \\ 1 \end{pmatrix}.$$

At $t = 0$ we have $\theta(0) = 0$ and so $L(0) = 0$, this implies that $L_d(0) = \dot{L}(0)$ simplifying our computations. It is easy to see that

$$\dot{L}(0) = \omega \begin{pmatrix} -E_2 & E_2\mathbf{r} \\ 0 & 0 \end{pmatrix}.$$

and hence,

$$\dot{\mathbf{p}} = -\omega E_2(\mathbf{p} - \mathbf{r}). \qquad (P1)$$

We can also perform this computation using the expansion for the exponential map that we found above:

$$e^{L(t)} = I_3 + \sin\theta \begin{pmatrix} -E_2 & E_2\mathbf{r} \\ 0 & 0 \end{pmatrix} + (1 - \cos\theta) \begin{pmatrix} -I_2 & \mathbf{r} \\ 0 & 0 \end{pmatrix}.$$

so that

$$\dot{M}(t) = \dot{\theta}\cos\theta \begin{pmatrix} -E_2 & E_2\mathbf{r} \\ 0 & 0 \end{pmatrix} + \dot{\theta}\sin\theta \begin{pmatrix} -I_2 & \mathbf{r} \\ 0 & 0 \end{pmatrix}$$
$$+ \sin\theta \begin{pmatrix} 0 & E_2\dot{\mathbf{r}} \\ 0 & 0 \end{pmatrix} + (1 - \cos\theta) \begin{pmatrix} 0 & \dot{\mathbf{r}} \\ 0 & 0 \end{pmatrix}.$$

Setting $t = 0$ and multiplying on the right by the point, we recover the same result as before.

Similar calculations for the second derivative give

$$\ddot{\mathbf{p}} = -\dot{\omega} E_2(\mathbf{p} - \mathbf{r}) + 2\omega E_2 \dot{\mathbf{r}} - \omega^2(\mathbf{p} - \mathbf{r}). \tag{P2}$$

Substituting these results into equation (C1), we get

$$\omega(\mathbf{p} - \mathbf{r})^T E_2(\mathbf{p} - \mathbf{c}) = 0.$$

This tells us that the vectors $(\mathbf{p} - \mathbf{r})$ and $(\mathbf{p} - \mathbf{c})$ are parallel since $\omega \neq 0$. Hence, the vectors $\mathbf{r}$, $\mathbf{p}$ and $\mathbf{c}$ are collinear.

Substituting into (C2) we obtain

$$-\dot{\omega}(\mathbf{p} - \mathbf{c})^T E_2(\mathbf{p} - \mathbf{r}) + 2\omega(\mathbf{p} - \mathbf{c})^T E_2 \dot{\mathbf{r}} - \omega^2(\mathbf{p} - \mathbf{c})^T(\mathbf{p} - \mathbf{r}) + \omega^2(\mathbf{p} - \mathbf{r})^2 = 0.$$

which simplifies to

$$-2(\mathbf{p} - \mathbf{c})^T E_2 \dot{\mathbf{r}} + \omega(\mathbf{r} - \mathbf{c})^T(\mathbf{p} - \mathbf{r}) = 0.$$

Recall that the effect of the matrix E_2 is to rotate vectors by $-\pi/2$ radians; hence, for any vectors $\mathbf{a}$ and $\mathbf{b}$ we have $\mathbf{a}^T E_2 \mathbf{b} = |\mathbf{a}||\mathbf{b}| \sin \phi$, where ϕ is the angle between the vectors. So if we write ψ for the angle between the vector $\mathbf{p} - \mathbf{c}$ and the velocity vector $\dot{\mathbf{r}}$ and write $v = |\dot{\mathbf{r}}|$ for the magnitude of velocity vector of the fixed centrode, the equation becomes

$$-2v|\mathbf{p} - \mathbf{c}| \sin \psi - \omega(\mathbf{r} - \mathbf{p}) \cdot (\mathbf{r} - \mathbf{c}) = 0.$$

Finally, since the vectors $(\mathbf{r} - \mathbf{p})$ and $(\mathbf{r} - \mathbf{c})$ are parallel, we can rearrange this equation to give the classical **Euler–Savaray relation**; see for example Hunt [54, p.125]:

$$\frac{2|\mathbf{p} - \mathbf{c}|}{|\mathbf{r} - \mathbf{p}||\mathbf{r} - \mathbf{c}|} \sin \psi = \frac{-\omega}{v}$$

The sign of the right-hand side here depends on whether or not $\mathbf{r}$ lies between $\mathbf{p}$ and $\mathbf{c}$. In the above we have assumed that it does not.

5.3.2   The Inflection Circle

At a point of inflection on a curve, the velocity and acceleration vectors are linearly dependent. The 2×2 matrix whose columns are the velocity and acceleration of a point of inflection will have zero determinant. We can use this to find all the points on the lamina that are points of inflection at some instant. In the last section, we found the following results for the velocity and acceleration of points on the lamina:

$$\dot{\mathbf{p}} = -\omega E_2(\mathbf{p} - \mathbf{r}), \qquad \ddot{\mathbf{p}} = -\dot{\omega} E_2(\mathbf{p} - \mathbf{r}) + 2\omega E_2 \dot{\mathbf{r}} - \omega^2(\mathbf{p} - \mathbf{r}).$$

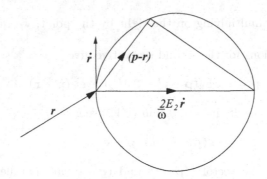

FIGURE 5.4. The Inflection Circle

For a 2×2 matrix whose columns are vectors $\mathbf{a}$ and $\mathbf{b}$, the determinant is given by $\mathbf{a}^T E_2 \mathbf{b}$. Hence, the equation for the points of inflection on the lamina is

$$0 = \dot{\mathbf{p}}^T E_2 \ddot{\mathbf{p}} = -2\omega^2 (\mathbf{p} - \mathbf{r})^T E_2 \dot{\mathbf{r}} + \omega^3 (\mathbf{p} - \mathbf{r})^2.$$

This simplifies further to

$$(\mathbf{p} - \mathbf{r})^T \left(\omega(\mathbf{p} - \mathbf{r}) - 2 E_2 \dot{\mathbf{r}} \right) = 0.$$

This is the equation of a circle, known as the **inflection circle**; see Figure 5.4. Notice that the inflection circle always passes through the point $\mathbf{r}$.

5.3.3   Ball's Point

Are there any points in the plane for which the velocity, acceleration, and third derivative are all in the same direction? Such a point on a curve is sometimes called an **undulation**. Clearly, such points must lie on the inflection circle. In equations (P1) and (P2) above, we found the first and second derivatives. We can compute the third derivative in the same way, and this gives

$$\mathbf{p}^{(3)} = (\omega^3 - \ddot{\omega}) E_2 (\mathbf{p} - \mathbf{r}) - 3\omega\dot{\omega}(\mathbf{p} - \mathbf{r}) + 3\omega^2 \dot{\mathbf{r}} + 3\dot{\omega} E_2 \dot{\mathbf{r}} + 3\omega E_2 \ddot{\mathbf{r}}. \qquad \text{(P3)}$$

So, the condition for the third derivative to be parallel to the tangent vector is given by

$$\dot{\mathbf{p}}^T E_2 \mathbf{p}^{(3)} = 0,$$
$$= 3\omega^2 \dot{\omega}(\mathbf{p} - \mathbf{r})^2 + 3\omega^3 (\mathbf{p} - \mathbf{r})^T \dot{\mathbf{r}} - 3\omega\dot{\omega}(\mathbf{p} - \mathbf{r})^T E_2 \dot{\mathbf{r}} - 3\omega^2 (\mathbf{p} - \mathbf{r})^T E_2 \ddot{\mathbf{r}}.$$

This equation is quadratic in the coordinates of $\mathbf{p}$, and any undulation will lie on the intersection of this conic and the inflection circle. We can make things a bit easier by subtracting a multiple of the equation for the inflection circle so as to make this equation linear. This gives

$$(\mathbf{p} - \mathbf{r})^T \left(\omega^2 \dot{\mathbf{r}} + \dot{\omega} E_2 \dot{\mathbf{r}} - \omega E_2 \ddot{\mathbf{r}} \right) = 0.$$

If we write $\mathbf{x} = (\omega^2 \dot{\mathbf{r}} + \dot{\omega} E_2 \dot{\mathbf{r}} - \omega E_2 \ddot{\mathbf{r}})$, then the general solution of the linear equation can be written

$$(\mathbf{p} - \mathbf{r}) = \lambda E_2 \mathbf{x},$$

where λ is arbitrary. We can fix λ by substituting this general solution into the equation for the inflection circle:

$$\lambda^2 \omega \mathbf{x}^2 - 2\lambda \mathbf{x}^T E_2 \dot{\mathbf{r}} = 0.$$

The solution $\lambda = 0$ corresponds to the undulation at $\mathbf{p} = \mathbf{r}$. The other solution, $\lambda = 2\mathbf{x}^T E_2 \dot{\mathbf{r}}/\omega \mathbf{x}^2$, corresponds to a second undulation called **Ball's point**, after R.S. Ball.

5.3.4 The Cubic of Stationary Curvature

Next we look at the curvature of the paths of points in the moving lamina. In particular, we look for points on the lamina for which the curvature is maximum or minimum. Actually, we will just require the curvature to be stationary. In differential geometry, such points are known as the vertices of the curve. The locus traced out by the centres of curvature as we move along a curve is known in classical geometry as the evolute of the original curve. The points of stationary curvature correspond to cusps on the evolute. More importantly here, these points are characterised by the fact that the curvature circle has at least four-point contact with the original curve at such points. This means that all the derivatives of the curve must agree with those for the circle up to the third, and all three equations (C1), (C2) and (C3) must be satisfied. These are three linear equations for the two unknowns: the components of $\mathbf{c}$, or equivalently those of $(\mathbf{p} - \mathbf{c})$. These can only be solved if the equations are consistent, the condition for consistency being given by the vanishing of the 3×3 determinant

$$\det \begin{pmatrix} \dot{\mathbf{p}}^T & 0 \\ \ddot{\mathbf{p}}^T & \dot{\mathbf{p}} \cdot \dot{\mathbf{p}} \\ \mathbf{p}^{(3)\,T} & 3\ddot{\mathbf{p}} \cdot \dot{\mathbf{p}} \end{pmatrix} = 0.$$

After expanding the above determinant, the consistency condition is found to be

$$(\dot{\mathbf{p}} \cdot \dot{\mathbf{p}})\dot{\mathbf{p}}^T E_2 \mathbf{p}^{(3)} - 3(\ddot{\mathbf{p}} \cdot \dot{\mathbf{p}})\dot{\mathbf{p}}^T E_2 \ddot{\mathbf{p}} = 0.$$

Substituting the results (P1), (P2) and (P3) from above into the consistency equation yields

$$3\omega^5 (\mathbf{p} - \mathbf{r})^2 (\mathbf{p} - \mathbf{r})^T \dot{\mathbf{r}} + 3\omega^3 \dot{\omega}(\mathbf{p} - \mathbf{r})^2 (\mathbf{p} - \mathbf{r})^T E_2 \dot{\mathbf{r}}$$
$$- 3\omega^4 (\mathbf{p} - \mathbf{r})^2 (\mathbf{p} - \mathbf{r})^T E_2 \ddot{\mathbf{r}} - 12\omega^4 (\mathbf{p} - \mathbf{r})^T \dot{\mathbf{r}}(\mathbf{p} - \mathbf{r})^T E_2 \dot{\mathbf{r}} = 0.$$

This is a cubic in the components of $\mathbf{p}$ and is called the **cubic of stationary curvature**. The equation is rather unwieldy, but we can simplify it by choosing suitable coordinates. Let us shift the origin of coordinates to the point $\mathbf{r}$ that is the instantaneous point on the centrode. Further, let us align the x-axis along the vector $\dot{\mathbf{r}}$, and hence the y-axis along the vector $-E_2\dot{\mathbf{r}}$. This choice of axes means that $(\mathbf{p} - \mathbf{r})^2 = (x^2 + y^2)$, $(\mathbf{p} - \mathbf{r})^T\dot{\mathbf{r}} = vx$ and $-(\mathbf{p} - \mathbf{r})^T E_2\dot{\mathbf{r}} = vy$, where $v = |\dot{\mathbf{r}}|$ as before, and x and y are the coordinates of the stationary points. The equation above simplifies to

$$(x^2 + y^2)\left(\frac{x}{\mu} + \frac{y}{\nu}\right) - xy = 0,$$

where

$$\mu = \frac{4v^2}{3a\sin\theta - \omega v} \quad\text{and}\quad \nu = \frac{4\omega v^2}{\dot{\omega}v - \omega a\cos\theta}$$

and we have written $\ddot{\mathbf{r}}^T\dot{\mathbf{r}} = av\cos\theta$ and $\ddot{\mathbf{r}}^T E_2\dot{\mathbf{r}} = -av\sin\theta$.

Notice that this cubic has a simple node at $x = y = 0$, that is, the point $\mathbf{r}$. The two tangents at this point are directed along the axes, that is, in the direction of $\dot{\mathbf{r}}$ and its perpendicular. The asymptote to the curve is the line

$$\frac{x}{\mu} + \frac{y}{\nu} + \frac{\mu\nu}{\mu^2 + \nu^2} = 0.$$

Finally, we note that the cubic may degenerate into a line and a conic under some circumstances. We will not pursue this here, but see Porteous [89, sect. 3.4].

5.3.5   The Burmester Points

Here we look for points that have five-point contact with their curvature circle. Certainly, such points must also lie on the cubic of stationary curvature. Extending the arguments above, we see that the 4×3 matrix

$$\begin{pmatrix} \dot{\mathbf{p}}^T & 0 \\ \ddot{\mathbf{p}}^T & \dot{\mathbf{p}}\cdot\dot{\mathbf{p}} \\ \mathbf{p}^{(3)\,T} & 3\ddot{\mathbf{p}}\cdot\dot{\mathbf{p}} \\ \mathbf{p}^{(4)\,T} & 4\mathbf{p}^{(3)}\cdot\dot{\mathbf{p}} + 3\ddot{\mathbf{p}}\cdot\ddot{\mathbf{p}} \end{pmatrix}$$

must have rank 2 at such points. If two of the four possible 3×3 determinants here vanish, then so will the remaining pair. Hence, we only need to consider two determinants here. For the first of these, we can take the first three rows as we did for the cubic of stationary curvature above. For the second, we can take the first two rows and the last to give the equation

$$\det\begin{pmatrix} \dot{\mathbf{p}}^T & 0 \\ \ddot{\mathbf{p}}^T & \dot{\mathbf{p}}\cdot\dot{\mathbf{p}} \\ \mathbf{p}^{(4)\,T} & 4\mathbf{p}^{(3)}\cdot\dot{\mathbf{p}} + 3\ddot{\mathbf{p}}\cdot\ddot{\mathbf{p}} \end{pmatrix} = 0,$$

which can be expanded to give

$$(\dot{\mathbf{p}}\cdot\dot{\mathbf{p}})\dot{\mathbf{p}}^T E_2\mathbf{p}^{(4)} - (4\mathbf{p}^{(3)}\cdot\dot{\mathbf{p}} + 3\ddot{\mathbf{p}}\cdot\ddot{\mathbf{p}})\dot{\mathbf{p}}^T E_2\ddot{\mathbf{p}} = 0.$$

Our next step is to express the derivatives of the point $\mathbf{p}$ by expressions involving only $\mathbf{p}$ itself and the instantaneous centrode point $\mathbf{r}$ and its derivatives:

$$\mathbf{p}^{(4)} = (6\dot{\omega}\omega^2 - \omega^{(3)})E_2(\mathbf{p} - \mathbf{r}) + (\omega^4 - 4\ddot{\omega}\omega - 3\dot{\omega}^2)(\mathbf{p} - \mathbf{r})$$
$$+ (4\ddot{\omega} - 2\omega^3)E_2\dot{\mathbf{r}} + 9\dot{\omega}\omega\dot{\mathbf{r}} + 6\dot{\omega}E_2\ddot{\mathbf{r}} + 3\omega^2\ddot{\mathbf{r}} + 4\omega E_2\mathbf{r}^{(3)}. \quad \text{(P4)}$$

Substituting this into the above and using the same coordinate system as above, we obtain another cubic:

$$(x^2 + y^2)(\alpha x + \beta y + \gamma) + y(\eta x + \zeta y + \xi) = 0$$

where

$$\alpha = \omega^3(6a\dot{\omega}\sin\theta - 12a\omega^2\cos\theta + 4b\omega^2\sin\phi - 15v\omega^2\dot{\omega}),$$
$$\beta = 2\omega^2(3v\dot{\omega}^2 + 2v\omega\ddot{\omega} - 72v^2\omega^8\sin\theta - 36v\omega^6\dot{\omega} - 2b\omega^2\cos\phi),$$
$$\gamma = 12v^2\omega^5,$$
$$\eta = -48v^2\omega(\dot{\omega} + 6v\omega^8\cos\theta),$$
$$\zeta = 288v^3\omega^9\sin\theta,$$
$$\xi = 24v^3\omega^4,$$

with $\mathbf{r}^{(3)} = (b\cos\phi, \, b\sin\phi)^T$.

Now, if we homogenise the two cubics by including a variable w, we can use Bézout's theorem to count the number of intersections. In $\mathbb{PC}^2$ the two cubics are thus

$$(x^2 + y^2)\left(\frac{x}{\mu} + \frac{y}{\mu}\right) - wxy = 0,$$
$$(x^2 + y^2)(\alpha x + \beta y + \gamma w) + wy(\eta x + \zeta y + \xi w) = 0.$$

We expect nine intersections between these two curves. However, there are some obvious common points. First, in the plane at infinity $w = 0$ we have two simple intersections, $x = 1$, $y = \pm i$. We also have an intersection at the pole $x = 0$, $y = 0$, $w = 1$. Recall that this is a simple node on the cubic of stationary curvature. On the other cubic, it is a regular point. However, a simple calculation reveals that the tangent at this regular point coincides with the tangent to one branch of the cubic of stationary curvature. Hence, the multiplicity of this intersection is 3. In total, these obvious intersections account for five of the nine crossings. Thus there are four other points in the plane that have five-point contact with a circle. These are the **Burmester points**. Pairs of these points may be complex conjugates, so there are at most four real Burmester points.

Note that the derivation of the above result in Porteous [89, sect. 3.5] is somewhat simpler. This is due to the fact that Porteous reparameterises the motion of the lamina so that its angular velocity is constant, $\omega = 1$. This has not been done in the above because although it is easy to reparameterise in theory, in practice it makes the expressions we need for particular cases very cumbersome.

5.4 The Planar 4-Bar

The standard example for the application of the above theory is to the 4-bar mechanism. In fact, reading some of the textbooks on mechanism theory one might be forgiven for concluding that the theory of the last section applies only to this example. On the other hand, the 4-bar is one of the oldest known mechanisms and it is also the most comprehensively studied and certainly the most widely used in practice. Hence, it is difficult to overstate its importance.

As the coupler bar of the mechanism moves, it traces out a curve in the proper Euclidean group $SE(2)$, that is, a one parameter family of rigid body motions. At each instant, we can find the inflection circle, Ball's point, cubic of stationary curvature, and Burmester points for the motion. Essentially this means that we must find the centrode $\mathbf{r}$, the angular velocity ω, the velocity of the centrode $\dot{\mathbf{r}}$, and some of the other constants that we used in the last section. There are several ways that we could proceed. We could, for example, use a parametric description of the motion of the coupler. It is well known that this motion is described by an elliptic curve (see [25]) and hence the parameterisation involves elliptic functions.

However, we don't need a parametric description of the coupler's motion since we can find all we need from the constraint equation

$$e^{\theta_1 C_1} e^{\theta_2 C_2} = e^{-\theta_4 C_4} e^{-\theta_3 C_3},$$

which simply expresses the fact that the coupler is rotating about the first two joints, C_1 and C_2, and at the same time it is rotating about the last two joints, C_3 and C_4.

The method most favoured by mechanical engineers is again different, but only slightly. Suppose we know the velocity and centre of curvature of two points on a moving lamina. From this information, we can find the position of the centrode, its velocity, and so forth. To apply this to the 4-bar, the moving lamina becomes the coupler bar, and the two points we know about are the positions of the second and third joints. These are attached to the coupler bar and simply circle the first and fourth joints, respectively.

From these considerations, some results are more or less immediate. The positions of the second and third joints, $\mathbf{c}_2$ and $\mathbf{c}_3$, lie on circular paths and hence have constant curvatures. So these points must lie on the cubic of stationary curvature and are also two of the Burmester points. We also know from the last section that the centrode, a point, and the centre of curvature of that point must lie on a line. Hence, $\mathbf{r}$, $\mathbf{c}_2$ and $\mathbf{c}_1$ are collinear, as are the points $\mathbf{r}$, $\mathbf{c}_3$ and $\mathbf{c}_4$. So the centrode $\mathbf{r}$ can be found by intersecting the lines that connect the first and second joints and the third and fourth joints; see Figure 5.5.

To make further progress, we will have to be more systematic. As usual, to simplify the calculations we can assume that the joint angles θ_1, θ_2, θ_3 and θ_4 are zero at the point we are interested in. The velocity of a general point $\mathbf{p}$ is

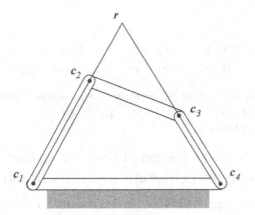

FIGURE 5.5. The Planar 4-Bar

given by the two equations (C1) and (R1) found above:

$$\dot{\mathbf{p}} = -\omega E_2(\mathbf{p} - \mathbf{r}) = -\omega' E_2(\mathbf{p} - \mathbf{c}).$$

Notice here that ω is the angular velocity of the lamina, while ω' is the angular velocity of the point $\mathbf{p}$ about its centre of curvature $\mathbf{c}$. Hence, for the two points $\mathbf{c}_2$ and $\mathbf{c}_3$ in the 4-bar we have

$$\omega(\mathbf{c}_2 - \mathbf{r}) = \omega_1(\mathbf{c}_2 - \mathbf{c}_1) \quad \text{and} \quad \omega(\mathbf{c}_3 - \mathbf{r}) = \omega_4(\mathbf{c}_3 - \mathbf{c}_4).$$

These equations can be rearranged to give

$$\mathbf{r} = \left(1 - \frac{\omega_1}{\omega}\right)\mathbf{c}_2 + \frac{\omega_1}{\omega}\mathbf{c}_1 = \mathbf{c}_2 + \frac{\omega_1}{\omega}(\mathbf{c}_1 - \mathbf{c}_2),$$

$$\mathbf{r} = \left(1 - \frac{\omega_4}{\omega}\right)\mathbf{c}_3 + \frac{\omega_3}{\omega}\mathbf{c}_4 = \mathbf{c}_3 + \frac{\omega_4}{\omega}(\mathbf{c}_4 - \mathbf{c}_3).$$

This confirms our earlier observation that $\mathbf{r}$ is the intersection of the lines connecting the two pairs of joints. It is usual for just one joint of a 4-bar mechanism to be driven, and hence we assume that it is the first joint that is attached to a motor. Since we know ω_1, the angular velocity of the first joint, the equations above can be combined and rearranged into

$$\omega_1(\mathbf{c}_1 - \mathbf{c}_2) + \omega_4(\mathbf{c}_3 - \mathbf{c}_4) + \omega(\mathbf{c}_2 - \mathbf{c}_3) = 0.$$

Thus, the ratios of these angular velocities can be found in terms of determinants:

$$\frac{\omega}{\omega_1} = \frac{(\mathbf{c}_3 - \mathbf{c}_4)^T E_2(\mathbf{c}_1 - \mathbf{c}_2)}{(\mathbf{c}_3 - \mathbf{c}_4)^T E_2(\mathbf{c}_3 - \mathbf{c}_2)} \quad \text{and} \quad \frac{\omega_4}{\omega_1} = \frac{(\mathbf{c}_2 - \mathbf{c}_3)^T E_2(\mathbf{c}_1 - \mathbf{c}_2)}{(\mathbf{c}_3 - \mathbf{c}_2)^T E_2(\mathbf{c}_3 - \mathbf{c}_4)}.$$

To find $\dot{\mathbf{r}}$ and $\dot{\omega}$ we look at the acceleration of the two joints. Using the equations (R2) and the corresponding equation for circular motion, we have the pair of equations

$$-\dot{\omega}E_2(\mathbf{c}_2 - \mathbf{r}) + 2\omega E_2\dot{\mathbf{r}} - \omega^2(\mathbf{c}_2 - \mathbf{r}) = -\dot{\omega}_1 E_2(\mathbf{c}_2 - \mathbf{c}_1) - \omega_1^2(\mathbf{c}_2 - \mathbf{c}_1),$$
$$-\dot{\omega}E_2(\mathbf{c}_3 - \mathbf{r}) + 2\omega E_2\dot{\mathbf{r}} - \omega^2(\mathbf{c}_3 - \mathbf{r}) = -\dot{\omega}_4 E_2(\mathbf{c}_3 - \mathbf{c}_4) - \omega_4^2(\mathbf{c}_3 - \mathbf{c}_4).$$

These equations can be rearranged to make $\dot{\mathbf{r}}$ the subject, and we can also use the substitutions $\omega(\mathbf{c}_2 - \mathbf{r}) = \omega_1(\mathbf{c}_2 - \mathbf{c}_1)$ and $\omega(\mathbf{c}_3 - \mathbf{r}) = \omega_4(\mathbf{c}_3 - \mathbf{c}_4)$ so that the pair of equations become

$$2\omega^2\dot{\mathbf{r}} = (\dot{\omega}\omega_1 - \omega\dot{\omega}_1)(\mathbf{c}_2 - \mathbf{c}_1) + (\omega\omega_1^2 - \omega^2\omega_1)E_2(\mathbf{c}_2 - \mathbf{c}_1),$$
$$2\omega^2\dot{\mathbf{r}} = (\dot{\omega}\omega_4 - \omega\dot{\omega}_4)(\mathbf{c}_3 - \mathbf{c}_4) + (\omega\omega_4^2 - \omega^2\omega_4)E_2(\mathbf{c}_3 - \mathbf{c}_4).$$

The first of these equations will give $\dot{\mathbf{r}}$ if we know $\dot{\omega}$. So we subtract the above equations and use the fact that $\omega_1(\mathbf{c}_1 - \mathbf{c}_2) + \omega_4(\mathbf{c}_3 - \mathbf{c}_4) + \omega(\mathbf{c}_2 - \mathbf{c}_3) = \mathbf{0}$ to produce the relation

$$\dot{\omega}_1(\mathbf{c}_1 - \mathbf{c}_2) + \dot{\omega}_4(\mathbf{c}_3 - \mathbf{c}_4) + \dot{\omega}(\mathbf{c}_2 - \mathbf{c}_3)$$
$$= \omega_1^2 E_2(\mathbf{c}_1 - \mathbf{c}_2) + \omega_4^2 E_2(\mathbf{c}_3 - \mathbf{c}_4) + \omega^2 E_2(\mathbf{c}_2 - \mathbf{c}_3).$$

From here, we can find $\dot{\omega}$ purely in terms of ω_1 and its derivative. To do this, we eliminate the $\dot{\omega}_4$ term by taking the scalar product with the vector $E_2(\mathbf{c}_4 - \mathbf{c}_2)$ and substituting for ω and ω_4. The result is

$$\dot{\omega} = \dot{\omega}_1 \frac{(\mathbf{c}_3 - \mathbf{c}_4)^T E_2(\mathbf{c}_1 - \mathbf{c}_2)}{(\mathbf{c}_2 - \mathbf{c}_3)^T E_2(\mathbf{c}_3 - \mathbf{c}_4)}$$
$$+ \frac{\omega_1^2}{\left((\mathbf{c}_2 - \mathbf{c}_3)^T E_2(\mathbf{c}_3 - \mathbf{c}_4)\right)^3} \Big((\mathbf{c}_3 - \mathbf{c}_4)^T(\mathbf{c}_1 - \mathbf{c}_2)\big((\mathbf{c}_2 - \mathbf{c}_3)^T E_2(\mathbf{c}_3 - \mathbf{c}_4)\big)^2$$
$$+ (\mathbf{c}_3 - \mathbf{c}_4)^T(\mathbf{c}_3 - \mathbf{c}_4)\big((\mathbf{c}_1 - \mathbf{c}_2)^T E_2(\mathbf{c}_2 - \mathbf{c}_3)\big)^2$$
$$+ (\mathbf{c}_2 - \mathbf{c}_3)^T(\mathbf{c}_3 - \mathbf{c}_4)\big((\mathbf{c}_1 - \mathbf{c}_2)^T E_2(\mathbf{c}_3 - \mathbf{c}_4)\big)^2\Big).$$

This information allows us to find the inflection circle. We could use similar methods to find the higher derivatives of $\mathbf{r}$ and hence locate Ball's point, the cubic of stationary curvature, and so on. Although reasonably straightforward, it is not difficult to see that the formulas will be be even longer and more complicated than the ones given above. Some slight simplification is achieved if we assume that the first joint is driven at a constant rate, which is reasonable for many practical machines. Before the advent of inexpensive computing, much emphasis was placed on technical drawing constructions for finding these important features. These days it is probably more relevant to find formulas that can be efficiently coded and incorporated in computer programs.

Many attempts have been made to extend these ideas of instantaneous kinematics to the spatial case, see for example [14, Chap VI] . Notice that the curves and points derived in the last section for the planar case arise from considering contact between paths of points and straight lines or circles. The significance of the straight lines and circles is that they are the paths of points generated by one-parameter subgroups of $SE(2)$. This suggests that in the spatial case we should look at the contact between point paths and helices. The difficulty is that helices are not algebraic curves and so cannot be represented as the zeros of some polynomial equations. This makes questions of contact rather more difficult to investigate.

6
Line Geometry

Line geometry is not as popular these days as it was even fifty years ago. This is perhaps because many of the original problems of the subject have been solved. Algebraic geometers think of ruled surfaces as line bundles over a curve or even more abstract descriptions. Differential geometers usually worry about the extrinsic geometry of ruled surfaces—that is, how such surfaces can sit in three dimensions—their curvature, and so forth. Symplectic geometers have all but forgotten that their subject began with the study of the symmetries of line complexes.

The subject is of great historical importance, since the space of all lines in three dimensions, the Klein quadric, which we discuss below, was perhaps the first concrete example of a non-Euclidean space.

Our interest in the subject is twofold, firstly because as robots and mechanisms move, any line attached to them will trace out a ruled surface or maybe some higher dimensional object. In particular, we might be interested in the surface traced out by the axis of some revolute joint. The second reason for our interest is that lines are zero-pitch screws. Line geometry is a special case of our more general interest in the geometry of the Lie algebra of the proper Euclidean group.

6.1   Lines in Three Dimensions

We begin by looking at some elementary geometry associated with lines in three dimensions. A line is determined by any two distinct points; see Figure 6.1.

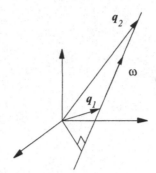

FIGURE 6.1. A Line in Three Dimensions

Suppose $\mathbf{q}_1 = (x_1, y_1, z_1)^T$ and $\mathbf{q}_2 = (x_2, y_2, z_2)^T$ are two points. A vector in the direction along the line joining the two points is

$$\omega = \mathbf{q}_2 - \mathbf{q}_1.$$

Now the position vector of any point on the line can be written as

$$\mathbf{q}_\lambda = \mathbf{q}_1 + \lambda\omega = (1 - \lambda)\mathbf{q}_1 + \lambda\mathbf{q}_2$$

where λ is a real parameter.

If $\mathbf{a}$ is some point not on the line, what is the shortest distance from this point to the line? The square of the distance to any point on the line is $(\mathbf{q}_\lambda - \mathbf{a}) \cdot (\mathbf{q}_\lambda - \mathbf{a})$. To find the minimum, we differentiate with respect to λ and set the result to zero:

$$(\mathbf{q}_\lambda - \mathbf{a}) \cdot \frac{d}{d\lambda}(\mathbf{q}_\lambda - \mathbf{a}) = 0 = (\mathbf{q}_\lambda - \mathbf{a}) \cdot \omega.$$

Notice that the minimum distance is along a vector perpendicular to the line. The equation has the solution

$$\lambda = (\mathbf{a} - \mathbf{q}_1) \cdot \omega/|\omega|^2.$$

The vector from the point to the line can be written as a triple product:

$$(\mathbf{q}_\lambda - \mathbf{a}) = \frac{1}{|\omega|^2}(|\omega|^2(\mathbf{q}_1 - \mathbf{a}) - ((\mathbf{q}_1 - \mathbf{a}) \cdot \omega)\omega) = \omega \times ((\mathbf{q}_1 - \mathbf{a}) \times \omega)/|\omega|^2.$$

Next consider two lines. In three dimensions, a pair of lines are generally skew. From the above, we know that the minimum distance between the lines must be along a vector perpendicular to both lines. Suppose the two lines are labelled A and B and are characterised by points on the lines $\mathbf{q}_a$ and $\mathbf{q}_b$ with vectors along the lines ω_a and ω_b. Now, $\omega_a \times \omega_b$ is a vector perpendicular to both lines. This vector vanishes only if the lines are parallel. How can we decide whether the lines actually meet? If they do meet, then the common point will be given by

$$\mathbf{q}_a + \lambda\omega_a = \mathbf{q}_b + \mu\omega_b.$$

This vector equation represents three linear equations in the two unknowns λ and μ. Such a system of equations has a solution if and only if the equations are consistent. That is, the lines meet if and only if the equations are consistent. When the lines are not parallel, the only possible linear dependency that the equations could have is given by dotting the equations with $\omega_a \times \omega_b$. Hence, the equations are consistent and the non-parallel lines meet if and only if the vectors for the lines satisfy

$$\mathbf{q}_a \cdot (\omega_a \times \omega_b) = \mathbf{q}_b \cdot (\omega_a \times \omega_b).$$

Instead of characterising the line by its direction ω and a point $\mathbf{q}$ on it, we could use the direction and the vector $\mathbf{v} = \mathbf{q} \times \omega$. This is sometimes called the moment of the line. Using these vectors for the two lines, the condition above can be rewritten as

$$\omega_a \cdot \mathbf{v}_b + \mathbf{v}_a \cdot \omega_b = 0.$$

For parallel lines, we could say that parallel lines meet at a point 'at infinity', and thus the relation just derived is satisfied if and only if the lines meet, possibly at infinity. Notice that any two lines that meet determine a plane, while two lines that lie in a plane must meet at infinity if the lines are parallel.

How can we tell whether the lines determined by their directions and a point on each actually determine the same line? Certainly, the direction vectors of the two lines must be proportional: $\omega_a = \gamma\omega_b$. Also, we must have

$$\mathbf{q}_a + \lambda\omega_a = \mathbf{q}_b$$

for some λ. These three equations for the single variable λ will be consistent if and only if we can find two linear dependencies. These can only be provided by taking the vector product of the equations with ω_a, and hence the condition for the lines to be the same is

$$\mathbf{q}_a \times \omega_a = \mathbf{q}_b \times \omega_a.$$

This can be rewritten very succinctly using the direction and moments for the lines. Now, two direction and moment pairs represent the same line if and only if

$$\omega_a = \gamma\omega_b \qquad \text{and} \qquad \mathbf{v}_a = \gamma\mathbf{v}_b$$

for some constant γ.

6.2   Plücker Coordinates

Imagine a space each of whose points corresponds to a line in three-dimensional space. How can we put coordinates on such a space? Each line must have a unique set of coordinates so that different lines have different coordinates.

We begin by looking at a single line in three-dimensional space determined by two points; see Figure 6.1 again. Let $\mathbf{q}_1 = (x_1, y_1, z_1)^T$ and $\mathbf{q}_2 = (x_2, y_2, z_2)^T$ be the two points. The direction and moment of the line are then given by

$$\boldsymbol{\omega} = \mathbf{q}_1 - \mathbf{q}_2, \qquad \text{and} \qquad \mathbf{v} = \mathbf{q}_1 \times \mathbf{q}_2.$$

The vector $\boldsymbol{\omega}$ is a vector in the direction of the line, while the moment $\mathbf{v}$ is normal to the plane through the origin containing the line. If we take two different points on the line, say

$$\mathbf{q}_1' = \lambda \mathbf{q}_1 + (1 - \lambda)\mathbf{q}_2 \qquad \text{and} \qquad \mathbf{q}_2' = \mu \mathbf{q}_1 + (1 - \mu)\mathbf{q}_2,$$

then the new points must be distinct, so $\lambda \neq \mu$. The direction and moment vectors become

$$\boldsymbol{\omega}' = \mathbf{q}_1' - \mathbf{q}_2' = (\lambda - \mu)\boldsymbol{\omega},$$
$$\mathbf{v}' = \mathbf{q}_1' \times \mathbf{q}_2' = (\lambda - \mu)\mathbf{v}.$$

So, the overall effect of using different points to determine the line is just to multiply these vectors by the same non-zero constant.

On the other hand, we saw above that two pairs of direction and moment vectors where the corresponding vectors are simply multiples of each other determine the same line.

If we use the components of $\boldsymbol{\omega}$ and $\mathbf{v}$ as the six homogeneous coordinates of five-dimensional projective space $\mathbb{PR}^5$, multiplying all the homogeneous coordinates by a non-zero scalar here gives the same point, and hence each line determines a unique point in this space. The six homogeneous coordinates are called the Plücker coordinates of the line; they are usually written as

$$
\begin{aligned}
p_{01} &= (x_1 - x_2), & p_{23} &= (y_1 z_2 - y_2 z_1), \\
p_{02} &= (y_1 - y_2), & p_{31} &= (x_2 z_1 - x_1 z_2), \\
p_{03} &= (z_1 - z_2), & p_{12} &= (x_1 y_2 - x_2 y_1).
\end{aligned}
$$

The notation becomes a little clearer if we use (x_1, x_2, x_3) as coordinates for a point in $\mathbb{R}^3$. The line through the two points $\mathbf{x} = (x_1, x_2, x_3)^T$ and $\mathbf{y} = (y_1, y_2, y_3)^T$ then has Plücker coordinates given by

$$p_{ij} = x_i y_j - x_j y_i,$$

with $x_0 = y_0 = 1$.

In terms of these Plücker coordinates, the two 3-dimensional vectors found above are

$$\boldsymbol{\omega} = \begin{pmatrix} p_{01} \\ p_{02} \\ p_{03} \end{pmatrix} \qquad \text{and} \qquad \mathbf{v} = \begin{pmatrix} p_{23} \\ p_{31} \\ p_{12} \end{pmatrix}.$$

But remember that the Plücker coordinates are homogeneous coordinates.

Not all points of $\mathbb{PR}^5$, however, represent lines. For a line, we have $\boldsymbol{\omega} = \mathbf{q}_1 - \mathbf{q}_2$ and $\mathbf{v} = \mathbf{q}_1 \times \mathbf{q}_2$ and hence $\boldsymbol{\omega} \cdot \mathbf{v} = 0$. In terms of the Plücker coordinates of the line, this gives

$$p_{01}p_{23} + p_{02}p_{31} + p_{03}p_{12} = 0.$$

This is a homogeneous equation of degree 2, and hence its solutions lie on a four-dimensional quadric hypersurface in $\mathbb{PR}^5$. This quadric is called the **Klein quadric**.

Points of $\mathbb{PR}^5$ not on the Klein quadric cannot be lines. On the other hand, no line in $\mathbb{R}^3$ has $\boldsymbol{\omega} = \mathbf{0}$ (we can have $\mathbf{v} = \mathbf{0}$; these are just lines through the origin). So there is a two-dimensional plane in the quadric, defined by $p_{01} = p_{02} = p_{03} = 0$ or equivalently $\boldsymbol{\omega} = \mathbf{0}$, whose points are not real lines. These points are usually referred to as lines "at infinity".

To summarise, we have shown that the lines in $\mathbb{R}^3$ are in one-to-one correspondence to points on a four-dimensional quadric excluding the points on a 2-plane.

Finally, suppose we are given the Plücker coordinates of a line. How do we find the points in three dimensions that lie on the line? Let $\boldsymbol{\omega}$ and $\mathbf{v}$ be the direction and moment of the line; then if $\mathbf{q}$ is a point on the line, it must satisfy

$$\boldsymbol{\omega} \times \mathbf{q} = \mathbf{v}.$$

To see this, write $\boldsymbol{\omega} = \mathbf{q}_1 - \mathbf{q}_2$, $\mathbf{v} = \mathbf{q}_1 \times \mathbf{q}_2$, and for the point on the line choose $\mathbf{q} = \mathbf{q}_1 + \lambda \boldsymbol{\omega}$. We can also describe the line parametrically. The points on the line can be described by

$$\mathbf{q} = (\mathbf{v} \times \boldsymbol{\omega})/|\boldsymbol{\omega}|^2 + \lambda\boldsymbol{\omega}$$

where λ is a parameter.

6.3   The Klein Quadric

We can think of the Klein quadric defined above in a slightly different way that we will find useful later. Consider the 2-planes through the origin in $\mathbb{R}^4$, and compare this with Figure 2.3. The connection between 2-planes in $\mathbb{R}^4$ and lines in $\mathbb{R}^3$ is as follows: Consider a 3-plane in $\mathbb{R}^4$ determined by the equation $x_0 = 1$. Notice this plane does not contain the origin. The intersection of a 2-plane through the origin with this 3-plane will be a line in the 3-plane. Different 2-planes give different lines, and the only 2-planes that do not intersect the 3-plane are the ones parallel to the 3-plane. If we include these as the 'lines at infinity', then there is a one-to-one correspondence between the 2-planes in $\mathbb{R}^4$ and the lines in $\mathbb{R}^3$.

The Klein quadric is a four-dimensional non-singular quadric in $\mathbb{PR}^5$. It contains two families of 2-planes. To see this, we perform the following change of coordinates:

$$p_{01} = X_a + Y_a, \qquad p_{23} = X_a - Y_a,$$
$$p_{02} = X_b + Y_b, \qquad p_{31} = X_b - Y_b,$$
$$p_{03} = X_c + Y_c, \qquad p_{12} = X_c - Y_c.$$

In these coordinates, the equation for the quadric becomes

$$p_{01}p_{23} + p_{02}p_{31} + p_{03}p_{12} = X_a^2 + X_b^2 + X_c^2 - Y_a^2 - Y_b^2 - Y_c^2 = 0.$$

Now, consider the points in $\mathbb{PR}^5$ satisfying the three homogeneous linear equations determined by the matrix equation $\mathbf{X} = M\mathbf{Y}$, where

$$\mathbf{X} = \begin{pmatrix} X_a \\ X_b \\ X_c \end{pmatrix} \quad \text{and} \quad \mathbf{Y} = \begin{pmatrix} Y_a \\ Y_b \\ Y_c \end{pmatrix}.$$

The points satisfying these equations lie in a 2-plane. If the constant matrix M is an element of the group $O(3)$, then the points also lie in the Klein quadric, since the equations imply that $\mathbf{X} \cdot \mathbf{X} = \mathbf{Y} \cdot \mathbf{Y}$ if M is orthogonal. This means that we have 2-planes lying entirely within the Klein quadric. In fact, we get a different 2-plane for each matrix $M \in O(3)$. There are two families of these 2-planes: planes determined by a matrix with determinant $\det(M) = +1$ are called α-planes, while planes given by matrices for which $\det(M) = -1$ are called β-planes.

What do these planes represent in terms of the lines in $\mathbb{R}^3$? If we write the Plücker coordinates as a pair of three-dimensional vectors $\boldsymbol{\omega}$ and $\mathbf{v}$ as in the last section, then the equations for a 2-plane can be written as

$$(I_3 - M)\boldsymbol{\omega} + (I_3 + M)\mathbf{v} = \mathbf{0}.$$

Now, we have two cases to consider: if $\det(M) = +1$, then $(M + I_3)$ is usually non-singular, so that we can write the equations as

$$(M + I_3)^{-1}(M - I_3)\boldsymbol{\omega} = \mathbf{v}.$$

The matrix $U_+ = (M + I_3)^{-1}(M - I_3)$ is anti-symmetric since

$$U_+^T = (M^T - I_3)(M^T + I_3)^{-1} = (I_3 - M)M^T M (I_3 + M)^{-1}$$
$$= -(M - I_3)(M + I_3)^{-1} = -U_+.$$

The final step here follows because $(M + I_3)$ and $(M - I_3)$ commute. Similarly, if $\det(M) = -1$, then $(M - I_3)$ is usually non-singular and the equation for the plane is

$$\boldsymbol{\omega} = (M - I_3)^{-1}(M + I_3)\mathbf{v}$$

and now the matrix $U_- = (M - I_3)^{-1}(M + I_3)$ is anti-symmetric.

Now, in the first case we can write the equation as

$$\mathbf{u}_+ \times \boldsymbol{\omega} = \mathbf{v}$$

where $\mathbf{u}_+$ is a point in three-dimensional space, and since we may interpret $\mathbf{v}$ as the moment of the line, we can see that the solutions to the equations consist of all the lines through the point $\mathbf{u}_+$.

When $\det(M) = +1$ the matrix $(M+I_3)$ is singular only when M is a rotation by 0, or π. In the case where the rotation is by 0 we have $M = I_3$, and this gives the set of lines through the origin. For rotations by π radians, the corresponding α-planes consist of parallel lines in $\mathbb{R}^3$. These could be thought of as sets of lines through a point at infinity.

In the second case, the equations are

$$\boldsymbol{\omega} = \mathbf{u}_- \times \mathbf{v}.$$

Here we have $\mathbf{u}_- \cdot \boldsymbol{\omega} = 0$, so the lines are all perpendicular to the vector $\mathbf{u}_-$. Moreover, if we write $\mathbf{v} = \mathbf{q} \times \boldsymbol{\omega}$, we can show that $\mathbf{u}_- \cdot \mathbf{q} = -1$. Hence, the β-planes consist of lines lying in a 2-plane in $\mathbb{R}^3$. Notice that the set of lines at infinity corresponds to the β-plane given by $M = -I_3$.

To summarise: α-planes are lines through a point in $\mathbb{R}^3$, while β-planes are lines in a 2-plane in $\mathbb{R}^3$.

Two different α-planes meet in a single point. This point represents the line joining the two points in $\mathbb{R}^3$ determined by the α-planes. Similarly, two different β-planes meet at a point, while this time the point represents the line in $\mathbb{R}^3$ that is the intersection of the two 2-planes determined by the β-planes. In general, an α-plane and a β-plane do not meet. This reflects the fact that in $\mathbb{R}^3$ a general point does not lie on a generally chosen plane, and hence there is no line lying on the plane that also passes through the point. On the other hand, if an α-plane and a β-plane do meet, then they meet in a line. In $\mathbb{R}^3$ this line in the Klein quadric corresponds to the set of lines in a plane passing through a point. Such a configuration of lines is sometimes called a plane star or a plane pencil of lines.

6.4 The Action of the Euclidean Group

To find the action of $SE(3)$ on the Plücker coordinates, we begin by looking at the action on a pair of points that determine a line. Given two points $\mathbf{q}_1$ and $\mathbf{q}_2$, an arbitrary rotation and translation takes the points to $\mathbf{q}_1' = R\mathbf{q}_1 + \mathbf{t}$ and $\mathbf{q}_2' = R\mathbf{q}_2 + \mathbf{t}$, respectively. The vector along the line through these points and the moment of the line are

$$\boldsymbol{\omega} = \mathbf{q}_1 - \mathbf{q}_2, \qquad \text{and} \qquad \mathbf{v} = \mathbf{q}_1 \times \mathbf{q}_2.$$

After the rotation and translation, these will be

$$\boldsymbol{\omega}' = \mathbf{q}_1' - \mathbf{q}_2' = R(\mathbf{q}_1 - \mathbf{q}_2)$$

and

$$\mathbf{v}' = \mathbf{q}_1' \times \mathbf{q}_2' = R(\mathbf{q}_1 \times \mathbf{q}_2) + R(\mathbf{q}_1 - \mathbf{q}_2) \times \mathbf{t} = R\mathbf{v} + R\boldsymbol{\omega} \times \mathbf{t}.$$

So, we can write the effect of a rotation and a translation on the Plücker coordinates in partitioned matrix form as

$$\begin{pmatrix} \boldsymbol{\omega}' \\ \mathbf{v}' \end{pmatrix} = \begin{pmatrix} R & 0 \\ TR & R \end{pmatrix} \begin{pmatrix} \boldsymbol{\omega} \\ \mathbf{v} \end{pmatrix}.$$

This is exactly the adjoint action of the group on its Lie algebra that we met in Section 4.2. Recall that $T = ad(\mathbf{t})$, that is, the anti-symmetric matrix with the property that $T\mathbf{x} = \mathbf{x} \times \mathbf{t}$ for any vector $\mathbf{x}$.

Notice that although we began by just looking at the effect of rigid motions on lines, we have found a linear representation on $\mathbb{R}^6$. We can think of this as an action on lines by taking the coordinates as Plücker coordinates on $\mathbb{PR}^5$. In this way, we obtain an action of the group on the Klein quadric. In other words the rigid transformations will preserve the Klein quadric. To see this, note that the equation defining the Klein quadric is invariant under a general rotation and translation:

$$\boldsymbol{\omega}' \cdot \mathbf{v}' = R\boldsymbol{\omega} \cdot (R\mathbf{v} + R\boldsymbol{\omega} \times \mathbf{t}) = \boldsymbol{\omega} \cdot \mathbf{v} = 0.$$

In fact, this shows that $\boldsymbol{\omega} \cdot \mathbf{v}$ is an invariant quadratic form. Given a quadratic form, we can always turn it into a symmetric bilinear form. In the case of the invariant quadratic form we saw above, the corresponding invariant symmetric bilinear form is

$$(\boldsymbol{\omega}_1^T, \mathbf{v}_1^T) \begin{pmatrix} 0 & I_3 \\ I_3 & 0 \end{pmatrix} \begin{pmatrix} \boldsymbol{\omega}_2 \\ \mathbf{v}_2 \end{pmatrix} = \boldsymbol{\omega}_1 \cdot \mathbf{v}_2 + \mathbf{v}_1 \cdot \boldsymbol{\omega}_2.$$

This form has been known for a century or more and is usually called the **reciprocal product** of screws. Compare this reciprocal product with the Killing form we met in Section 4.7.

The fact that this form is invariant makes it easy to interpret geometrically. All we need to do is to look at it in a simple situation. The group invariance then takes care of more complicated cases. So assume we have two lines—a line through the origin along the x-axis

$$\begin{pmatrix} \boldsymbol{\omega}_1 \\ \mathbf{v}_1 \end{pmatrix} = \begin{pmatrix} \mathbf{i} \\ \mathbf{0} \end{pmatrix}$$

and a second line at an angle ϕ to the first and displaced l units along the z-axis:

$$\begin{pmatrix} \boldsymbol{\omega}_2 \\ \mathbf{v}_2 \end{pmatrix} = \begin{pmatrix} \cos \phi \mathbf{i} + \sin \phi \mathbf{j} \\ l\mathbf{k} \times (\cos \phi \mathbf{i} + \sin \phi \mathbf{j}) \end{pmatrix} = \begin{pmatrix} \cos \phi \mathbf{i} + \sin \phi \mathbf{j} \\ l \cos \phi \mathbf{j} - l \sin \phi \mathbf{i} \end{pmatrix}.$$

The reciprocal product of these two lines is given by

$$\boldsymbol{\omega}_1 \cdot \mathbf{v}_2 + \mathbf{v}_1 \cdot \boldsymbol{\omega}_2 = \mathbf{i} \cdot (l \cos \phi \mathbf{j} - l \sin \phi \mathbf{i}) = -l \sin \phi.$$

That is, the reciprocal product of these two lines is minus the product of the minimum distance between the lines with the sine of the angle between them. The group invariance tells us that this must be true for any pair of lines. Notice also that a pair of lines meet at a point (possibly at infinity) if and only if their reciprocal product is zero, in agreement with Section 6.1. When the reciprocal product of two screws is zero, we say that the screws are reciprocal to each other. The Killing form evaluated on a pair of lines gives the cosine of the angle between the lines.

It is possible to show that the Killing form and the form of the reciprocal product are the only invariant bilinear forms on $se(3)$, in the sense that any invariant bilinear form must be some linear combination of these two. To see this, we must consider the equations

$$\begin{pmatrix} R^T & -R^T T \\ 0 & R^T \end{pmatrix} \begin{pmatrix} A & B \\ C & D \end{pmatrix} \begin{pmatrix} R & 0 \\ TR & R \end{pmatrix} = \begin{pmatrix} A & B \\ C & D \end{pmatrix}.$$

The bilinear form is given by the matrix on the right-hand side. The equation expresses the fact that the form should be invariant under the action of the group, that is, the action of an arbitrary rotation and translation.

Multiplying out the above equation gives four equations for the four unknown 3×3 matrices A, B, C and D:

$$A = R^T AR + R^T BTR - R^T TCR - R^T TDTR,$$
$$B = R^T BR - R^T TDR,$$
$$C = R^T CR + R^T DTR,$$
$$D = R^T DR.$$

The last of these equations has the solution $D = \lambda I_3$ for some scalar λ. It is clear that this a solution. The fact that the only matrices that commute with a rotation matrix are multiples of the identity means that it is the only solution. Substituting this result into the second and third equations gives

$$R^T BR - \lambda R^T TR = B \qquad \text{and} \qquad R^T CR + \lambda R^T TR = C.$$

These have to be satisfied for any rotation and translation, and hence we must have $\lambda = 0$. Hence, the solution for B and C is $B = bI_3$ and $C = cI_3$ for some scalars b and c. Turning to the first equation, we have

$$R^T AR + bR^T TR - cR^T TR = A.$$

Again, this must be satisfied for arbitrary rotations and translations, so we conclude that $b = c = \beta$ say, and finally that $A = \alpha I_3$. To summarise, the invariant bilinear forms on the Lie algebra $se(3)$ are given by

$$(\omega_1^T, \mathbf{v}_1^T) \begin{pmatrix} -2\alpha I_3 & \beta I_3 \\ \beta I_3 & 0 \end{pmatrix} \begin{pmatrix} \omega_2 \\ \mathbf{v}_2 \end{pmatrix} = -2\alpha(\omega_1 \cdot \omega_2) + \beta(\omega_1 \cdot \mathbf{v}_2 + \mathbf{v}_1 \cdot \omega_2).$$

The factor of -2 here is for consistency with the Killing form and will tidy up some relations a little later. These results can also be proved using the representation theory that will be introduced in Chapter 7.

Each symmetric bilinear form determines a quadratic form. Setting each of these to zero gives a degree-two homogeneous equation in the Plücker coordinates:

$$-\alpha(\boldsymbol{\omega} \cdot \boldsymbol{\omega}) + \beta(\boldsymbol{\omega} \cdot \mathbf{v}) = 0.$$

In other words the solution forms a pencil of quadrics. The set of solutions to one of these equations, that is, with fixed values of α and β, forms a quadric hypersurface in the projective space $\mathbb{PR}^5$. When $\alpha = 0$ we recover the Klein quadric. On the other hand, if $\beta = 0$ we obtain the quadric $\boldsymbol{\omega} \cdot \boldsymbol{\omega} = 0$; this is a singular quadric. The only real solutions lie on the 2-plane $\boldsymbol{\omega} = \mathbf{0}$; this is the plane of lines at infinity that we met in Section 6.2.

More generally, a point in $\mathbb{PR}^5$ that does not lie on the 2-plane $\boldsymbol{\omega} = \mathbf{0}$ can be written as

$$\begin{pmatrix} \boldsymbol{\omega} \\ \mathbf{r} \times \boldsymbol{\omega} + p\boldsymbol{\omega} \end{pmatrix}.$$

Substituting this into the quadratic form gives

$$(-\alpha + \beta p)\boldsymbol{\omega} \cdot \boldsymbol{\omega} = 0.$$

Solutions are points for which $p = \alpha/\beta$. So each point in $\mathbb{PR}^5$ lies on one of these invariant quadrics. Furthermore, the only points that lie on more than one are the points on the 2-plane $\boldsymbol{\omega} = \mathbf{0}$ that lie in every invariant quadric.

The scalar p is thus an invariant for the group action; in fact it is the pitch of the corresponding Lie algebra element. For a point $(\boldsymbol{\omega}, \mathbf{v})^T$ its pitch is given by

$$p = \frac{\boldsymbol{\omega} \cdot \mathbf{v}}{\boldsymbol{\omega} \cdot \boldsymbol{\omega}}.$$

Conventionally, points with $\boldsymbol{\omega} = \mathbf{0}$ are assigned an infinite pitch. Lines in $\mathbb{R}^3$ correspond to points in $\mathbb{PR}^5$ with pitch zero. We will denote the symmetric matrices representing these forms by

$$Q_p = \alpha Q_\infty + \beta Q_0$$

where $p = \alpha/\beta$ and

$$Q_0 = \begin{pmatrix} 0 & I_3 \\ I_3 & 0 \end{pmatrix} \quad \text{and} \quad Q_\infty = \begin{pmatrix} -2I_3 & 0 \\ 0 & 0 \end{pmatrix}.$$

The quadrics defined by these matrices will be referred to as pitch quadrics.

It is not too difficult to show that points in $\mathbb{PR}^5$ with the same pitch can always be brought into coincidence by some rotation and translation. So we can now describe the orbits of the group $SE(3)$ on $\mathbb{PR}^5$. Each orbit is either the 2-plane $\boldsymbol{\omega} = \mathbf{0}$ or the part of a pitch quadric that excludes $\boldsymbol{\omega} = \mathbf{0}$. The isotropy group of points can also be determined for points with infinite pitch, that is, points of the form $(\mathbf{0}, \mathbf{v})^T$. We have

$$\begin{pmatrix} R(\mathbf{v}) & 0 \\ TR(\mathbf{v}) & R(\mathbf{v}) \end{pmatrix} \begin{pmatrix} \mathbf{0} \\ \mathbf{v} \end{pmatrix} = \begin{pmatrix} \mathbf{0} \\ \mathbf{v} \end{pmatrix}$$

where $R(\mathbf{v})$ is a rotation about the vector $\mathbf{v}$. This group of matrices is isomorphic to $SO(2) \ltimes \mathbb{R}^3$. For a point with a finite pitch, the isotropy group is the symmetry group of a cylinder $SO(2) \times \mathbb{R}$. The transformations that leave points of the form $(\boldsymbol{\omega}, \mathbf{r} \times \boldsymbol{\omega} + p\boldsymbol{\omega})^T$ fixed consist of a rotation about $\boldsymbol{\omega}$ followed by a translation of the form $\mathbf{t} = (R(\boldsymbol{\omega}) - I_3)\mathbf{r} + \lambda\boldsymbol{\omega}$ with λ and the angle of rotation arbitrary.

We can look at all of this in a slightly different way. Rather than passing to projective space, consider the points in $\mathbb{R}^6$ satisfying

$$\boldsymbol{\omega} \cdot \mathbf{v} = 0.$$

As we have seen above, these points can be thought of as representing lines in $\mathbb{R}^3$ up to an overall multiplicative factor. Now suppose we normalise by imposing the extra condition

$$\boldsymbol{\omega} \cdot \boldsymbol{\omega} = 1.$$

The only possible multiplicative factors are now ± 1, so to each line in $\mathbb{R}^3$ there are two such points in $\mathbb{R}^6$,

$$\begin{pmatrix} \boldsymbol{\omega} \\ \mathbf{r} \times \boldsymbol{\omega} \end{pmatrix} \quad \text{and} \quad \begin{pmatrix} -\boldsymbol{\omega} \\ -\mathbf{r} \times \boldsymbol{\omega} \end{pmatrix}.$$

corresponding to the same line. We can interpret points in this affine variety as **directed lines**, that is, lines together with a given direction along the line. The space of these directed lines double covers the space of undirected lines. We must be a little bit careful however, since the Klein quadric includes points representing lines at infinity. So, the directed lines double cover the open set of finite lines in the Klein quadric. Many relationships between lines are best pictured using this directed representation. Moreover, since the double covering map is a local homeomorphism, if we are only interested in local properties of the space of lines we lose nothing if we work with the directed representation and then project to the finite lines in the Klein quadric.

6.5   Ruled Surfaces

In this short section, we cannot hope to do justice to this vast subject. Entire books have been written devoted to the topic, for example Edge [31]. We content ourselves with a couple of examples. We will return to the subject in Chapter 9 when we have developed some more efficient tools to study the differential geometry of these surfaces.

A **ruled surface** is a one-parameter family of lines. The parameterisation must be at least continuous, so that the family of lines forms a surface. The idea is that these surfaces can be made up of a collection of lines. An obvious example is the plane, which can be ruled by a family of parallel lines. A degenerate example would be the set of lines tangent to a curve. A ruled surface formed in this way is called a **developable**.

A one-parameter family of lines can be considered as a curve in the Klein quadric. Hence, much of the study of these surfaces can be reduced to the study of curves in the Klein quadric.

We will confine our attentions to a few simple examples that arise frequently in robotics and mechanism theory. We begin with the regulus.

6.5.1   The Regulus

Consider the cylindrical hyperboloid in $\mathbb{R}^3$ given by the equation

$$x^2 + y^2 - z^2 - r^2 = 0.$$

As we saw with the Klein quadric, this quadric contains two families of linear spaces of half the dimension of the quadric itself; see Section 6.4. In this case, since the quadric is two-dimensional the linear subspaces are lines. Each family is a ruled surface called a regulus; see Figure 6.2. The lines are given by pairs of linear equations.

Family 1	Family 2
$x = z\cos\theta - w\sin\theta,$	$x = z\cos\theta + w\sin\theta,$
$y = z\sin\theta + w\cos\theta,$	$y = z\sin\theta - w\cos\theta.$

These equations come from the relation

$$\begin{pmatrix} x \\ y \end{pmatrix} = M \begin{pmatrix} z \\ w \end{pmatrix}$$

where M is a matrix in $O(2)$. The two families are distinguished by the sign of M's determinant; see Section 2.2.2. Different lines in a single family correspond to different values of the parameter θ. Notice that we can turn one regulus into the other by reflecting in the plane $z = 0$. This involves changing the parameter to $\theta + \pi$ in one of the families.

We can find the Plücker coordinates of the lines by looking at a pair of points on each line. A convenient pair of points is given by $z = 1$, $w = 0$ and $z = 0$, $w = 1$. These give the points

$$(0, \cos\theta, \sin\theta, 1), \qquad (1, -\sin\theta, \cos\theta, 0),$$

for lines on the first regulus and

$$(0, \cos\theta, \sin\theta, 1), \qquad (1, \sin\theta, -\cos\theta, 0),$$

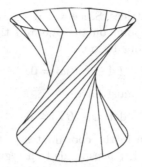

FIGURE 6.2. The Regulus of a Hyperboloid

for lines on the second regulus. These give the following Plücker coordinates for the lines:

$$
\begin{aligned}
p_{01} &= -\cos\theta, & p_{23} &= -\cos\theta, \\
p_{02} &= -\sin\theta, & p_{31} &= -\sin\theta, \\
p_{03} &= -1, & p_{12} &= 1,
\end{aligned}
$$

for the first regulus and

$$
\begin{aligned}
p_{01} &= -\cos\theta, & p_{23} &= \cos\theta, \\
p_{02} &= -\sin\theta, & p_{31} &= \sin\theta, \\
p_{03} &= -1, & p_{12} &= -1,
\end{aligned}
$$

for the second.

In each case, the lines all lie on a 2-plane in $\mathbb{PR}^5$. In each case, the 2-plane is determined by three linear equations.

$$
\begin{aligned}
p_{01} - p_{23} &= 0, & p_{01} + p_{23} &= 0, \\
p_{02} - p_{31} &= 0, \quad \text{or} & p_{02} + p_{31} &= 0, \\
p_{03} + p_{12} &= 0, & p_{03} - p_{12} &= 0.
\end{aligned}
$$

The lines of a regulus are points in the Klein quadric that also lie on a 2-plane; hence, as a curve in the Klein quadric, the regulus corresponds to a conic. All of the above also applies to any one-sheeted hyperboloid, since by Sylvester's theorem we can always find a projective transformation that will turn a one-sheeted hyperboloid into a cylindrical one. Hence, on any one-sheeted hyperboloid in $\mathbb{R}^3$ there is a pair of reguli each corresponding to a conic in the Klein quadric.

Looking at things the other way around, suppose we had a conic curve in the Klein quadric. What surface does that correspond to in $\mathbb{R}^3$? The ruled surface corresponding to a conic in the Klein quadric will always be a quadric in $\mathbb{R}^3$. To see this, first observe that the conic will be the intersection of the Klein quadric with a 2-plane in $\mathbb{PR}^5$. Now, assume that the equations for the 2-plane can be written as

$$
A\boldsymbol{\omega} + B\mathbf{v} = \mathbf{0},
$$

where A and B are 3×3 matrices. Since we are dealing with lines, we can write $\mathbf{v} = \mathbf{t} \times \boldsymbol{\omega}$, where $\mathbf{t}$ is any point on the line and thus a point on the surface. Using $T = ad(\mathbf{t})$, the equations become

$$(A - BT)\boldsymbol{\omega} = \mathbf{0}.$$

For non-trivial solutions, we must have

$$\det(A - BT) = 0.$$

This is a quadratic equation in the coordinates of $\mathbf{t}$ since $\det(T) = 0$. Hence, any conic in the Klein quadric corresponds to a regulus of some quadric in $\mathbb{R}^3$.

Finally, notice that, given a pair of skew lines, if we rotate one of the lines about the other, we generate a regulus of a cylindrical hyperboloid. Such a situation might occur in practice if the lines are the axes of a pair of joints attached to a rigid link.

6.5.2 The Cylindroid

Consider a line in $\mathbb{PR}^5$. In general, such a linear subspace will intersect the Klein quadric in just two points. The points of the Klein quadric correspond to lines in $\mathbb{R}^3$, but the other points of the linear subspace can be interpreted as elements of the Lie algebra $se(3)$, or more precisely, as rays through the origin in $se(3)$; see Section 6.4. Almost all elements of $se(3)$ generate screw motions; the axis of the screw is again a line in $\mathbb{R}^3$. If we take the axes of all the screws in the linear subspace, we obtain a one-parameter family of lines, a ruled surface in $\mathbb{R}^3$.

We investigate the cylindroid by considering the ruled surface generated by a pair of lines. By a rigid transformation, we can move the pair of lines so that one line lies along the x-axis and the common perpendicular between the lines lies along the z-axis. The two lines will then have Plücker coordinates

$$
\begin{aligned}
p_{01} &= 1 \\
p_{02} &= 0 \\
p_{03} &= 0 \\
p_{23} &= 0 \\
p_{31} &= 0 \\
p_{12} &= 0
\end{aligned}
\qquad \text{and} \qquad
\begin{aligned}
p_{01} &= \cos\phi \\
p_{02} &= \sin\phi \\
p_{03} &= 0 \\
p_{23} &= -l,\sin\phi \\
p_{31} &= l\cos\phi \\
p_{12} &= 0
\end{aligned}
$$

where l is the length of the common perpendicular between the lines, sometimes called the offset distance, and ϕ is the angle between the lines, the twist angle. A point on the linear system joining the lines is given by

$$
\begin{aligned}
p_{01} &= \mu + \lambda\cos\phi \\
p_{02} &= \lambda\sin\phi \\
p_{03} &= 0 \\
p_{23} &= -\lambda l\sin\phi \\
p_{31} &= \lambda l\cos\phi \\
p_{12} &= 0
\end{aligned}
$$

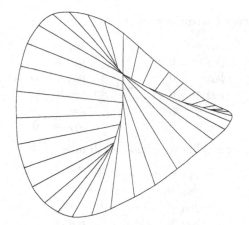

FIGURE 6.3. The Cylindroid

where μ and λ are parameters. The axes of these screws are given by

$$
\begin{aligned}
p_{01} &= \mu + \lambda \cos \phi \\
p_{02} &= \lambda \sin \phi \\
p_{03} &= 0 \\
p_{23} &= -\lambda l \sin \phi - p\mu - p\lambda \cos \phi \\
p_{31} &= \lambda l \cos \phi - p\lambda \sin \phi \\
p_{12} &= 0.
\end{aligned}
$$

The pitch p is given as usual by $\boldsymbol{\omega} \cdot \mathbf{v}/\boldsymbol{\omega} \cdot \boldsymbol{\omega}$, which simplifies to

$$
p = \frac{-\mu \lambda l \sin \phi}{\mu^2 + \lambda^2 + 2\mu\lambda \cos \phi}.
$$

Multiplying through by the denominator, we obtain a cubic rational parameterisation of the curve in the Klein quadric. We can also find the equation of the ruled surface in $\mathbb{R}^3$. Using the fact that points on the lines are given by $\mathbf{q} = (\mathbf{v} \times \boldsymbol{\omega})/|\omega|^2 + \gamma\boldsymbol{\omega}$ (see Section 6.2) we get a parameterisation of the points on the surface:

$$
x = \mu\gamma + \lambda\gamma \cos \phi, \qquad y = \lambda\gamma \sin \phi, \qquad z = -l\frac{\lambda^2 + \lambda\mu \cos \phi}{\lambda^2 + \mu^2 + 2\mu\lambda \cos \phi}.
$$

Eliminating the parameters, we obtain the cubic equation

$$
z(x^2 + y^2) = -l(y^2 + xy \cot \phi).
$$

In the kinematics literature, this ruled surface is called the cylindroid; in the differential geometry literature it goes by the name of Plücker's conoid. The surface is illustrated in Figure 6.3.

To find out a little more about the disposition of lines in the cylindroid, we will look at a particular case where the twist angle between the generating lines is $\pi/2$ and neither of the lines passes through the origin.

That is, the Plücker coordinates of the lines are

$$
\begin{aligned}
p_{01} &= & 1 \\
p_{02} &= & 0 \\
p_{03} &= & 0 \\
p_{23} &= & 0 \\
p_{31} &= & r_x \\
p_{12} &= & 0
\end{aligned}
\qquad \text{and} \qquad
\begin{aligned}
p_{01} &= & 0 \\
p_{02} &= & 1 \\
p_{03} &= & 0 \\
p_{23} &= & r_y \\
p_{31} &= & 0 \\
p_{12} &= & 0.
\end{aligned}
$$

So, a general line of the cylindroid has coordinates

$$
\begin{aligned}
p_{01} &= & \lambda(\lambda^2 + \mu^2) \\
p_{02} &= & \mu(\lambda^2 + \mu^2) \\
p_{03} &= & 0 \\
p_{23} &= & -\mu(\lambda^2 r_x - \mu^2 r_y) \\
p_{31} &= & \lambda(\lambda^2 r_x - \mu^2 r_y) \\
p_{12} &= & 0.
\end{aligned}
$$

Remembering that we can multiply Plücker coordinates by a non-zero constant without effect, if we normalise by setting $\lambda = \cos\theta$ and $\mu = \sin\theta$, then we can see that the lines in the cylindroid with direction

$$
\boldsymbol{\omega} = \begin{pmatrix} \cos\theta \\ \sin\theta \\ 0 \end{pmatrix}
$$

pass through the point

$$
\mathbf{r} = \begin{pmatrix} 0 \\ 0 \\ r_x \cos^2\theta + r_y \sin^2\theta \end{pmatrix}.
$$

On the other hand, through each point on the z-axis between the extreme values of $z = r_x$ and r_y pass exactly two lines of the cylindroid.

In general, if we have a one-parameter sequence of rigid motions, such a motion will be generated, via the exponential map, by a curve in the Lie algebra $se(3)$. In the projective space $\mathbb{PR}^5$, we would also get a curve unless the motion was just a pure rotation, pure translation, or helical motion, which corresponds to lines through the origin in $se(3)$. So, generically, by taking the axes of the screws in the motion, we obtain a ruled surface. This ruled surface of screw axes is called the **fixed axode** of the motion.

6.5.3   Curvature Axes

The final example of a ruled surface comes from the theory of surfaces.

At most points on a curved surface, the surface can be approximated locally by a quadric:

$$
k_x x^2 + k_y y^2 - 2z = 0,
$$

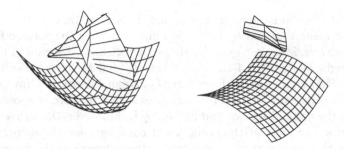

FIGURE 6.4. Lines of Curvature for Two Surface Points

where k_x and k_y are the principal curvatures; see O'Neill [81, Chap. V sect. 3]. Here the z-direction is along the surface normal at the point and the x- and y-directions are aligned with the directions of principal curvature. The exceptions, where this approximation is not valid, are umbilic points where all directions are directions of principal curvature. We will not consider umbilics here.

The curvature at an angle ϕ to the x-axis is given by the Euler relation

$$k_\phi = k_x \cos^2 \phi + k_y \sin^2 \phi;$$

see [81, Chap. V sect. 2], for example. We can think of the possible motion along the surface in this direction as an infinitesimal rotation. So there will be a rotation axis perpendicular to the direction of motion through the centre of curvature. As we move around the point on the surface, that is as ϕ changes, we will generate a sequence of lines, a ruled surface.

The Plücker coordinates of these curvature axes will be

$$\mathbf{s}(\phi) = \begin{pmatrix} -\mathbf{i}\sin\phi + \mathbf{j}\cos\phi \\ (\mathbf{i}\cos\phi + \mathbf{j}\sin\phi)/k_\phi \end{pmatrix}.$$

Writing $\lambda = \cos\phi$ and $\mu = \sin\phi$, we obtain a rational cubic parameterisation of these points in the Klein quadric:

$$\mathbf{s}(\lambda, \mu) = \begin{pmatrix} -\mu(k_x\lambda^2 + k_y\mu^2) \\ \lambda(k_x\lambda^2 + k_y\mu^2) \\ 0 \\ \lambda(\lambda^2 + \mu^2) \\ \mu(\lambda^2 + \mu^2) \\ 0 \end{pmatrix}.$$

So, the curvature axes at a point on the surface form a rational cubic ruled surface.

As above we can find a parameterisation of the points on this surface,

$$x = -\mu\gamma(k_x\lambda^2 + k_y\mu^2), \qquad y = \lambda\gamma(k_x\lambda^2 + k_y\mu^2), \qquad z = \frac{(\lambda^2 + \mu^2)}{(k_x\lambda^2 + k_y\mu^2)}.$$

Eliminating the parameters gives an equation for the ruled surface,

$$z(k_x y^2 + k_y x^2) = (x^2 + y^2).$$

These surfaces are illustrated in Figure 6.4. It is clear that when both of the principal curvatures k_x and k_y have the same sign, then the surface formed by the curvature axes is again a cylindroid; see the left-hand diagram in Figure 6.4.

On the right of Figure 6.4 the case where k_x and k_y have opposite signs is illustrated. That is, a point with negative Gaussian curvature. Only part of the ruled surface generated by the curvature axes is shown, there is another branch underneath the original surface and hidden by it. Moreover, the surface contains a pair of lines that cross at the point under consideration; these correspond to directions with zero curvature. Approaching these directions the curvature axes get further and further away from the original surface. Hence, only a small part of one branch of the ruled surface is shown in the figure.

6.6   Line Complexes

A **line complex** is a set of lines all of which obey a linear relation, that is, the set of lines whose Plücker coordinates satisfy an equation of the form

$$b_1 p_{01} + b_2 p_{02} + b_3 p_{03} + \alpha_1 p_{23} + \alpha_2 p_{31} + \alpha_3 p_{12} = 0$$

where the b_is and α_is are constants. The lines of a complex lie in the intersection of the Klein quadric with a hyperplane in $\mathbb{PR}^5$. If we write the Plücker coordinates as $\omega^T = (p_{01}, p_{02}, p_{03})$ and $\mathbf{v}^T = (p_{23}, p_{31}, p_{12})$ and collect the constants into a pair of three-dimensional vectors $\boldsymbol{\alpha}^T = (\alpha_1, \alpha_2, \alpha_3)$ and $\mathbf{b}^T = (b_1, b_2, b_3)$, then the equation can be written as

$$\boldsymbol{\omega} \cdot \mathbf{b} + \mathbf{v} \cdot \boldsymbol{\alpha} = 0.$$

This is reminiscent of the reciprocal product that we met in Section 6.4. In fact, we can see that another way of looking at the line complex is as the set of lines reciprocal to a fixed screw:

$$(\boldsymbol{\omega}^T, \mathbf{v}^T) \begin{pmatrix} 0 & I_3 \\ I_3 & 0 \end{pmatrix} \begin{pmatrix} \boldsymbol{\alpha} \\ \mathbf{b} \end{pmatrix} = 0.$$

If the fixed screw is in fact a line, that is, if $\boldsymbol{\alpha} \cdot \mathbf{b} = 0$, then the complex consists of all possible lines that meet the fixed line determined by $\boldsymbol{\alpha}$ and $\mathbf{b}$. This is a singular situation, however. It corresponds to the case when the hyperplane of the complex is tangent to the Klein quadric. If we write

$$\Phi = p_{01} p_{23} + p_{02} p_{31} + p_{03} p_{12},$$

then the partial derivatives are

$$\frac{\partial \Phi}{\partial p_{01}} = p_{23}, \qquad \frac{\partial \Phi}{\partial p_{02}} = p_{31}, \qquad \frac{\partial \Phi}{\partial p_{03}} = p_{12},$$

$$\frac{\partial \Phi}{\partial p_{23}} = p_{01}, \qquad \frac{\partial \Phi}{\partial p_{31}} = p_{02}, \qquad \frac{\partial \Phi}{\partial p_{12}} = p_{03}.$$

Hence, the equation of the tangent hyperplane to the Klein quadric at the point $\boldsymbol{\alpha}^T = (p_{01}, p_{02}, p_{03})$, $\mathbf{b}^T = (p_{23}, p_{31}, p_{12})$ is

$$b_1 p_{01} + b_2 p_{02} + b_3 p_{03} + \alpha_1 p_{23} + \alpha_2 p_{31} + \alpha_3 p_{12} = 0.$$

To study the disposition of lines in a non-singular complex, we make use of another representation of line complexes. Suppose (x_0, x_1, x_2, x_3) and (y_0, y_1, y_2, y_3) are a pair of points in $\mathbb{PR}^3$. Now the line joining these two points is in the complex if and only if the following equation is satisfied:

$$(y_1, y_2, y_3, y_0) \begin{pmatrix} 0 & -\alpha_3 & \alpha_2 & b_1 \\ \alpha_3 & 0 & -\alpha_1 & b_2 \\ -\alpha_2 & \alpha_1 & 0 & b_3 \\ -b_1 & -b_2 & -b_3 & 0 \end{pmatrix} \begin{pmatrix} x_1 \\ x_2 \\ x_3 \\ x_0 \end{pmatrix} = 0.$$

This is because the left-hand side of the above equation expands to

$$b_1(x_0 y_1 - x_1 y_0) + b_2(x_0 y_2 - x_2 y_0) + b_3(x_0 y_3 - x_3 y_0)$$
$$+ \alpha_3(x_1 y_2 - x_2 y_1) + \alpha_2(x_1 y_3 - x_3 y_1) + \alpha_1(x_2 y_3 - x_3 y_2)$$
$$= b_1 p_{01} + b_2 p_{02} + b_3 p_{03} + \alpha_1 p_{23} + \alpha_2 p_{31} + \alpha_3 p_{12}.$$

From this, we can see that if we fix a point (y_0, y_1, y_2, y_3) say, then the points that lie on lines in the complex through that fixed point must lie in a two-dimensional plane. More specifically, the equation of the 2-plane is given by

$$x_0 C_0 + x_1 C_1 + x_2 C_2 + x_3 C_3 = 0,$$

where the constant C_is are given by

$$\begin{aligned} C_0 &= b_1 y_1 + b_2 y_2 + b_3 y_3, \\ C_1 &= -b_1 y_0 + \alpha_3 y_2 + \alpha_2 y_3, \\ C_2 &= -b_2 y_0 - \alpha_3 y_1 + \alpha_1 y_3, \\ C_3 &= -b_3 y_0 - \alpha_2 y_1 - \alpha_1 y_2. \end{aligned}$$

This means that through any point in $\mathbb{PR}^3$ the lines of the complex that pass through that point lie in a plane. That is, the lines form a plane star; see Section 6.3.

In the above notation, the line complex is singular if and only if the determinant of the anti-symmetric matrix is zero. This is because we have

$$\det \begin{pmatrix} 0 & -\alpha_3 & \alpha_2 & b_1 \\ \alpha_3 & 0 & -\alpha_1 & b_2 \\ -\alpha_2 & \alpha_1 & 0 & b_3 \\ -b_1 & -b_2 & -b_3 & 0 \end{pmatrix} = (\boldsymbol{\alpha} \cdot \mathbf{b})^2.$$

Writing the equation for the line complex as an anti-symmetric pairing of points in $\mathbb{PR}^3$ has another useful consequence. Consider what happens to the equation if we make a linear change of coordinates in $\mathbb{PR}^3$. The new coordinates of points are given by

$$\begin{pmatrix} x_1' \\ x_2' \\ x_3' \\ x_0' \end{pmatrix} = M \begin{pmatrix} x_1 \\ x_2 \\ x_3 \\ x_0 \end{pmatrix}$$

where $M \in GL(4)$. Now, writing the anti-symmetric matrix that defines the complex as

$$X = \begin{pmatrix} 0 & -\alpha_3 & \alpha_2 & b_1 \\ \alpha_3 & 0 & -\alpha_1 & b_2 \\ -\alpha_2 & \alpha_1 & 0 & b_3 \\ -b_1 & -b_2 & -b_3 & 0 \end{pmatrix}$$

in the new coordinates, we get a new matrix X' that satisfies the relation

$$X = M^T X' M.$$

A transformation for which $X = M^T X M$ is a symmetry of the complex; lines of the complex are transformed into other lines in the complex. The group of matrices M that preserves a 4×4 anti-symmetric matrix is the symplectic group $Sp(4, \mathbb{R})$; see Section 3.1.

Which of these symmetries are rigid body motions? To answer this, we could look to see which matrices M in the above equation have the form

$$M = \begin{pmatrix} R & \mathbf{t} \\ 0 & 1 \end{pmatrix}.$$

Alternatively, and perhaps more simply, we could study the equation $\boldsymbol{\omega} \cdot \mathbf{b} + \mathbf{v} \cdot \boldsymbol{\alpha} = 0$ under rigid coordinate changes. This leads to the equation

$$\begin{pmatrix} R & 0 \\ TR & R \end{pmatrix} \begin{pmatrix} \boldsymbol{\alpha} \\ \mathbf{b} \end{pmatrix} = \begin{pmatrix} \boldsymbol{\alpha} \\ \mathbf{b} \end{pmatrix}$$

which has solution

$$R = e^{\theta\alpha} \quad \text{and} \quad \mathbf{t} = \frac{1}{|\boldsymbol{\alpha}|^2} \boldsymbol{\alpha} \times (I_3 - R)\mathbf{b} - \lambda\boldsymbol{\alpha}$$

for some θ and some λ. Hence, we have shown that

$$Sp(4, \mathbb{R}) \cap SE(3) = SO(2) \times \mathbb{R}.$$

The rigid symmetries of a line complex are the symmetries of a cylinder. The axis of the cylinder is given by the line with Plücker coordinates

$$\begin{pmatrix} \boldsymbol{\alpha} \\ \mathbf{b} - \boldsymbol{\alpha}(\boldsymbol{\alpha} \cdot \mathbf{b}/|\boldsymbol{\alpha}|^2) \end{pmatrix}.$$

Notice that this is the axis of the screw $(\boldsymbol{\alpha}, \mathbf{b})^T$.

A different application of these ideas comes from geometric optics. Light rays can be thought of as lines. It turns out that the effect of an optical system in turning one light ray into another can be described as an element of the group $Sp(4, \mathbb{R})$; see Bamberg and Sternberg [7, Chap.9].

Lastly here, we observe that there is an accidental isomorphism between the 10-dimensional groups $Sp(4, \mathbb{R})$ and $SO(3,2)$. The latter group is the group of conformal transformations of $\mathbb{R}^3$; that is, the group of transformations that preserve the usual scalar product on $\mathbb{R}^3$, but only up to multiplication by a positive constant. This discovery by Lie led to the idea of 'Sphere geometry' and was one of the early results that inspired his work on 'continuous groups'; see Hawkins [47].

6.7   Inverse Robot Jacobians

In Section 4.5 the Jacobians of serial robots were discussed. In many practical applications, especially the control of such manipulators, it is the inverse of the Jacobian that is key. In many algorithms the inversion of the Jacobian matrix is performed numerically; this is computationally expensive. Hunt pointed out that for many designs of practical machines the inverse Jacobian can be found symbolically; see [55]. In Hunt's method a convenient coordinate system is chosen to write down the Jacobian and then 5×5 cofactors are computed to form the adjugate of the Jacobian matrix. A little later, but seemingly independently, Fijany and Bejczy [34] showed that if the robot has a spherical wrist then a 3×3 diagonal block of the of the Jacobian vanishes and then only inverses of 3×3 matrices need to be calculated.

The method outlined here relies on some properties of line complexes and the fact that the joint screws are lines; that is, the joints are either revolute or prismatic. This means that some of the rows of the inverse Jacobian can be found by inspection with very little effort.

Suppose the joint axes of the robot and hence the columns of the robot's Jacobian matrix are $\mathbf{s}_1, \mathbf{s}_2, \ldots, \mathbf{s}_6$. If it is possible to find a line $\mathbf{z}_1$ say, that meets the five lines $\mathbf{s}_2, \mathbf{s}_3, \ldots, \mathbf{s}_6$ then we know that $\mathbf{z}_1$ will be reciprocal to these lines. That is, $\mathbf{z}_1^T Q_0 \mathbf{s}_i = 0$, $i = 2, 3, \ldots, 6$; in other words the five joint axes lie in a singular line complex. In this case we will generally have $\mathbf{z}_1^T Q_0 \mathbf{s}_1 \neq 0$ and hence we can divide by this factor and set

$$\mathcal{W}_1 = \frac{1}{\mathbf{z}_1^T Q_0 \mathbf{s}_1} Q_0 \mathbf{z}_1.$$

The first row of the inverse Jacobian matrix is then $\mathcal{W}_1^T$, since by construction $\mathcal{W}_1^T \mathbf{s}_1 = 1$ and $\mathcal{W}_1^T \mathbf{s}_i = 0$ for $i = 2, 3, \ldots, 6$. The rows of the inverse Jacobian matrix are, in fact, wrenches; see Chapter 12 below.

When the five joint axes do not lie in a singular complex, it is often still possible to find a screw reciprocal to all five without having to calculate large determinants. These ideas are best understood from an example.

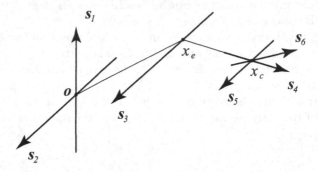

FIGURE 6.5. The Arrangement of Joints in a 6R Robot

Consider the robot illustrated in Figure 6.5; the diagram shows the disposition of the joint axes in some general configuration. This robot is similar too, but a little simpler than the standard PUMA design. We may assume that

$$\mathbf{s}_1 = \begin{pmatrix} \boldsymbol{\omega}_1 \\ \mathbf{0} \end{pmatrix}, \qquad \mathbf{s}_2 = \begin{pmatrix} \boldsymbol{\omega}_2 \\ \mathbf{0} \end{pmatrix}, \qquad \mathbf{s}_3 = \begin{pmatrix} \boldsymbol{\omega}_2 \\ \mathbf{x}_e \times \boldsymbol{\omega}_2 \end{pmatrix},$$

and

$$\mathbf{s}_4 = \begin{pmatrix} \boldsymbol{\omega}_4 \\ \mathbf{x}_c \times \boldsymbol{\omega}_4 \end{pmatrix}, \qquad \mathbf{s}_5 = \begin{pmatrix} \boldsymbol{\omega}_5 \\ \mathbf{x}_c \times \boldsymbol{\omega}_5 \end{pmatrix}, \qquad \mathbf{s}_6 = \begin{pmatrix} \boldsymbol{\omega}_6 \\ \mathbf{x}_c \times \boldsymbol{\omega}_6 \end{pmatrix}.$$

Notice that the fourth joint axis passes through $\mathbf{x}_c$ and $\mathbf{x}_e$ so we could also have written $\mathbf{s}_4^T = \left(\boldsymbol{\omega}_4^T, (\mathbf{x}_e \times \boldsymbol{\omega}_4)^T \right)$.

Now if we choose $\mathbf{z}_1$ to pass through the wrist centre $\mathbf{x}_c$, then it will meet the last three joint axes; further if it is parallel to the second and third axes it will be reciprocal to $\mathbf{s}_2$ and $\mathbf{s}_3$. In fact, we can see that

$$\mathbf{z}_1 = \begin{pmatrix} \boldsymbol{\omega}_2 \\ \mathbf{x}_c \times \boldsymbol{\omega}_2 \end{pmatrix}.$$

The line generating the second row of the inverse Jacobian must be reciprocal to all the joints except the second. This can be done by taking $\mathbf{z}_2$ as the line joining the points $\mathbf{x}_c$ and $\mathbf{x}_e$, the elbow. Clearly, this line meets the last four joint axes and because $\mathbf{s}_1$, $\mathbf{x}_c$ and $\mathbf{x}_e$ lie in a plane; $\mathbf{z}_2$ will also meet the first joint axis. In this robot, unlike the PUMA, there is no offset between the first joint and the elbow. It is clear that $\mathbf{z}_2 = \mathbf{s}_4$ the fourth joint axis. Notice that if the three points, $\mathbf{o}$, $\mathbf{x}_e$ and $\mathbf{x}_c$, are co-linear, then $\mathbf{z}_2$ will also be reciprocal to $\mathbf{s}_2$ and the Jacobian will be singular.

For $\mathbf{z}_3$ we can use the line joining $\mathbf{o}$ and $\mathbf{x}_c$. That is $\mathbf{z}_3 = \left(\mathbf{x}_c^T, \mathbf{0}^T \right)$. This is not one of the joints in the robot, so in general we would have to keep track of $\mathbf{x}_c$ to find this line. Notice that we don't have to normalise these lines since they will be divided by a factor $\mathbf{z}_3^T Q_0 \mathbf{s}_3$ when we compute the row in the inverse Jacobian, $\mathcal{W}_3^T$.

To find the last three rows is just a little more involved. Let

$$\mathbf{z}_\alpha = \begin{pmatrix} \mathbf{x}_e \\ \mathbf{0} \end{pmatrix}, \qquad \mathbf{z}_\beta = \begin{pmatrix} \omega_2(\omega_1 \cdot (\omega_2 \times \omega_4)) \\ \omega_1 \times \omega_2(\mathbf{x}_e \cdot (\omega_2 \times \omega_4)) \end{pmatrix}, \qquad \mathbf{z}_\gamma = \begin{pmatrix} \mathbf{0} \\ \omega_1 \times \omega_2 \end{pmatrix}.$$

These three screws are linearly independant and each is reciprocal to $\mathbf{s}_1$, $\mathbf{s}_2$ and $\mathbf{s}_3$. The form of $\mathbf{z}_\beta$ has been chosen so that $\mathbf{z}_\alpha$ and $\mathbf{z}_\beta$ are also reciprocal to $\mathbf{s}_4$. Now we can find a screw $\mathbf{z}_4$, reciprocal to all the joint screws but $\mathbf{s}_4$ as a linear combination of the three screws given above:

$$\mathbf{z}_4 = \alpha \mathbf{z}_\alpha + \beta \mathbf{z}_\beta + \mathbf{z}_\gamma.$$

The coefficient of $\mathbf{z}_\gamma$ cannot be zero, since then $\mathbf{z}_4$ would be reciprocal to $\mathbf{s}_4$, so it does not harm to fix this coefficient to be 1. The constants α and β can be found by setting the reciprocal products of $\mathbf{z}_4$ with $\mathbf{s}_5$ and $\mathbf{s}_6$ to zero,

$$\mathbf{z}_4^T Q_0 \mathbf{s}_5 = \alpha(\mathbf{z}_\alpha^T Q_0 \mathbf{s}_5) + \beta(\mathbf{z}_\beta^T Q_0 \mathbf{s}_5) + (\mathbf{z}_\gamma^T Q_0 \mathbf{s}_5) = 0$$

and

$$\mathbf{z}_4^T Q_0 \mathbf{s}_6 = \alpha(\mathbf{z}_\alpha^T Q_0 \mathbf{s}_6) + \beta(\mathbf{z}_\beta^T Q_0 \mathbf{s}_6) + (\mathbf{z}_\gamma^T Q_0 \mathbf{s}_6) = 0.$$

The solution to this is straightforward but not very instructive.

To find $\mathbf{z}_5$ and $\mathbf{z}_6$ we can compare linear combinations of $\mathbf{z}_\alpha$ and $\mathbf{z}_\beta$ to $\mathbf{s}_5$ and $\mathbf{s}_6$,

$$\mathbf{z}_5 = \mathbf{z}_\alpha(\mathbf{z}_\beta^T Q_0 \mathbf{s}_6) - \mathbf{z}_\beta(\mathbf{z}_\alpha^T Q_0 \mathbf{s}_6)$$

and

$$\mathbf{z}_6 = \mathbf{z}_\alpha(\mathbf{z}_\beta^T Q_0 \mathbf{s}_5) - \mathbf{z}_\beta(\mathbf{z}_\alpha^T Q_0 \mathbf{s}_5).$$

Clearly it would be a simple matter to substitute values for the joint screws $\mathbf{s}_1, \ldots, \mathbf{s}_6$ parameterised by the joint angles and hence obtain a symbolic expression for the inverse Jacobian, but this will not be pursued here.

6.8   Grassmannians

The space of all n-planes through the origin in $\mathbb{R}^m$ is known as a **Grassmann manifold** or **Grassmannian**, and is often written $G(n, m)$. So, for rays through the origin, we have $G(1, m) = \mathbb{PR}^m$, that is, the Grassmannian of lines through the origin is simply a projective space. The space we are interested in here is the Grassmannian $G(2, 4)$ of 2-planes in $\mathbb{R}^4$; see Section 6.3. Each 2-plane can be determined by a pair of vectors, say $\mathbf{x} = (x_0, x_1, x_2, x_3)^T$ and $\mathbf{y} = (y_0, y_1, y_2, y_3)^T$. We get the same 2-plane if we use a pair of vectors that are linearly dependent on these. We may eliminate this ambiguity by looking at the anti-symmetric product of the two vectors

$$\mathbf{x} \wedge \mathbf{y} = \begin{pmatrix} x_0 y_1 - x_1 y_0 \\ x_0 y_2 - x_2 y_0 \\ x_0 y_3 - x_3 y_0 \\ x_2 y_3 - x_3 y_2 \\ x_3 y_1 - x_1 y_3 \\ x_1 y_2 - x_2 y_1 \end{pmatrix} = \begin{pmatrix} p_{01} \\ p_{02} \\ p_{03} \\ p_{23} \\ p_{31} \\ p_{12} \end{pmatrix}.$$

This is the Plücker embedding again, but now we see how to generalise it to other Grassmannians. For $G(n, m)$ we simply take the n-fold anti-symmetric product of m-dimensional vectors. In this way $G(n, m)$ is embedded in projective space of dimension $\binom{m}{n} - 1$. Each coordinate is given by an $n \times n$ determinant that is a minor of the $m \times n$ matrix

$$\begin{pmatrix} x_1 & y_1 & \cdots & z_1 \\ x_2 & y_2 & \cdots & z_2 \\ \vdots & \vdots & \cdots & \vdots \\ x_m & y_m & \cdots & z_m \end{pmatrix}.$$

So, for example, for the Grassmannian $G(3, 6)$ of 3-planes in $\mathbb{R}^6$ we have 20 Plücker coordinates of the form p_{ijk} with $1 \leq i < j < k \leq 6$. The first of these is given by

$$p_{123} = \det \begin{pmatrix} x_1 & y_1 & z_1 \\ x_2 & y_2 & z_2 \\ x_3 & y_3 & z_3 \end{pmatrix}.$$

Not every possible combination of these coordinates corresponds to planes. For $G(2, 4)$, the Klein quadric, only Plücker coordinates satisfying the quadratic relation

$$p_{01}p_{23} + p_{02}p_{31} + p_{03}p_{12} = 0.$$

correspond to 2-planes. This relation can now be seen as a Pfaffian, that is, the square root of the determinant of an anti-symmetric matrix:

$$\text{Pfaff} \begin{pmatrix} 0 & p_{01} & p_{02} & p_{03} \\ p_{10} & 0 & p_{12} & p_{13} \\ p_{20} & p_{21} & 0 & p_{23} \\ p_{30} & p_{31} & p_{32} & 0 \end{pmatrix} = 0$$

where $p_{ji} = -p_{ij}$.

Alternatively, the relation can be derived by expanding the following matrix in terms of its 2×2 minors:

$$\det \begin{pmatrix} x_0 & y_0 & x_0 & y_0 \\ x_1 & y_1 & x_1 & y_1 \\ x_2 & y_2 & x_2 & y_2 \\ x_3 & y_3 & x_3 & y_3 \end{pmatrix} = 0.$$

For other Grassmannians, we get several quadratic relations like this. For example, for the Grassmannian $G(2,5)$ of 2-planes in $\mathbb{R}^5$ we get five quadratic equations by computing the five 4×4 minors of the rectangular matrix

$$\begin{pmatrix} x_0 & y_0 & x_0 & y_0 \\ x_1 & y_1 & x_1 & y_1 \\ x_2 & y_2 & x_2 & y_2 \\ x_3 & y_3 & x_3 & y_3 \\ x_4 & y_4 & x_4 & y_4 \end{pmatrix}.$$

Or we could compute the Pfaffians of minors associated with the diagonals in the matrix

$$\begin{pmatrix} 0 & p_{01} & p_{02} & p_{03} & p_{04} \\ p_{10} & 0 & p_{12} & p_{13} & p_{14} \\ p_{20} & p_{21} & 0 & p_{23} & p_{24} \\ p_{30} & p_{31} & p_{32} & 0 & p_{34} \\ p_{40} & p_{41} & p_{42} & p_{43} & 0 \end{pmatrix}.$$

Remember that the determinant of an odd-order anti-symmetric matrix is always zero. The equations are

$$p_{01}p_{23} - p_{02}p_{13} + p_{03}p_{12} = 0$$
$$p_{01}p_{24} - p_{02}p_{14} + p_{04}p_{12} = 0$$
$$p_{01}p_{34} - p_{03}p_{14} + p_{04}p_{13} = 0$$
$$p_{02}p_{34} - p_{03}p_{24} + p_{04}p_{23} = 0$$
$$p_{12}p_{34} - p_{13}p_{24} + p_{14}p_{23} = 0.$$

Remarkably, these equations, the Plücker relations, are enough to specify the Grassmannian. That is, points in the projective space $\mathbb{PR}^9$ whose coordinates satisfy this linear system of quadratic equations, and only these points, correspond to 2-planes in $\mathbb{R}^5$. Notice that these relations are not independent, this reflects the fact that 4×4 determinants of the 5×4 above are not all independent. See Hodge and Pedoe [52, sect. 1.5], or Semple and Roth [109, chap. X].

In general, the topology of a Grassmannian manifold is given by the fact that we can realise $G(n,m)$ as the quotient of a pair of groups

$$G(n,m) = \frac{O(m)}{O(n) \times O(m-n)}.$$

That is, Grassmannian manifolds are homogeneous spaces. To see this, consider a general n-plane in $\mathbb{R}^m$. The group $O(m)$ acts transitively on these planes, moving them around in general. To find the isotropy group, we set up a system of orthogonal coordinates so that the first n coordinate vectors span the n-plane, while the other $(m-n)$ coordinates are orthogonal to it. Consider the n-plane spanned by the columns of the $m \times n$ matrix

$$\begin{pmatrix} I_n \\ 0 \end{pmatrix}.$$

The action of the group $O(m)$ is by multiplication. The elements of $O(m)$ that preserve the n-plane have the form

$$\begin{pmatrix} A_n & 0 \\ 0 & B_{m-n} \end{pmatrix}$$

where $A_n \in O(n)$ and $B_{m-n} \in 0(m-n)$. This generalises to any n-plane: an $O(n)$ transformation just moves around the coordinates in the n-plane but leaves the plane's position unchanged. Likewise, an $O(m-n)$ transformation just affects the coordinates orthogonal to the plane but leaves the plane fixed. The isotropy group for the action of $O(m)$ on n-planes is just the direct product of these two groups; any other group element changes the position of the plane. See Porteous [88, Chap. 12], or Husemöller [58, sect. 8.2]. Notice that we can find the dimension of a Grassmannian from this. The dimension of the group $O(n)$ is $n(n-1)/2$; see Section 4.1; hence

$$\dim(G(n,m)) = \frac{1}{2}m(m-1) - \frac{1}{2}n(n-1) - \frac{1}{2}(m-n)(m-n-1) = (m-n)n.$$

The space of oriented n-planes in $\mathbb{R}^m$ can be given a similar description using orientation-preserving groups:

$$G^+(n,m) = \frac{SO(m)}{SO(n) \times SO(m-n)},$$

where $G^+(n,m)$ is the Grassmannian of oriented or directed n-planes in $\mathbb{R}^m$. Compare this with the space of directed lines in $\mathbb{R}^3$ that we met at the end of Section 6.4. The difference between that space and $G^+(2,4)$ is that the oriented Grassmannian contains lines at infinity. These are oriented 2-planes through the origin in $\mathbb{R}^4$ that do not meet the 3-plane $x_0 = 1$; see Section 6.3. Hence, $G^+(2,4)$ double covers the Klein quadric $G(2,4)$, and the finite directed lines form an open set in the oriented Grassmannian.

7
Representation Theory

In the latter half of the twentieth century a large part of group theory was concerned with the theory of group representations. This followed from the nineteenth century's concentration on invariants and covariants. Crudely speaking invariants are trivial representations and covariants are just elements of some non-trivial representation. The significance of these ideas is that if we want to write down equations and relations in terms of coordinates then we expect that if we change coordinates then the geometry or mechanics expressed in our equations should not be altered. The simplest way to do this is to make sure that the expressions are equalities between invariants, but we can also use relations between covariants—so long as the covariants we compare correspond to the same representation. The upshot of this that whenever we want to study some new kind of object, a line, an ellipsoid or an inertia matrix perhaps, then we should always ask how the new object transforms under a change of coordinates, that is, which representation does the object belong to?

7.1 Definitions

A representation of a Lie group is nothing more than a linear action of the group. We have seen several linear actions of the groups $SO(3)$ and $SE(3)$ already. However, we want to be more systematic here and apply some of the large modern subject of representation theory to robotics. An excellent introduction to representation theory can be found in Fulton and Harris [36]. We begin by making some formal definitions.

A **representation of a Lie group** G on a vector space V is a smooth map

$$R : G \times V \longrightarrow V$$

satisfying certain axioms. However, it is more usual to think of a representation as a collection of maps from V to V, one for each element of the group G. In this way, we think of the group elements as providing symmetries of the vector space. Let us write these maps as $R_g = R(g, *)$. With this notation the axioms are

$$R_g(R_h \mathbf{v})) = R_{gh}(\mathbf{v}). \tag{R1}$$

The map respects the group product for all g, $h \in G$ and all $\mathbf{v} \in V$. The map for the identity element must always be the identity map on the vector space:

$$R_e(\mathbf{v}) = \mathbf{v}. \tag{R2}$$

Lastly, the maps must be linear:

$$R_g(a\mathbf{v}_1 + b\mathbf{v}_2) = aR_g(\mathbf{v}_1) + bR_g(\mathbf{v}_2) \tag{R3}$$

for any $g \in G$, all $\mathbf{v}_1$, $\mathbf{v}_2 \in V$, and all scalars a, b. Compare this with Section 2.4.

So, the maps R_g must be linear and invertible. Linearity comes from (R3), and we can use (R1) and (R2) to show that the inverse of a map R_g will be given by $R_{g^{-1}}$. Such maps on a vector space are called endomorphisms. Given a basis for a finite dimensional vector space, an endomorphism can be written as a non-singular matrix; hence, these representations are sometimes called matrix representations. However, a change of basis should not affect the representation, so we consider two matrix representations to be equivalent if they are related by a coordinate transformation, that is, by a similarity

$$R' = R \Longleftrightarrow R'_g = MR_gM^{-1} \text{ for all } g \in G, \text{ and some non-singular matrix } M.$$

For every group, there is at least one representation: If we map every element of the group to the identity matrix, then all the axioms above are satisfied. However, this is not a very exciting representation, it is called the **trivial representation** of the group. If every different element of the group is represented by a different symmetry of the vector space, then we say that the representation is **faithful**. Another way of saying this is that a faithful representation is given by an injective map from the group to the space of endomorphisms of a vector space.

As examples of different and more interesting representations of the same group, consider the two representations of $SE(3)$ that we have already seen. The 4×4 representation that we met in Section 2.5 with typical element

$$P_g = \begin{pmatrix} R & \mathbf{t} \\ 0 & 1 \end{pmatrix}$$

and the adjoint representation we met in Section 4.2. This is a 6×6 representation with corresponding typical element

$$Ad_g = \begin{pmatrix} R & 0 \\ TR & R \end{pmatrix}.$$

For Lie algebras, we have a similar notion of a representation. To each Lie algebra element X, a representation L associates an endomorphism of V, which we will label L_X or sometimes $L(X)$. As usual, the endomorphisms are required to respect the structure of the algebra, which here means that for any pair of Lie algebra elements X and Y we must have

$$L_{[X,Y]} = L_X L_Y - L_Y L_X.$$

That is, the commutator of the Lie algebra elements must be represented by the commutator of the endomorphisms. In terms of the action of these endomorphisms on the vector space V, we could write the above relation as

$$L_{[X,Y]}(\mathbf{v}) = L_X(L_Y(\mathbf{v})) - L_Y(L_X(\mathbf{v}))$$

for any $\mathbf{v} \in V$. Once again, if we write the endomorphisms as matrices, then equivalent representations are given by similar matrices.

Given a representation of a Lie group, we get a representation of its Lie algebra. To do this, we think of the representation as a homomorphism from the group to the group of endomorphisms of V; then we take the Jacobian of this map at the identity in the group. Recall that this is how we produced the adjoint representation of a Lie algebra from the adjoint representation of its group in Section 4.2. Notice that the trivial representation of a Lie algebra sends every element to zero.

On the other hand, given a representation of a Lie algebra, we can exponentiate to get a representation of the simply connected covering group; see Section 4.6. Sometimes the representation of the covering group will also be a representation for one of its quotients. This happens when the representation is constant on the kernel of the quotient projection.

Thus, we see that if we want to study the representations of a Lie group we can study the representations of its Lie algebra instead. We lose nothing at all if the group is simply connected, and even if the group is not simply connected the worst that can happen is that we may find representations of the simply connected cover of the group that do not project to the group itself. We have already seen an example of this phenomenon: in Section 2.3 we saw that the group $SU(2)$ double covers $SO(3)$, and hence their Lie algebras are isomorphic. The 2×2 adjoint representation of $SU(2)$ cannot be a representation of $SO(3)$, since both

$$\begin{pmatrix} 1 & 0 \\ 0 & 1 \end{pmatrix} \quad \text{and} \quad \begin{pmatrix} -1 & 0 \\ 0 & -1 \end{pmatrix}$$

are mapped to the identity by the projection from $SU(2)$ to $SO(3)$, but each element of $SO(3)$ must be represented by a unique matrix. On the other hand, the 3×3 representation of $SO(3)$ is a perfectly good representation of $SU(2)$.

It is usually easier to study the representations of a Lie algebra rather than a Lie group. This is because of the linear (vector space) structure of the Lie algebra. Hence, in the following we will study in some detail the representation theory of $so(3)$ and $se(3)$ rather than that of the corresponding Lie groups. We will then see which of these representations can pass to the groups $SO(3)$ and $SE(3)$. One final complication concerns the field of scalars we can use. Things will be far easier if we work with complex scalars. Now, any real representation of a Lie algebra, that is, a representation by matrices with real entries, is also a complex representation since the complex numbers include the reals. So, we will have all the real representations if we know all the complex ones. However, it may happen that there are two real representations that are equivalent under a complex similarity, but for which there is no real similarity that links them. Hence, a complex representation may split into two or more real representations. Again, we will address this point when we come to study specific examples. First, we look at how representations can be combined.

7.2   Combining Representations

We can manufacture new representations from ones we already know in a number of ways. Perhaps the simplest is the **dual of a representation**. Given a vector space V, its dual is the space V^* of linear functionals on V, that is, the space of linear functions $\mathbf{f} : V \longrightarrow \mathbb{R}$ such that

$$\mathbf{f}(a\mathbf{v}_1 + b\mathbf{v}_2) = a\mathbf{f}(\mathbf{v}_1) + b\mathbf{f}(\mathbf{v}_2)$$

for any constants a and b. The map $\mathbf{f}(\mathbf{v})$ is called the **evaluation map** of the functional. Notice that the linearity property of the map means that to specify a dual vector we only have to give the result of the evaluation map on a basis for the original space. Hence, a finite vector space and its dual will have the same dimension. Given some basis for V, we can think of vectors and dual vectors as $n \times 1$ column vectors. The evaluation map would then be given by the matrix product

$$\mathbf{f}(\mathbf{v}) = \mathbf{f}^T \mathbf{v}.$$

The dual of a group representation is then a representation on the vector space dual to the original one. It is defined in such a way that the evaluation map is independent of the group. If we write the dual representation as R_g^*, then this means that

$$(R_g^* \mathbf{f})(R_g \mathbf{v}) = \mathbf{f}(\mathbf{v}).$$

So, in terms of matrices we have $R_g^* = R_{g^{-1}}^T$. Sometimes the dual representation is equivalent to the original. For example, for the standard representation of the orthogonal groups $O(n)$ the dual representation is exactly the same as the original since all the matrices in the group satisfy $M^T M = I_n$; see Section 2.2.1. This is not usually the case, however.

The definition of the dual representation of a Lie algebra is a little different. This is because we want the definitions to be consistent with the way we pass from a group representation to a representation of its Lie algebra as outlined above. This means that the dual representation L_X^* must satisfy

$$(L_X^* \mathbf{f})(\mathbf{v}) + \mathbf{f}(L_X \mathbf{v}) = 0.$$

Consequently, we must have $L_X^* = -L_X^T$; hence, the eigenvalues of the operator L_X^* are of the opposite sign to those of L_X.

Our next operations combine representations on two vector spaces. To understand these constructions, we first combine the vector spaces that the representations act on and then give the action of the group or algebra on the combined space.

Given a pair of vector spaces V and W, their **direct sum**, $V \oplus W$, is a vector space whose elements are given by the disjoint union of the components $V \cup W$. The vector space operations, multiplication by scalars, and addition of vectors is then defined componentwise. That is, if we have bases for the two spaces $\{\mathbf{v}_1, \mathbf{v}_2, \ldots, \mathbf{v}_m\}$ for V and $\{\mathbf{w}_1, \mathbf{w}_2, \ldots, \mathbf{w}_n\}$ for W, then a basis for $V \oplus W$ would be $\{\mathbf{v}_1, \mathbf{v}_2, \ldots, \mathbf{v}_m, \mathbf{w}_1, \mathbf{w}_2, \ldots, \mathbf{w}_n\}$. Notice that $V \oplus W$ has two obvious subspaces, one isomorphic to V and the other isomorphic to W.

Now we can define the **direct sum** of two representations, $L_1 \oplus L_2$. Given two representations L_1 on V and L_2 on W, their direct sum is a representation on $V \oplus W$ such that when we restrict to the subspace isomorphic to V we recover L_1 and restricting to W gives L_2. In terms of matrices, this means we can partition the matrices as

$$(L_1 \oplus L_2)(X) = \begin{pmatrix} L_1(X) & 0 \\ 0 & L_2(X) \end{pmatrix}$$

for any X in the Lie algebra. From this, we can see that the eigenvalues of $(L_1 \oplus L_2)(X)$ are simply the eigenvalues $L_1(X)$ together with those of $L_2(X)$.

If a representation can be written as a direct sum of two other representations, then it is said to be decomposable. A representation that is not the direct sum of two others is called **indecomposable**. Clearly, to get all the representations of a Lie algebra it is enough to find only the indecomposable ones, since the others can be constructed as direct sums of these.

It may happen that in a vector space V there is a proper subspace $U \subset V$ that is preserved by the representation. That is, $L_X(\mathbf{u}) \in U$ for all $\mathbf{u} \in U$ and all X in the Lie algebra. Such a subspace is called an **invariant subspace** of the representation. Notice that for the direct product $V \oplus U$ both the subspaces V and W are invariant. A representation that has no invariant subspace (except $\mathbf{0}$) is called **irreducible**. Irreducible representations, then, are also indecomposable, but the converse is not necessarily true. For $se(3)$ we can have indecomposable representations that are not irreducible. For example, the standard 4×4 representation with general form

$$\begin{pmatrix} \Omega & \mathbf{v} \\ 0 & 0 \end{pmatrix}$$

that we found in Section 4.3 clearly has an invariant subspace generated by vectors of the form $(x, y, z, 0)^T$, but there is no complementary invariant subspace, and so this representation is indecomposable but not irreducible.

For semi-simple Lie algebras like $so(3)$, all indecomposable representations are also irreducible. This is because semi-simple Lie algebras always have a positive definite, group invariant, bilinear form defined on them, the negative of the Killing form. Hence, so do their representations. For any invariant subspace $U \subset V$, we can find the orthogonal complement $U^\perp$ with respect to the form. Since the form is invariant under the Lie algebra, $U^\perp$ is also an invariant subspace, and hence we have the decomposition $V = U \oplus U^\perp$. So, for these algebras we need only find the irreducible representations.

The other main way to combine vector spaces and representations is by the tensor product. For a pair of vector spaces V and W with bases as above, their **tensor product**, $V \otimes W$, has a basis given by the symbols $\{\mathbf{v}_1 \otimes \mathbf{w}_1, \ldots, \mathbf{v}_1 \otimes \mathbf{w}_n, \mathbf{v}_2 \otimes \mathbf{w}_1, \ldots, \mathbf{v}_m \otimes \mathbf{w}_n\}$. We can think of the tensor product as a projection from the Cartesian product of the constituent vector spaces. In order for the tensor product to be a linear space, this projection must have the property that $(a\mathbf{v}_i, b\mathbf{w}_j) \longmapsto ab(\mathbf{v}_i \otimes \mathbf{w}_j)$ for constants a, b. Another way of putting this is

$$a\mathbf{v}_i \otimes \mathbf{w}_j = \mathbf{v}_i \otimes a\mathbf{w}_j = a(\mathbf{v}_i \otimes \mathbf{w}_j).$$

A simple consequence of these definitions is that the tensor product distributes over the direct sum:

$$(U \oplus V) \otimes W = (U \otimes W) \oplus (V \otimes W).$$

The tensor product of two group representations is defined slightly differently from the tensor product of two Lie algebra representations. Given two representations of a group, R_1 and R_2, on the spaces V and W, respectively, the tensor product of the representations acts on the tensor product of the vector spaces and is defined by its action on the basis elements:

$$(R_1 \otimes R_2)(g)\mathbf{v}_i \otimes \mathbf{w}_j = (R_1(g)\mathbf{v}_i) \otimes (R_2(g)\mathbf{w}_j)$$

for all g in the group.

Now, we want the action of the Lie algebra to be consistent with the processes of taking derivatives and exponentiating that we mentioned in the last section. To see what this means, we set $g = e^{tX}$ and take derivatives at $t = 0$:

$$\frac{d}{dt}(R_1 \otimes R_2)(g)\Big|_{t=0} \mathbf{v}_i \otimes \mathbf{w}_j = \left(\frac{dR_1}{dt}\Big|_{t=0} \mathbf{v}_i\right) \otimes \mathbf{w}_j + \mathbf{v}_i \otimes \left(\frac{dR_2}{dt}\Big|_{t=0} \mathbf{w}_j\right).$$

Hence, we define the action of the tensor product of two representations of a
Lie algebra on the tensor product of their vector spaces by

$$(L_1 \otimes L_2)(X)\mathbf{v}_i \otimes \mathbf{w}_j = (L_1(X)\mathbf{v}_i) \otimes \mathbf{w}_j + \mathbf{v}_i \otimes (L_2(X)\mathbf{w}_j).$$

Suppose that $\mathbf{e}$ is an eigenvector of $L_1(X)$ with eigenvalue λ and that $\mathbf{f}$ is an
eigenvector of $L_2(X)$ with eigenvalue μ. Now,

$$(L_1 \otimes L_2)(X)\mathbf{e} \otimes \mathbf{f} = (L_1(X)\mathbf{e}) \otimes \mathbf{f} + \mathbf{e} \otimes (L_2(X)\mathbf{f}) = \lambda \mathbf{e} \otimes \mathbf{f} + \mathbf{e} \otimes \mu \mathbf{f} = (\lambda + \mu)\mathbf{e} \otimes \mathbf{f}.$$

That is, the tensor product of two eigenvectors is an eigenvector for the tensor
product representation whose eigenvalue is the sum of the original eigenvalues.
    There is no reason why we cannot take the tensor product of a vector space
with itself. Usually, we write this as

$$V \otimes V = V^{\otimes 2}.$$

Inductively, we can define the **tensor power** of a vector space as $V^{\otimes n} = V \otimes V^{\otimes n-1}$. Sitting inside the tensor powers are two important subspaces. The first
is the space of all symmetric vectors, that is, vectors that are invariant under
permutation of factors. For example, if the vector space V is three-dimensional
with a basis $\{\mathbf{v}_1, \mathbf{v}_2, \mathbf{v}_3\}$, then the symmetric square is a vector subspace of
$V^{\otimes 2}$ with a basis $\{\mathbf{v}_1 \otimes \mathbf{v}_1, \mathbf{v}_2 \otimes \mathbf{v}_2, \mathbf{v}_3 \otimes \mathbf{v}_3, \mathbf{v}_1 \otimes \mathbf{v}_2 + \mathbf{v}_2 \otimes \mathbf{v}_1, \mathbf{v}_1 \otimes \mathbf{v}_3 + \mathbf{v}_3 \otimes \mathbf{v}_1, \mathbf{v}_2 \otimes \mathbf{v}_3 + \mathbf{v}_3 \otimes \mathbf{v}_2\}$. We will denote this space $\mathrm{Sym}^2 V$ and write the
symmetric product of a pair of vectors as $\mathbf{v}_i \mathbf{v}_j$.
    In general, $\mathrm{Sym}^n V$ has generators

$$\mathbf{v}_i \mathbf{v}_j \cdots \mathbf{v}_k = \sum_\sigma \mathbf{v}_{\sigma(i)} \otimes \mathbf{v}_{\sigma(j)} \otimes \cdots \otimes \mathbf{v}_{\sigma(k)}, \qquad i \leq j \leq \cdots \leq k$$

where the sum is taken over all permutations σ of the indices $(i, j, \ldots, k)$. Notice
that if the vector space V has dimension d, then the dimension of the space
$\mathrm{Sym}^n V$ is given by the binomial coefficient $\binom{n+d-1}{n}$.
    The other subspace is the space of anti-symmetric vectors. These are invariant
under even permutations of factors but change sign under an odd permutation.
For the three-dimensional example above, the anti-symmetric vectors in $V^{\otimes 2}$
are generated by $\{\mathbf{v}_1 \otimes \mathbf{v}_2 - \mathbf{v}_2 \otimes \mathbf{v}_1, \mathbf{v}_2 \otimes \mathbf{v}_3 - \mathbf{v}_3 \otimes \mathbf{v}_2, \mathbf{v}_3 \otimes \mathbf{v}_1 - \mathbf{v}_1 \otimes \mathbf{v}_3\}$. We
will denote the space of degree n anti-symmetric vectors by $\bigwedge^n V$. The space
has generators

$$\mathbf{v}_i \wedge \mathbf{v}_j \wedge \cdots \wedge \mathbf{v}_k = \sum_\sigma \mathrm{sign}(\sigma) \mathbf{v}_{\sigma(i)} \otimes \mathbf{v}_{\sigma(j)} \otimes \cdots \otimes \mathbf{v}_{\sigma(k)}, \qquad i < j < \cdots < k$$

where the sum is taken over all permutations σ of the indices $(i, j, \ldots, k)$. For
a d-dimensional vector space V we can see that the dimension of $\bigwedge^n V$ is given
by the binomial coefficient $\binom{d}{n}$.

As usual, we extend the definitions of symmetric and anti-symmetric powers of vector spaces to symmetric and anti-symmetric powers of representations. These new representations have very useful geometric interpretations.

The dual V^* to a vector space V can be thought of as the space of coordinate functions on V. For any basis of V, $\{\mathbf{v}_1, \ldots, \mathbf{v}_n\}$, we can always find a basis of V^*, $\{\mathbf{v}_1^*, \ldots, \mathbf{v}_n^*\}$, with the property that

$$\mathbf{v}_i^*(\mathbf{v}_j) = \begin{cases} 1, & \text{if } i = j, \\ 0, & \text{if } i \neq j. \end{cases}$$

So, given any vector $\mathbf{v} \in V$ we can find its coordinates $x_i = \mathbf{v}_i^*(\mathbf{v})$.

Endomorphisms of a vector space V can be thought of as elements of $V \otimes V^*$. We can see this by taking a basis $\{\mathbf{v}_1, \mathbf{v}_2, \ldots, \mathbf{v}_n\}$ for V and a dual basis $\{\mathbf{v}_1^*, \mathbf{v}_2^*, \ldots, \mathbf{v}_n^*\}$ for V^*. A general element of $V \otimes V^*$ will be given by $\sum_{ij} m_{ij} \mathbf{v}_i \otimes \mathbf{v}_j^*$. This acts on a typical vector $\mathbf{v} = a_1 \mathbf{v}_1 + \cdots + a_n \mathbf{v}_n$ from V:

$$\sum_{ij} m_{ij} \mathbf{v}_i \otimes \mathbf{v}_j^*(\mathbf{v}) = \sum_{ij} m_{ij} a_j \mathbf{v}_i.$$

Hence, we can think of the m_{ij} as elements of the matrix M representing the endomorphism:

$$\mathbf{v} \longmapsto M\mathbf{v}.$$

Elements of the space $\mathrm{Sym}^2 V^*$ can be thought of as quadratic forms on V. As above, a general element of this space can be represented by a matrix, a symmetric matrix this time since the coefficients of $\mathbf{v}_i^* \otimes \mathbf{v}_j^*$ and $\mathbf{v}_j^* \otimes \mathbf{v}_i^*$ are the same. The evaluation of this form on a vector from V is given by the matrix product $\mathbf{v}^T M \mathbf{v}$, where M is the symmetric matrix defined by the element of $\mathrm{Sym}^2 V^*$. If we concentrate on the coordinates of V, then $\mathbf{v}^T M \mathbf{v}$ gives a homogeneous quadratic polynomial in the coordinates of V. So we can also think of $\mathrm{Sym}^2 V^*$ as the space of homogeneous quadratic polynomials in the coordinates of V. Also, by polarising the quadratic form we can think of $\mathrm{Sym}^2 V^*$ as the space of symmetric bilinear forms on V. For a pair of vectors $\mathbf{u}$ and $\mathbf{v}$ in V the bilinear form is given by $\mathbf{u}^T M \mathbf{v}$.

The corresponding representation of the group (or Lie algebra) gives the action of the group (or algebra) on the quadratic forms, polynomials, or bilinear forms under coordinate changes.

Generally, the space $\mathrm{Sym}^k V^*$ can be interpreted as the space of degree k homogeneous polynomials in the coordinates. Hence, the representation $\mathrm{Sym}^k R^*$ gives the action of the group on these polynomials under basis changes in V.

The interpretation of the anti-symmetric powers of a representation has to do with the action of the group on linear subspaces. Suppose we have a representation of a group or its Lie algebra on some m-dimensional vector space. The group acts via the representation on the vectors, moving them around in general. Now, think of the set of n-planes through the origin in the vector space. The representation will also act on these. In Section 6.8 we saw that the set of

n-planes in $\mathbb{R}^m$ was a Grassmannian, $G(n,m)$. We can label each n-plane by an n-fold anti-symmetric (or Grassmann) product of linearly independent vectors in the plane. Hence, the Grassmannian $G(n,m)$ sits inside the vector space $\bigwedge^n \mathbb{R}^m$. Not only that, but the n-fold anti-symmetric product of the original representation acts on $\bigwedge^n \mathbb{R}^m$ and when restricted to $G(n,m)$ gives the action of the group or Lie algebra on the n-planes.

Finally, here we introduce the notion of **plethyism**. This is the study of how representations combine. We saw above that any representation of a Lie algebra can be written as a direct sum of indecomposable representations or, if we are lucky, of irreducible representations. The question now arises, given two representations with a known decomposition what is the decomposition of their tensor product? Because the tensor product distributes over the direct sum, we only need to find the decompositions for tensor products of indecomposable (irreducible) representations to solve the general problem. Finding the decompositions for symmetric and anti-symmetric powers of representations is also important. For example, consider the standard representation M of $GL(n)$ on $\mathbb{R}^n$. As we saw above, the symmetric powers of the dual of this representation give the action of $GL(n)$ on the homogeneous polynomials in the coordinates of $\mathbb{R}^n$. Now suppose that the representation $\mathrm{Sym}^k M^*$ contains the trivial representation as a component. This would mean that there was a homogeneous polynomial of degree k in the coordinates of $\mathbb{R}^n$ that is left unchanged by linear coordinate changes. In classical language, such a polynomial is called an invariant. In this way, the classical invariant theory of Sylvester and Cayley can be described in terms of representation theory; see Dieudonné and Carrell [26]. Further, as mentioned at the beginning of this chapter, since the coordinates of a vector space are artificial constructs, any equation or relation that we write concerning some physical phenomenon in the space should be independent of the coordinate basis we choose. More precisely, we expect that physically meaningful expressions should be elements of some representation of the group of basis changes. If there is a symmetry in the problem, then we should expect to deal with elements from representations of the symmetry group.

This view even extends to infinite-dimensional representations. A linear differential operator on $\mathbb{R}^n$ can be thought of as a linear operator acting on the infinite-dimensional vector space of smooth functions on $\mathbb{R}^n$. We can look for representations of groups where, instead of a matrix, the group elements are represented by linear differential operators. There is a very strong connection between the special functions of mathematical physics and this view of representation theory. In mathematical physics, we usually have a single differential equation to solve. However, if we can find other differential operators such that the complete set forms a Lie algebra, then the analysis is greatly simplified. Solutions of the original equations are eigenfunctions that span a finite-dimensional sub-representation of the Lie algebra. The elements of these representations involve the special functions. A clearer picture of this approach can be found in Miller [76], Talman [119], or Vilenkin [121].

In the case of robotics, if we assume rigidity, then the symmetry group is $SE(3)$, and we expect to be dealing with elements from representations of this group. Hence, whenever any physical property is introduced we always want to know how it transforms under rigid coordinate changes, that is, to which representation of $SE(3)$ does it belong?

7.3   Representations of $SO(3)$

We begin here by looking at the irreducible representations of the Lie algebra $so(3)$. This is one of the simplest examples of a Lie algebra, and hence its representation theory is a standard example in most textbooks. Moreover, the group is important in nuclear physics for describing spin and isospin. Therefore the representation theory of $so(3)$ or the isomorphic Lie algebra $su(2)$ can be found in many nuclear and quantum physics texts.

Our first step is to perform a change of variables. Rather than the X, Y and Z that we met in Section 4.3, we will use

$$J_3 = Z, \qquad J_+ = Y + iX, \qquad J_- = Y - iX.$$

Remember that we are working with complex scalars, and i is the imaginary unit here. With this new basis the commutation relations of the algebra become

$$[J_3, J_+] = iJ_+, \qquad [J_3, J_-] = -iJ_-, \qquad [J_+, J_-] = 2iJ_3.$$

Next, we will assume that we have a representation L, and we will study the eigenvectors of $L(J_3)$. To simplify the notation a little, we will write $J_*(\mathbf{v})$ rather than $L(J_*)\mathbf{v}$. Now, suppose that $\mathbf{e}$ is an eigenvector of J_3 with eigenvalue λ. We will use the commutation relations to show that $J_+(\mathbf{e})$ and $J_-(\mathbf{e})$ are also eigenvectors of J_3:

$$[J_3, J_+](\mathbf{e}) = iJ_+(\mathbf{e}) = J_3 J_+(\mathbf{e}) - J_+ J_3(\mathbf{e}),$$
$$= J_3(J_+(\mathbf{e})) - \lambda J_+(\mathbf{e}).$$

Hence, $J_3(J_+(\mathbf{e})) = (\lambda + i)J_+(\mathbf{e})$, so either $J_+(\mathbf{e})$ is zero or it is another eigenvector of J_3. The eigenvalue of $J_+(\mathbf{e})$ is then $(\lambda + i)$. Similarly, we find that the eigenvalue of $J_-(\mathbf{e})$ is $(\lambda - i)$. Repeating the above, we get many more eigenvectors of J_3: the vectors $J_+^k(\mathbf{e})$ have eigenvalues $(\lambda + ki)$ and the $J_-^k(\mathbf{e})$ have eigenvectors $(\lambda - ki)$. All the eigenvalues are different, so the corresponding eigenvectors are different. However, if the vector space is finite-dimensional we cannot keep producing eigenvectors in this fashion indefinitely. There must be some for which $J_+(\mathbf{e}) = \mathbf{0}$. Let $\mathbf{e}_n$ be an eigenvector of J_3 for which $J_+(\mathbf{e}_n) = \mathbf{0}$ and assume that the eigenvalue of $\mathbf{e}_n$ is n. From this eigenvalue, we can produce a string of eigenvalues $J_-^k(\mathbf{e}_n)$ with eigenvalues $(n - ki)$. Again, we cannot continue this process indefinitely; there must be a non-zero eigenvector $J_-^m(\mathbf{e}_n)$

for which $J_-J_-^m(\mathbf{e}_n) = \mathbf{0}$. So we have a finite chain of eigenvectors for J_3, namely $\mathbf{e}_n$, $J_-(\mathbf{e}_n)$, $J_-^2(\mathbf{e}_n), \ldots, J_-^m(\mathbf{e}_n)$. We know the action of J_3 and J_- on these vectors, so now we look at the action of J_+, again using the commutation relations:

$$[J_+, J_-](\mathbf{e}_n) = 2iJ_3(\mathbf{e}_n),$$
$$J_+J_-(\mathbf{e}_n) - J_-J_+(\mathbf{e}_n) = 2in(\mathbf{e}_n),$$
$$J_+(J_-(\mathbf{e}_n)) = 2in(\mathbf{e}_n).$$

The last line follows because we know that $J_+(\mathbf{e}_n) = \mathbf{0}$. To find the action of J_+ on $J_-^2(\mathbf{e}_n)$ we look at

$$[J_+, J_-]J_-(\mathbf{e}_n) = 2iJ_3J_-(\mathbf{e}_n),$$
$$J_+J_-^2(\mathbf{e}_n) - J_-J_+J_-(\mathbf{e}_n) = 2i(n - i)J_-(\mathbf{e}_n),$$
$$J_+(J_-^2(\mathbf{e}_n)) = 2i(2n - i)J_-(\mathbf{e}_n).$$

In general, we can see that $J_+(J_-^k(\mathbf{e}_n)) = f(k)J_-^{k-1}(\mathbf{e}_n)$, where $f(n)$ satisfies the recurrence relation

$$f(k + 1) = f(k) + 2i(n - ki); \qquad f(1) = 2in.$$

This has solution $f(k) = (2in + k - 1)k$. That is,

$$J_+(J_-^k(\mathbf{e}_n)) = (2in + k - 1)kJ_-^{k-1}(\mathbf{e}_n).$$

In particular, we have

$$J_+(J_-^{m+1}(\mathbf{e}_n)) = (2in + m)(m + 1)J_-^m(\mathbf{e}_n) = \mathbf{0}$$

since we know that $J_-^{m+1}(\mathbf{e}_n) = \mathbf{0}$. But $J_-^m(\mathbf{e}_n) \neq \mathbf{0}$ and m is a positive integer; hence, we conclude that $(2in + m) = 0$ and thus $n = mi/2$.

To recap, we have shown above that the vector space generated by

$$\mathbf{e}_n, \ J_-(\mathbf{e}_n), \ J_-^2(\mathbf{e}_n), \ldots, J_-^m(\mathbf{e}_n)$$

is preserved by the Lie algebra, that is, it is an invariant subspace of the representation. On the other hand, it is not too difficult to see that the space generated by these vectors is irreducible—just try to find a vector invariant under the action of J_-. The generality of our constructions means that every irreducible representation of $so(3)$ has the above form; the only difference between them is the number of eigenvectors in the chain. Thus we have a different irreducible representation for each different positive integer m. The eigenvalues for a particular irreducible representation are then

$$\frac{m}{2}i, \ \frac{m-2}{2}i, \ \frac{m-4}{2}i, \ldots, \ \frac{4-m}{2}i, \ \frac{2-m}{2}i, \ \frac{-m}{2}i.$$

That is, they are symmetrically disposed about 0. Notice that 0 is an eigenvalue only if m is even. So, the irreducible representations of $so(3)$ are classified by the eigenvalues of the matrix J_3. For the two representations that we found in Section 4.3, we have

$$J_3 = \begin{pmatrix} 0 & i/2 \\ i/2 & 0 \end{pmatrix}$$

(remember $su(2)$ is isomorphic to $so(3)$), which has eigenvalues $i/2$ and $-i/2$, and

$$J_3 = \begin{pmatrix} 0 & -1 & 0 \\ 1 & 0 & 0 \\ 0 & 0 & 0 \end{pmatrix}$$

which has eigenvalues i, 0 and $-i$.

Which of these representations exponentiate to representations of $SO(3)$, not just its double cover $SU(2)$? We can think of the kernel of the projection from $SU(2)$ to $SO(3)$ as consisting of the identity and the element $e^{2\pi J_3}$; see Section 2.3. The eigenvalues of $e^{2\pi J_3}$ can be found by exponentiating the eigenvalues of J_3. There are two cases to consider. When m is an odd integer, then the eigenvalues of J_3 are of the form $in/2$ for some integer n; hence, the eigenvalues of the group element are $e^{\pi in}$, that is, $+1$ for even values of n and -1 for odd n. In this case, the representation takes different values on the elements in the kernel, and thus these representations do not pass to the quotient $SO(3)$. In the other case, the eigenvalues of J_3 are all integer multiples of i, so the eigenvalues of the corresponding group element are all $+1$, just as for the identity matrix. In fact, since J_3 is always diagonalisable, we have that $e^{2\pi J_3}$ is the identity matrix. So these representations do pass to $SO(3)$. To summarise, the irreducible representations of $SO(3)$ are the exponentials of the representations given above where $m = 2n$, that is, where the element J_3 has eigenvalues $ni, (n-1)i, \ldots, i, 0, -i, \ldots, (1-n)i, -ni$.

Finally, we are in a position to say something about real representations as opposed to the complex representations we have been studying so far. In Section 7.1, it was mentioned that two real representations might be equivalent via a complex similarity but there might be no real similarity that relates them. In fact, for $so(3)$ this cannot happen. We can show this by contradiction. Assume we have two real representations of $so(3)$, K and L, similar by a complex matrix $M = A + iB$, where A and B are real. The fact that both representations are real means that we can look at the real and imaginary parts of the similarity separately, so that

$$K_X A = A L_X, \quad \text{and} \quad K_X B = B L_X$$

for all $X \in so(3)$. We can eliminate K from these equations to get

$$(B^{-1}A)L_X - L_X(B^{-1}A) = 0.$$

That is, the real matrix $(B^{-1}A)$ must commute with all the elements of the representation. However, it is known that the only non-zero matrices with this property are multiples of the identity; hence, we have $A = \lambda B$ as the only solutions. This, however, means that the complex matrix M is simply a complex multiple of the real matrix A. So, in this case, a complex similarity can always be turned into a real similarity.

All that remains is to see which irreducibles actually do contain a real representation. Clearly, the L_1 defining representation is real. The tensor product of two real representations is also real; hence, all the representations L_k, where k is an integer, are real since they are contained in tensor powers of L_1; see the next section. In fact, these are the only real representations of $so(3)$; see Fulton and Harris [36, Lecture 11].

7.4 $SO(3)$ Plethyism

Having found all the irreducible representations, our next task is to see how they combine, that is, to study plethyisms of the algebra. To begin with, let's make life a bit easier by simplifying our notation. Let us label the irreducible representation of $so(3)$ where J_3 has largest eigenvalue in by L_n. Then n can take positive integer values and half-odd-integer values. We also write L_0 for the one-dimensional trivial representation.

As a first example, consider the tensor product of two copies of the L_1 representation. If we write the eigenvectors of J_3 in the two representations as $\mathbf{u}_1, \mathbf{u}_0, \mathbf{u}_{-1}$ and $\mathbf{v}_1, \mathbf{v}_0, \mathbf{v}_{-1}$ then the tensor product $L_1 \otimes L_1$ has basis

$$\mathbf{u}_1 \otimes \mathbf{v}_1, \qquad \text{with eigenvalue } 2i,$$
$$\mathbf{u}_1 \otimes \mathbf{v}_0, \ \mathbf{u}_0 \otimes \mathbf{v}_1, \qquad \text{with eigenvalues } i,$$
$$\mathbf{u}_1 \otimes \mathbf{v}_{-1}, \ \mathbf{u}_0 \otimes \mathbf{v}_0, \ \mathbf{u}_{-1} \otimes \mathbf{v}_1, \qquad \text{with eigenvalues } 0,$$
$$\mathbf{u}_0 \otimes \mathbf{v}_{-1}, \ \mathbf{u}_{-1} \otimes \mathbf{v}_0, \qquad \text{with eigenvalues } -i,$$
$$\mathbf{u}_{-1} \otimes \mathbf{v}_{-1}, \qquad \text{with eigenvalue } -2i.$$

Remember, we get the eigenvalues of the composite vector by adding the eigenvalues of the components. From the list of eigenvalues and their multiplicities, we can find the irreducible components of this representation. The largest eigenvalue is $2i$, and hence L_2 must be a component. This will account for vectors with eigenvalues $2i$, i, 0, $-i$, $-2i$, thus leaving two vectors with eigenvalue 0 and one each with eigenvalues i and $-i$. The remaining eigenvalues can only come from a copy of L_1 and a trivial component L_0. So we have the result

$$L_1 \otimes L_1 = L_2 \oplus L_1 \oplus L_0.$$

Notice that we have not said anything about which eigenvectors belong in which component. Simply to find the irreducible components, this information isn't necessary, but we can use the theory above to show the decomposition explicitly. For example, in the above calculation, the vector with the largest eigenvalue was $\mathbf{u}_1 \otimes \mathbf{v}_1$. We can find the basis for the rest of this L_2 representation by repeatedly applying the J_- operator. This gives

$$J_-(\mathbf{u}_1 \otimes \mathbf{v}_1) = \mathbf{u}_1 \otimes \mathbf{v}_0 + \mathbf{u}_0 \otimes \mathbf{v}_1,$$
$$J_-^2(\mathbf{u}_1 \otimes \mathbf{v}_1) = \mathbf{u}_1 \otimes \mathbf{v}_{-1} + 2\mathbf{u}_0 \otimes \mathbf{v}_0 + \mathbf{u}_{-1} \otimes \mathbf{v}_1,$$
$$J_-^3(\mathbf{u}_1 \otimes \mathbf{v}_1) = 3\mathbf{u}_0 \otimes \mathbf{v}_{-1} + 3\mathbf{u}_{-1} \otimes \mathbf{v}_0,$$
$$J_-^4(\mathbf{u}_1 \otimes \mathbf{v}_1) = 6\mathbf{u}_{-1} \otimes \mathbf{v}_{-1}.$$

The L_1 component then has a basis $\mathbf{u}_1 \otimes \mathbf{v}_0 - \mathbf{u}_0 \otimes \mathbf{v}_1$, $\mathbf{u}_1 \otimes \mathbf{v}_1 - \mathbf{u}_1 \otimes \mathbf{v}_1$ and $\mathbf{u}_0 \otimes \mathbf{v}_{-1} - \mathbf{u}_{-1} \otimes \mathbf{v}_0$. This can be recognised as essentially the vector product of two 3-vectors. Finally, the invariant L_0 is generated by $\mathbf{u}_1 \otimes \mathbf{v}_{-1} + \mathbf{u}_0 \otimes \mathbf{v}_0 + \mathbf{u}_{-1} \otimes \mathbf{v}_1$, which is the scalar product in these coordinates.

For the general case, we consider a pair of representations L_m and L_n. If we take the tensor product $L_m \otimes L_n$, we will get $2m \times 2n$ eigenvectors. However, we are really only interested in the eigenvalues and their multiplicity. Since we must tensor every eigenvector from L_m with every one from L_n, think of laying these products out at the vertices of a $2m \times 2n$ grid. The eigenvalues of the components add to give the eigenvalue of the composite; hence, the diagonals in the grid will contain terms with the same eigenvalue, so the largest eigenvalue we get is $(m + n)i$ for the term $\mathbf{u}_m \otimes \mathbf{v}_n$ at the corner of the grid. Next we get two terms with eigenvalue $(m + n - 1)i$, then three terms with eigenvalue one less, and so on, until we reach the next corner of the grid. If $m > n$, then we will reach the next corner after $2n$ steps. Here we will have $2n$ terms with eigenvalue $(m - n)i$. Continuing the process, the number of terms remains constant as the eigenvalues decrease to the next corner, which occurs after $2m$ steps. Finally, the number of terms decreases to one at the final corner, where the final term is $\mathbf{u}_{-m} \otimes \mathbf{v}_{-n}$. We see that L_{m+n} is a component, and so is L_{m+n-1}, and L_{m+n-2}, and so on until L_{m-n}. Now we have accounted for all the eigenvalues, so we have the result

$$L_m \otimes L_n = L_{m+n} \oplus L_{m+n-1} \oplus \cdots \oplus L_{|m-n|}.$$

For our next example, we look at the symmetric square of the $L_{1/2}$ representation. Using $\mathbf{v}_{1/2}$ and $\mathbf{v}_{-1/2}$ as the eigenvectors of J_3 in $L_{1/2}$, we see that the symmetric square has eigenvectors

$$\mathbf{v}_{1/2} \otimes \mathbf{v}_{1/2}, \quad \mathbf{v}_{1/2} \otimes \mathbf{v}_{-1/2} + \mathbf{v}_{-1/2} \otimes \mathbf{v}_{1/2}, \quad \mathbf{v}_{-1/2} \otimes \mathbf{v}_{-1/2}.$$

The eigenvalues of these vectors are i, 0, and $-i$, respectively. Hence, we have that $\mathrm{Sym}^2 L_{1/2} = L_1$. More generally, suppose we take the n-th symmetric power of $L_{1/2}$. The largest eigenvalue is now $in/2$ from the term $\mathbf{v}_{1/2} \otimes \cdots \otimes \mathbf{v}_{1/2}$ with n factors. Due to the symmetrisation, we get only one term for each smaller eigenvalue down to $-in/2$. So we have

$$L_{n/2} = \mathrm{Sym}^n L_{1/2}.$$

In other words, every irreducible representation of $so(3)$ is a symmetric power of $L_{1/2}$.

Our final example is the anti-symmetric square of L_1. This has eigenvectors

$$\mathbf{v}_1 \wedge \mathbf{v}_0, \quad \mathbf{v}_1 \wedge \mathbf{v}_{-1}, \quad \mathbf{v}_0 \wedge \mathbf{v}_{-1}$$

with eigenvalues i, 0 and $-i$ once more. So we see that

$$\textstyle\bigwedge^2 L_1 = L_1.$$

Again, this can be interpreted as the vector product of 3-vectors.

There are many more relations like the ones we have derived above. Since the rotation group $SO(3)$ acts on functions on $\mathbb{R}^3$ as linear differential operators as we mentioned above, there is a close connection with the special functions of mathematical physics, in this case with the spherical harmonic functions. Thus, many of the relations between representations of $so(3)$ can be simply translated into relations between spherical harmonics.

7.5 Representations of $SE(3)$

The representation theory for $se(3)$ is less complete than for $so(3)$. This is because $se(3)$ is not semi-simple, and hence indecomposable representations are not necessarily irreducible, as we saw above. It is possible to describe all irreducible representations of $se(3)$; see for example Miller [76, Chap. 6], Talman [119], or Vilenkin [121]. Unfortunately, for our purposes this is not enough; the standard representation and the adjoint representation are both indecomposable but not irreducible. Fortunately, we can get a fair way without a general theory, but it does mean that we are limited to studying the subject example by example.

We can make a couple of general observations, however. First, any representation of $se(3)$ will restrict to a representation of the subgroup $so(3)$. However, even if the representation of $se(3)$ is irreducible, when restricted to the subgroup the representation of $so(3)$ may reduce. This is because elements from the $\mathbb{R}^3$ subgroup may mix up vectors from different $so(3)$-invariant subspaces.

As with $so(3)$, things become clearer if we use a convenient basis for the Lie algebra. In Section 4.3 we used $\boldsymbol{\omega}_i$, $\boldsymbol{\omega}_j$, $\boldsymbol{\omega}_k$ and $\mathbf{v}_i$, $\mathbf{v}_j$, $\mathbf{v}_k$ as the generators of $se(3)$. We now make the following change of basis:

$$J_3 = \boldsymbol{\omega}_k, \quad J_+ = \boldsymbol{\omega}_j + i\boldsymbol{\omega}_i, \quad J_- = \boldsymbol{\omega}_j - i\boldsymbol{\omega}_i,$$
$$P_3 = \mathbf{v}_k, \quad P_+ = \mathbf{v}_j + i\mathbf{v}_i, \quad P_- = \mathbf{v}_j - i\mathbf{v}_i.$$

With this basis, the commutation relations become

$$[J_3, J_+] = iJ_+, \quad [J_3, J_-] = -iJ_-, \quad [J_+, J_-] = 2iJ_3,$$
$$[J_3, P_+] = iP_+, \quad [J_3, P_-] = -iP_-, \quad [J_3, P_3] = 0,$$
$$[J_+, P_+] = 0, \quad [J_+, P_-] = 2iP_3, \quad [J_+, P_3] = -iP_+,$$
$$[J_-, P_+] = -2iP_3, \quad [J_-, P_-] = 0, \quad [J_-, P_3] = iP_-,$$
$$[P_3, P_+] = 0, \quad [P_3, P_-] = 0, \quad [P_+, P_-] = 0.$$

A simple consequence of this is that in any faithful representation R of $se(3)$ on a vector space V, the matrices $R(P_+)$, $R(P_3)$ and $R(P_-)$ span an L_1 representation of $so(3)$. Of course, so do the matrices $R(J_+)$, $R(J_3)$ and $R(J_-)$, that is, if we think of the matrices as elements of the vector space $V \otimes V^*$. In fact, if we restrict our attention to the subspace spanned by the six vectors, then the representation $R \otimes R^*$ must contain a subrepresentation isomorphic to the adjoint representation, or equivalently, the vector space has an invariant subspace isomorphic to $se(3)$.

The adjoint representation of $se(3)$ is six-dimensional—when restricted to the $so(3)$ subgroup the representation is $L_1 \oplus L_1$. Hence, the operator $ad(J_3)$ has six eigenvectors, which are just J_+, J_3, J_-, P_+, P_3 and P_-. However, it is slightly easier to use the basis generated by J_- as in Section 7.3. Hence, we will use the vectors $\boldsymbol{\omega}_1 = J_+$, $\boldsymbol{\omega}_0 = ad(J_-)(\boldsymbol{\omega}_1)$, $\boldsymbol{\omega}_{-1} = ad(J_-)(\boldsymbol{\omega}_0)$, $\mathbf{v}_1 = P_+$, $\mathbf{v}_0 = ad(J_-)(\mathbf{v}_1)$ and $\mathbf{v}_{-1} = ad(J_-)(\mathbf{v}_0)$. Notice that the subscript here is the multiple of i that is the vector's eigenvalue. An advantage of this notation is that it emphasises that we are thinking of the matrices as vectors.

The action of $ad(P_+)$ on these vectors is as follows:

$$ad(P_+)\boldsymbol{\omega}_1 = 0, \quad ad(P_+)\boldsymbol{\omega}_0 = -2\mathbf{v}_1, \quad ad(P_+)\boldsymbol{\omega}_{-1} = -2\mathbf{v}_0,$$
$$ad(P_+)\mathbf{v}_1 = 0, \quad ad(P_+)\mathbf{v}_0 = 0, \quad ad(P_+)\mathbf{v}_{-1} = 0.$$

We can find all the other actions using the commutation relations. For example,

$$ad(P_3)\boldsymbol{\omega}_1 = \frac{1}{2i} ad([P_+, J_-])\boldsymbol{\omega}_1,$$
$$= ad(J_-)ad(P_+)\boldsymbol{\omega}_1 - ad(P_+)ad(J_-)\boldsymbol{\omega}_1,$$
$$= ad(J_-)0 + ad(P_+)\boldsymbol{\omega}_0,$$
$$= -2\mathbf{v}_1.$$

The dual of the adjoint representation is sometimes called the coadjoint representation. Let us denote the space of linear functionals on $se(3)$ by $se(3)^*$ and take as a basis for this space $\{\mathbf{a}_1, \mathbf{a}_0, \mathbf{a}_{-1}, \mathbf{b}_1, \mathbf{b}_0, \mathbf{b}_{-1}\}$, such that

$$\mathbf{a}_1(\boldsymbol{\omega}_{-1}) = 1, \quad \mathbf{a}_0(\boldsymbol{\omega}_0) = 1, \quad \mathbf{a}_{-1}(\boldsymbol{\omega}_1) = 1,$$
$$\mathbf{b}_1(\mathbf{v}_{-1}) = 1, \quad \mathbf{b}_0(\mathbf{v}_0) = 1, \quad \mathbf{b}_{-1}(\mathbf{v}_1) = 1,$$

and all other evaluations are zero. In Section 7.2, we saw that to find the dual of a representation we simply take the negative of the transpose of the original matrices in the representation. Here we can compute the action of the Lie algebra elements on the given basis. For example, we have

$$(ad^*(J_3)\mathbf{a}_1)(\boldsymbol{\omega}_{-1}) = -\mathbf{a}_1(ad(J_3)\boldsymbol{\omega}_{-1}) = \mathbf{a}_1(\boldsymbol{\omega}_{-1}) = 1,$$

that is, $ad^*(J_3)\mathbf{a}_1 = \mathbf{a}_1$. In fact, it is easy to see that when restricted to the $so(3)$ subalgebra, ad^* is again $L_1 \oplus L_1$. The action of $ad^*(P_+)$ can be calculated in a similar fashion; for example

$$(ad^*(P_+)\mathbf{b}_{-1})(\boldsymbol{\omega}_0) = -\mathbf{b}_{-1}(ad(P_+)\boldsymbol{\omega}_0) = 2\mathbf{b}_{-1}(\mathbf{v}_1) = 2,$$

so that $ad^*(P_+)\mathbf{b}_{-1} = 2\mathbf{a}_0$. Further computations reveal that

$$ad^*(P_+)\mathbf{a}_1 = 0, \quad ad^*(P_+)\mathbf{a}_0 = 0, \quad ad^*(P_+)\mathbf{a}_{-1} = 0,$$
$$ad^*(P_+)\mathbf{b}_1 = 0, \quad ad^*(P_+)\mathbf{b}_0 = 2\mathbf{a}_1, \quad ad^*(P_+)\mathbf{b}_{-1} = 2\mathbf{a}_0.$$

The isomorphism from $se(3)$ to its dual $se(3)^*$, given on basis elements by $\boldsymbol{\omega}_i \longmapsto \mathbf{b}_i$ and $\mathbf{v}_i \longmapsto -\mathbf{a}_i$, transforms the adjoint representation into the coadjoint representation. Hence, the adjoint representation and the coadjoint representation are similar.

Our next example is the symmetric square of the coadjoint representation. Let's call this representation Q. As in Section 7.2, we will simply write $\mathbf{a}_i\mathbf{a}_j$ to denote the symmetric product of a pair of vectors. The symmetric square of the coadjoint representation is 21-dimensional. Restricting to the $so(3)$ subalgebra we have the decomposition

$$Sym^2(L_1 \oplus L_1) = L_2 \oplus L_2 \oplus L_2 \oplus L_1 \oplus L_0 \oplus L_0 \oplus L_0.$$

We will content ourselves with finding the invariants, that is, the trivial components of the representation. As far as the $so(3)$ action is concerned, there are three trivial L_0 components; they are determined by the vectors

$$\mathbf{a}_0\mathbf{a}_0 - 2\mathbf{a}_1\mathbf{a}_{-1}, \quad \mathbf{b}_0\mathbf{b}_0 - 2\mathbf{b}_1\mathbf{b}_{-1}, \quad \mathbf{a}_0\mathbf{b}_0 - \mathbf{a}_1\mathbf{b}_{-1} - \mathbf{a}_{-1}\mathbf{b}_1.$$

It is clear that these vectors have 0 as eigenvalue. After some simple calculations, we can also check that $Q(J_+)$ gives zero when applied to each of these three vectors. To find the elements invariant under all of $se(3)$, we further require the vectors to be invariant with respect to $Q(P_+)$. A short calculation shows that only the first and last of the above vectors are invariants for the full group of rotations and translations. For example, the middle of the three gives

$$Q(P_+)\big(\mathbf{b}_0\mathbf{b}_0 - 2\mathbf{b}_1\mathbf{b}_{-1}\big) = ad^*(P_+)(\mathbf{b}_0)\mathbf{b}_0 + \mathbf{b}_0 ad^*(P_+)(\mathbf{b}_0)$$
$$- 2ad^*(P_+)(\mathbf{b}_1)\mathbf{b}_{-1} - 2\mathbf{b}_1 ad^*(P_+)(\mathbf{b}_{-1}),$$
$$= 4\mathbf{a}_1\mathbf{b}_0 - 4\mathbf{a}_0\mathbf{b}_1,$$
$$\neq 0.$$

This calculation gives the same result as the one we performed in Section 6.4 showing that there are just two invariant bilinear forms on the Lie algebra $se(3)$. To make the correspondence explicit, we must go back to the original basis for the Lie algebra. In terms of the original basis for $se(3)$, we have

$$\omega_i = \frac{1}{2i}(J_+ - J_-) = \frac{1}{2i}\left(\omega_1 - \frac{1}{2}\omega_{-1}\right),$$

$$\omega_j = \frac{1}{2}(J_+ + J_-) = \frac{1}{2}\left(\omega_1 + \frac{1}{2}\omega_{-1}\right),$$

$$\omega_k = J_3 = \frac{-1}{2i}\omega_0,$$

$$\mathbf{v}_i = \frac{1}{2i}(P_+ - P_-) = \frac{1}{2i}\left(\mathbf{v}_1 - \frac{1}{2}\mathbf{v}_{-1}\right),$$

$$\mathbf{v}_j = \frac{1}{2}(P_+ + P_-) = \frac{1}{2}\left(\mathbf{v}_1 + \frac{1}{2}\mathbf{v}_{-1}\right),$$

$$\mathbf{v}_k = P_3 = \frac{-1}{2i}\mathbf{v}_0.$$

and hence we have

$$\mathbf{a}_1(\omega_i) = \frac{-1}{4i}, \quad \mathbf{a}_1(\omega_j) = \frac{1}{4}, \quad \mathbf{a}_0(\omega_k) = \frac{-1}{2i}, \quad \mathbf{a}_{-1}(\omega_i) = \frac{1}{2i}, \quad \mathbf{a}_{-1}(\omega_j) = \frac{1}{2},$$

$$\mathbf{b}_1(\mathbf{v}_i) = \frac{-1}{4i}, \quad \mathbf{b}_1(\mathbf{v}_j) = \frac{1}{4}, \quad \mathbf{b}_0(\mathbf{v}_k) = \frac{-1}{2i}, \quad \mathbf{b}_{-1}(\mathbf{v}_i) = \frac{1}{2i}, \quad \mathbf{b}_{-1}(\mathbf{v}_j) = \frac{1}{2}.$$

with all other pairings zero. So, given a pair of general elements in the Lie algebra

$$\mathbf{s}_1 = x_1\omega_i + y_1\omega_j + z_1\omega_k + p_1\mathbf{v}_i + q_1\mathbf{v}_j + r_1\mathbf{v}_k$$

and

$$\mathbf{s}_2 = x_2\omega_i + y_2\omega_j + z_2\omega_k + p_2\mathbf{v}_i + q_2\mathbf{v}_j + r_2\mathbf{v}_k,$$

the results of the two invariant bilinear forms are

$$\left(\mathbf{a}_0\mathbf{a}_0 - 2\mathbf{a}_1\mathbf{a}_{-1}\right)(\mathbf{s}_1, \mathbf{s}_2) = 2\mathbf{a}_0(\mathbf{s}_1)\mathbf{a}_0(\mathbf{s}_2) - 2\mathbf{a}_1(\mathbf{s}_1)\mathbf{a}_{-1}(\mathbf{s}_2) - 2\mathbf{a}_1(\mathbf{s}_2)\mathbf{a}_{-1}(\mathbf{s}_1),$$

$$= \frac{-1}{2}z_1z_2 - 2\left(\frac{y_1}{4} - \frac{x_1}{4i}\right)\left(\frac{y_2}{2} + \frac{x_2}{2i}\right) - 2\left(\frac{y_1}{2} + \frac{x_1}{2i}\right)\left(\frac{y_2}{4} - \frac{x_2}{4i}\right),$$

$$= \frac{-1}{2}(x_1x_2 + y_1y_2 + z_1z_2)$$

and

$$\left(\mathbf{a}_0\mathbf{b}_0 - \mathbf{a}_1\mathbf{b}_{-1} - \mathbf{a}_{-1}\mathbf{b}_1\right)(\mathbf{s}_1, \mathbf{s}_2) = \mathbf{a}_0(\mathbf{s}_1)\mathbf{b}_0(\mathbf{s}_2) + \mathbf{a}_0(\mathbf{s}_2)\mathbf{b}_0(\mathbf{s}_1) - \mathbf{a}_1(\mathbf{s}_1)\mathbf{b}_{-1}(\mathbf{s}_2)$$

$$- \mathbf{a}_1(\mathbf{s}_2)\mathbf{b}_{-1}(\mathbf{s}_1) - \mathbf{a}_{-1}(\mathbf{s}_1)\mathbf{b}_1(\mathbf{s}_2) - \mathbf{a}_{-1}(\mathbf{s}_2)\mathbf{b}_1(\mathbf{s}_1),$$

$$= \frac{-z_1r_2}{4} - \frac{z_2r_1}{4} - \left(\frac{y_1}{4} - \frac{x_1}{4i}\right)\left(\frac{q_2}{2} + \frac{p_2}{2i}\right) - \left(\frac{y_2}{4} - \frac{x_2}{4i}\right)\left(\frac{q_1}{2} + \frac{p_1}{2i}\right)$$

$$- \left(\frac{y_1}{2} + \frac{x_1}{2i}\right)\left(\frac{q_2}{4} - \frac{p_2}{4i}\right) - \left(\frac{y_2}{2} + \frac{x_2}{2i}\right)\left(\frac{q_1}{4} - \frac{p_1}{4i}\right),$$

$$= \frac{-1}{4}(x_1p_2 + y_1q_2 + z_1r_2 + p_1x_2 + q_1y_2 + r_1z_2).$$

That is, we recover the Killing form and the reciprocal product of Lie algebra elements up to multiplication by a constant. As usual, the pitch of a single element is given by the ratio of these two forms.

All this extends to any faithful representation of $se(3)$. We can always find the pitch of an element in such a representation. Given a matrix X, we can always compute its trace $Tr(X)$, which is the sum of its diagonal elements. Thinking of the matrix as an element of $V \times V^*$ for some vector space V, we can think of the trace as evaluating V^* on V. The trace has many useful properties. Certainly it is a linear map, and also $Tr(XY) = Tr(YX)$ for any matrices X and Y. This last relation implies that the trace is invariant under similarity transforms:

$$Tr(MXM^{-1}) = Tr(XM^{-1}M) = Tr(X).$$

In Section 4.7, we met the Killing form as the trace of matrices in the adjoint representation of a Lie algebra. However, we need not restrict ourselves to the adjoint representation; we will get an invariant bilinear form using any representation. The result of such a form evaluated on a pair of elements from the representation X and Y, say, will be $Tr(XY)$. We can see that this form is invariant by looking at the action of another element in the representation:

$$Z(Tr(XY)) = Tr([Z,X]Y + X[Z,Y]) = Tr(ZXY - XZY + XZY - XYZ) = 0$$

since $Tr(XYZ) = Tr(ZYX)$. Further, it is not difficult to see that we also get symmetric, bilinear, invariant forms on any matrix representation from $Tr(XMY)$ if M is a matrix that commutes with all the matrices in the representation.

For $se(3)$, we have two matrices

$$J_t = J_3^2 + \frac{1}{2}J_+J_- + \frac{1}{2}J_-J_+ \qquad \text{and} \qquad P_t = P_3J_3 + \frac{1}{2}P_+J_- + \frac{1}{2}P_-J_+$$

that commute with all the elements of the algebra. Here, for brevity, we will simply write P_3 for the matrix representing P_3 and so forth. Thus, for example we have

$$[P_t, J_+] = P_3J_3J_+ - J_+P_3J_3$$
$$+ \frac{1}{2}P_+J_-J_+ - \frac{1}{2}J_+P_+J_- + \frac{1}{2}P_-J_+J_+ - \frac{1}{2}J_+P_-J_+,$$
$$= iP_3J_+ + [P_3, J_+]J_3 + \frac{1}{2}P_+[J_-, J_+] + \frac{1}{2}[P_-, J_+]J_+,$$
$$= iP_3J_+ + iP_+J_3 - iP_+J_3 - iP_3J_+,$$
$$= 0.$$

where we have made liberal use of the commutation relations.

So, for any matrix representation of $se(3)$ we have a pair of symmetric, bilinear, invariant forms, namely

$$K < X, Y >= Tr(XJ_tY), \qquad \text{and} \qquad R < X, Y >= Tr(XP_tY).$$

In the adjoint representation, these reduce to the Killing form and the reciprocal product, respectively.

In any representation, the pitch of an element will therefore be given by

$$p(X) = \frac{R < X, X >}{K < X, X >} = \frac{Tr(XP_tX)}{Tr(XJ_tX)}.$$

This is particularly useful in robotics, where it is common to use several different representations of the group of rigid motions.

7.6   The Principle of Transference

We can generate a large number of representations of $se(3)$ using the following trick. Let L be an n-dimensional representation of $so(3)$. Now, we form a $2n$-dimensional representation, D, of $se(3)$. For the elements in $so(3)$ we use the partitioned matrices

$$D(J_3) = \begin{pmatrix} L(J_3) & 0 \\ 0 & L(J_3) \end{pmatrix},$$

$$D(J_+) = \begin{pmatrix} L(J_+) & 0 \\ 0 & L(J_+) \end{pmatrix},$$

$$D(J_-) = \begin{pmatrix} L(J_-) & 0 \\ 0 & L(J_-) \end{pmatrix},$$

and for the translations we use

$$D(P_3) = \begin{pmatrix} 0 & 0 \\ L(J_3) & 0 \end{pmatrix}, \quad D(P_+) = \begin{pmatrix} 0 & 0 \\ L(J_+) & 0 \end{pmatrix}, \quad D(P_-) = \begin{pmatrix} 0 & 0 \\ L(J_-) & 0 \end{pmatrix}.$$

It is straightforward to check that these matrices satisfy the commutation relations. Another way of looking at this trick is to set

$$I_2 = \begin{pmatrix} 1 & 0 \\ 0 & 1 \end{pmatrix} \quad \text{and} \quad E = \begin{pmatrix} 0 & 0 \\ 1 & 0 \end{pmatrix}.$$

Then we have

$$D(J_*) = L(J_*) \otimes I_2 \quad \text{and} \quad D(P_*) = L(J_*) \otimes E.$$

There is yet another way of looking at this. Notice that linear sums of the two matrices I_2 and E close to form a commutative ring, essentially because $E^2 = 0$, and E and I_2 commute. We can think of this as a representation of the ring of **dual numbers**. This ring is generated by a single symbol, usually written ε, subject to the single condition $\varepsilon^2 = 0$. We can think of the dual numbers as the quotient of a polynomial ring:

$$\mathbb{D} = \mathbb{R}[\varepsilon]/\varepsilon^2.$$

In this way, the dual numbers are very similar to the complex numbers. In the complex case, however, the generator i squares to -1, not zero. So, we can write a general dual number as $\check{z} = a + \varepsilon b$. We add and multiply them just like complex numbers except, that we replace ε^2 and all higher powers by 0. To avoid confusion with unit vectors, we will write dual quantities as $\check{z}$ rather than the more common $\hat{z}$. Now we have an isomorphism between the algebras $se(3)$ and $so(3) \otimes \mathbb{D}$; hence, any representation of $so(3) \otimes \mathbb{D}$ will also be a representation of $se(3)$.

This, in essence, is the Principle of Transference. In the original formulation of the nineteenth century, this principle takes the form

> *Any identity between three-dimensional vectors becomes a valid identity between lines in space when the scalars are considered to be dual numbers.*

(See Rooney [95].) According to Ball, this or something similar was proved by Kotelnikov. However, modern folklore has it that all traces of this work were lost during the Bolshevik revolution of 1917.

A modern statement of the principle would be:

> *All representations of the group $SO(3)$ become representations of $SE(3)$ when tensored with the dual numbers.*

(See Selig [102].) The key point is that any identity or equation must relate quantities that are elements of some representation, as we saw at the end of Section 7.2.

One final way of looking at this principle is to consider the isomorphism $so(3) \otimes \mathbb{D} = so(3, \mathbb{D})$. That is, we can think of elements of $se(3)$ as anti-symmetric matrices with dual rather than real entries. There are some subtleties to this since the dual numbers are only a ring, and not a field, there are dual numbers that are zero divisors. Hence, representations of such algebras act on modules rather than vector spaces. However, many properties of Lie theory persist under this change of ground ring.

As an example we will look at invariant 3-tensors. We begin with rotationally invariant 3-tensors. A general 3-tensor here is an array of $3^3 = 27$ numbers which transforms according to the $L_1 \otimes L_1 \otimes L_1$ representation of $SO(3)$. Now,

$$L_1 \otimes L_1 \otimes L_1 = (L_2 \oplus L_1 \oplus L_0) \otimes L_1 = (L_3 \oplus L_2 \oplus L_1) \oplus (L_2 \oplus L_1 \oplus L_0) \oplus L_1;$$

hence there is just one rotationally invariant 3-tensor.

If we write the eigenvectors of J_3 as $\mathbf{u}_*$, $\mathbf{v}_*$ and $\mathbf{w}_*$, then the invariant element can be found to be

$$\mathcal{I} = \mathbf{u}_0 \otimes \mathbf{v}_+ \otimes \mathbf{w}_- - \mathbf{u}_0 \otimes \mathbf{v}_- \otimes \mathbf{w}_+ - \mathbf{u}_+ \otimes \mathbf{v}_0 \otimes \mathbf{w}_-$$
$$+ \mathbf{u}_- \otimes \mathbf{v}_0 \otimes \mathbf{w}_+ + \mathbf{u}_+ \otimes \mathbf{v}_- \otimes \mathbf{w}_0 - \mathbf{u}_- \otimes \mathbf{v}_+ \otimes \mathbf{w}_0,$$

where we have used subscripts $+$ and $-$ rather than 1 and -1. This is to avoid confusion when we change the basis using

$$\mathbf{u}_0 = \mathbf{u}_3, \quad \mathbf{u}_+ = \mathbf{u}_1 + i\mathbf{u}_2, \quad \mathbf{u}_- = \mathbf{u}_1 - i\mathbf{u}_2$$

and similar for $\mathbf{v}_*$ and $\mathbf{w}_*$. Now the rotationally invariant tensor is given by

$$\mathcal{I} = -2i \Big(\mathbf{u}_1 \otimes \mathbf{v}_2 \otimes \mathbf{w}_3 + \mathbf{u}_3 \otimes \mathbf{v}_1 \otimes \mathbf{w}_2 + \mathbf{u}_2 \otimes \mathbf{v}_3 \otimes \mathbf{w}_1$$
$$- \mathbf{u}_2 \otimes \mathbf{v}_1 \otimes \mathbf{w}_3 - \mathbf{u}_3 \otimes \mathbf{v}_2 \otimes \mathbf{w}_1 - \mathbf{u}_1 \otimes \mathbf{v}_3 \otimes \mathbf{w}_2 \Big).$$

It is clear that this is a multiple of the alternating tensor, usually written:

$$\mathcal{E}_{ijk} = \begin{cases} 1, & \text{if } ijk \text{ is an even permutation of } 123, \\ -1, & \text{if } ijk \text{ is an odd permutaion of } 123, \\ 0, & \text{otherwise.} \end{cases}$$

(See the end of Section 4.1.)

Now we dualise to find the invariants with respect to $SE(3)$. To do this we write the vectors as

$$\check{\mathbf{u}}_i = \mathbf{u}_i + \varepsilon \mathbf{u}_{i+3}, \qquad i = 1, 2, 3$$

and similar for $\mathbf{v}_i$ and $\mathbf{w}_i$. The vectors $\mathbf{u}_4$, $\mathbf{u}_5$ and $\mathbf{u}_6$ are just copies of the original vectors with subscript reduced by 3. Substituting this into the expression for the invariant above gives a dual invariant,

$$\check{\mathcal{I}} \propto \mathcal{Q}_\infty + \varepsilon \mathcal{Q}_0.$$

So we get two real $SE(3)$-invariant 3-tensors, the first $\mathcal{Q}_\infty$ is the alternating tensor extended with zeros. That is,

$$\left(\mathcal{Q}_\infty \right)_{ijk} = \begin{cases} -2, & \text{if } ijk \text{ is an even permutation of } 123, \\ +2, & \text{if } ijk \text{ is an odd permutaion of } 123, \\ 0, & \text{otherwise; } i, j, k = 1, 2, \ldots, 6. \end{cases}$$

The other real invariant has 18 non-zero entries,

$$\mathcal{Q}_0 = \mathbf{u}_4 \otimes \mathbf{v}_2 \otimes \mathbf{w}_3 + \mathbf{u}_6 \otimes \mathbf{v}_1 \otimes \mathbf{w}_2 + \mathbf{u}_5 \otimes \mathbf{v}_3 \otimes \mathbf{w}_1$$
$$- \mathbf{u}_5 \otimes \mathbf{v}_1 \otimes \mathbf{w}_3 - \mathbf{u}_6 \otimes \mathbf{v}_2 \otimes \mathbf{w}_1 - \mathbf{u}_4 \otimes \mathbf{v}_3 \otimes \mathbf{w}_2$$
$$\mathbf{u}_1 \otimes \mathbf{v}_5 \otimes \mathbf{w}_3 + \mathbf{u}_3 \otimes \mathbf{v}_4 \otimes \mathbf{w}_2 + \mathbf{u}_2 \otimes \mathbf{v}_6 \otimes \mathbf{w}_1$$
$$- \mathbf{u}_2 \otimes \mathbf{v}_4 \otimes \mathbf{w}_3 - \mathbf{u}_3 \otimes \mathbf{v}_5 \otimes \mathbf{w}_1 - \mathbf{u}_1 \otimes \mathbf{v}_6 \otimes \mathbf{w}_2$$
$$\mathbf{u}_1 \otimes \mathbf{v}_2 \otimes \mathbf{w}_6 + \mathbf{u}_3 \otimes \mathbf{v}_1 \otimes \mathbf{w}_5 + \mathbf{u}_2 \otimes \mathbf{v}_3 \otimes \mathbf{w}_4$$
$$- \mathbf{u}_2 \otimes \mathbf{v}_1 \otimes \mathbf{w}_6 - \mathbf{u}_3 \otimes \mathbf{v}_2 \otimes \mathbf{w}_4 - \mathbf{u}_1 \otimes \mathbf{v}_3 \otimes \mathbf{w}_5.$$

If we look at the linear functions on the 3-tensors, the vector-space dual, we get a more familiar picture of these invariants. As we saw above, representations of $SO(3)$ are essentially unaffected by taking the vector-space dual, so we can consider the above invariants as invariant functions of 3-tensors, or of three screws. With the above definitions we have

$$(\mathcal{Q}_\infty)_{ijk}(\mathbf{s_u})_i(\mathbf{s_v})_j(\mathbf{s_w})_k = \mathbf{s_u}^T Q_\infty[\mathbf{s_v}, \mathbf{s_w}]$$

and

$$(\mathcal{Q}_0)_{ijk}(\mathbf{s_u})_i(\mathbf{s_v})_j(\mathbf{s_w})_k = \mathbf{s_u}^T Q_0[\mathbf{s_v}, \mathbf{s_w}],$$

where summation over repeated indices is to be assumed and the indices i, j and k range from 1 to 6. So, we can write these functions in terms of the invariant 6×6 symmetric matrices we have already met in Section 6.4, and the commutators of screws, see the end of Section 4.3.

Other properties that carry over into the dual number case include the Killing form

$$Tr(\check{S}_1\check{S}_2) = Tr(\Omega_1\Omega_2) + \varepsilon\Big(Tr(\Omega_1 T_2 + T_1\Omega_2)\Big).$$

Hence, the dual Killing form combines the ordinary Killing form of $so(3)$ with the reciprocal product.

To find the pitch and axis of an $SO(3, \mathbb{D})$ matrix, we can copy the real case in Section 4.4. We get

$$Tr(\check{R}) = 1 + 2\cos(\check{\theta}) = 1 + 2\cos(\theta) - 2\varepsilon d\sin(\theta)$$

and

$$\check{R} - \check{R}^T = 2\sin(\check{\theta})\check{V} = 2\sin(\theta)V + 2\varepsilon\Big(\sin(\theta)U + d\cos(\theta)V\Big).$$

See Samuel et al. [97].

In fact, the only properties that cannot be extended to dual number matrices are those that involve dividing by scalars or what amounts to the same thing, solving equations.

8
Screw Systems

8.1 Generalities

At the beginning of Chapter 4, we stated that Ball's instantaneous screws are rays through the origin in the Lie algebra $se(3)$. Clifford referred to the elements of the Lie algebra as motors. In the following, the practice of referring to Lie algebra elements as screws will continue, with the hope that no confusion will arise.

Consider a serial robot arm in a position where its joint screws are given by $s_1, s_2, \ldots, s_6$. Now, concentrating on the n-th link of the robot, what is the range of possible velocities that this link can have? In Section 4.5 we found that the velocity of such a link is given by

$$\dot{q}_n = \dot{\theta}_1 s_1 + \dot{\theta}_2 s_2 + \cdots + \dot{\theta}_n s_n.$$

So if we assume that we can drive the joints at any velocity, then the range of possible velocities is just the linear span of the joint screws, span $(s_1, s_2, \ldots, s_n)$, that is, the set of all vectors of the form

$$q = c_1 s_1 + c_2 s_2 + \cdots + c_n s_n$$

where the c_is range over the real numbers. Notice that span $(s_1, s_2, \ldots, s_n)$ is a vector subspace of $se(3)$. If all the joint screws are linearly independent, then the dimension of this subspace is n.

A **screw system** is simply a vector subspace of $se(3)$. In addition to the application mentioned above, there are many other uses for this concept; see

Phillips [85, 86], for example. Perhaps Klein [63] was the first to consider such ideas, but it was Hunt [54] who first gave a heuristic classification of screw systems. The first rigourous classification appears in Gibson and Hunt [37] and was then tidied up in Donelan and Gibson [28]. In this chapter, we will rederive Gibson and Hunt's classification but using a method with more of a group-theoretic flavour.

Given an n-system, that is, an n-dimensional vector subspace, we can always find n linearly independent screws that generate it, say, $s_1, s_2, \ldots, s_n$. The Grassmannian $G(n, 6)$ of all n-planes in $se(3)$ lies in the projective space of dimension $\binom{6}{n} - 1$. We may think of a point on the Grassmannian as being given by the anti-symmetric product $s_1 \wedge s_2 \wedge \cdots \wedge s_n$, where the $\binom{6}{n}$ components of the n-fold anti-symmetric product are taken as homogeneous coordinates. Hence, two sets of generators define the same screw system if the n-fold anti-symmetric products of their elements are multiples of each other. We will consider two screw systems to be equivalent if there is a rigid transformation that can transform one of the screw systems into the other. The action of $SE(3)$ on the n-fold anti-symmetric products is just the n-th anti-symmetric power of the adjoint representation, $\wedge^n Ad$. Thus, the action of the group on the Grassmannian is via the projection and restriction of $\wedge^n Ad$. Notice that this all means that a different set of generators, $s_1', s_2', \ldots, s_n'$, will generate the same (equivalent) n-system as before, provided the new generators are linear functions of the old ones, which may also have been subject to an overall rigid transformation. That is, we may transform the generators as

$$s_i' = \sum_{j=1}^{n} m_{ij} Ad(g) s_j$$

with m_{ij} constants and $g \in SE(3)$ without affecting the screw system. The only restriction is that the constants m_{ij} must be such that the new screws are linearly independent—the $n \times n$ matrix M, with elements m_{ij} must have non-zero determinant. To put it another way, M has to be an element of the general linear group $GL(n)$.

The task of classifying screw systems can now be seen to be equivalent to describing the different orbits of the group in the Grassmannian. Suppose that each different orbit represents a point in an abstract space. Such a space is usually called a **moduli space**. We can expect that the moduli space for any classification problem decomposes into a number of disjoint pieces, and each connected component will usually be an interesting topological space, though not a manifold in general.

Our first example is very simple: we consider 1-systems. These are just rays in $se(3)$, that is, lines through the origin. Such a ray can be thought of as a point in the projective space $\mathbb{PR}^5$. In Section 6.4 we saw that the orbits of the group on this space are a pencil of quadrics, that is, a one-dimensional linear system of quadrics. A general quadric in the pencil, one of the pitch quadrics, can be associated with its symmetric matrix

$$Q_p = \begin{pmatrix} -2\alpha I_3 & \beta I_3 \\ \beta I_3 & 0 \end{pmatrix}$$

where $p = \alpha/\beta$ and $p = \infty$ when $\beta = 0$. So Q_0 is the Klein quadric and Q_∞ the degenerate quadric, which is the plane of lines at infinity. We also have that $\mathbf{s}^T Q_p \mathbf{s} = 0$ if the screw $\mathbf{s}$ has pitch p. There is one quadric for each pitch p, including the degenerate case when the pitch is infinite. Hence, the moduli space for 1-systems is a single projective line, parameterised by p. Topologically, this space is a circle. Unfortunately, for the higher screw systems it is much harder to describe the moduli space.

The Gibson–Hunt classification of higher-dimensional systems turns on how a system, now a projective $(n-1)$ plane, lies with respect to this system of pitch quadrics. A general $(n-1)$ plane does not lie completely in any of the quadrics; Gibson and Hunt call these systems type I systems. If n is 3 or less, then it is possible for the system to lie entirely within one of the quadrics: these are the type II systems. A further subdivision occurs when we consider how a system can intersect the degenerate, infinite pitch quadric. Recall that this is a projective 2-plane that lies in all the other quadrics. The systems that miss this 2-plane are A systems; those that intersect it at a single point are B systems, and so forth. Hence, a 2-system of type IIC lies in one of the pitch quadrics and intersects the plane of infinite pitch screws in a line. This, in broad outline, is the classification scheme developed by Gibson and Hunt. There are some extra subtleties, but we will discuss these when we come to them.

The amount of work that we have to do can be considerably reduced if we use the following: For any n-system, call it Δ, there is a unique $6 - n$ system called the **reciprocal system** to Δ. A screw $\mathbf{s}$ belongs to the reciprocal system of Δ if and only if it satisfies the linear equations

$$\mathbf{s}_i^T Q_0 \mathbf{s} = 0$$

for all $\mathbf{s}_i \in \Delta$. Since Δ contains n linearly independent screws and Q_0 is non-degenerate, the above equations constitute n independent linear equations for the six unknown coordinates of $\mathbf{s}$. The reciprocal system thus has $6 - n$ linearly independent screws. Thought of as projective spaces, the reciprocal screw system is just the polar plane with respect to the Klein quadric of the original system. Notice that it is possible for some screws to be in Δ and also in the reciprocal system. Such screws are necessarily lines.

If we take the reciprocal of a system Δ and then take the reciprocal of the result, we will end up with the original system Δ. If we write the reciprocal of Δ as $\overline{\Delta}$, we can write this as

$$\overline{\overline{\Delta}} = \Delta.$$

The bilinear form Q_0 is invariant with respect to rigid transformations:

$$(Ad(g)\mathbf{s}_i)^T Q_0 \mathbf{s}_j = \mathbf{s}_i^T Q_0 (Ad(g)\mathbf{s}_j)$$

for any $g \in SE(3)$ and any pair of screws $\mathbf{s}_i$ and $\mathbf{s}_j$. Hence, if we apply a rigid transformation to the whole system Δ and then find the reciprocal of the result, we will get the transform of $\overline{\Delta}$. That is

$$\overline{\wedge^n Ad(g)\Delta} = \wedge^{6-n} Ad(g)\overline{\Delta}$$

for any $g \in SE(3)$. Hence, to classify 4-systems it is enough to classify their reciprocal 2-systems, and to classify 5-systems; all we need is the pitch of the reciprocal 1-system. All that remains is to classify 2- and 3-systems. To help with this we review a little more Lie theory.

Suppose that we have a vector subspace of $se(3)$ generated by the screws $\mathbf{s}_1$, $\mathbf{s}_2, \ldots, \mathbf{s}_n$. Generally, such a subspace will not be a subalgebra. Let us denote

$$\Delta_1 = \mathrm{span}\,(\mathbf{s}_1, \mathbf{s}_2, \ldots, \mathbf{s}_n).$$

If we take commutators of pairs of elements from Δ_1, then we may get screws that lie outside Δ_1. So we will write

$$\Delta_2 = \Delta_1 \oplus [\Delta_1, \Delta_1]$$

where

$$[\Delta_1, \Delta_1] = \mathrm{span}\,([\mathbf{s}_i, \mathbf{s}_j] : \mathbf{s}_i \in \Delta_1, \mathbf{s}_j \in \Delta_1).$$

If $[\Delta_1, \Delta_1] = 0$, then we know that Δ_1 is a commutative subalgebra. If, more generally, $\Delta_2 = \Delta_1$, then the subspace Δ_1 is a subalgebra, not necessarily commutative. When these relations do not hold we can say only that $\Delta_1 \subset \Delta_2$. The screw system Δ_2 does have a physical significance. We started this section by looking at the velocities of links in a robot arm. Suppose we now look at the time derivative of the n-th link's velocity screw. Using the results of Section 4.5, we have

$$\ddot{\mathbf{q}}_n = \ddot{\theta}_1 \mathbf{s}_1 + \ddot{\theta}_2 \mathbf{s}_2 + \cdots + \ddot{\theta}_n \mathbf{s}_n + \dot{\theta}_1 \dot{\theta}_2 [\mathbf{s}_1, \mathbf{s}_2] + \dot{\theta}_1 \dot{\theta}_3 [\mathbf{s}_1, \mathbf{s}_3] + \cdots + \dot{\theta}_{n-1} \dot{\theta}_n [\mathbf{s}_{n-1}, \mathbf{s}_n].$$

Once again, if we assume that the joint rates and their accelerations can take any value, then the screw $\ddot{\mathbf{q}}_n$ can range over all of Δ_2.

We can iterate the above process by defining

$$\Delta_{i+1} = \Delta_i \oplus [\Delta_1, \Delta_i].$$

However, these screw systems cannot keep growing indefinitely, since $se(3)$ is only six-dimensional. Eventually there will be a stage where $\Delta_{i+1} = \Delta_i$. When this happens, it is easy to see that $\Delta_{i+k} = \Delta_i$ for any positive integer k. The screw system that we end up with, Δ_i, will be a subalgebra.

To see this, we have to show that the commutator of any pair of elements of Δ_i is again in Δ_i. Linearity of the Lie bracket means that we only have to consider the monomials of the form $[\mathbf{s}_j, [\mathbf{s}_k, [\cdots, [\mathbf{s}_l, \mathbf{s}_m] \cdots]]]$, which generate Δ_i. In fact, we can show that $[\Delta_i, \Delta_j] \subseteq \Delta_{i+j}$. We do this by induction. First, it is simple to see that $[\Delta_1, \Delta_j] \subseteq \Delta_{j+1}$ for any positive j. Now assume that $[\Delta_i, \Delta_j] \subseteq \Delta_{i+j}$ is true for all j and all $i \leq k$ for some positive integer k. We will show that $[\Delta_{k+1}, \Delta_j] \subseteq \Delta_{j+k+1}$, thus proving the inductive step. Suppose that $\sigma_k \in \Delta_k$ and $\sigma_j \in \Delta_j$. Then a general monomial in Δ_{k+1} will have the form $[\mathbf{s}, \sigma_k]$. So we have

$$[[\mathbf{s}, \sigma_k], \sigma_j] = [\mathbf{s}, [\sigma_k, \sigma_j]] - [\sigma_k, [\mathbf{s}, \sigma_j]]$$

using the Jacobi identity. By hypothesis, $[\sigma_k, \sigma_j] \subseteq \Delta_{j+k}$. Hence the first right-hand side term above is in Δ_{j+k+1}. The second term is also in this vector space since $[\mathbf{s}, \sigma_j] \in \Delta_{j+1}$.

So we have shown that $[\Delta_i, \Delta_j] \subseteq \Delta_{i+j}$ in general. Thus, if $\Delta_{i+k} = \Delta_i$, then $[\Delta_i, \Delta_i] \subseteq \Delta_i$, which is the condition for Δ_i to be a subalgebra.

The subalgebra Δ_i that we end up with after this process of iteration can only be one of a finite list, one of the Lie algebras to the connected subgroups of $SE(3)$ that we found in Section 3.5. We will call the subgroup produced in this way the **completion group** of the screw system Δ_1 and write its Lie algebra as Δ_∞. It is clear that two screw systems with different completion groups cannot be equivalent. Indeed, transforming a screw system into an equivalent one using an element of $SE(3)$ simply conjugates the completion group of the system. The completion group provides a coarse classification for screw systems. For a given dimension of screw system, we will determine which completion groups can arise and then which different screw systems have the same completion group.

8.2   2-systems

A 2-system can be thought of as a point in the 8-dimensional Grassmannian $G(2,6)$. Our task here is to describe all the orbits of $SE(3)$ in the Grassmannian. The action of an element $\mathbf{s}$ of $se(3)$ on a point $\mathbf{s}_1 \wedge \mathbf{s}_2$ in $G(2,6)$ is given by

$$\mathbf{s}(\mathbf{s}_1 \wedge \mathbf{s}_2) = (ad(\mathbf{s})\mathbf{s}_1) \wedge \mathbf{s}_2 + \mathbf{s}_1 \wedge (ad(\mathbf{s})\mathbf{s}_2) = [\mathbf{s}, \mathbf{s}_1] \wedge \mathbf{s}_2 + \mathbf{s}_1 \wedge [\mathbf{s}, \mathbf{s}_2];$$

see Section 7.2. Now, suppose $[\mathbf{s}, \mathbf{s}_1] \wedge \mathbf{s}_2 + \mathbf{s}_1 \wedge [\mathbf{s}, \mathbf{s}_2] = 0$. This means that $\mathbf{s}$ has no effect on the point, and hence $e^{t\mathbf{s}}$ is an element of the isotropy group of the point. The Lie algebra of a point's isotropy group is given by all elements $\mathbf{s}$ that satisfy the linear equations

$$[\mathbf{s}, \mathbf{s}_1] \wedge \mathbf{s}_2 + \mathbf{s}_1 \wedge [\mathbf{s}, \mathbf{s}_2] = 0.$$

So, it is a simple matter of linear algebra to compute the Lie algebra of the isotropy group given a point on $G(2,6)$.

Elements of the isotropy group itself are elements g of $SE(3)$ satisfying

$$(Ad(g)\mathbf{s}_1) \wedge (Ad(g)\mathbf{s}_2) = \lambda \mathbf{s}_1 \wedge \mathbf{s}_2$$

where λ is some non-zero constant. Remember that we are looking at points in a projective space, so multiplication by a non-zero constant has no effect. This condition is harder to work with, but the isotropy group is not necessarily just the exponential of this Lie algebra since there may also be discrete symmetries of the system. However, we can show that the constant in the equation above can only be $\lambda = \pm 1$, and this severely limits the kind of discrete symmetries that a 2-system can have. To see this, consider projecting the system onto the rotation invariant subspace determined by the first three components of each screw. On this subspace, the condition becomes

$$(R\boldsymbol{\omega}_1) \wedge (R\boldsymbol{\omega}_2) = R(\boldsymbol{\omega}_1 \wedge \boldsymbol{\omega}_2) = \lambda \boldsymbol{\omega}_1 \wedge \boldsymbol{\omega}_2$$

where R is a rotation matrix. Now $\boldsymbol{\omega}_1 \wedge \boldsymbol{\omega}_2$ is another 3-vector (see Section 7.4) so we see that $\lambda = 1$ if R is a rotation about that vector, or $\lambda = -1$ if R is a rotation of π about a perpendicular vector. This argument fails if the projection to the subspace is trivial, which can happen if either or both of the original screws have infinite pitch. When both screws have infinite pitch, we can look at the action of the group restricted to the subspace of infinite pitch screws. The argument is the same as above but using the last three components:

$$(R\mathbf{v}_1) \wedge (R\mathbf{v}_2) = R(\mathbf{v}_1 \wedge \mathbf{v}_2) = \lambda \mathbf{v}_1 \wedge \mathbf{v}_2.$$

Since the screws have infinite pitch, the action of the translations is trivial. When only one of the screws has infinite pitch, say $\mathbf{s}_2$, then we can look at the subspace in which $\boldsymbol{\omega}_1 \wedge \mathbf{v}_2$ lies. Again, the translations act trivially on this space, so the argument above applies again. Hence, to find the discrete symmetries of a 2-system we only have to check rotations about $\boldsymbol{\omega}_1 \times \boldsymbol{\omega}_2$ and rotation of π perpendicular to this.

In Section 3.3 we saw that any two points on the same orbit of a group have conjugate isotropy groups, so henceforth we will just talk of the isotropy group of an orbit. Notice that two points in $G(2,6)$ cannot be on the same orbit if they have different isotropy groups. So these groups provide another important characteristic of the orbits.

Returning to the study of the completion groups, we can see that only seven subgroups can occur as the completion groups of 2-systems. The subgroups must be at least two-dimensional, and the group $\mathbb{R}^3$ cannot occur, as it cannot be generated by two elements. The list of possible completion subgroups is therefore

$$\mathbb{R}^2, \; SO(2) \times \mathbb{R}, \; H_p \ltimes \mathbb{R}^2, \; SO(3), \; SE(2), \; SE(2) \times \mathbb{R}, \text{ and } SE(3).$$

We will examine each case in turn. Assume that the 2-system we are considering is generated by the two screws

$$\Delta_1 = \operatorname{span}(\mathbf{s}_1, \mathbf{s}_2).$$

The first case to consider is that when the two generators commute. So, $[\Delta_1, \Delta_1] = 0$, and hence Δ_1 is now a commutative subalgebra.

8.2.1   The Case $\mathbb{R}^2$

In this case Δ_1 is a commutative subalgebra. Notice also that all the elements of this screw system have infinite pitch; hence, in the Gibson–Hunt classification would assign this 2-system the type IIC. Now in any 2-system with this completion group we can always find a rigid transformation which makes

$$s_1 = \begin{pmatrix} 0 \\ i. \end{pmatrix}$$

Then we can rotate about the x-axis until the second screw is of the form

$$s_2' = \begin{pmatrix} 0 \\ ai + bj \end{pmatrix}.$$

Finally, we can add multiples of the first screw to this and multiply by a constant to give a normal form for this 2-system:

$$s_1 = \begin{pmatrix} 0 \\ i \end{pmatrix}, \qquad s_2 = \begin{pmatrix} 0 \\ j \end{pmatrix}.$$

Since any 2-system of this type can be reduced to this single normal form, the Grassmannian contains just a single orbit of this type. A short computation shows that the Lie algebra to the orbit's isotropy group is $se(2) \times \mathbb{R}$. For the normal form, the isotropy group consists of rotations about the z-axis together with any translation. Now any rotation of π radians about an axis perpendicular to the z-axis will give a discrete symmetry. For example, a rotation about the x-axis will not change s_1 and will only change the sign of s_2; hence $s_1 \wedge s_2$ will change sign. All these symmetries, however, can be thought of as a fixed π rotation conjugated with some rotation about the z-axis. The isotropy group is thus $E(2) \times \mathbb{R}$. That is, we can think of the rotations by π as reflections in the xy-plane.

8.2.2   The Case $SO(2) \times \mathbb{R}$

Again, Δ_1 is a commutative subalgebra, this time containing screws of finite pitch. The line corresponding to this 2-system in $\mathbb{PR}^5$ will intersect each of the pitch quadrics twice, except the plane Q_∞, which it meets just once. As generators of the system, let us take the two intersections with the Klein quadric Q_0. This gives us one pitch zero screw and one infinite pitch screw since Q_∞ is contained in every other pitch quadric. By a suitable rigid transformation, we can align the zero pitch screw, a line, along the x-axis. The infinite pitch screw commutes with this line and hence must have the same axis. Thus, we have produced the following normal form for this system:

$$s_1 = \begin{pmatrix} i \\ 0 \end{pmatrix}, \qquad s_2 = \begin{pmatrix} 0 \\ i \end{pmatrix}.$$

This is the 2-system with Gibson–Hunt type IB^0. Again there is just one orbit of this type. This time the Lie algebra of the isotropy group of the orbit is $so(2) \times \mathbb{R}$, generated by rotations and translations about the x-axis for the normal form. Now, the discrete symmetries consist of any rotation of π radians about an axis perpendicular to the x-axis. Any of these symmetries will change the sign of both $\mathbf{s}_1$ and $\mathbf{s}_2$ and hence $\mathbf{s}_1 \wedge \mathbf{s}_2$ is unchanged. Again, these discrete symmetries can be thought of as a fixed one and a conjugation by some rotation about the x-axis. So the isotropy group for this orbit is $O(2) \times \mathbb{R}$.

It is not possible to have $[\mathbf{s}_1, \mathbf{s}_2] \subset \Delta_1$ with $[\mathbf{s}_1, \mathbf{s}_2] \neq 0$. So, for the following cases we have $\Delta_2 \neq \Delta_1$.

8.2.3   The Case $SO(3)$

The next cases have three-dimensional completion groups. That is, we look at the cases where $\Delta_3 = \Delta_2$. There are three possibilities.

All the elements of the $so(3)$ subalgebra have pitch zero; hence, so do elements of any 2-system with this completion group. Such a system has Gibson–Hunt type IIA^0. In fact, Gibson and Hunt do not single out this system; for them it is simply a system of type IIA for which the parameter takes the value zero. Using rigid transformations and linear changes of basis, we can maneuver the generators of any such system into the normal form

$$\mathbf{s}_1 = \begin{pmatrix} \mathbf{i} \\ \mathbf{0} \end{pmatrix}, \qquad \mathbf{s}_2 = \begin{pmatrix} \mathbf{j} \\ \mathbf{0} \end{pmatrix}.$$

So once again there is just one orbit of this type; its isotropy group is $SO(2)$, generated by rotations about $[\mathbf{s}_1, \mathbf{s}_2]$. The discrete symmetries are π rotations about axes perpendicular to $[\mathbf{s}_1, \mathbf{s}_2]$ hence the isotropy group is $O(2)$.

8.2.4   The Case $H_p \ltimes \mathbb{R}^2$

The Lie algebra of this group consists of screws of pitch p with parallel axes together with infinite pitch screws along axes perpendicular to the first. This subalgebra cannot be generated by two infinite pitch screws; hence, the 2-system must have contained one infinite pitch screw and screws of pitch p. That is, it is a system of Gibson–Hunt type IIB. We can produce a normal form for this system by transforming the pitch p screw to lie along the x-axis and then rotating about this axis until the infinite pitch screw lies along the y-axis:

$$\mathbf{s}_1 = \begin{pmatrix} \mathbf{i} \\ p\mathbf{i} \end{pmatrix}, \qquad \mathbf{s}_2 = \begin{pmatrix} \mathbf{0} \\ \mathbf{j} \end{pmatrix}.$$

The isotropy group's Lie algebra can now be found; it is $\mathbb{R}^2$. Notice that we get a single orbit for each different value of the pitch p. However, the pitch cannot be infinite since then the system would be of type IIC. Also, zero pitch is disallowed since this gives the next case. This system does have discrete symmetries. We can think of these as being generated by a π rotation about the z-axis and a π rotation about the x-axis. Combining these would give a π rotation about the y-axis; hence, the discrete symmetries form a group of order 4, $\mathbb{Z}_2 \times \mathbb{Z}_2$, where $\mathbb{Z}_2$ is the group of integers modulo 2. So the system's isotropy group is $\mathbb{Z}_2 \times \mathbb{Z}_2 \times \mathbb{R}^2$. Note that the group of order 4 $\mathbb{Z}_2 \times \mathbb{Z}_2$ can also be thought of as the order 4 dihedral group D_2.

8.2.5   The Case $SE(2)$

As mentioned above, this case can be thought of as the same as the last case but with $p = 0$. So there is just a single orbit with the normal form

$$s_1 = \begin{pmatrix} i \\ 0 \end{pmatrix}, \qquad s_2 = \begin{pmatrix} 0 \\ j \end{pmatrix}.$$

The Lie algebra to the isotropy group is still $\mathbb{R}^2$, and the isotropy group itself is still $\mathbb{Z}_2 \times \mathbb{Z}_2 \times \mathbb{R}^2$. Gibson and Hunt do not separate out this case. For them it is simply the system of type IIB with $p = 0$.

8.2.6   The Case $SE(2) \times \mathbb{R}$

There is just one four-dimensional subgroup. This group can also be written as $SO(2) \ltimes \mathbb{R}^3$, from which it is clear that the Lie algebra must contain a pure rotation. Without loss of generality, we may take this rotation as a generator of the 2-system. We may also take a pure translation as the other generator. However, if the axis of the translation is perpendicular to the axis of rotation, then the action of the rotation on this element would only generate an $\mathbb{R}^2$, as in the IIB $(p = 0)$ case above. On the other hand, if the axis of the translation is parallel to that of the rotation, then the translation is unaffected by the rotation, and the closure group generated would only be $SO(2) \times \mathbb{R}$ as in the IB^0 case above. So the axis of the translation generator must be at some general angle to the rotation axis. A normal form for the generators could be

$$s_1 = \begin{pmatrix} i \\ 0 \end{pmatrix}, \qquad s_2 = \begin{pmatrix} 0 \\ i + pj \end{pmatrix}.$$

Here p is essentially the tangent of the angle between the axes of the two generators. Now suppose that we perform a rotation of π radians about the x-axis. The only change is the sign of p. Hence, any pair of these systems that differ only by the sign of p are equivalent. We cannot affect p with any other rigid motion, so we get a single orbit for each different finite but positive value of p. In each case, the isotropy group of the orbit is $\mathbb{Z}_2 \times \mathbb{Z}_2 \times \mathbb{R}$. The Gibson–Hunt type of these systems is IB.

8.2.7 The Case $SE(3)$

The final possible completion group is the whole six-dimensional group. To generate the whole group from just two elements, those elements must contain two independent rotations and two independent translations. A general pair of generators would thus have the form

$$s_1' = \begin{pmatrix} \boldsymbol{\omega}_1 \\ \mathbf{v}_1 + p_1\boldsymbol{\omega}_1 \end{pmatrix}, \qquad s_2' = \begin{pmatrix} \boldsymbol{\omega}_2 \\ \mathbf{v}_2 + p_2\boldsymbol{\omega}_2 \end{pmatrix}.$$

The 2-system generated by these screws contains a different pair of generators such that the axes of the two screws meet orthogonally. The condition for two screws to meet orthogonally is

$$s_1^T Q_p s_2 = 0$$

for all pitch quadrics Q_p. By the process of Gramm–Schmidt orthogonalisation, we can bring the generators to orthonormal form with respect to one of the pitch quadrics, say Q_∞. So we may assume that the generators s_1' and s_2' satisfy

$$(s_1')^T Q_\infty s_1' = (s_2')^T Q_\infty s_2' = -2 \qquad \text{and} \qquad (s_1')^T Q_\infty s_2' = 0.$$

An arbitrary change of basis for the screw system would have the form

$$s_1 = m_{11} s_1' + m_{12} s_2',$$
$$s_2 = m_{21} s_1' + m_{22} s_2',$$

where m_{11}, m_{12}, m_{21} and m_{22} are constants. But in order to preserve the orthonormality under Q_∞, the matrix

$$M = \begin{pmatrix} m_{11} & m_{12} \\ m_{21} & m_{22} \end{pmatrix}$$

must be orthogonal, that is $M \in O(2)$. If we can choose a matrix M such that

$$s_1^T Q_p s_2 = m_{11} m_{21} q_{11}' + (m_{11} m_{22} + m_{12} m_{22}) q_{12}' + m_{12} m_{22} q_{22}' = 0$$

with $q_{ij}' = (s_i')^T Q_p s_j'$ for some other p, then we are done. Notice, however, that the left-hand side of this equation is just the off-diagonal element in the symmetric matrix

$$\Upsilon_p = M \begin{pmatrix} q_{11}' & q_{12}' \\ q_{12}' & q_{22}' \end{pmatrix} M^T.$$

By Sylvester's theorem, there is always an orthogonal matrix that will make Υ_0, say, diagonal, that is, will reduce the off-diagonal element to zero.

So, after a final rigid transformation of both generators we can bring the original generators to the normal form

$$s_1 = \begin{pmatrix} \mathbf{i} \\ p_a \mathbf{i} \end{pmatrix}, \qquad s_2 = \begin{pmatrix} \mathbf{j} \\ p_b \mathbf{j} \end{pmatrix}.$$

In this case, we will perform the calculation for the isotropy group explicitly. So, let $\mathbf{s} = (x, y, z, t_x, t_y, t_z)^T$ be a general element of the Lie algebra. Now, for this element to be in the Lie algebra of the isotropy group we must have $[\mathbf{s}, \mathbf{s}_1] \wedge \mathbf{s}_2 + \mathbf{s}_1 \wedge [\mathbf{s}, \mathbf{s}_2] = 0$, that is,

$$
\begin{pmatrix} 0 \\ z \\ -y \\ 0 \\ p_a z + t_z \\ -p_a y - t_y \end{pmatrix} \wedge \begin{pmatrix} 0 \\ 1 \\ 0 \\ 0 \\ p_b \\ 0 \end{pmatrix} + \begin{pmatrix} 1 \\ 0 \\ 0 \\ p_a \\ 0 \\ 0 \end{pmatrix} \wedge \begin{pmatrix} -z \\ 0 \\ x \\ -p_b z - t_z \\ 0 \\ p_b x + t_x \end{pmatrix} = 0.
$$

Expanding the anti-symmetric products, we get 15 linear equations for the components of a general element of the isotropy group's Lie algebra. The equations are very simple, and after a little rearrangement we get

$$
x = 0, \quad y = 0, \quad t_x = 0, \quad t_y = 0, \quad t_z = 0, \quad \text{and} \quad (p_a - p_b)z = 0.
$$

So, if $p_a \neq p_b$ we also have $z = 0$, and the isotropy group's Lie algebra is trivial. The discrete symmetries are once again $\mathbb{Z}_2 \times \mathbb{Z}_2$, consisting of rotations of π about any of the coordinate axes for the normal form. In this case, the 2-system contains screws with different pitches but no infinite pitch screws, so these systems correspond to the Gibson–Hunt type IA. However, if $p_a = p_b$, then we can have any value for z, and the isotropy group contains the elements $e^{z\mathbf{k}}$, that is, a copy of $SO(2)$. There is also a discrete symmetry, a π rotation about any axis perpendicular to the z-axis. Hence, the isotropy group of these orbits is $O(2)$. Screws in these 2-systems all have the same pitch, and hence these systems have Gibson–Hunt type IIA.

Notice that the Gibson–Hunt classification distinguishes precisely those orbits that have different isotropy groups. In other words, a pair of 2-systems will have different Gibson–Hunt type if and only if the orbits defined by the systems are topologically different. The normal forms that we have used are taken from Donelan and Gibson [29], who also classify the screw systems up to the action of the group of similarities rather than just rigid transformations. The results are summarised in Table 8.1.

The IA systems are closely connected with the cylindroid that we met in Section 6.5.2. Any IA system will meet the Klein quadric Q_0 in two points neither of which will be on the degenerate quadric Q_∞. If the two points are real, then by suitable transformations we can bring the screw system into the normal form given in Section 6.5.2, that is,

$$
\mathbf{s}_1 = \begin{pmatrix} \mathbf{i} \\ \mathbf{0} \end{pmatrix}, \qquad \mathbf{s}_2 = \begin{pmatrix} \cos\phi\mathbf{i} + \sin\phi\mathbf{j} \\ -l\sin\phi\mathbf{i} + l\cos\phi\mathbf{j} \end{pmatrix}.
$$

A general screw in a IA system has the form

$$
\mathbf{s} = \begin{pmatrix} \lambda\mathbf{i} + \mu\mathbf{j} \\ \lambda p_a\mathbf{i} + \mu p_b\mathbf{j} \end{pmatrix}.
$$

TABLE 8.1. The 2-Systems

Completion Group	Gibson–Hunt Type	Normal Form	Isotropy Group
$\mathbb{R}^2$	IIC	$\mathbf{s}_1 = (0,0,0,1,0,0)^T$ $\mathbf{s}_2 = (0,0,0,0,1,0)^T$	$E(2) \times \mathbb{R}$
$SO(2) \times \mathbb{R}$	IB^0	$\mathbf{s}_1 = (1,0,0,0,0,0)^T$ $\mathbf{s}_2 = (0,0,0,1,0,0)^T$	$O(2) \times \mathbb{R}$
$SO(3)$	IIA $(p = 0)$	$\mathbf{s}_1 = (1,0,0,0,0,0)^T$ $\mathbf{s}_2 = (0,1,0,0,0,0)^T$	$O(2)$
$H_p \ltimes \mathbb{R}^2$	IIB $(p \neq 0)$	$\mathbf{s}_1 = (1,0,0,p,0,0)^T$ $\mathbf{s}_2 = (0,0,0,0,1,0)^T$	$\mathbb{Z}_2 \times \mathbb{Z}_2 \times \mathbb{R}^2$
$SE(2)$	IIB $(p = 0)$	$\mathbf{s}_1 = (1,0,0,0,0,0)^T$ $\mathbf{s}_2 = (0,0,0,0,1,0)^T$	$\mathbb{Z}_2 \times \mathbb{Z}_2 \times \mathbb{R}^2$
$SE(2) \times \mathbb{R}$	IB $(p \neq 0)$	$\mathbf{s}_1 = (1,0,0,0,0,0)^T$ $\mathbf{s}_2 = (0,0,0,1,p,0)^T$	$\mathbb{Z}_2 \times \mathbb{Z}_2 \times \mathbb{R}$
$SE(3)$	IIA $(p \neq 0)$	$\mathbf{s}_1 = (1,0,0,p,0,0)^T$ $\mathbf{s}_2 = (0,1,0,0,p,0)^T$	$O(2)$
$SE(3)$	IA $(p_a \neq p_b)$	$\mathbf{s}_1 = (1,0,0,p_a,0,0)^T$ $\mathbf{s}_2 = (0,1,0,0,p_b,0)^T$	$\mathbb{Z}_2 \times \mathbb{Z}_2$

The system contains two pitch zero screws if p_a and p_b have opposite signs. Assuming $p_a > 0$ and $p_b < 0$, the two lines are

$$\mathbf{s}_1' = \left(\frac{\sqrt{-p_b}\mathbf{i} + \sqrt{p_a}\mathbf{j}}{p_a\sqrt{-p_b}\mathbf{i} + p_b\sqrt{p_a}\mathbf{j}} \right) \quad \text{and} \quad \mathbf{s}_2' = \left(\frac{-\sqrt{-p_b}\mathbf{i} + \sqrt{p_a}\mathbf{j}}{-p_a\sqrt{-p_b}\mathbf{i} + p_b\sqrt{p_a}\mathbf{j}} \right).$$

From the directions of these two lines, we have

$$\cos\phi = \frac{p_a + p_b}{p_a - p_b}.$$

That is, from the Killing form of the two screws, the reciprocal product gives

$$-l\sin\phi = \frac{4p_a p_b}{p_a - p_b}.$$

So, we see that the axes of the screws in IA 2-systems form a cylindroid if the pitches in the normal form have opposite signs.

8.3   3-systems

3-systems can be thought of as points in the 9-dimensional Grassmannian $G(3,6)$. Taking the quotient by the 6-dimensional group of rigid motions, we expect at most $9 - 6 = 3$ parameter families of orbits. That is, the moduli space will be at most three-dimensional. In this case, elements of the Lie algebra of an orbit's isotropy group satisfy the relation

$$[\mathbf{s}, \mathbf{s}_1] \wedge \mathbf{s}_2 \wedge \mathbf{s}_3 + \mathbf{s}_1 \wedge [\mathbf{s}, \mathbf{s}_2] \wedge \mathbf{s}_3 + \mathbf{s}_1 \wedge \mathbf{s}_2 \wedge [\mathbf{s}, \mathbf{s}_3] = 0$$

while elements of the isotropy group itself satisfy

$$(Ad(g)\mathbf{s}_1) \wedge (Ad(g)\mathbf{s}_2) \wedge (Ad(g)\mathbf{s}_3) = \mathbf{s}_1 \wedge \mathbf{s}_2 \wedge \mathbf{s}_3.$$

Arguments similar to those given at the beginning of the last section show that this is the only possibility; that is, the only possible non-zero constant here is $+1$.

Thought of projectively, a 3-system is a 2-plane in $\mathbb{PR}^5$. If this plane does not lie entirely in one of the pitch quadrics, that is, if the system has Gibson–Hunt type I, then it will intersect all the pitch quadrics. These intersections determine a pencil of conic curves on the plane, and the projective type of this pencil is certainly invariant under the action of $SE(3)$. For regular pencils the projective type is given by the Segre symbol of the pencil; see Section 3.4. Another way to think of this is as the number and type of singular quadrics in the pencil. Notice that, in this problem, some of the possibile types may not arise. The Gibson–Hunt classification of 3-systems takes account of this, and so the type of a 3-system will depend on the number and type of singular quadrics in the pencil. Notice also that type II systems will be either α- or β-planes of the pitch quadric that the system lies in; see Section 6.3.

Turning to the possible completion groups, there are six possible subgroups of dimension 3 or greater:

$$\mathbb{R}^3, \; SO(3), \; SE(2), \; H_p \ltimes \mathbb{R}^2, \; SE(2) \times \mathbb{R} \text{ and } SE(3).$$

Once again, we will look at each case in turn. To begin with, the first four cases are all subalgebras. That is, $\Delta_2 = \Delta_1$.

8.3.1   The Case $\mathbb{R}^3$

Clearly, any three independent infinite pitch screws will generate this 3-system. After a rotation and a Gramm–Schmidt orthogonalisation, we can assume a normal form for the generators:

$$\mathbf{s}_1 = \begin{pmatrix} \mathbf{0} \\ \mathbf{i} \end{pmatrix}, \qquad \mathbf{s}_2 = \begin{pmatrix} \mathbf{0} \\ \mathbf{j} \end{pmatrix}, \qquad \mathbf{s}_3 = \begin{pmatrix} \mathbf{0} \\ \mathbf{k} \end{pmatrix}.$$

The isotropy group of the system is the whole group $SE(3)$, since this system is exactly the invariant subspace of all infinite pitch screws. Hence, the Gibson–Hunt type of this system is IID.

8.3.2   The Case $SO(3)$

This system can be generated by any three linearly independent zero pitch screws. By a process similar to the one described immediately above, we can transform any set of generators to the normal form

$$\mathbf{s}_1 = \begin{pmatrix} \mathbf{i} \\ \mathbf{0} \end{pmatrix}, \qquad \mathbf{s}_2 = \begin{pmatrix} \mathbf{j} \\ \mathbf{0} \end{pmatrix}, \qquad \mathbf{s}_3 = \begin{pmatrix} \mathbf{k} \\ \mathbf{0} \end{pmatrix}.$$

The isotropy group of such a 3-system is $SO(3)$, the same as the completion group. Gibson and Hunt do not separate out this particular system; it is just one of the type IIA systems.

8.3.3   The Case $SE(2)$

To generate a 3-system with this completion group, we need a pitch zero screw and two independent infinite pitch screws. The axes of the infinite pitch screws must be perpendicular to the axis of the zero pitch screw, and hence we can always transform such a system into the normal form

$$\mathbf{s}_1 = \begin{pmatrix} \mathbf{i} \\ \mathbf{0} \end{pmatrix}, \qquad \mathbf{s}_2 = \begin{pmatrix} \mathbf{0} \\ \mathbf{j} \end{pmatrix}, \qquad \mathbf{s}_3 = \begin{pmatrix} \mathbf{0} \\ \mathbf{k} \end{pmatrix}.$$

The isotropy group of such a system is $E(2) \times \mathbb{R}$. Again, for Gibson and Hunt this is just a particular case of type IIC.

8.3.4   The Case $H_p \ltimes \mathbb{R}^2$

These systems are almost the same as above, except that we have a pitch p generator rather than zero pitch. Any such system can be transformed to the normal form

$$s_1 = \begin{pmatrix} \mathbf{i} \\ p\mathbf{i} \end{pmatrix}, \qquad s_2 = \begin{pmatrix} \mathbf{0} \\ \mathbf{j} \end{pmatrix}, \qquad s_3 = \begin{pmatrix} \mathbf{0} \\ \mathbf{k} \end{pmatrix}.$$

where p is non-zero. On the other hand, it is not possible to transform a system of this type into another if the values of p are different. Hence, we have an orbit of this type for each different non-zero value of p. The isotropy group of these orbits is still $E(2) \times \mathbb{R}$. The Gibson–Hunt type is IIC again.

8.3.5   The Case $SE(2) \times \mathbb{R}$

This group can also be written as $SO(2) \ltimes \mathbb{R}^3$. A 3-system with this completion group must contain a pure rotation, which we can take as a rotation about the x-axis. The other generators must be pure translations. At least one of these must have some component in the x-direction. Other than this requirement, we only need the translations to be linearly independent. If one of the translations has a component in the x-direction, we can take a multiple of it away from the other translation to produce a translation with no component in the x-direction. We may then combine the translations so that they are orthogonal, and finally we can rotate about the x-axis to get the normal form

$$s_1 = \begin{pmatrix} \mathbf{i} \\ \mathbf{0} \end{pmatrix}, \qquad s_2 = \begin{pmatrix} \mathbf{0} \\ \mathbf{j} \end{pmatrix}, \qquad s_3 = \begin{pmatrix} \mathbf{0} \\ \mathbf{i} + p\mathbf{k} \end{pmatrix}.$$

Now, a rotation of π about the y-axis reverses sign of p; hence, we get a different orbit for each different value of $|p|$. The Lie algebra of the isotropy group for any of these orbits is $\mathbb{R}^2$. There are no discrete symmetries unless $p = 0$, in which case a π rotation about the y-axis is a symmetry. Hence, if $p \neq 0$ the isotropy group is $\mathbb{R}^2$ and the Gibson–Hunt type is IC. But if $p = 0$, the isotropy group is $\mathbb{Z}_2 \times \mathbb{R}^2$ with Gibson–Hunt type IC^0.

8.3.6   The Case $SE(3)$

In the last section, we saw that we can generate $SE(3)$ with just two screws. A 3-system with this completion group will contain an extra screw. If this extra screw has finite pitch, then we transform it into a normal form that has three screws whose axes are mutually orthogonal and all meet at a common point. The condition for three lines to have these properties is $s_1^T Q_0[s_2, s_3] = 0$. However, if three lines meet orthogonally in pairs, then the three meeting points of the pairs must coincide. Hence, the conditions that we must satisfy are $s_i^T Q_p s_j = 0$ for all p and $1 \leq i < j \leq 3$. We can use much the same argument as we did for

the IA 2-systems in the last section to show that such a system can be reduced
to the normal form

$$\mathbf{s}_1 = \begin{pmatrix} \mathbf{i} \\ p_a\mathbf{i} \end{pmatrix}, \qquad \mathbf{s}_2 = \begin{pmatrix} \mathbf{j} \\ p_b\mathbf{j} \end{pmatrix}, \qquad \mathbf{s}_3 = \begin{pmatrix} \mathbf{k} \\ p_c\mathbf{k} \end{pmatrix}.$$

First, we produce an orthonormal basis with respect to Q_∞ using Gramm–
Schmidt. To preserve the orthonormal properties of the basis, we use an orthogonal matrix $M \in O(3)$ for a further basis change:

$$\mathbf{s}_1 = m_{11}\mathbf{s}_1' + m_{12}\mathbf{s}_2' + m_{13}\mathbf{s}_3',$$
$$\mathbf{s}_2 = m_{21}\mathbf{s}_1' + m_{22}\mathbf{s}_2' + m_{23}\mathbf{s}_3',$$
$$\mathbf{s}_3 = m_{31}\mathbf{s}_1' + m_{32}\mathbf{s}_2' + m_{33}\mathbf{s}_3'.$$

The three conditions that pairs of screws meet orthogally are the vanishing of
the off-diagonal elements of the symmetric matrix

$$\Upsilon_p = M \begin{pmatrix} q_{11}' & q_{12}' & q_{13}' \\ q_{12}' & q_{22}' & q_{23}' \\ q_{13}' & q_{23}' & q_{33}' \end{pmatrix} M^T$$

where $q_{ij}' = (\mathbf{s}_i')^T Q_p \mathbf{s}_j'$, and we can choose $p = 0$. Sylvester's theorem assures
us that we can always find an M that diagonalises Υ_0 as required. Finally, we
can translate the common point on the three axes to the origin and rotate the
axes into coincidence with the coordinate axes.

If all three pitches are different here, the isotropy group consists of the π-
radian rotations about the coordinate axes, $\mathbb{Z}_2 \times \mathbb{Z}_2$. Gibson and Hunt denote
this system IA_1, the subscript here referring to the projective type of its intersection with pitch quadrics. When two of the pitches p_a, p_b and p_c are the
same, the isotropy group of the system becomes $O(2)$ and the Gibson–Hunt
type is IA_2. It is also possible to have all three pitches the same, in which case
the isotropy group is $SO(3)$ and the Gibson–Hunt type IIA, since the system
now lies entirely within one of the pitch quadrics.

If the extra screw has infinite pitch, things are slightly different. We cannot
completely orthonormalise the three screws, since an infinite pitch screw has
$\mathbf{s}^T Q_p \mathbf{s} = 0$ for any p. The best we can do is to produce a set of generators that
satisfy

$$(\mathbf{s}_i')^T Q_\infty \mathbf{s}_j' = 0, \qquad 1 \le i < j \le 3$$

and

$$(\mathbf{s}_1')^T Q_\infty \mathbf{s}_1' = (\mathbf{s}_2')^T Q_\infty \mathbf{s}_2' = -2, \qquad (\mathbf{s}_3')^T Q_\infty \mathbf{s}_3' = 0.$$

To preserve these relations, any further change of basis can only have the form

$$\mathbf{s}_1 = m_{11}\mathbf{s}_1' + m_{12}\mathbf{s}_2' + m_{13}\mathbf{s}_3',$$
$$\mathbf{s}_2 = m_{21}\mathbf{s}_1' + m_{22}\mathbf{s}_2' + m_{23}\mathbf{s}_3',$$
$$\mathbf{s}_3 = \phantom{m_{11}\mathbf{s}_1' + m_{12}\mathbf{s}_2' +} \; m_{33}\mathbf{s}_3'.$$

where the 2×2 block

$$\tilde{M} = \begin{pmatrix} m_{11} & m_{12} \\ m_{21} & m_{22} \end{pmatrix}$$

is in $O(2)$, and $m_{33} \neq 0$ but m_{13} and m_{23} can be any real numbers. After such a transformation, the terms $q_{ij} = \mathbf{s}_i^T Q_p \mathbf{s}_j$ will be the ij-th element in the symmetric matrix

$$\Upsilon_p = M \begin{pmatrix} q'_{11} & q'_{12} & q'_{13} \\ q'_{12} & q'_{22} & q'_{23} \\ q'_{13} & q'_{23} & q'_{33} \end{pmatrix} M^T$$

where, as before, $q'_{ij} = (\mathbf{s}'_i)^T Q_p \mathbf{s}'_j$, that is, the values for the old set of generators. For definiteness, we will take $p = 0$. Now, if $q'_{13} = 0$ and $q'_{23} = 0$, then we can choose an $\tilde{M}$ to diagonalise the top left-hand corner of the matrix Υ_0. A final overall rigid motion will bring the common intersection of the two finite screws' axes to the origin and align the axes of the three screws along the coordinate axes. Thus, we have reduced these 3-systems to the normal form

$$\mathbf{s}_1 = \begin{pmatrix} \mathbf{i} \\ p_a \mathbf{i} \end{pmatrix}, \qquad \mathbf{s}_2 = \begin{pmatrix} \mathbf{j} \\ p_b \mathbf{j} \end{pmatrix}, \qquad \mathbf{s}_3 = \begin{pmatrix} \mathbf{0} \\ \mathbf{k} \end{pmatrix}.$$

When the two moduli p_a and p_b are different, the isotropy group of the system has Lie algebra $\mathbb{R}^2$. The discrete symmetries are given by rotations of π about the coordinate axes; hence, the isotropy group of these systems is $\mathbb{Z}_2 \times \mathbb{Z}_2 \times \mathbb{R}^2$. Gibson and Hunt denote systems like this IB$_3$. It is also possible for p_a and p_b to be equal. In this case, the isotropy group of the system is $E(2)$ and the Gibson–Hunt type IIB.

When one or both of q'_{13} or q'_{23} are non-zero we may pick $\tilde{M}$ and m_{33} so that $q_{13} = \mathbf{s}_1^T Q_0 \mathbf{s}_3 = 1$ and $q_{23} = \mathbf{s}_2^T Q_0 \mathbf{s}_3 = 0$. We may then choose m_{13} and m_{23} so that the top left-hand corner of the matrix Υ_0 is diagonal, and we can also make the two diagonal elements the same. A final overall rigid transformation brings the system to the form

$$\mathbf{s}_1 = \begin{pmatrix} \mathbf{i} \\ p_a \mathbf{i} \end{pmatrix}, \qquad \mathbf{s}_2 = \begin{pmatrix} \mathbf{j} \\ p_a \mathbf{j} \end{pmatrix}, \qquad \mathbf{s}_3 = \begin{pmatrix} \mathbf{0} \\ \mathbf{i} + p_b \mathbf{k} \end{pmatrix}.$$

Now, the isotropy group of this system is $\mathbb{Z}_2$, consisting of the single rotation of π about the y-axis. A rotation of π radians about the x-axis simply changes the sign of p_b; hence, two such systems that only differ by the sign of p_b are equivalent and determine the same orbit. The Gibson–Hunt type of these orbits is IB$_0$.

The results are summarised as Table 8.2. Notice that we have exactly reproduced the Gibson–Hunt classification. The distinction made between orbits with different completion groups is only a slight ornamentation; they were separated by the Gibson–Hunt classification anyway. However, the concept of the completion group of a screw system is useful in its own right. Consider a serial robot arm as we did at the beginning of this chapter. How do the screw systems defined by the joints of the robot change as the robot moves? A detailed description can probably be given only for specific examples. However, we can find at least one general result. For two joints, clearly the Gibson–Hunt type is invariant since the joints are rigidly attached to each other. For three and more joints, it is not hard to see that the Gibson–Hunt type of the system can change as the robot moves. The new positions of the joints will be related to the original ones by the adjoint representation; hence, the new joint screws will lie in the Lie algebra of the completion group of the original screws. Hence, the Lie algebra generated by the joint screws in their new positions will be a subalgebra of the Lie algebra of the original completion group. Reversing the movement of the robot from the new position back to the original one, we see that the original joint screws must lie in the Lie algebra generated by the new ones. Hence, the completion groups at the two positions must be the same; that is, the completion group is an invariant of the robot. This observation is perhaps most relevant to three joint-structures where there are several different possibilities for completion groups.

The type I systems have a strong connection with the reguli that we met in Section 6.5.1. Since any type I system will meet the Klein quadric in a conic, we see that the lines of a I system comprise a regulus. Using the results of Section 6.5.1, we can find the equation for the quadric in $\mathbb{R}^3$ that the regulus lies on. For example, the normal form for the type IA_1 system is defined by the linear equations

$$\begin{pmatrix} p_a & 0 & 0 \\ 0 & p_b & 0 \\ 0 & 0 & p_c \end{pmatrix} \begin{pmatrix} \omega_x \\ \omega_y \\ \omega_z \end{pmatrix} + \begin{pmatrix} -1 & 0 & 0 \\ 0 & -1 & 0 \\ 0 & 0 & -1 \end{pmatrix} \begin{pmatrix} v_x \\ v_y \\ v_z \end{pmatrix} = \begin{pmatrix} 0 \\ 0 \\ 0 \end{pmatrix}.$$

Writing $\mathbf{v} = \mathbf{x} \times \boldsymbol{\omega}$, we obtain a set of three homogeneous linear equations in the components of $\boldsymbol{\omega}$. The equation of the quadric is given by the condition for these linear equations to have non-trivial solutions

$$\det \left(\begin{pmatrix} p_a & 0 & 0 \\ 0 & p_b & 0 \\ 0 & 0 & p_c \end{pmatrix} + \begin{pmatrix} 0 & -z & y \\ z & 0 & -x \\ -y & x & 0 \end{pmatrix} \right) = 0$$

which simplifies to

$$p_a x^2 + p_b y^2 + p_c z^2 + p_a p_b p_c = 0.$$

Notice that if two of the moduli are the same, that is, for a type IA_2 system, the quadric is cylindrical. The lines in a IB_0 system lie on a paraboloid with equation

$$p_b z^2 + xz + p_a y + p_a^2 p_b = 0$$

and the lines of a IB_3 system lie on a pair of parallel planes $z^2 - p_a p_b = 0$.

TABLE 8.2. The 3-Systems

Completion Group	Gibson–Hunt Type	Normal Form	Isotropy Group
$\mathbb{R}^3$	IID	$\mathbf{s}_1 = (0,0,0,1,0,0)^T$ $\mathbf{s}_2 = (0,0,0,0,1,0)^T$ $\mathbf{s}_3 = (0,0,0,0,0,1)^T$	$SE(3)$
$SO(3)$	IIA $(p=0)$	$\mathbf{s}_1 = (1,0,0,0,0,0)^T$ $\mathbf{s}_2 = (0,1,0,0,0,0)^T$ $\mathbf{s}_3 = (0,0,1,0,0,0)^T$	$SO(3)$
$SE(2)$	IIC $(p=0)$	$\mathbf{s}_1 = (1,0,0,0,0,0)^T$ $\mathbf{s}_2 = (0,0,0,0,1,0)^T$ $\mathbf{s}_3 = (0,0,0,0,0,1)^T$	$E(2) \times \mathbb{R}$
$H_p \ltimes \mathbb{R}^2$	IIC $(p \neq 0)$	$\mathbf{s}_1 = (1,0,0,p,0,0)^T$ $\mathbf{s}_2 = (0,0,0,0,1,0)^T$ $\mathbf{s}_3 = (0,0,0,0,0,1)^T$	$E(2) \times \mathbb{R}$
$SE(2) \times \mathbb{R}$	IC $(p \neq 0)$	$\mathbf{s}_1 = (1,0,0,0,0,0)^T$ $\mathbf{s}_2 = (0,0,0,0,1,0)^T$ $\mathbf{s}_3 = (0,0,0,1,0,p)^T$	$\mathbb{R}^2$
$SE(2) \times \mathbb{R}$	IC^0	$\mathbf{s}_1 = (1,0,0,0,0,0)^T$ $\mathbf{s}_2 = (0,0,0,0,1,0)^T$ $\mathbf{s}_3 = (0,0,0,1,0,0)^T$	$\mathbb{Z}_2 \times \mathbb{R}^2$
$SE(3)$	IIB	$\mathbf{s}_1 = (1,0,0,p,0,0)^T$ $\mathbf{s}_2 = (0,1,0,0,p,0)^T$ $\mathbf{s}_3 = (0,0,0,0,0,1)^T$	$E(2)$
$SE(3)$	IB$_0$ $(p_a \neq p_b)$	$\mathbf{s}_1 = (1,0,0,p_a,0,0)^T$ $\mathbf{s}_2 = (0,1,0,0,p_a,0)^T$ $\mathbf{s}_3 = (0,0,0,1,0,p_b)^T$	$\mathbb{Z}_2$
$SE(3)$	IB$_3$ $(p_a \neq p_b)$	$\mathbf{s}_1 = (1,0,0,p_a,0,0)^T$ $\mathbf{s}_2 = (0,1,0,0,p_b,0)^T$ $\mathbf{s}_3 = (0,0,0,0,0,1)^T$	$\mathbb{Z}_2 \times \mathbb{Z}_2 \times \mathbb{R}^2$
$SE(3)$	IIA $(p \neq 0)$	$\mathbf{s}_1 = (1,0,0,p,0,0)^T$ $\mathbf{s}_2 = (0,1,0,0,p,0)^T$ $\mathbf{s}_3 = (0,0,1,0,0,p)^T$	$SO(3)$
$SE(3)$	IA$_2$ $(p_a \neq p_b)$	$\mathbf{s}_1 = (1,0,0,p_a,0,0)^T$ $\mathbf{s}_2 = (0,1,0,0,p_b,0)^T$ $\mathbf{s}_3 = (0,0,1,0,0,p_b)^T$	$O(2)$
$SE(3)$	IA$_1$, $(p_a \neq p_b \neq p_c \neq p_a)$	$\mathbf{s}_1 = (1,0,0,p_a,0,0)^T$ $\mathbf{s}_2 = (0,1,0,0,p_b,0)^T$ $\mathbf{s}_3 = (0,0,1,0,0,p_c)^T$	$\mathbb{Z}_2 \times \mathbb{Z}_2$

In Section 6.3 we met the α-planes and β-planes that lie in the Klein quadric. Clearly these are type II systems, as they lie entirely in one of the pitch quadrics. Recall that they satisfy the equations

$$(I_3 - M)\boldsymbol{\omega} + (I_3 + M)\mathbf{v} = \mathbf{0}$$

where $M \in O(3)$ and $\det(M) = +1$ for an α-plane and $\det(M) = -1$ for a β-plane. In terms of the Gibson–Hunt classification, the α-planes are given by the IIA ($p = 0$) and IIC ($p = 0$) systems. For the normal forms the matrix M is given by

$$M = \begin{pmatrix} 1 & 0 & 0 \\ 0 & 1 & 0 \\ 0 & 0 & 1 \end{pmatrix}$$

for the IIA ($p = 0$) system and

$$M = \begin{pmatrix} 1 & 0 & 0 \\ 0 & -1 & 0 \\ 0 & 0 & -1 \end{pmatrix}$$

for the IIC ($p = 0$) system. Remember that in general, an α-plane consisted of the set of lines through a point in $\mathbb{R}^3$. Clearly these are the IIA ($p = 0$) systems. The other α-planes can be thought of as lines through 'points at infinity'; these are the IIC ($p = 0$) systems. If we specify an α-plane by giving a special orthogonal matrix M, then we can distinguish the two cases by looking at the rank of the matrix $I_3 + M$. When $\text{Rank}(I_3 + M) = 3$, then the α-plane does not intersect the infinite pitch quadric. The only other possibility is $\text{Rank}(I_3 + M) = 1$, in which case the α-plane intersects the infinite pitch quadric in a line, and so the system is of type IIC ($p = 0$).

The β-planes correspond to the IIB ($p = 0$) systems and the IID system. For the normal forms of these systems, we have

$$M = \begin{pmatrix} 1 & 0 & 0 \\ 0 & 1 & 0 \\ 0 & 0 & -1 \end{pmatrix}$$

for the IIB ($p = 0$) system and

$$M = \begin{pmatrix} -1 & 0 & 0 \\ 0 & -1 & 0 \\ 0 & 0 & -1 \end{pmatrix},$$

for the IID system. Again, we have an interpretation in terms of lines in $\mathbb{R}^3$: the IIB ($p = 0$) systems are systems of lines lying in a plane. The IID system can be thought of as the system of lines lying in the 'plane at infinity'. If we are specifying the β-plane by an orthogonal matrix M with determinant -1, then if $M \neq -I_3$ the plane has type IIB $(p = 0)$. Notice that $\text{Rank}(I_3 + M) = 2$ for the IIB $(p = 0)$ systems and $\text{Rank}(I_3 + M) = 0$ for the IID system.

The reciprocal of a 3-system is also a 3-system. Moreover, the isotropy group of any 3-system is also the isotropy group of its reciprocal system. This can be seen by considering the relation

$$\overline{\wedge^3 Ad(g)\Delta} = \wedge^3 Ad(g)\overline{\Delta};$$

see Section 8.1. This means that the Gibson–Hunt type of a 3-system is unchanged if we take the reciprocal. The moduli of a 3-system, however, may change if we take the reciprocal systems. For example, taking the normal form of the IA_1, a short calculation reveals that the reciprocal system has moduli $-p_a$, $-p_b$ and $-p_c$. The lines of the reciprocal system lie on the same quadric as those from the original system. This change of sign in the moduli also applies to the other I systems. Hence, the relationship between the two reguli on any quadric in $\mathbb{PR}^3$ is that they are determined by the intersection of the Klein quadric with a pair of reciprocal 3-planes in $\mathbb{PR}^5$.

8.4   Identification of Screw Systems

Given a number of linearly independent screws, how can we find the Gibson–Hunt type of the n-system generated by the screws? The classification scheme above gives an implicit algorithm that involves computing isotropy groups and completion groups, and transforming to normal form. Although relatively straightforward, this method is computationally rather intensive, since it involves finding and solving several linear equations, among other things. Our original question should really be sharpened to, How can we find the Gibson–Hunt type of an n-system efficiently?

One way of finding the type of a screw system would be to have a finite list of polynomial invariants that we could compute easily and that would distinguish the different cases. We could try to find these invariants in a systematic way by studying the ring of invariants for the representations $\wedge^n Ad$; see Donelan and Gibson [28]. Here, in the interest of brevity, an ad hoc approach is taken.

8.4.1   1-systems and 5-systems

By the remarks in Section 8.1, the identification of 1-systems amounts to finding the pitch of the generator.

Formally we have two invariants, if $\mathbf{s}$ is the generator of the 1-system, then we have the two quantities

$$\mathbf{s}^T Q_0 \mathbf{s}, \quad \text{and} \quad \mathbf{s}^T Q_\infty \mathbf{s}.$$

These are only relative invariants, they are unaffected by rigid transformations but under a scaling, $\mathbf{s} \longrightarrow \lambda \mathbf{s}$, they are both multiplied by λ^2. Hence the ratio gives a single absolute invariant,

$$p = -\mathbf{s}^T Q_0 \mathbf{s} / \mathbf{s}^T Q_\infty \mathbf{s}.$$

This is the pitch of the system; if $\mathbf{s}^T Q_\infty \mathbf{s} = 0$, then the pitch is said to be infinite.

According to Section 8.1, the classification of 5-systems proceeds via their reciprocal 1-systems. That is, the modulus of a 5-system is simply the pitch of the reciprocal 1-system. But how can we find that pitch? Suppose the 5-system is generated by the five independent screws, $\mathbf{s}_1$, $\mathbf{s}_2$, $\mathbf{s}_3$, $\mathbf{s}_4$ and $\mathbf{s}_5$. Then the anti-symmetric product of these is an element of the coadjoint representation of $se(3)$:

$$\mathcal{W} = \mathbf{s}_1 \wedge \mathbf{s}_2 \wedge \mathbf{s}_3 \wedge \mathbf{s}_4 \wedge \mathbf{s}_5.$$

Notice that to compute the coefficients of $\mathcal{W}$ we have to evaluate six 5×5 determinants. To turn $\mathcal{W}$ into a screw, that is, an element of the adjoint representation, we must multiply by an element of $ad \otimes ad$. Using the element Q_0, that is, $\mathbf{s} = Q_0 \mathcal{W}$ gives a screw reciprocal to the 5-system since

$$\mathbf{s}_i^T Q_0 \mathbf{s} = \mathbf{s}_i^T \mathcal{W} = \mathbf{s}_i \wedge \mathbf{s}_1 \wedge \mathbf{s}_2 \wedge \mathbf{s}_3 \wedge \mathbf{s}_4 \wedge \mathbf{s}_5 = 0, \qquad i = 1, 2 \ldots, 5.$$

Alternatively, we can perform our computations in the coadjoint representation, since

$$\mathbf{s}^T Q_0 \mathbf{s} = \mathcal{W}^T Q_0 Q_0 Q_0 \mathcal{W} \qquad \text{and} \qquad \mathbf{s}^T Q_\infty \mathbf{s} = \mathcal{W}^T Q_0 Q_\infty Q_0 \mathcal{W}.$$

Hence, the pitch of a 5-system is given by

$$p = \frac{\mathcal{W}^T Q_0^* \mathcal{W}}{\mathcal{W}^T Q_\infty^* \mathcal{W}}$$

with

$$Q_0^* = Q_0 Q_0 Q_0 = \begin{pmatrix} 0 & I_3 \\ I_3 & 0 \end{pmatrix}, \qquad \text{and} \qquad Q_\infty^* = Q_0 Q_\infty Q_0 = \begin{pmatrix} 0 & 0 \\ 0 & -2I_3 \end{pmatrix}.$$

A computationally more efficient procedure would be to find $\mathbf{s}$, up to a constant scale factor, by solving the five homogeneous linear equations, $\mathbf{s}_i^T Q_0 \mathbf{s} = 0$.

8.4.2   2-systems

Here we can look for the invariants of the pencil of quadratics, $\alpha \Upsilon_0 + \beta \Upsilon_\infty$. That is,

$$\det(\alpha \Upsilon_0 + \beta \Upsilon_\infty) = \alpha^2 i_1 + \alpha\beta i_2 + \beta^2 i_3$$

Clearly the functions i_1, i_2 and i_3 are invariant with respect to rigid motions since the components of the matrices are, but how do they transform under a change of basis in the pencil? If the new basis of the pencil is

$$\mathbf{s}_1' = m_{11}\mathbf{s}_1 + m_{12}\mathbf{s}_2, \qquad \mathbf{s}_2' = m_{21}\mathbf{s}_1 + m_{22}\mathbf{s}_2,$$

then it is not difficult to see that the new invariants will be given by, $i'_n = \det(M)^2 i_n$, where M is the matrix with elements m_{ij}. Hence these functions are not absolute invariants of the screw system but only relative invariants. We can form absolute invariants from ratios of these functions. In detail, we have

$$i_1 = \det \begin{pmatrix} \mathbf{s}_1^T Q_0 \mathbf{s}_1 & \mathbf{s}_1^T Q_0 \mathbf{s}_2 \\ \mathbf{s}_2^T Q_0 \mathbf{s}_1 & \mathbf{s}_2^T Q_0 \mathbf{s}_2 \end{pmatrix},$$

$$i_2 = \det \begin{pmatrix} \mathbf{s}_1^T Q_0 \mathbf{s}_1 & \mathbf{s}_1^T Q_\infty \mathbf{s}_2 \\ \mathbf{s}_2^T Q_0 \mathbf{s}_1 & \mathbf{s}_2^T Q_\infty \mathbf{s}_2 \end{pmatrix} + \det \begin{pmatrix} \mathbf{s}_1^T Q_\infty \mathbf{s}_1 & \mathbf{s}_1^T Q_0 \mathbf{s}_2 \\ \mathbf{s}_2^T Q_\infty \mathbf{s}_1 & \mathbf{s}_2^T Q_0 \mathbf{s}_2 \end{pmatrix},$$

$$i_3 = \det \begin{pmatrix} \mathbf{s}_1^T Q_\infty \mathbf{s}_1 & \mathbf{s}_1^T Q_\infty \mathbf{s}_2 \\ \mathbf{s}_2^T Q_\infty \mathbf{s}_1 & \mathbf{s}_2^T Q_\infty \mathbf{s}_2 \end{pmatrix}.$$

Notice also that we have

$$i_2 = -2[\mathbf{s}_1, \mathbf{s}_2]^T Q_0 [\mathbf{s}_1, \mathbf{s}_2] \quad \text{and} \quad i_3 = -2[\mathbf{s}_1, \mathbf{s}_2]^T Q_\infty [\mathbf{s}_1, \mathbf{s}_2],$$

which can be verified by direct computation, or by using the partitioned form for $\mathbf{s}_1$ and $\mathbf{s}_2$ and then simplifying using the formula for the vector triple product. So, for example, if we write

$$\mathbf{s}_1 = \begin{pmatrix} \boldsymbol{\omega}_1 \\ \mathbf{v}_1 \end{pmatrix} \quad \text{and} \quad \mathbf{s}_2 = \begin{pmatrix} \boldsymbol{\omega}_2 \\ \mathbf{v}_2 \end{pmatrix},$$

then

$$[\mathbf{s}_1, \mathbf{s}_2]^T Q_\infty [\mathbf{s}_1, \mathbf{s}_2] = -2(\boldsymbol{\omega}_1 \times \boldsymbol{\omega}_2) \cdot (\boldsymbol{\omega}_1 \times \boldsymbol{\omega}_2)$$

and

$$i_3 = 4 \det \begin{pmatrix} (\boldsymbol{\omega}_1 \cdot \boldsymbol{\omega}_1) & (\boldsymbol{\omega}_1 \cdot \boldsymbol{\omega}_2) \\ (\boldsymbol{\omega}_2 \cdot \boldsymbol{\omega}_1) & (\boldsymbol{\omega}_2 \cdot \boldsymbol{\omega}_2) \end{pmatrix} = 4\{(\boldsymbol{\omega}_1 \cdot \boldsymbol{\omega}_1)(\boldsymbol{\omega}_2 \cdot \boldsymbol{\omega}_2) - (\boldsymbol{\omega}_1 \cdot \boldsymbol{\omega}_2)(\boldsymbol{\omega}_2 \cdot \boldsymbol{\omega}_1)\}.$$

This does not give us enough invariants to separate out the different cases. To produce more invariants consider the following construction. Form the 6×6 antisymmetric matrix

$$A = \mathbf{s}_1 \mathbf{s}_2^T - \mathbf{s}_2 \mathbf{s}_1^T.$$

Now the coefficients of the polynomial

$$\det(A - \lambda Q_0 - \mu Q_\infty)$$

are again clearly invariants under rigid motions. Under a change of basis in the pencil it is simple to calculate that

$$A' = \det(M)^2 A.$$

Expanding the determinant gives

$$\det(A - \lambda Q_0 - \mu Q_\infty) = -\lambda^6 - i_1 \lambda^4 - i_4 \lambda^3 \mu - i_5 \lambda^2 \mu^2.$$

TABLE 8.3. Invariants of 2-Systems

Gibson–Hunt Type	i_1	i_2	i_3	i_4	i_5
IIC	0	0	0	0	4
IB^0	-1	0	0	0	0
IIA $(p = 0)$	0	0	4	0	0
IIB $(p \neq 0)$	0	0	0	$4p$	$4p^2$
IIB $(p = 0)$	0	0	0	0	0
IB $(p \neq 0)$	-1	0	0	0	0
IIA $(p \neq 0)$	$4p^2$	$-8p$	4	$8p^3$	$4p^4$
IA $(p_a \neq p_b)$	$4p_a p_b$	$-4(p_a + p_b)$	4	$4(p_a^2 p_b + p_b^2 p_a)$	$4p_a^2 p_b^2$

Notice we only get two new invariants, which could also be written as determinants but now the formulas are rather more cumbersome.

Once again we can see that i_4 and i_5 are relative invariants that transform according to $i'_n = \det(M)^2 i_n$ under a change of basis in the pencil.

If we evaluate these invariants on the normal forms for the 2-systems that we derived above we obtain the results found in Table 8.3.

These five invariants are almost enough to distinguish between the different cases. Certainly the vanishing of any of the invariants is independent of the basis chosen for the screw system. Also, although the exact numerical value of the invariants may change under a change of basis, the fact that an invariant is non-zero is significant. For example, if all the invariants are non-zero, then the 2-system must be of type IIA $(p \neq 0)$ or IA $(p_a \neq p_b)$. Now we look at the quadratic equation

$$i_3 x^2 + i_2 x + i_1 = 0.$$

In the case IA $(p_a \neq p_b)$, the roots of this equation are the moduli $x = p_a, p_b$. But if this quadratic is a perfect square, so that the discriminant vanishes, $i_2^2 - 4i_1 i_3 = 0$ then we have the IIA $(p \neq 0)$ case and the modulus is given by $p = -i_2/2i_3$.

The only difficulty with the invariants given above would be to distinguish between the IB^0 and IB $(p \neq 0)$ 2-systems. A simple way to do this would be to examine the Lie bracket, $[\mathbf{s}_1, \mathbf{s}_2]$; this is a covariant of the screw system, under a rigid transformation it transforms according to the adjoint representation of

$SE(3)$. Under a change of basis in the screw system it is simple to calculate that

$$[s_1', s_2'] = \det(M)[s_1, s_2].$$

On the normal forms we have

$$[s_1, s_2] = \begin{pmatrix} 0 \\ 0 \\ 0 \\ 0 \\ 0 \\ p \end{pmatrix}$$

for the IB $(p \neq 0)$ system and $[s_1, s_2] = \mathbf{0}$ for the IB^0 system. Hence, the IB^0 system can be distinguished by the vanishing of the Lie Bracket of its basis screws. Finally, we need to recover the modulus from the IB $(p \neq 0)$ system. To do this we can take

$$i_\infty = [s_1, s_2]^T \begin{pmatrix} 0 & 0 \\ 0 & -2I_3 \end{pmatrix} [s_1, s_2].$$

The symmetric matrix here is simply Q^*_∞ that we met in the previous section. This expression would not normally be an invariant, since it is only invariant under rotations but not translations. However in this case we know that $[s_1, s_2]$ has infinite pitch and for infinite pitch screws this function is clearly invariant, so at last we have that

$$p^2 = i_\infty/2i_1.$$

As an example, let's look at the screw system determined by a pair of lines, a fairly common situation in robotics:

$$s_1 = \begin{pmatrix} \mathbf{i} \\ \mathbf{0} \end{pmatrix} \quad \text{and} \quad s_2 = \begin{pmatrix} \mathbf{i} \cos \phi + \mathbf{j} \sin \phi \\ \mathbf{j} l \cos \phi - \mathbf{i} l \sin \phi \end{pmatrix}.$$

That is, one line is along the x-axis while the other is at an angle ϕ to the first and displaced a distance l along the z-axis. The reduced pitch quadrics for this system are

$$\Upsilon_0 = \begin{pmatrix} 0 & -l \sin \phi \\ -l \sin \phi & 0 \end{pmatrix} \quad \text{and} \quad \Upsilon_\infty = \begin{pmatrix} -2 & -2 \cos \phi \\ -2 \cos \phi & -2 \end{pmatrix}.$$

The first three invariants are therefore

$$i_1 = -l^2 \sin^2 \phi, \quad i_2 = -4l \sin \phi \cos \phi, \quad i_3 = 4 \sin^2 \phi.$$

So in general this is a type IA $(p_a \neq p_b)$ system with moduli

$$p_a, p_b = l \left(\frac{\cos \phi \pm 1}{2 \sin \phi} \right) = \frac{l}{2} \tan \frac{\phi}{2}, \frac{-l}{2} \cot \frac{\phi}{2}.$$

When $l = 0$ the lines meet and $i_1 = i_2 = 0$ but $i_3 \neq 0$, so in this case we have a IIA $(p = 0)$ system. On the other hand, if $l \neq 0$ but $\phi = 0, \pi$, that is the lines are parallel or anti-parallel, then the first three invariants all vanish. The other invariants also vanish $i_4 = i_5 = 0$. So referring to Table 8.3 again, we see that this is a IIB $(p = 0)$ system.

8.4.3   4-systems

Let the 4-system be generated by the four independent screws, s_1, s_2, s_3 and s_4. Now to find the reciprocal 2-system we could solve the four homogeneous linear equations $s_i^T Q_0 s = 0$ with $i = 1, 2, 3, 4$.

A better approach may be to find the Plücker coordinates of the 2-system directly. Suppose that the screws

$$s_a = \begin{pmatrix} x_1 \\ x_2 \\ x_3 \\ x_4 \\ x_5 \\ x_6 \end{pmatrix}, \qquad s_b = \begin{pmatrix} y_1 \\ y_2 \\ y_3 \\ y_4 \\ y_5 \\ y_6 \end{pmatrix}$$

form a basis for the reciprocal 2-system, then the 15 Plücker coordinates of this 2-system have the form

$$p_{ij} = x_i y_j - x_j y_i.$$

When studying 2-systems above, a 6×6 anti-symmetric matrix was introduced. Now the elements of this matrix are simply the Plücker coordinates of the 2-system

$$A = s_a s_b^T - s_b s_a^T = \begin{pmatrix} 0 & p_{12} & p_{13} & \cdots & p_{16} \\ -p_{12} & 0 & p_{23} & \cdots & p_{26} \\ \vdots & \vdots & \ddots & \vdots & \vdots \\ -p_{16} & -p_{26} & -p_{36} & \cdots & 0 \end{pmatrix}.$$

So it is easy to see that the Plücker coordinates of the reciprocal 2-system must satisfy the linear equations

$$AQ_0 s_i = \mathbf{0}, \qquad i = 1, 2, 3, 4.$$

However, the coordinates also have to satisfy the Plücker relations, see Section 6.8. There will be fifteen relations of the form

$$p_{12}p_{34} - p_{13}p_{24} + p_{14}p_{23} = 0;$$

not all of these relations are independent so it will be possible to find the Plücker coordinates of the 2-system reciprocal to the given 4-system.

Having found the Plücker coordinates of the 2-system we next compute the invariants. It should be possible to express these invariants directly in terms of the Plücker coordinates. Indeed we may derive the results:

$$i_1 = -(p_{14}^2 + p_{25}^2 + p_{36}^2)$$
$$\quad + 2(p_{12}p_{45} + p_{13}p_{46} + p_{23}p_{56} - p_{15}p_{24} - p_{16}p_{34} - p_{26}p_{35}),$$
$$i_2 = 4(p_{23}(p_{35} - p_{26}) + p_{13}(p_{34} - p_{16}) + p_{12}(p_{24} - p_{15})),$$
$$i_3 = 4(p_{12}^2 + p_{23}^2 + p_{13}^2),$$
$$i_4 = 4((p_{15} - p_{24})p_{45} + (p_{16} - p_{34})p_{46} + (p_{26} - p_{35})p_{56}),$$
$$i_5 = 4(p_{45}^2 + p_{46}^2 + p_{56}^2),$$
$$i_\infty = -((p_{15} - p_{24})^2 + (p_{26} - p_{35})^2 + (p_{34} - p_{16})^2).$$

These results can be found quite simply as follows. For i_1, i_4 and i_5 we can expand the determinant

$$\det(A - \lambda Q_0 - \mu Q_\infty) = -\lambda^6 - i_1\lambda^4 - i_4\lambda^3\mu - i_5\lambda^2\mu^2$$

where A is written in terms of Plücker coordinates as above. Notice that this procedure will produce more terms than just the invariants. However these extra terms must vanish for 2-systems, that is when the coordinates satisfy the Plücker relations.

To find the other invariants we can use the easily verifiable relation

$$[s_a, s_b] = \begin{pmatrix} p_{23} \\ -p_{13} \\ p_{12} \\ p_{26} - p_{35} \\ p_{34} - p_{16} \\ p_{15} - p_{24} \end{pmatrix}$$

and substitute into the expressions for the invariants in terms of the commutator of the basis elements.

8.4.4 3-systems

Next, we turn to the problem of finding the type of a 3-system given a triple of generators. Here fewer invariants are known—although presumably there are several more to be found.

As with the 2-systems, we begin by forming the two symmetric matrices that represent restriction of the pitch quadrics to the 3-system. These are now 3×3 matrices, though.

$$\Upsilon_\infty = \begin{pmatrix} s_1^T Q_\infty s_1 & s_1^T Q_\infty s_2 & s_1^T Q_\infty s_3 \\ s_2^T Q_\infty s_1 & s_2^T Q_\infty s_2 & s_2^T Q_\infty s_3 \\ s_3^T Q_\infty s_1 & s_3^T Q_\infty s_2 & s_3^T Q_\infty s_3 \end{pmatrix}, \quad \Upsilon_0 = \begin{pmatrix} s_1^T Q_0 s_1 & s_1^T Q_0 s_2 & s_1^T Q_0 s_3 \\ s_2^T Q_0 s_1 & s_2^T Q_0 s_2 & s_2^T Q_0 s_3 \\ s_3^T Q_0 s_1 & s_3^T Q_0 s_2 & s_3^T Q_0 s_3 \end{pmatrix}.$$

Now consider the determinant

$$\det(\alpha \Upsilon_\infty + \beta \Upsilon_0) = \alpha^3 \det(\Upsilon_\infty) + \alpha^2 \beta \Theta + \alpha \beta^2 \Phi + \beta^3 \det(\Upsilon_0).$$

The coefficients of the monomials in α and β are clearly invariants. Moreover, as in the case of the 2-systems, a linear change of basis in the system simply multiplies the determinant by $\det(M)^2$ where M is the general linear matrix representing the basis change. That is,

$$\det(\Upsilon_p') = (\det(M))^2 \det(\Upsilon_p).$$

For type I systems the plane determined by the system intersects the pitch quadrics in a pair of conic curves. Hence these systems determine a pencil of conics. The invariants above are exactly the classical invariants for pencils of conics, see Salmon [96, sect. 370].

We have one more obvious invariant, $i_0 = s_1^T Q_0 [s_2, s_3]$. It is not hard to see that this is invariant with respect to cyclic interchanges of the three screws. Only a little more work is required to show that under a change of basis in the system, we have

$$(s_1')^T Q_0 [s_2', s_3'] = \det(M) s_1^T Q_0 [s_2, s_3]$$

where M is the matrix of the general linear transformation.

The quantity $s_1^T Q_\infty [s_2, s_3]$ is also an invariant, but we have already seen it since

$$\det(\Upsilon_\infty) = -2 \big(s_1^T Q_\infty [s_2, s_3] \big)^2.$$

So we have five invariants, the result of evaluating these invariants on the normal forms for the 3-systems is given in Table 8.4.

Notice that the five invariants can distinguish between A-systems and can distinguish B-systems from C- and D-systems.

For the A-systems, if the invariants all vanish except, $\det(\Upsilon_\infty)$ then the 3-system in a IIA ($p = 0$) system. Otherwise we set up the cubic equation

$$\det(\Upsilon_\infty)x^3 + \Theta x^2 + \Phi x + \det(\Upsilon_0) = 0.$$

If this cubic has three distinct solutions, then we have a IA$_1$ systems and the roots of the cubic are the moduli p_a, p_b and p_c. When the cubic has a repeated root we have a IA$_2$ system again with the two roots of the cubic as moduli. Lastly, if the equation is a perfect cube we have a IIA ($p \neq 0$) system with the moduli p as the single root of the equation.

If $i_0 = s_1^T Q_0 [s_2, s_3]$ is non-zero, then we have a B-system. The IB$_0$ systems are easy to distinguish by the non-vanishing of Φ. The moduli, in this case, are given by

$$p_a = -\det(\Upsilon_0)/\Phi, \qquad \text{and} \qquad p_b^2 = 2i_0^2/\Phi.$$

For the IB$_3$ system consider the reduced pitch quadrics for the normal form

$$\alpha \Upsilon_\infty + \beta \Upsilon_0 = \begin{pmatrix} 2\beta p_a - 2\alpha & 0 & 0 \\ 0 & 2\beta p_b - 2\alpha & 0 \\ 0 & 0 & 0 \end{pmatrix}.$$

TABLE 8.4. Invariants of 3-Systems

Gibson–Hunt Type	$\det(\Upsilon_0)$	Φ	Θ	$\det(\Upsilon_\infty)$	i_0
IID	0	0	0	0	0
IIC $(p = 0)$	0	0	0	0	0
IIC $(p \neq 0)$	0	0	0	0	0
IC $(p \neq 0)$	0	2	0	0	0
IC^0	0	2	0	0	0
IIB	0	0	0	0	1
IB$_3$ $(p_a \neq p_b)$	0	0	0	0	1
IB$_0$ $(p_a \neq p_b)$	$-2p_a$	2	0	0	p_b
IIA $(p = 0)$	0	0	0	-8	0
IIA $(p \neq 0)$	$8p^3$	$-24p^2$	$24p$	-8	$3p$
IA$_2$ $(p_a \neq p_b)$	$8p_a p_b^2$	$-16p_a p_b - 8p_b^2$	$8p_a + 16p_b$	-8	$p_a + 2p_b$
IA$_1$, $(p_a \neq p_b \neq p_c \neq p_a)$	$8p_a p_b p_c$	$-8(p_a p_b + p_b p_c + p_c p_a)$	$8(p_a + p_b + p_c)$	-8	$p_a + p_b + p_c$

In general this matrix has rank 2, but this drops to rank 1 precisely when $\alpha/\beta = p_a$ or p_b. So we can find the moduli by solving

$$\mathrm{Rank}(\alpha \Upsilon_\infty + \beta \Upsilon_0) = 1$$

for α/β. Moreover, when $p_a = p_b$ we have a IIB system.

For the IIC systems we can perform a similar calculation. Now $\alpha \Upsilon_\infty + \beta \Upsilon_0$ normally has rank 1, so we can find the value of α/β which makes this matrix vanish. This value $p = \alpha/\beta$ will be the modulus of the IIC system and if it is 0, then we have a IIC $(p=0)$ system.

The IC systems are easy to spot since the only non-zero invariant we have is Φ. However there doesn't seem to be a simple way to compute the modulus. The best we can do here seems to be to convert the system to normal form. By a suitable change of basis we can find two infinite pitch generators; in partitioned form the generators will be

$$\mathbf{s}_1 = \begin{pmatrix} \boldsymbol{\omega}_1 \\ \mathbf{v}_1 \end{pmatrix}, \quad \mathbf{s}_2 = \begin{pmatrix} \mathbf{0} \\ \mathbf{v}_2 \end{pmatrix} \quad \text{and} \quad \mathbf{s}_3 = \begin{pmatrix} \mathbf{0} \\ \mathbf{v}_3 \end{pmatrix}.$$

We also require that the vectors $\mathbf{v}_2$ and $\mathbf{v}_3$ are orthogonal. Now for the normal form we have that $\boldsymbol{\omega}_1 \cdot (\mathbf{v}_2 \times \mathbf{v}_3)$ is proportional to p, so if this quantity vanishes then we must have an IC^0 system.

When this quantity is non-zero we must take account of the magnitudes of these vectors

$$\frac{\boldsymbol{\omega}_1 \cdot (\mathbf{v}_2 \times \mathbf{v}_3)}{|\boldsymbol{\omega}_1||\mathbf{v}_2||\mathbf{v}_3|} = \pm p\sqrt{1 + p^2}.$$

The square root factor here is because in the normal form, $|\mathbf{v}_3| = \sqrt{1 + p^2}$. The above is essentially a quadratic in p^2. The solution will always have a positive and a negative root. The negative root should be discarded to avoid an imaginary pitch.

For the IID system we don't need to compute the invariants since both Υ_0 and Υ_∞ both vanish.

In particular cases it may be simpler not to compute all these invariants. For example consider the following. Let $\mathbf{s}_a$ be an arbitrary screw and $\mathbf{s}_b$ a pitch zero screw about an axis perpendicular to $\mathbf{s}_a$. Now form the 3-system with generators

$$\mathbf{s}_1 = \mathbf{s}_a,$$
$$\mathbf{s}_2 = [\mathbf{s}_b, \mathbf{s}_a],$$
$$\mathbf{s}_3 = [\mathbf{s}_b, [\mathbf{s}_b, \mathbf{s}_a]].$$

Although this arrangement looks rather contrived at present, we will meet it again in Section 13.6.1. Without loss of generality, we can take

$$\mathbf{s}_a = \begin{pmatrix} \mathbf{i} \\ p\mathbf{i} \end{pmatrix} \quad \text{and} \quad \mathbf{s}_b = \begin{pmatrix} \mathbf{j} \\ -l\mathbf{i} \end{pmatrix}$$

so that the generators are

$$\mathbf{s}_1 = \begin{pmatrix} \mathbf{i} \\ p\mathbf{i} \end{pmatrix}, \quad \mathbf{s}_2 = \begin{pmatrix} -\mathbf{k} \\ -p\mathbf{k} \end{pmatrix}, \quad \mathbf{s}_3 = \begin{pmatrix} -\mathbf{i} \\ -p\mathbf{i} - l\mathbf{j} \end{pmatrix}.$$

The restricted pitch quadrics are thus given by

$$\Upsilon_\infty = \begin{pmatrix} -2 & 0 & 2 \\ 0 & -2 & 0 \\ 2 & 0 & -2 \end{pmatrix} \quad \text{and} \quad \Upsilon_0 = \begin{pmatrix} 2p & 0 & -2p \\ 0 & 2p & 0 \\ -2p & 0 & 2p \end{pmatrix}.$$

Now, the rank of Υ_∞ is 2, so this is a B system. We also have

$$p\Upsilon_\infty + \Upsilon_0 = 0$$

so this is a II system. That is, this system has type IIB with pitch p.

8.5   Operations on Screw Systems

Thinking of screw systems as linear subspaces in $se(3)$, there are two natural binary operations we can perform.

For two screw systems, the union is the linear span of the screws from both of them. For two screw systems $\mathbf{s}_1 \wedge \cdots \wedge \mathbf{s}_m$ and $\mathbf{s}_{m+1} \wedge \cdots \wedge \mathbf{s}_{m+n}$ that do not intersect, the union is given by

$$\mathbf{s}_1 \wedge \cdots \wedge \mathbf{s}_m \wedge \mathbf{s}_{m+1} \wedge \cdots \wedge \mathbf{s}_{m+n}.$$

This defines a linear mapping between the anti-symmetric powers of $se(3)$:

$$\cup : \wedge^m se(3) \times \wedge^n se(3) \longrightarrow \wedge^{m+n} se(3).$$

The dual operation to the union is the intersection of screw systems. The intersection of two screw systems is simply the linear subspace common to both. This operation has applications in gripping (see Section 12.5) and to parallel robots. Suppose that a rigid body is constrained by a finger or some mechanism. To first order, the velocities of the body are restricted to some linear subspace of $se(3)$, a screw system. The body may be constrained by several mechanisms, as for example in a multi-fingered gripper, the body of a Stewart platform, or even the body of a walking robot when the contact points between the feet and the ground do not move. The effect of imposing several constraints, again to first order, will be the intersection of the screw systems determined by each. That is, allowed movements must be consistent with all the constraints.

Algebraically the intersection or meet of a pair of screw systems is given by the **shuffle product**. Consider two screw systems, $\mathbf{s}_1 \wedge \cdots \wedge \mathbf{s}_j$ and $\mathbf{z}_1 \wedge \cdots \wedge \mathbf{z}_k$

where $j + k \geq 6$. The shuffle product of these two systems will be

$$\frac{1}{(6-k)!(j+k-6)!} \times$$

$$\sum_\sigma \mathrm{sign}(\sigma) \det(\mathbf{s}_{\sigma(1)}, \dots, \mathbf{s}_{\sigma(6-k)}, \mathbf{z}_1, \cdots, \mathbf{z}_k)\mathbf{s}_{\sigma(6-k+1)} \wedge \cdots \wedge \mathbf{s}_{\sigma(j)}.$$

The sum here ranges over all permutations σ of $\{1, 2, \dots, j\}$. The determinant can be interpreted as follows: Suppose that we choose a basis $X_1, X_2, \dots, X_6$ for $se(3)$, so any screw can be written $\mathbf{s}_i = c_{i1}X_1 + c_{i2}X_2 + \cdots + c_{i6}X_6$. where c_{ij} are numerical coefficients. Given six screws $\mathbf{s}_1, \dots, \mathbf{s}_6$, the determinant

$$\det(\mathbf{s}_1, \mathbf{s}_2, \dots, \mathbf{s}_6) = \det(c_{ij})$$

is the determinant of the 6×6 matrix of coefficients. This determinant is often written as

$$\det(\mathbf{s}_1, \mathbf{s}_2, \dots, \mathbf{s}_6) = [\mathbf{s}_1, \mathbf{s}_2, \dots, \mathbf{s}_6]$$

and called the **bracket** of the screws.

This product can be thought of as a map,

$$\cap : \wedge^i se(3) \times \wedge^j se(3) \longrightarrow \wedge^{6-i-j} se(3).$$

If we collect together the exterior powers of $se(3)$ we get a vector space usually written

$$\wedge se(3) = \wedge^0 se(3) \oplus \wedge^1 se(3) \oplus \wedge^2 se(3) \oplus \cdots \oplus \wedge^6 se(3).$$

Here $\wedge^0 se(3)$ is to be interpreted as a copy of the coefficient field $\mathbb{R}$ and $\wedge^1 se(3)$ is just a copy of $se(3)$ itself. This space, together with the wedge product is a **Grassmann algebra**. If we include the shuffle product as well we get a **Grassmann–Cayley algebra**, see [127].

There are several useful applications of Grassmann–Cayley algebra in robotics, see for example White [127] and Downing, Samuel and Hunt [24] . However, this algebra only takes account of the vector space structure of $se(3)$; the action of the group and the Lie algebra structure are ignored. For example, the isotropy group of the union of two screw systems is the intersection of the individual isotropy groups, since the isotropy group of the union consists of group elements that preserve both systems. However, to find the isotropy group of an intersection we must use the precise isotropy group of the screw systems, not just their conjugacy class. More generally, the Gibson–Hunt type of the intersection or union of a pair of screw systems cannot be found using Grassmann–Cayley algebra alone.

Finally, some observations on dual vector spaces. The dual to the space $\wedge^n se(3)$ can be thought of as $\wedge^{6-n} se(3)$, that is, $\left(\wedge^n se(3) \right)^* = \wedge^{6-n} se(3)$. Suppose we take an n-system $\Delta_a = \mathbf{s}_1 \wedge \mathbf{s}_2 \wedge \cdots \wedge \mathbf{s}_n$ and a $(6-n)$-system $\Delta_b = \mathbf{s}_{n+1} \wedge \mathbf{s}_{n+2} \wedge \cdots \wedge \mathbf{s}_6$. Then using a basis $X_1, X_2, \dots, X_6$ for $se(3)$ as above, the union of these two systems will be given by

$$\Delta_a \cup \Delta_b = \mathbf{s}_1 \wedge \cdots \mathbf{s}_n \wedge \mathbf{s}_{n+1} \wedge \cdots \mathbf{s}_6 = [\mathbf{s}_1, \dots, \mathbf{s}_6]X_1 \wedge X_2 \wedge \cdots \wedge X_6$$

using the bracket determinant defined above. So we can define the evaluation map of Δ_a on a dual vector Δ_b by

$$\Delta_a(\Delta_b) = [\mathbf{s}_1, \ldots, \mathbf{s}_6].$$

Although this formula looks as if it should depend on the basis we choose for $se(3)$, the transformation properties of the construction ensure that the result of the evaluation is in fact independent of the basis. Note that if the two screw systems intersect, that is, if they have a common screw, then the evaluation will give zero.

The dual to a 1-system is a 5-system. In Chapter 11 we call these vectors co-screws. If we write a 5-system as

$$\mathcal{W} = \mathbf{s}_1 \wedge \mathbf{s}_2 \wedge \mathbf{s}_3 \wedge \mathbf{s}_4 \wedge \mathbf{s}_5,$$

we can treat these vectors in much the same ways as screws. There are six-dimensional vectors that transform according to the coadjoint representation of $se(3)$. Hence, we could talk about co-screw systems. However, we saw above that the reciprocal gives an isomorphism from screw systems to their dual systems, and hence the classification of co-screw systems is the same as the classification of their dual screw systems. Note that the intersection and union of reciprocal systems obey the familiar de Morgan's laws:

$$\overline{(\Delta_a \cup \Delta_b)} = \overline{\Delta_a} \cap \overline{\Delta_b} \qquad \text{and} \qquad \overline{(\Delta_a \cap \Delta_b)} = \overline{\Delta_a} \cup \overline{\Delta_b}.$$

This is a straightforward consequence of the definition of the reciprocal.

9
Clifford Algebra

At the turn on the nineteenth century there was a vituperative dispute about which was the 'correct' notation to use in modern geometry. The matrix-vector methods promoted by Gibbs won and the quaternion-Clifford algebra methods lost. This is why modern students in science and engineering no longer learn about quaternions. However, news of this revolution was slow to spread in some areas, particularly in kinematics. So Study and latter Blaschke [12] and Dimentberg [27] continued to develop 'dual quaternions' and applied them to the theory of mechanisms. Mathematicians never really forgot about these things, although the real impetus to look at these structures afresh came when physicists rediscovered them. Pauli's σ-matrices and Dirac's γ-matrices turned out to be generators of Clifford algebras.

In the last twenty years there has been something of a backlash against vector-matrix methods and in favour of Clifford algebra led mainly by David Hestenes, see [50] . The Clifford algebra viewpoint is gaining some ground, at least with computer scientists; many commercial computer graphics systems use quaternions to represent rotations. There are good reasons for this; for example, suppose a rotation is produced as the result of a computation. Any computational procedure will introduce errors due to finite precision arithmetic. If the rotation is represented an orthogonal matrix, then the errors will mean the result is probably not orthogonal. To recover an orthogonal matrix requires a time consuming Gramm–Schmidt process. By contrast, if the rotation is represented by a unit modulus quaternion, all that is needed to recover the normalisation is to divide by the sum of the squares of the components.

There are many cases in kinematics where using the Clifford algebras described below lead to simple formulas and straightforward algorithms. In particular, the realisation of the group of rigid body motions, $SE(3)$, in the Study quadric is probably the neatest and most useful way of picturing the group manifold.

Clifford algebras are associative algebras similar to Grassmann algebras, which we have more or less met in Chapter 7. Grassmann algebras are just the algebras of anti-symmetric tensors.

Originally, Grassmann developed his "Extension theory" to turn geometry into algebra and hence facilitate geometrical computations. Clifford extended this work and combined it with ideas from Hamilton's quaternions by introducing a new product. In this way Clifford was able to describe an algebra for metric geometries, not just projective geometry. In fact, as we will see a little later, the Grassmann algebra can be thought of as sitting in the Clifford algebra and the Grassmann product can be written in terms of the Clifford product. More recently, Rota introduced the shuffle product into Grassmann algebras to produce what he called Grassmann–Cayley algebras. This gives us algebraic operations for both the 'meet' and 'join' of linear subspaces. In a non-degenerate Clifford algebra this operation is easy to model using the Clifford product and a particular element in the algebra, (the so-called unit pseudo-scalar). However, in a degenerate algebra, like the one we need for Euclidean geometry, we have to introduce this operation as a separate idea derived from the Clifford product. A brief history of Clifford algebra can be found in [69].

The reason for the utility of these Clifford algebras is that they contain many representations of the orthogonal group. That is, given a vector space $\mathbb{R}^n$ we can construct the corresponding Clifford or Grassmann algebra containing various representations of $O(n)$. In particular, we have already seen that the Grassmann algebra of anti-symmetric tensors on $\mathbb{R}^n$ contains all the representations $\wedge^k \mathbb{R}^n$ corresponding to the action of $O(n)$ on k-planes. Clifford algebras have an even richer structure.

Here we will quickly specialise to the Clifford algebra for the group of proper rigid motions. This construction is usually attributed to Clifford himself by the kinematic community citing [20]. However, it seems more likely that the idea is due to Study, see [118]. The confusion probably arises because in [20] Clifford introduces the biquaternions also known as double quaternions; the algebra relevant to the Euclidean group is the dual quaternion algebra, but this is also sometimes called the biquaternion algebra.

This algebra is often a very efficient vehicle for computations. We illustrate this by taking the promised closer look at the differential geometry of ruled surfaces. In the following chapter these ideas will be developed more fully for use in Euclidean geometry and applied to some robot kinematic problems.

9.1   Geometric Algebra

Given a symmetric bilinear form Q on the vector space $V = \mathbb{R}^n$, we can form a Clifford algebra. Suppose we have a basis for $\mathbb{R}^n$ given by the vectors $\mathbf{x}_1, \mathbf{x}_2, \ldots, \mathbf{x}_n$, and that in this basis the bilinear form has the matrix Q_{ij}. That is, evaluating the form on a pair of basis elements gives $Q(\mathbf{x}_i, \mathbf{x}_j) = Q_{ij}$. The Clifford algebra is constructed by imposing the following relations on the free associative algebra generated by the basis vectors:

$$\mathbf{x}_i\mathbf{x}_j + \mathbf{x}_j\mathbf{x}_i = 2Q_{ij}, \qquad 1 \leq i, j \leq n.$$

Notice that the Clifford product is simply denoted by juxtaposing the elements. So, the condition says that the anti-commutator of generators gives a scalar.

Now suppose we pick a new set of generators for our algebra, where the new generators are related to the original ones by a general linear transformation. That is, we let

$$\mathbf{e}_i = \sum_{j=1}^n m_{ij}\mathbf{x}_j,$$

where m_{ij} are the elements of an $n \times n$ general linear matrix M. In this new basis, we have

$$\mathbf{e}_i\mathbf{e}_j + \mathbf{e}_j\mathbf{e}_i = \sum_{k=1}^n \sum_{j=1}^n m_{ik}m_{jl}(\mathbf{x}_k\mathbf{x}_l + \mathbf{x}_l\mathbf{x}_k)$$

$$= \sum_{k=1}^n \sum_{j=1}^n 2m_{ik}m_{jl}Q_{kl} = 2\left(MQM^T\right)_{ij}.$$

That is, a general linear change of basis in $\mathbb{R}^n$ corresponds to a congruence transform of the symmetric matrix Q. Hence, by Sylvester's theorem, we can always find a set of generators for which the anti-commutator is diagonal. Moreover, the diagonal elements will be 1, -1, or 0. So, any Clifford algebra has a set of generators that anti-commute,

$$\mathbf{e}_i\mathbf{e}_j + \mathbf{e}_j\mathbf{e}_i = 0, \qquad i \neq j,$$

and that square to 1, -1, or 0. Thus, Clifford algebras are completely determined by the number of generators that square to 1, the number that square to -1, and how many square to 0. So, we will denote a Clifford algebra by $C\ell(p,q,r)$, where p is the number of generators that square to 1, q the number squaring to -1, and r the number squaring to 0.

The simplest example of a Clifford algebra is probably $C\ell(0,1,0)$, which is isomorphic to the complex numbers, $\mathbb{C}$. Elements of $C\ell(0,1,0)$ have the form $x + y\mathbf{e}_1$, where x and y are real numbers. Addition of these elements is componentwise, and multiplication is associative:

$$(x + y\mathbf{e}_1) + (w + z\mathbf{e}_1) = (x + w) + (y + z)\mathbf{e}_1$$

and

$$(x + y\mathbf{e}_1)(w + z\mathbf{e}_1) = (xw - yz) + (xz + yw)\mathbf{e}_1,$$

since $\mathbf{e}_1\mathbf{e}_1 = -1$. Hence, we see that the Clifford generator $\mathbf{e}_1$ plays exactly the role of the complex unit i. Observe that, by similar arguments, the algebra $C\ell(0, 0, 1)$ is the ring of dual numbers, $\mathbb{D}$, that we met in Section 7.6.

Our next example is $C\ell(0, 2, 0)$, which turns out to be isomorphic to the quaternions, $\mathbb{H}$. A general element of this algebra has the form

$$w + x\mathbf{e}_1 + y\mathbf{e}_2 + z\mathbf{e}_1\mathbf{e}_2.$$

We can identify the unit quaternions as $i \mapsto \mathbf{e}_1, j \mapsto \mathbf{e}_2$ and $k \mapsto \mathbf{e}_1\mathbf{e}_2$. Now, we can easily verify that the generators behave as expected. First, we look at the squares:

$$i^2 = \mathbf{e}_1^2 = -1, \quad j^2 = \mathbf{e}_2^2, \quad k^2 = \mathbf{e}_1\mathbf{e}_2\mathbf{e}_1\mathbf{e}_2 = -\mathbf{e}_1\mathbf{e}_1\mathbf{e}_2\mathbf{e}_2 = -1.$$

They all square to -1 as expected. Next we look at the products:

$$ij = \mathbf{e}_1\mathbf{e}_2 = k, \quad jk = \mathbf{e}_2\mathbf{e}_1\mathbf{e}_2 = -\mathbf{e}_1\mathbf{e}_2\mathbf{e}_2 = i, \quad ki = \mathbf{e}_1\mathbf{e}_2\mathbf{e}_1 = -\mathbf{e}_1\mathbf{e}_1\mathbf{e}_2 = j.$$

Finally, we check that these elements anti-commute. The fact that $ij + ji = 0$ is clear, so we only have to check

$$jk + kj = \mathbf{e}_2\mathbf{e}_1\mathbf{e}_2 + \mathbf{e}_1\mathbf{e}_2\mathbf{e}_2 = 0, \quad ki + ik = \mathbf{e}_1\mathbf{e}_2\mathbf{e}_1 + \mathbf{e}_1\mathbf{e}_1\mathbf{e}_2 = 0.$$

Notice that whenever we have a monomial in the generators $\mathbf{e}_\alpha \mathbf{e}_\beta \cdots \mathbf{e}_\gamma$ we can bring the term to the form $\pm\mathbf{e}_i\mathbf{e}_j \cdots \mathbf{e}_k$ with $i \leq j \leq \cdots \leq k$, by commuting pairs of generators and multiplying by -1 every time we make a swap. If any generators are repeated, we can simplify the pair to 1, -1, or 0 depending on the relevant value of Q_{ii}. In this way, we can bring any monomial to the normal form; $\pm\mathbf{e}_i\mathbf{e}_j \cdots \mathbf{e}_k$ with $i < j < \cdots < k$, strict inequalities this time. Together with 1, these monomials form a basis of the algebra as a vector space. That is, any element of the algebra can be written as a sum of terms, each term being a scalar or a scalar multiplied by one of the monomials in the generators. Hence, the dimension of the algebra generated by $n = p + q + r$ elements will be 2^n.

The degrees of the monomials give us a grading on the Clifford algebra. This means we can decompose a Clifford algebra into vector subspaces

$$C\ell(p, q, r) = V_0 \oplus V_1 \oplus V_2 \oplus \cdots \oplus V_n,$$

where each subspace V_k has a basis given by the degree k monomials and $V_0 = \mathbb{R}$, generated by 1. This grading is dependent on the choice of generators for the algebra; a different choice gives a different decomposition. However, if we split the algebra into even and odd degree subspaces,

$$C\ell(p, q, r) = C\ell^+(p, q, r) \oplus C\ell^-(p, q, r),$$

then this decomposition is independent of the choice of basis. This can be seen by observing that the even part, $C\ell^+(p, q, r)$, is a subalgebra—the product of even degree monomials is always even, since generators can only cancel in pairs. It is easy to see that this even subalgebra has dimension $2^{p+q+r-1}$. In fact, any Clifford algebra is isomorphic to the even subalgebra of a Clifford algebra with one more generator:

$$C\ell(p, q, r) = C\ell^+(p, q+1, r).$$

Explicitly, this isomorphism is given by sending a generator $\mathbf{e}_i$ of $C\ell(p, q, r)$ to the element $\mathbf{e}_i\mathbf{e}_0$ of $C\ell^+(p, q+1, r)$. All we really need to do here is to check that this mapping preserves the relations on the algebra:

$$\mathbf{e}_i\mathbf{e}_j + \mathbf{e}_j\mathbf{e}_i \longmapsto \mathbf{e}_i\mathbf{e}_0\mathbf{e}_j\mathbf{e}_0 + \mathbf{e}_j\mathbf{e}_0\mathbf{e}_i\mathbf{e}_0 = -(\mathbf{e}_i\mathbf{e}_j + \mathbf{e}_j\mathbf{e}_i)\mathbf{e}_0^2.$$

We see that the relations will be preserved so long as $\mathbf{e}_0$ squares to -1.

On any Clifford algebra there is conjugation, that is, a bijective map from the algebra to itself that reverses the order of products. Denoting the conjugation by an asterisk and a pair of Clifford algebra elements by $\mathbf{c}_1$ and $\mathbf{c}_2$, this means

$$(\mathbf{c}_1\mathbf{c}_2)^* = \mathbf{c}_2^*\mathbf{c}_1^*.$$

A mapping with this property is called an anti-involution.

We can define the conjugation by giving its action on the generators and then use linearity and the anti-involution property to generalise the action to arbitrary elements in the algebra. We have $\mathbf{e}_i^* = -\mathbf{e}_i$ for any generator. The conjugation has no effect on scalars.

So, on the complex numbers $\mathbb{C} = C\ell(0, 1, 0)$, the conjugation is the familiar complex conjugate. The complex numbers are commutative, so the anti-involution property is of no consequence here. On the quaternions $\mathbb{H} = C\ell(0, 2, 0)$, we obtain the quaternionic conjugate:

$$(a + bi + cj + dk)^* = a - bi - cj - dk.$$

The conjugations of $i = \mathbf{e}_1$ and $j = \mathbf{e}_2$ are clear from the definition of the conjugate of generators. To find the conjugate of $k = \mathbf{e}_1\mathbf{e}_2$ we use the anti-involution property:

$$k^* = (\mathbf{e}_1\mathbf{e}_2)^* = (-\mathbf{e}_2)(-\mathbf{e}_1) = -\mathbf{e}_1\mathbf{e}_2 = -k.$$

More generally, we can write

$$(\mathbf{e}_1\mathbf{e}_2\cdots\mathbf{e}_k)^* = (-1)^k\mathbf{e}_k\cdots\mathbf{e}_2\mathbf{e}_1.$$

Consider the subspace spanned by the generators. A typical element from $C\ell(p, q, r)$ would have the form

$$\mathbf{x} = x_1\mathbf{e}_1 + x_2\mathbf{e}_2 + \cdots + x_n\mathbf{e}_n,$$

where $n = p+q+r$. Let us label the n-dimensional vector space of these elements by V. For elements of V we have the relation

$$\mathbf{x}\mathbf{x}^* = (-x_1^2 - x_2^2 - \cdots - x_p^2 + x_{p+1}^2 + \cdots + x_{p+q}^2).$$

Notice that the result of multiplying such an element by its conjugate is a scalar.

Another mapping from the algebra to itself is the **main involution**, denoted by α. It is defined by

$$\alpha(\mathbf{e}_1\mathbf{e}_2\cdots\mathbf{e}_k) = (-1)^k \mathbf{e}_1\mathbf{e}_2\cdots\mathbf{e}_k.$$

Although this looks superficially similar to the conjugation, it is rather different. It is, in fact, a homomorphism since it preserves products rather than reversing them. On the even subalgebra, the main involution is the identity. However, the main involution reverses the sign of monomials of odd order. It is not hard to see that for any element of the algebra we have $\alpha(\mathbf{c})^* = \alpha(\mathbf{c}^*)$; that is, the main involution commutes with conjugation.

In general, we cannot invert every element in a Clifford algebra. For an element $\mathbf{c}$ it is not always possible to find another element $\mathbf{c}^{-1}$ such that $\mathbf{c}\mathbf{c}^{-1} = \mathbf{c}^{-1}\mathbf{c} = 1$. This is related to the fact that Clifford algebras generally have zero divisors. This means that it is possible to have a pair of non-zero elements $\mathbf{a}$ and $\mathbf{b}$, say, that satisfy $\mathbf{ab} = 0$. So, the existence of an inverse for either $\mathbf{a}$ or $\mathbf{b}$ would imply that the other was zero, contradicting our hypothesis. It is not hard to find examples. For instance, in the Clifford algebras $C\ell(0, n, 0)$ where $n \geq 3$, a small calculation reveals that

$$(\mathbf{e}_1 + \mathbf{e}_2\mathbf{e}_3)(\mathbf{e}_2 + \mathbf{e}_1\mathbf{e}_3) = 0.$$

Hence, neither of the elements $(\mathbf{e}_1 + \mathbf{e}_2\mathbf{e}_3)$ or $(\mathbf{e}_2 + \mathbf{e}_1\mathbf{e}_3)$ has an inverse. Moreover, we see that this phenomenon is unrelated to the possibility that the symmetric form was degenerate, although in such a case we would have generators that square to zero and hence possess no inverse. Elements of a Clifford algebra that can be inverted are called **units**. The set of all units in any Clifford algebra forms a group. For the cases $C\ell(0, 1, 0) = \mathbb{C}$ and $C\ell(0, 2, 0) = \mathbb{H}$ it is well known that neither the complex numbers nor the quaternions possess zero divisors, so in these cases the group of units is the whole of the Clifford algebra. But this, of course, is not generally the case.

In the following, we will confine our attention to the Clifford algebras $C\ell(0, n, 0)$ and look at two subgroups of the group of units in these algebras. The first subgroup is called Pin(n) and is defined by

$$\text{Pin}(n) = \{\mathbf{g} \in C\ell(0, n, 0) : \mathbf{g}\mathbf{g}^* = 1 \text{ and } \alpha(\mathbf{g})\mathbf{x}\mathbf{g}^* \in V \text{ for all } \mathbf{x} \in V\}.$$

Notice that the condition $\mathbf{g}\mathbf{g}^* = 1$ ensures that $\mathbf{g}$ is a unit in the algebra. Also notice that the intersection of this group with the n-dimensional space V is an

$(n-1)$-dimensional sphere. If we write an element of V as $\mathbf{x} = x_1\mathbf{e}_1 + x_2\mathbf{e}_2 + \cdots + x_n\mathbf{e}_n$, then the first condition for such an element to be in the group is

$$\mathbf{x}\mathbf{x}^* = 1 = x_1^2 + x_2^2 + \cdots + x_n^2.$$

The second condition, $\mathbf{x}\mathbf{x}'\mathbf{x}^* \in V$, is automatically satisfied. To see this, we need only investigate the case where $\mathbf{x}' = \mathbf{e}_i$ since the map $\mathbf{x}' \mapsto \mathbf{x}\mathbf{x}'\mathbf{x}^*$ is a linear mapping. Now, when $\mathbf{x}' = \mathbf{e}_i$ there are three possible types of terms in the product $\mathbf{x}\mathbf{x}'\mathbf{x}^*$. First, we have terms of the form $-\mathbf{e}_i\mathbf{e}_i\mathbf{e}_i = \mathbf{e}_i$, clearly in V. Next we can get terms like $-\mathbf{e}_i\mathbf{e}_i\mathbf{e}_j = \mathbf{e}_j$, again in V. Lastly, we will have terms of the form $-\mathbf{e}_j\mathbf{e}_i\mathbf{e}_k$, but for every such term there will also be a term of the form $-\mathbf{e}_k\mathbf{e}_i\mathbf{e}_j$, and these terms will cancel.

The group $\mathrm{Pin}(n)$ acts on the subspace $V = \mathbb{R}^n$. This action is given by

$$\mathbf{g} : \mathbf{x} \longmapsto \alpha(\mathbf{g})\mathbf{x}\mathbf{g}^*.$$

This action preserves the usual scalar product on $V = \mathbb{R}^n$. To see this, first note that the scalar product can be written as

$$\mathbf{x} \cdot \mathbf{y} = \frac{1}{2}(\mathbf{x}\mathbf{y}^* + \mathbf{y}\mathbf{x}^*).$$

So, transforming a pair of vectors $\mathbf{x}$ and $\mathbf{y}$ and then taking the scalar product gives

$$\frac{1}{2}\big(\alpha(\mathbf{g})\mathbf{x}\mathbf{g}^*(\alpha(\mathbf{g})\mathbf{y}\mathbf{g}^*)^* + \alpha(\mathbf{g})\mathbf{y}\mathbf{g}^*(\alpha(\mathbf{g})\mathbf{x}\mathbf{g}^*)^*\big)$$

$$= \frac{1}{2}\big(\alpha(\mathbf{g})\mathbf{x}\mathbf{g}^*\mathbf{g}\mathbf{y}^*\alpha(\mathbf{g}^*) + \alpha(\mathbf{g})\mathbf{y}\mathbf{g}^*\mathbf{g}\mathbf{x}^*\alpha(\mathbf{g}^*)\big)$$

$$= \frac{1}{2}\big(\alpha(\mathbf{g})\mathbf{x}\mathbf{y}^*\alpha(\mathbf{g}^*) + \alpha(\mathbf{g})\mathbf{y}\mathbf{x}^*\alpha(\mathbf{g}^*)\big)$$

$$= \frac{1}{2}(\mathbf{x}\mathbf{y}^* + \mathbf{y}\mathbf{x}^*)\alpha(\mathbf{g}\mathbf{g}^*)$$

$$= \frac{1}{2}(\mathbf{x}\mathbf{y}^* + \mathbf{y}\mathbf{x}^*).$$

Linear transformations of $V = \mathbb{R}^n$ that preserve the scalar product are just elements of the orthogonal group $O(n)$. Thus, we can think of the action as defining a homomorphism from $\mathrm{Pin}(n)$ to $O(n)$. In fact this homomorphism is the double covering map. It is not hard to see that the kernel of this homomorphism consists of the two elements 1 and -1. It is harder to see that the homomorphism is onto; see for example Curtis [23], Porteous [88], or Fulton and Harris [36]. However, we can easily see that the reflections of V correspond to the group elements in V. Recall that the intersection of the group $\mathrm{Pin}(n)$ with V is the $(n-1)$-sphere of elements of the form $\mathbf{v} = a_1\mathbf{e}_1 + \cdots + a_n\mathbf{e}_n$, subject to the relation $a_1^2 + \cdots + a_n^2 = 1$. The action of such an element is a reflection in the hyperplane perpendicular to $\mathbf{v}$. Elements in the direction of $\mathbf{v}$ have their sign reversed,

$$\alpha(\mathbf{v})\mathbf{v}\mathbf{v}^* = \alpha(\mathbf{v}) = -\mathbf{v},$$

while elements perpendicular to $\mathbf{v}$ are unchanged. An element $\mathbf{v}^\perp$ perpendicular to $\mathbf{v}$ will satisfy $\mathbf{v}^\perp \mathbf{v}^* = \mathbf{v}\mathbf{v}^\perp$ so that

$$\alpha(\mathbf{v})\mathbf{v}^\perp \mathbf{v}^* = \alpha(\mathbf{v})\mathbf{v}\mathbf{v}^\perp = \mathbf{v}^\perp.$$

The second group we look at is defined as

$$\mathrm{Spin}(n) = \{\mathbf{g} \in C\ell^+(0,n,0) : \mathbf{g}\mathbf{g}^* = 1 \text{ and } \mathbf{g}\mathbf{x}\mathbf{g}^* \in V \text{ for all } \mathbf{x} \in V\}.$$

These are the even elements of $\mathrm{Pin}(n)$. The action of $\mathrm{Spin}(n)$ on $V = \mathbb{R}^n$ can be written

$$\mathbf{g} : \mathbf{x} \longmapsto \mathbf{g}\mathbf{x}\mathbf{g}^*.$$

It is unnecessary here to include the main involution since it has no effect on elements of the even subalgebra. Certainly $\mathrm{Spin}(n)$ acts by orthogonal transformations on $V = \mathbb{R}^n$, but these transformations are in fact rotations, since they are generated by elements that are pairs of reflections.

Historically, $\mathrm{Spin}(n)$ was described before $\mathrm{Pin}(n)$. Since $\mathrm{Spin}(n)$ double covers $SO(n)$, the double cover of $O(n)$ was called $\mathrm{Pin}(n)$. This joke is usually attributed to J-P. Serre.

As an example, we will look at the group $\mathrm{Spin}(3)$. A general element of the even algebra $C\ell^+(0,3,0)$ has the form

$$\mathbf{g} = a_0 + a_1\mathbf{e}_2\mathbf{e}_3 + a_2\mathbf{e}_3\mathbf{e}_1 + a_3\mathbf{e}_1\mathbf{e}_2.$$

The conjugate of this element is

$$\mathbf{g}^* = a_0 - a_1\mathbf{e}_2\mathbf{e}_3 - a_2\mathbf{e}_3\mathbf{e}_1 - a_3\mathbf{e}_1\mathbf{e}_2,$$

so the condition $\mathbf{g}\mathbf{g}^* = 1$ becomes

$$a_0^2 + a_1^2 + a_2^2 + a_3^2 = 1.$$

The condition $\mathbf{g}\mathbf{x}\mathbf{g}^* \in V$ is automatically satisfied for these elements, so the elements of $\mathrm{Spin}(3)$ lie on a three-dimensional sphere. The algebra $C\ell^+(0,3,0)$ is isomorphic to the quaternions $\mathbb{H} = C\ell(0,2,0)$; hence, the group $\mathrm{Spin}(3)$ is isomorphic to the group of unit quaternions, yet another accidental isomorphism.

As mentioned above, the group acts on $\mathbb{R}^3$ by rotations. In fact, a rotation of θ about one of the coordinate axes is given by

$$\cos\frac{\theta}{2} + \sin\frac{\theta}{2}\mathbf{e}_i\mathbf{e}_j,$$

where $i \neq j$. For example, a rotation about e_3 is

$$\left(\cos \frac{\theta}{2} + \sin \frac{\theta}{2} e_1 e_2 \right) e_3 \left(\cos \frac{\theta}{2} - \sin \frac{\theta}{2} e_1 e_2 \right) = e_3$$

$$\left(\cos \frac{\theta}{2} + \sin \frac{\theta}{2} e_1 e_2 \right) e_1 \left(\cos \frac{\theta}{2} - \sin \frac{\theta}{2} e_1 e_2 \right) = \cos \theta e_1 + \sin \theta e_2$$

$$\left(\cos \frac{\theta}{2} + \sin \frac{\theta}{2} e_1 e_2 \right) e_2 \left(\cos \frac{\theta}{2} - \sin \frac{\theta}{2} e_1 e_2 \right) = -\sin \theta e_1 + \cos \theta e_2.$$

As previewed above, we now look at Grassmann algebras and their relation to Clifford algebras. Recall that Grassmann algebras were briefly introduced in Section 8.5. Given a vector space $V = \mathbb{R}^n$, with a basis $x_1, x_2, \ldots, x_n$, the Grassmann algebra is constructed by imposing the following relations on the free associative algebra generated by the basis vectors:

$$x_i \wedge x_j + x_j \wedge x_i = 0, \qquad 1 \leq i, j \leq n,$$

the Grassmann, or exterior, product is denoted $\wedge$. We could consider this as the Clifford algebra $C\ell(0, 0, n)$. However, it is probably more enlightening to think of the Grassmann algebras as the equivalent of Clifford algebras in the case where there in no metric Q_{ij} on V. A Grassmann algebra has a natural grading defined on it. If we write the Grassmann algebra on V as $\wedge V$, then the grading is given by

$$\wedge V = \wedge^0 V \oplus \wedge^1 V \oplus \wedge^2 V \oplus \cdots \wedge^n V,$$

where $\wedge^i V$ is the anti-symmetric tensor product as defined in Section 7.2. Remember that we interpret $\wedge^0 V = \mathbb{R}$ as the field of scalars and $\wedge^1 V = V$. If $b \in \wedge^i V$ and $c \in \wedge^j V$, then we have $b \wedge c \in \wedge^{i+j} V$.

On any Clifford algebra, we can define a Grassmann algebra by defining the Grassmann product in terms of the Clifford product. First we define the product of a generator $x \in V$ with an arbitrary element of the algebra c,

$$x \wedge c = \frac{1}{2} (xc + \alpha(c)x), \quad \text{and} \quad c \wedge x = \frac{1}{2} (cx + x\alpha(c)).$$

This can then be extended to the whole of the algebra by assuming the exterior product to be linear and associative. On the generators, we get

$$e_i \wedge e_j = \frac{1}{2} (e_i e_j - e_j e_i) = e_i e_j - Q_{ij}.$$

When the generators are orthogonal, that is when Q is diagonal, the exterior product and the Clifford product agree except that the exterior product of any element with itself vanishes,

$$\mathbf{e}_i \wedge \mathbf{e}_j = \mathbf{e}_i \mathbf{e}_j, \qquad \mathbf{e}_i \wedge \mathbf{e}_i = 0;$$

see Lounesto [69, Chap 3.].

Specialising to the Clifford algebras $C\ell(0, n, 0)$ again, we see that since the group Pin(n) acts linearly on V, the Clifford algebra contains all the anti-symmetric powers of this representation. For the orthogonal groups $O(n)$, and hence also for their double covers Pin(n), we have that the adjoint representation of the group is equivalent to the anti-symmetric square of the standard representation

$$Ad(R) = \wedge^2 R.$$

This is because the Lie algebra of $O(n)$ is given by $n \times n$ anti-symmetric matrices. Hence, we can identify the Lie algebra of the groups Pin(n) or Spin(n) as the subspace $\wedge^2 V$. In the standard basis for $C\ell(0, n, 0)$, we see that this corresponds to the degree-2 elements of the form $\frac{1}{2}\mathbf{e}_i \mathbf{e}_j$. The Lie bracket is given by the commutator of these degree-2 elements. On the basis for this space, this gives

$$\left[\frac{1}{2}\mathbf{e}_i \mathbf{e}_j, \frac{1}{2}\mathbf{e}_k \mathbf{e}_l\right] = \frac{1}{4}(\mathbf{e}_i \mathbf{e}_j \mathbf{e}_k \mathbf{e}_l - \mathbf{e}_k \mathbf{e}_l \mathbf{e}_i \mathbf{e}_j).$$

If i, j, k and l are all different, then the elements commute and the Lie bracket is zero. This is also the case if $i = k$, $j = l$ or $i = l$, $j = k$. On the other hand, if there is only one coincidence between i, j and k, l, then the result will be another degree-2 element.

9.2   Clifford Algebra for the Euclidean Group

The Clifford algebras for the Euclidean groups turn out to be $C\ell(0, n, 1)$, with n generators that square to -1 and a single generator that squares to 0. Let us label the first n generators $\mathbf{e}_1, \mathbf{e}_2, \ldots, \mathbf{e}_n$ in the usual way and call the generator that squares to 0 simply $\mathbf{e}$. Now, by analogy with the Spin groups, we look at a subgroup of the units in $C\ell(0, n, 1)$ and examine its action on $\mathbb{R}^n$. This time the $\mathbb{R}^n$ we look at will be the elements of the form $1 + \mathbf{x}\mathbf{e}$ where $\mathbf{x} = (x_1 \mathbf{e}_1 + x_2 \mathbf{e}_2 + \cdots + x_n \mathbf{e}_n)$. The group we consider is given by the elements of the form

$$\left\{\left(\mathbf{g} + \frac{1}{2}\mathbf{t}\mathbf{g}\mathbf{e}\right) \in C\ell(0, n, 1) : \mathbf{g} \in \mathrm{Spin}(n), \mathbf{t} = t_1 \mathbf{e}_1 + \cdots + t_n \mathbf{e}_n\right\}.$$

This is clearly a subgroup of the even subalgebra $C\ell^+(0, n, 1)$ of $C\ell(0, n, 1)$. The action of the group on the subspace is given by

$$\left(\mathbf{g} + \frac{1}{2}\mathbf{t}\mathbf{g}\mathbf{e}\right)(1 + \mathbf{x}\mathbf{e})\left(\mathbf{g} - \frac{1}{2}\mathbf{t}\mathbf{g}\mathbf{e}\right)^* = 1 + (\mathbf{g}\mathbf{x}\mathbf{g}^* + \mathbf{t})\mathbf{e},$$

since we have

$$\left(\mathbf{g} - \frac{1}{2}\mathbf{tge}\right)^* = \left(\mathbf{g}^* + \frac{1}{2}\mathbf{g}^*\mathbf{te}\right).$$

The action on the points of $\mathbb{R}^n$, that is , $\mathbf{x} \mapsto \mathbf{gxg}^* + \mathbf{t}$ is certainly a rigid motion.

The product of two group elements is given by

$$\left(\mathbf{g} + \frac{1}{2}\mathbf{tge}\right)\left(\mathbf{g}' + \frac{1}{2}\mathbf{t}'\mathbf{g}'\mathbf{e}\right) = \left(\mathbf{gg}' + \frac{1}{2}(\mathbf{t} + \mathbf{gt}'\mathbf{g}^*)\mathbf{gg}'\mathbf{e}\right).$$

Clearly, the elements of Spin(n) act on the vectors $\mathbf{t}$, so the group is a semi-direct product. In fact, we can see that this action coincides with the standard action of $SO(n)$ on $\mathbb{R}^n$; hence, we can identify the group as Spin(n) $\ltimes \mathbb{R}^n$. Note that if we had chosen $\mathbf{g} \in Pin(n)$ the corresponding group defined as above would give the double cover of the full Euclidean group $E(n)$. However, we must be careful here to use the action of Pin(n) on $\mathbb{R}^n$ as defined above in Section 9.1, the one that includes the main involution.

The group we have defined above, Spin(n) $\ltimes \mathbb{R}^n$, double covers the group of proper rigid motions $SE(n)$. The action of Spin(n) $\ltimes \mathbb{R}^n$ on $\mathbb{R}^n$ defined above can be thought of as a homomorphism of Spin(n) $\ltimes \mathbb{R}^n$ onto $SE(n)$. The isotropy group of any point in $\mathbb{R}^n$ under the action consists of the two elements $(\mathbf{g} + \frac{1}{2}\mathbf{tge}) = 1 \,\text{or} - 1$; hence, this subgroup is also the kernel of the homomorphism. The above closely follows the treatment given by Porteous [88, p.267].

In the next section, we will look, in some detail, at the case $n = 3$. Before we do that, however, we look at the Clifford algebra for planar motions, $SE(2)$. Clifford algebra does not seem to have been exploited much for solving problems in planar mechanisms theory, perhaps because the subject is simple enough not to need such fancy methods; however, see McCarthy [73].

The group Spin(2) is isomorphic to the group of unit modulus complex numbers. A typical element of Spin(2) has the form

$$\mathbf{g} = \cos\frac{\theta}{2} + \sin\frac{\theta}{2}\mathbf{e}_1\mathbf{e}_2.$$

The isomorphism follows from identifying $\frac{1}{2}\mathbf{e}_1\mathbf{e}_2$ with the imaginary unit i.

A typical element of the group Spin(2) $\ltimes \mathbb{R}^2$ thus has the form $\mathbf{g} + \frac{1}{2}\mathbf{tge}$, where $\mathbf{g} \in$ Spin(2) as above and $\mathbf{t} = x\mathbf{e}_1 + y\mathbf{e}_2$. This gives an element of the form

$$\mathbf{g} + \frac{1}{2}\mathbf{tge} =$$

$$\cos\frac{\theta}{2} + \sin\frac{\theta}{2}\mathbf{e}_1\mathbf{e}_2 + \frac{1}{2}\left(x\cos\frac{\theta}{2} + y\sin\frac{\theta}{2}\right)\mathbf{e}_1\mathbf{e} - \frac{1}{2}\left(x\sin\frac{\theta}{2} - y\cos\frac{\theta}{2}\right)\mathbf{e}_2\mathbf{e}.$$

A rotation about a point $\mathbf{c} = c_x \mathbf{e}_1 + c_y \mathbf{e}_2$ is given by the conjugation

$$\left(1 + \frac{1}{2}\mathbf{ce}\right)\mathbf{g}\left(1 - \frac{1}{2}\mathbf{ce}\right) = \mathbf{g} + \frac{1}{2}(\mathbf{cg} - \mathbf{gc})\mathbf{e}$$

$$= \cos\frac{\theta}{2} + \sin\frac{\theta}{2}(\mathbf{e}_1\mathbf{e}_2 + c_y\mathbf{e}_1\mathbf{e} - c_x\mathbf{e}_2\mathbf{e}).$$

Hence, it is simple to find the centre of rotation of group elements.

The Lie algebra of the group is given by the degree-2 elements, as in the case of the groups Pin(n) in the last section. In this case, we can see that the degree-2 elements are closed under the commutator and that the group preserves the space of degree-2 elements. Moreover, we can see that the action of the group on the degree-2 elements also preserves the commutator. Since the dimension of the space of degree-2 elements is the same as the dimension of the Lie algebra it is not difficult to see that these spaces must be isomorphic. That is, the degree-2 elements give the Lie algebra again, and this applies to any size Euclidean group.

However, for the Euclidean group in two dimensions it is simple to see that the elements $\frac{1}{2}\mathbf{e}_1\mathbf{e}_2$, $\frac{1}{2}\mathbf{e}_1\mathbf{e}$, and $\frac{1}{2}\mathbf{e}_2\mathbf{e}$ satisfy the correct commutation relations:

$$\left[\frac{1}{2}\mathbf{e}_1\mathbf{e}_2, \frac{1}{2}\mathbf{e}_1\mathbf{e}\right] = \frac{1}{2}\mathbf{e}_2\mathbf{e}, \quad \left[\frac{1}{2}\mathbf{e}_1\mathbf{e}_2, \frac{1}{2}\mathbf{e}_2\mathbf{e}\right] = -\frac{1}{2}\mathbf{e}_1\mathbf{e}, \quad \left[\frac{1}{2}\mathbf{e}_1\mathbf{e}, \frac{1}{2}\mathbf{e}_2\mathbf{e}\right] = 0.$$

In order to calculate the exponential map, we need to know about the powers of the degree-2 elements. Since the element $\mathbf{e}$ squares to zero, we see that all powers greater than one of $\mathbf{e}_1\mathbf{e}$ and $\mathbf{e}_2\mathbf{e}$ vanish. The element $\mathbf{e}_1\mathbf{e}_2$ behaves like the imaginary unit:

$$(\mathbf{e}_1\mathbf{e}_2)^n = \begin{cases} 1, & \text{if } n = 0 \bmod 4, \\ \mathbf{e}_1\mathbf{e}_2, & \text{if } n = 1 \bmod 4, \\ -1, & \text{if } n = 2 \bmod 4, \\ -\mathbf{e}_1\mathbf{e}_2, & \text{if } n = 3 \bmod 4. \end{cases}$$

For more general elements, we have

$$(\mathbf{e}_1\mathbf{e}_2 + c_y\mathbf{e}_1\mathbf{e} - c_x\mathbf{e}_2\mathbf{e})^n = \begin{cases} 1, & \text{if } n = 0 \bmod 4, \\ \mathbf{e}_1\mathbf{e}_2 + c_y\mathbf{e}_1\mathbf{e} - c_x\mathbf{e}_2\mathbf{e}, & \text{if } n = 1 \bmod 4, \\ -1, & \text{if } n = 2 \bmod 4, \\ -\mathbf{e}_1\mathbf{e}_2 - c_y\mathbf{e}_1\mathbf{e} + c_x\mathbf{e}_2\mathbf{e}, & \text{if } n = 3 \bmod 4. \end{cases}$$

which can be verified by induction. Hence, the exponential of such an element is given by

$$\exp\left(\frac{\theta}{2}(\mathbf{e}_1\mathbf{e}_2 + c_y\mathbf{e}_1\mathbf{e} - c_x\mathbf{e}_2\mathbf{e})\right) = \cos\frac{\theta}{2} + \sin\frac{\theta}{2}(\mathbf{e}_1\mathbf{e}_2 + c_y\mathbf{e}_1\mathbf{e} - c_x\mathbf{e}_2\mathbf{e}).$$

So we can associate Lie algebra elements with centres of rotation, except of course the pure translations. These are given by

$$\exp\left(\frac{1}{2}(t_x\mathbf{e}_1\mathbf{e} + t_y\mathbf{e}_2\mathbf{e})\right) = \left(1 + \frac{1}{2}t_x\mathbf{e}_1\mathbf{e} + \frac{1}{2}t_y\mathbf{e}_2\mathbf{e}\right).$$

These relations can be used, for example, to derive the Campbell–Baker–Hausdorff formula for the planar group given in Section 5.3. Let the two centres of rotation be given by the Lie algebra elements

$$C_a = \frac{1}{2}(e_1 e_2 + a_y e_1 e - a_x e_2 e)$$

and

$$C_b = \frac{1}{2}(e_1 e_2 + b_y e_1 e - b_x e_2 e)$$

Now we can compute the product of the two exponentials

$$\exp\left(\theta_1 C_a\right) \exp\left(\theta_2 C_b\right)$$

$$= \left(\cos\frac{\theta_1}{2} + \sin\frac{\theta_1}{2}(e_1 e_2 + a_y e_1 e - a_x e_2 e)\right)\left(\cos\frac{\theta_2}{2} + \sin\frac{\theta_2}{2}(e_1 e_2 + b_y e_1 e - b_x e_2 e)\right)$$

$$= \left(\cos\frac{\theta_1}{2}\cos\frac{\theta_2}{2} - \sin\frac{\theta_1}{2}\sin\frac{\theta_2}{2}\right) + \left(\cos\frac{\theta_1}{2}\sin\frac{\theta_2}{2} + \sin\frac{\theta_1}{2}\cos\frac{\theta_2}{2}\right)e_1 e_2$$

$$+ \left(\cos\frac{\theta_1}{2}\sin\frac{\theta_2}{2}b_y + \sin\frac{\theta_1}{2}\cos\frac{\theta_2}{2}a_y + \sin\frac{\theta_1}{2}\sin\frac{\theta_2}{2}(b_x - a_x)\right)e_1 e$$

$$- \left(\cos\frac{\theta_1}{2}\sin\frac{\theta_2}{2}b_x + \sin\frac{\theta_1}{2}\cos\frac{\theta_2}{2}a_x - \sin\frac{\theta_1}{2}\sin\frac{\theta_2}{2}(b_y - a_y)\right)e_2 e.$$

This can be written as

$$\exp\left(\theta_1 C_a\right) \exp\left(\theta_2 C_b\right) = \cos\left(\frac{\theta_1 + \theta_2}{2}\right) + \sin\left(\frac{\theta_1 + \theta_2}{2}\right)(e_1 e_2 + y\, e_1 e - x\, e_2 e),$$

where

$$x = \alpha a_x + \beta b_x + \gamma(b_y - a_y),$$
$$y = \alpha a_y + \beta b_y + \gamma(b_x - a_x),$$

with

$$\alpha = \tan\frac{\theta_1}{2} \Big/ \left(\tan\frac{\theta_1}{2} + \tan\frac{\theta_2}{2}\right),$$

$$\beta = \tan\frac{\theta_2}{2} \Big/ \left(\tan\frac{\theta_1}{2} + \tan\frac{\theta_2}{2}\right),$$

$$\gamma = \tan\frac{\theta_1}{2}\tan\frac{\theta_2}{2} \Big/ \left(\tan\frac{\theta_1}{2} + \tan\frac{\theta_2}{2}\right).$$

Since the commutator is

$$[C_a, C_b] = \frac{1}{2}(b_x - a_x)e_1 e + \frac{1}{2}(b_y - a_y)e_2 e,$$

we can write the product as a single exponential

$$\exp\left(\theta_1 C_a\right) \exp\left(\theta_2 C_b\right) = \exp\left(\phi(\alpha C_a + \beta C_b + \gamma[C_a, C_b])\right),$$

where the angle of rotation is $\phi = \theta_1 + \theta_2$.

9.3   Dual Quaternions

Here we look at the Clifford algebra for $SE(3)$, but first we can make some more general comments. For even dimensions, we have the isomorphism

$$Cℓ(0, 2k, 1) = Cℓ(0, 2k, 0) \otimes \mathbb{D};$$

here $\mathbb{D}$ is the ring of dual numbers, as usual. Writing ε for $1 \otimes \varepsilon$ and $\mathbf{e}_i$ for $\mathbf{e}_i \otimes 1$ the isomorphism is given by sending $\mathbf{e}_i \mapsto \mathbf{e}_i$ and $\mathbf{e}_1 \cdots \mathbf{e}_{2k}\mathbf{e} \mapsto \varepsilon$. This is an algebra isomorphism, so the image of any element can be found from the images of the generators of the algebra. It is only an isomorphism in even dimensions, since only then does each $\mathbf{e}_i$ commute with $\mathbf{e}_1 \cdots \mathbf{e}_{2k}\mathbf{e}$. Now, the even subalgebra of a Clifford algebra is isomorphic to the Clifford algebra with one less generator, so

$$Cℓ^+(0, 2k + 1, 1) = Cℓ(0, 2k, 1).$$

So, when n is odd the double cover of the group $SE(n)$ lies in the dual algebra $Cℓ(0, n - 1, 0) \otimes \mathbb{D}$.

When $n = 3$, that is, $SE(3)$, the relevant algebra is $Cℓ(0, 2, 0) \otimes \mathbb{D} = \mathbb{H} \otimes \mathbb{D}$. This algebra is called the **dual quaternion algebra**. We can look at the isomorphism in detail in this case. The double cover of the group of proper isometries of $\mathbb{R}^3$ lies in the Clifford algebra $Cℓ(0, 3, 1)$. A typical element of this group is given by $(\mathbf{g} + \frac{1}{2}\mathbf{tge})$, where $\mathbf{g} \in \text{Spin}(3)$ and $\mathbf{t} \in \mathbb{R}^3$. Taking $\{\mathbf{e}_1, \mathbf{e}_2, \mathbf{e}_3, \mathbf{e}\}$ as a basis of $Cℓ(0, 3, 1)$, we may write

$$\mathbf{g} = a_0 + a_1\mathbf{e}_2\mathbf{e}_3 + a_2\mathbf{e}_3\mathbf{e}_1 + a_3\mathbf{e}_1\mathbf{e}_2.$$

This is an element of $\text{Spin}(3)$ so long as $\mathbf{g}\mathbf{g}^* = 1$, which means that the scalars a_i, must satisfy

$$a_0^2 + a_1^2 + a_2^2 + a_3^2 = 1.$$

Similarly, we may write

$$\mathbf{t} = b_1\mathbf{e}_1 + b_2\mathbf{e}_2 + b_3\mathbf{e}_3,$$

where the b_is are scalars. So our typical group element looks like

$$\left(\mathbf{g} + \frac{1}{2}\mathbf{tge}\right) = a_0 + a_1\mathbf{e}_2\mathbf{e}_3 + a_2\mathbf{e}_3\mathbf{e}_1 + a_3\mathbf{e}_1\mathbf{e}_2$$

$$+ \frac{1}{2}(a_0b_1 - a_2b_3 + a_3b_2)\mathbf{e}_1\mathbf{e} + \frac{1}{2}(a_0b_2 + a_1b_3 - a_3b_1)\mathbf{e}_2\mathbf{e}$$

$$+ \frac{1}{2}(a_0b_3 - a_1b_2 + a_2b_1)\mathbf{e}_3\mathbf{e} + \frac{1}{2}(a_1b_1 + a_2b_2 + a_3b_3)\mathbf{e}_1\mathbf{e}_2\mathbf{e}_3\mathbf{e}.$$

The group elements lie in the even subalgebra

$$Cℓ^+(0, 3, 1) = Cℓ(0, 2, 1) = \mathbb{H} \otimes \mathbb{D}.$$

We can give the isomorphism explicitly as follows:

$$\mathbf{e}_2\mathbf{e}_3 \mapsto i,$$
$$\mathbf{e}_3\mathbf{e}_1 \mapsto j,$$
$$\mathbf{e}_1\mathbf{e}_2 \mapsto k,$$
$$-\mathbf{e}_1\mathbf{e}_2\mathbf{e}_3\mathbf{e} \mapsto \varepsilon,$$

where i, j and k are the unit quaternions. As a consequence, we also have

$$\mathbf{e}_1\mathbf{e} \mapsto i\varepsilon,$$
$$\mathbf{e}_2\mathbf{e} \mapsto j\varepsilon,$$
$$\mathbf{e}_3\mathbf{e} \mapsto k\varepsilon.$$

Any group element can now be written as a dual quaternion:

$$\check{h} = a_0 + a_1 i + a_2 j + a_3 k + \frac{1}{2}(a_0 b_1 - a_2 b_3 + a_3 b_2)i\varepsilon$$
$$+ \frac{1}{2}(a_0 b_2 + a_1 b_3 - a_3 b_1)j\varepsilon + \frac{1}{2}(a_0 b_3 - a_1 b_2 + a_2 b_1)k\varepsilon$$
$$- \frac{1}{2}(a_1 b_1 + a_2 b_2 + a_3 b_3)\varepsilon.$$

These elements have the general shape $\check{h} = h_0 + h_1\varepsilon$, where h_0 and h_1 are ordinary quaternions. The condition that $\mathbf{g} \in \mathrm{Spin}(3)$ becomes

$$\check{h}\check{h}^* = 1,$$

for the dual quaternions, where $\check{h}^* = h_0^* + h_1^*\varepsilon$ is the dual-quaternionic conjugate. This condition can be written as a pair of quaternionic equations:

$$h_0 h_0^* = 1,$$
$$(h_0 h_1^* + h_1 h_0^*) = 0.$$

These dual quaternions are elements of the group $\mathrm{Spin}(3) \ltimes \mathbb{R}^3$, which double covers $SE(3)$. To get elements of $SE(3)$, we must take the quotient by the $\mathbb{Z}_2$ subgroup, that is, we must identify the elements $\check{h}$ and $-\check{h}$. The elements of $SE(3)$ can be represented by points in the projective space $\mathbb{PR}^7$. Write a general dual quaternion as

$$\check{h} = h_0 + h_1\varepsilon = (a_0 + a_1 i + a_2 j + a_3 k) + (c_0 + c_1 i + c_2 j + c_3 k)\varepsilon$$

and take $(a_0 : a_1 : a_2 : a_3 : c_0 : c_1 : c_2 : c_3)$ as homogeneous coordinates in $\mathbb{PR}^7$. Now, $\check{h}$ and $\lambda\check{h}$ correspond to the same point of $\mathbb{PR}^7$, so this identifies $\check{h}$ and $-\check{h}$ as required. The relation $h_0 h_0^* = 1$ is redundant, and we are left with a single quadratic relation for group elements:

$$h_0 h_1^* + h_1 h_0^* = 0, \qquad \text{or} \qquad a_0 c_0 + a_1 c_1 + a_2 c_2 + a_3 c_3 = 0,$$

in the present coordinates. This six-dimensional non-singular quadric is the **Study quadric**. The only points on the quadric not representing group elements are the points satisfying

$$h_0 h_0^* = 0, \qquad \text{that is} \qquad a_0^2 + a_1^2 + a_2^2 + a_3^2 = 0.$$

This corresponds to a 3-plane in the quadric.

The action on points in $\mathbb{R}^3$, which we met at the beginning of Section 9.2, now becomes

$$(1 + \mathbf{x}\varepsilon) \longmapsto (h_0 + h_1\varepsilon)(1 + \mathbf{x}\varepsilon)(h_0^* - h_1^*\varepsilon).$$

In the above, pure quaternions have been confused with three-dimensional vectors; that is, $\mathbf{x} = xi + yj + zk$. This will be done regularly in what follows. Notice that the translation part of the transformation can be recovered from the relations

$$h_1 h_0^* = -h_0 h_1^* = \mathbf{t}/2$$

so that

$$h_1 h_0^* - h_0 h_1^* = \mathbf{t}.$$

The Lie algebra is given, as usual, by the degree-2 elements. The standard basis is given by the elements

$$\frac{1}{2}\mathbf{e}_2\mathbf{e}_3, \ \frac{1}{2}\mathbf{e}_3\mathbf{e}_1, \ \frac{1}{2}\mathbf{e}_1\mathbf{e}_2, \ \frac{1}{2}\mathbf{e}_1\mathbf{e}, \ \frac{1}{2}\mathbf{e}_2\mathbf{e}, \ \frac{1}{2}\mathbf{e}_3\mathbf{e}.$$

Notice that if we use the isomorphism with the dual quaternions given above, then an arbitrary element of the Lie algebra will be a pure dual quaternion, that is, an element of the form

$$\check{\mathbf{v}} = \frac{1}{2}(xi + yj + zk) + \frac{1}{2}\varepsilon(t_x i + t_y j + t_z k).$$

The commutators of the unit quaternions are twice their quaternionic product; that is, $[i, j] = 2ij = 2k$ for example. Remember that the unit dual number ε commutes with all other elements.

We can also write elements of the Lie algebra as pairs of three-dimensional vectors, or **dual vectors**

$$\check{\mathbf{v}} = \mathbf{v}_0 + \varepsilon\mathbf{v}_1,$$

where $\mathbf{v}_0^T = (x/2, \ y/2, \ z/2)$ and $\mathbf{v}_1^T = (t_x/2, \ t_y/2, \ t_z/2)$. In terms of these dual vectors the Lie bracket is given by twice the dual vector product, that is, the vector product extended to dual vectors. For a pair of dual vectors, we have

$$\frac{1}{2}[\check{\mathbf{v}}, \check{\mathbf{u}}] = \check{\mathbf{v}} \times \check{\mathbf{u}} = (\mathbf{v}_0 \times \mathbf{u}_0) + \varepsilon(\mathbf{v}_0 \times \mathbf{u}_1 + \mathbf{v}_1 \times \mathbf{u}_0).$$

The vector product has simply been distributed over the expressions for the dual vectors $(\mathbf{v}_0 + \varepsilon\mathbf{v}_1) \times (\mathbf{u}_0 + \varepsilon\mathbf{u}_1)$ and the square of the dual unit set to zero.

In a similar fashion, we can define the **dual scalar product** of a pair of dual vectors; this is given by

$$\check{\mathbf{v}} \cdot \check{\mathbf{u}} = (\mathbf{v}_0 \cdot \mathbf{u}_0) + \varepsilon(\mathbf{v}_0 \cdot \mathbf{u}_1 + \mathbf{v}_1 \cdot \mathbf{u}_0).$$

The result is, in general, a dual number. It combines the Killing form and the reciprocal product of the Lie algebra elements. In other words, it is simply a multiple of the dual Killing form; see Section 7.6.

The action of $SE(3)$ on elements of the Lie algebra, that is, the adjoint action of the group, can also be described in terms of the dual quaternion algebra. Dual vectors transform according to

$$(\mathbf{v}_0 + \varepsilon \mathbf{v}_1) \longmapsto (h_0 + \varepsilon h_1)(\mathbf{v}_0 + \varepsilon \mathbf{v}_1)(h_0 + \varepsilon h_1)^*.$$

This works because we can think of $h_0 \mathbf{v}_0 h_0^*$ as a rotation $R\mathbf{v}_0$. Then using the relation $h_1 h_0^* = -h_0 h_1^* = \mathbf{t}/2$, we can see that the dual part of the relation simplifies to

$$h_0 \mathbf{v}_1 h_0^* + h_0 \mathbf{v}_0 h_1^* + h_1 \mathbf{v}_0 h_0^* = h_0 \mathbf{v}_1 h_0^* + [h_0 \mathbf{v}_1 h_0^*, \mathbf{t}]/2$$

and by the remarks above this can be written in terms of three-dimensional vectors as $R\mathbf{v}_0 + R\mathbf{v}_1 \times \mathbf{t}$.

Note that the above action has a slightly different form from the action on points, a matter of a sign in the right-most term.

The standard 'trick' for multiplying quaternions extends to dual quaternions as expected from the principle of transference. If we write a pair of quaternions as $a_0 + \mathbf{a}$ and $b_0 + \mathbf{b}$ where $\mathbf{a} = a_x i + a_y j + a_z k$ and similarly for $\mathbf{b}$, then their product is given by

$$(a_0 + \mathbf{a})(b_0 + \mathbf{b}) = a_0 b_0 - \mathbf{a} \cdot \mathbf{b} + a_0 \mathbf{b} + b_0 \mathbf{a} + \mathbf{a} \times \mathbf{b},$$

where we have again confused three-dimensional vectors with pure quaternions. The above equation also holds when we extend it to the dual numbers:

$$(\check{a}_0 + \check{\mathbf{a}})(\check{b}_0 + \check{\mathbf{b}}) = \check{a}_0 \check{b}_0 - \check{\mathbf{a}} \cdot \check{\mathbf{b}} + \check{a}_0 \check{\mathbf{b}} + \check{b}_0 \check{\mathbf{a}} + \check{\mathbf{a}} \times \check{\mathbf{b}},$$

using the dual scalar and dual vector products.

The exponential map from the Lie algebra to the group is simple to find. First, we look at the ordinary quaternion case. The Lie algebra is given by pure quaternions, any of which can be written in the form

$$\frac{\theta}{2}\mathbf{v} = \frac{\theta}{2}(xi + yj + zk),$$

where $x^2 + y^2 + z^2 = 1$. Now, since $(xi + yj + zk)^2 = -1$, the exponential is simply

$$e^{\theta \mathbf{v}/2} = \cos\left(\frac{\theta}{2}\right) + \sin\left(\frac{\theta}{2}\right)\mathbf{v} = \cos\left(\frac{\theta}{2}\right) + \sin\left(\frac{\theta}{2}\right)(xi + yj + zk).$$

Turning now to the dual quaternions, we see that a pure dual quaternion can be written in the form

$$\check{\mathbf{s}} = \frac{1}{2}\check{\theta}\check{\mathbf{v}},$$

where $\check{\mathbf{v}}^2 = -1$ and the dual angle $\check{\theta} = \theta + \varepsilon p$. Notice here that the dual vector $\check{\mathbf{v}}$ is a directed line, since the dual equation $\check{\mathbf{v}}^2 = -1$ implies the two scalar equations $\mathbf{v}_0 \cdot \mathbf{v}_0 = 1$ and $\mathbf{v}_0 \cdot \mathbf{v}_1 = 0$; see Section 6.4. Now, by direct computation, or by invoking the principle of transference, we see that the exponential of such an element is given by

$$e^{\check{\theta}\check{\mathbf{v}}/2} = \cos\left(\frac{1}{2}\check{\theta}\right) + \sin\left(\frac{1}{2}\check{\theta}\right)\check{\mathbf{v}}.$$

Writing the directed line as $\check{\mathbf{v}} = \mathbf{v}_0 + \varepsilon \mathbf{v}_1$ and expanding the trigonometric functions of the dual angle (see Section 7.6) this becomes

$$e^{\check{\theta}\check{\mathbf{v}}/2} =$$
$$\left(\cos\left(\frac{\theta}{2}\right) + \sin\left(\frac{\theta}{2}\right)\mathbf{v}_0\right) + \varepsilon\left(\sin\left(\frac{\theta}{2}\right)\mathbf{v}_1 + \frac{p}{2}\cos\left(\frac{\theta}{2}\right)\mathbf{v}_0 - \frac{p}{2}\sin\left(\frac{\theta}{2}\right)\right).$$

Comparing this with the general group element $\mathbf{g} + \mathbf{tge}/2$, we see that

$$\mathbf{g} = \cos\left(\frac{\theta}{2}\right) + \sin\left(\frac{\theta}{2}\right)\mathbf{v}_0.$$

To find $\mathbf{t}$, we can post-multiply by $\mathbf{g}^*$ to obtain

$$\mathbf{t} = \sin(\theta)\mathbf{v}_1 + (1 - \cos(\theta))\mathbf{v}_0 \times \mathbf{v}_1 + p\mathbf{v}_0,$$

remembering that $\mathbf{v}_0 \cdot \mathbf{v}_0 = 1$ and $\mathbf{v}_0 \cdot \mathbf{v}_1 = 0$.

9.4   Geometry of Ruled Surfaces

In this section, we look a little more closely at the differential geometry of ruled surfaces as promised in Chapter 6. We could view this section as an application of the algebraic methods outlined above. However, ruled surfaces are important in robotics anyway.

We can use Clifford algebra to study lines in space because the algebra contains the anti-symmetric square of the standard representation of $SE(3)$. This representation is essentially the Lie algebra; hence, we may represent lines in space as the pitch zero elements, that is, dual vectors. We normalise the line vectors by requiring that they satisfy the single dual condition

$$\check{\mathbf{v}} \cdot \check{\mathbf{v}} = 1$$

which, if we write the dual vector as $\check{\mathbf{v}} = \mathbf{v}_0 + \mathbf{v}_1\varepsilon$, is equivalent to the two vector conditions

$$\mathbf{v}_0 \cdot \mathbf{v}_0 = 1,$$
$$\mathbf{v}_0 \cdot \mathbf{v}_1 = 0.$$

Notice that these dual vectors correspond to directed lines; see Section 6.4.

A ruled surface is a smoothly parameterised family of lines. Using the parameter λ we can write the lines as

$$\check{\mathbf{l}}(\lambda) = \mathbf{v}(\lambda) + \varepsilon\mathbf{r}(\lambda) \times \mathbf{v}(\lambda).$$

The surface that this corresponds to is parameterised as

$$\mathbf{x}(\lambda, \mu) = \mathbf{r}(\lambda) + \mu\mathbf{v}(\lambda).$$

With a fixed value for λ in the above, varying the parameter μ simply takes us along a generating line of the surface.

A **non-cylindric** ruled surface is one in which the direction of the generating lines is continuously changing, in other words, one for which

$$\frac{d\mathbf{v}}{d\lambda} \neq \mathbf{0}.$$

On a non-cylindric ruled surface, there is a special curve known as the striction curve. The geometry of the surface is very closely related to the geometry of this curve.

There are many equivalent ways to define the striction curve. We will begin with a definition that characterises the striction curve as the solution to an optimisation problem. The **striction curve** of a ruled surface is the curve of minimal length that meets all the generating lines of the surface. Another way of looking at this is that the striction curve is the curve described by a stretched rubber band constrained to lie on the surface.

Before proceeding, we briefly look at the common perpendicular to a pair of lines, since in Section 6.1 we saw that the shortest distance between a pair of lines is along the common perpendicular. For two non-parallel lines $\check{\mathbf{l}}_1$ and $\check{\mathbf{l}}_2$, the common perpendicular is the axis of the screw, $\check{\mathbf{l}}_1 \times \check{\mathbf{l}}_2$. To see this, suppose that the dual vector product gives, $\check{\mathbf{l}}_1 \times \check{\mathbf{l}}_2 = \check{\mu}\check{\mathbf{n}}$, where $\check{\mathbf{n}}$ is a line $\check{\mathbf{n}} \cdot \check{\mathbf{n}} = 1$. Now, when the lines are not parallel, the number $\check{\mu}$ is not pure dual. Hence, from the relations $\check{\mathbf{l}}_1 \cdot \check{\mathbf{l}}_1 \times \check{\mathbf{l}}_2 = 0$ and $\check{\mathbf{l}}_2 \cdot \check{\mathbf{l}}_1 \times \check{\mathbf{l}}_2 = 0$ we may infer

$$\check{\mathbf{l}}_1 \cdot \check{\mathbf{n}}, \quad \text{and} \quad \check{\mathbf{l}}_2 \cdot \check{\mathbf{n}} = 0,$$

which means that $\check{\mathbf{n}}$ meets both $\check{\mathbf{l}}_1$ and $\check{\mathbf{l}}_2$ at right angles. Moreover, the principle of transference tells us that we may write

$$\check{\mathbf{l}}_1 \times \check{\mathbf{l}}_2 = \sin(\check{\alpha})\check{\mathbf{n}}.$$

The dual angle $\breve{\alpha}$ here can easily be seen to be $\breve{\alpha} = \alpha + \varepsilon d$, where α is the angle between the lines and d the minimum distance between them.

Now, suppose that the lines are two nearby lines in a ruled surface, $\breve{l}(\lambda)$ and $\breve{l}(\lambda + \delta)$. To a first approximation, the second of these lines is given by

$$\breve{l}(\lambda + \delta) \approx \breve{l}(\lambda) + \delta \frac{d\breve{l}}{d\lambda}(\lambda).$$

Hence, the common perpendicular to these lines is given approximately by the axis of the screw:

$$\breve{l}(\lambda) \times \frac{d\breve{l}}{d\lambda}(\lambda).$$

In the limit $\delta \to 0$, this gives a line tangential to the surface through a point on the line $\breve{l}(\lambda)$. This point is called the **striction point** on the line. The locus of the striction points forms the striction curve.

In Section 6.5, developable ruled surfaces were introduced. These surfaces are formed from the tangent lines to a curve. That is, given a unit speed curve $\mathbf{r}(\lambda)$, that is, one parameterised in such a way that $|\dot{\mathbf{r}}(\lambda)| = 1$, its tangent developable is given by

$$\breve{l}(\lambda) = \dot{\mathbf{r}}(\lambda) + \varepsilon \mathbf{r}(\lambda) \times \dot{\mathbf{r}}(\lambda).$$

The derivative of such a ruled surface is always a line:

$$d\breve{l}/d\lambda = \ddot{\mathbf{r}}(\lambda) + \varepsilon \mathbf{r}(\lambda) \times \ddot{\mathbf{r}}(\lambda).$$

In general, the derivative $d\breve{l}/d\lambda$ will not be a line, since

$$\frac{d\breve{l}}{d\lambda} = \dot{\mathbf{v}} + \varepsilon(\dot{\mathbf{r}} \times \mathbf{v} + \mathbf{r} \times \dot{\mathbf{v}}).$$

The pitch of this screw is zero only if $\dot{\mathbf{v}} \cdot (\dot{\mathbf{r}} \times \mathbf{v}) = 0$. Since we cannot have $\mathbf{v} = \mathbf{0}$ or $\dot{\mathbf{v}} = \mathbf{0}$, the only solutions are $\dot{\mathbf{r}} = \mathbf{0}$, $\dot{\mathbf{v}} \propto \dot{\mathbf{r}}$ or $\mathbf{v} \propto \dot{\mathbf{r}}$. The first two cases correspond to a family of lines all passing through a single point. Such a surface is usually called a **cone**. The third solution corresponds to a developable surface.

We can write the axis of the screw, $d\breve{l}/d\lambda$, as $\breve{\mathbf{n}}$ so that

$$\frac{d\breve{l}}{d\lambda} = \frac{1}{|\dot{\mathbf{v}}|}(\dot{\mathbf{v}} \cdot \dot{\mathbf{v}} + \varepsilon \dot{\mathbf{v}} \cdot (\dot{\mathbf{r}} \times \mathbf{v}))\breve{\mathbf{n}}.$$

This can be written as

$$\frac{d\breve{l}}{d\lambda} = \breve{\kappa}\breve{\mathbf{n}} \qquad \text{where} \qquad \breve{\kappa} = \frac{1}{|\dot{\mathbf{v}}|}(\dot{\mathbf{v}} \cdot \dot{\mathbf{v}} + \varepsilon \dot{\mathbf{v}} \cdot (\dot{\mathbf{r}} \times \mathbf{v})),$$

introducing the dual quantity $\breve{\kappa}$.

By the remarks above, the lines $\check{\mathbf{l}}$ and $\check{\mathbf{n}}$ meet at right angles at the point of striction. A third line through the point of striction and perpendicular to the two other lines is then given by the dual vector product

$$\check{\mathbf{t}} = \check{\mathbf{l}} \times \check{\mathbf{n}}.$$

Using the triple product formulas, it is simple to verify that this is indeed a line. We can also see that $\check{\mathbf{n}} \times \check{\mathbf{t}} = \check{\mathbf{l}}$ and $\check{\mathbf{t}} \times \check{\mathbf{l}} = \check{\mathbf{n}}$. That is, these three lines form an orthonormal basis with the striction point as origin. The line $\check{\mathbf{t}}$ is tangential to the surface and the line $\check{\mathbf{n}}$ can be seen to be normal to the surface.

The derivatives of these lines can be written in terms of the above basis, for example

$$\frac{d\check{\mathbf{n}}}{d\lambda} = \left(\frac{d\check{\mathbf{n}}}{d\lambda} \cdot \check{\mathbf{l}} \right) \check{\mathbf{l}} + \left(\frac{d\check{\mathbf{n}}}{d\lambda} \cdot \check{\mathbf{n}} \right) \check{\mathbf{n}} + \left(\frac{d\check{\mathbf{n}}}{d\lambda} \cdot \check{\mathbf{t}} \right) \check{\mathbf{t}}.$$

The coefficients can be evaluated by considering the dual scalar products of the lines. For instance, if we differentiate $\check{\mathbf{l}} \cdot \check{\mathbf{t}} = 0$ we obtain, after a little rearrangement,

$$\frac{d\check{\mathbf{t}}}{d\lambda} \cdot \check{\mathbf{l}} = -\check{\mathbf{t}} \cdot \frac{d\check{\mathbf{l}}}{d\lambda} = -\check{\kappa} \check{\mathbf{t}} \cdot \check{\mathbf{n}} = 0.$$

The only pairs that do not simplify are

$$\frac{d\check{\mathbf{n}}}{d\lambda} \cdot \check{\mathbf{t}} = -\frac{d\check{\mathbf{t}}}{d\lambda} \cdot \check{\mathbf{n}}.$$

So we define $\check{\tau} = (d\check{\mathbf{t}}/d\lambda) \cdot \check{\mathbf{t}}$. In this way, we can produce the following formulas:

$$\frac{d\check{\mathbf{l}}}{d\lambda} = \check{\kappa}\check{\mathbf{n}}$$

$$\frac{d\check{\mathbf{n}}}{d\lambda} = -\check{\kappa}\check{\mathbf{l}} \qquad +\check{\tau}\check{\mathbf{t}}$$

$$\frac{d\check{\mathbf{t}}}{d\lambda} = -\check{\tau}\check{\mathbf{n}}.$$

These formulas look like dual versions of the Frenet–Serret relations for a curve in $\mathbb{R}^3$. However, they are not simply the result of applying the principle of transference to the Frenet–Serret relations. The Frenet–Serret relations apply to unit speed curves in $\mathbb{R}^3$, but the above applies to curves on the 'unit sphere' in $\mathbb{D}^3$. As a consequence, the dual numbers $\check{\kappa}$ and $\check{\tau}$ are not differential invariants of the surface; see Guggenheimer [43, sect. 8.2]. This means that two different parameterisations of a ruled surface may have different values of $\check{\kappa}$ and $\check{\tau}$.

However, if we write $\check{\kappa} = \kappa_0 + \varepsilon\kappa_1$, then since the surface is non-cylindric we must have $\kappa_0 \neq 0$. Hence, we can reparameterise the surface so that $\kappa_0 = 1$. With this parameterisation κ_1 and $\check{\tau} = \tau_0 + \varepsilon\tau_1$ will be differential invariants. The quantity κ_1 is usually called the **distribution parameter** of the surface. Notice that $\kappa_1 = 0$ implies that $d\check{\mathbf{l}}/d\lambda$ is a line, and hence the surface is a cone or a developable.

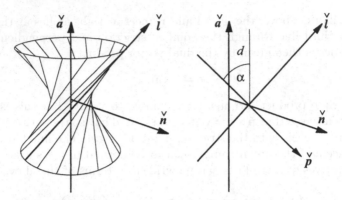

FIGURE 9.1. A Line Sweeps out a Cylindrical Regulus

Since the three lines $\check{\mathbf{l}}$, $\check{\mathbf{n}}$ and $\check{\mathbf{t}}$ form an orthogonal system, we may write vectors in terms of their directions. For example, the tangent vector to the striction curve can be written

$$\frac{d\mathbf{s}}{d\lambda} = \alpha\mathbf{v} + \beta\mathbf{u} + \gamma\mathbf{w},$$

where

$$\check{\mathbf{l}} = \mathbf{v} + \varepsilon\mathbf{s} \times \mathbf{v},$$
$$\check{\mathbf{n}} = \mathbf{u} + \varepsilon\mathbf{s} \times \mathbf{u},$$
$$\check{\mathbf{t}} = \mathbf{w} + \varepsilon\mathbf{s} \times \mathbf{w}.$$

Now, if we differentiate $\mathbf{s} \times \mathbf{v}$, $\mathbf{s} \times \mathbf{u}$ and $\mathbf{s} \times \mathbf{w}$ and rearrange using the relations derived above, we obtain three equations:

$$\frac{d\mathbf{s}}{d\lambda} \times \mathbf{v} = \kappa_1\mathbf{u}, \qquad \frac{d\mathbf{s}}{d\lambda} \times \mathbf{u} = -\kappa_1\mathbf{v} + \tau_1\mathbf{w}, \qquad \frac{d\mathbf{s}}{d\lambda} \times \mathbf{w} = -\tau_1\mathbf{u}.$$

Using the relations $\mathbf{v} \times \mathbf{u} = \mathbf{w}$, $\mathbf{u} \times \mathbf{w} = \mathbf{v}$ and $\mathbf{w} \times \mathbf{v} = \mathbf{u}$, derived from the relations between the lines, we can identify the coefficients α, β and γ. Thus, we arrive at the relation

$$\frac{d\mathbf{s}}{d\lambda} = \tau_1\mathbf{v} + \kappa_1\mathbf{w}.$$

Notice also that we have

$$\frac{d\mathbf{s}}{d\lambda} \cdot \frac{d\mathbf{v}}{d\lambda} = 0,$$

which gives another characterisation of the striction curve.

As an example, let's look at the regulus of a cylindrical hyperboloid. Notice that we can easily parameterise this surface as

$$\check{\mathbf{l}}(\lambda) = e^{ad(\check{\mathbf{a}})\lambda}\,\check{\mathbf{l}}(0),$$

where $\breve{a}$ is a line, the axis of the cylindrical hyperboloid. This is because we can think of the surface as generated by swinging a line $\breve{l}(0)$ about the axis. From this, we can compute the derivative easily:

$$\frac{d\breve{l}}{d\lambda} = \breve{a} \times \breve{l} = \sin(\breve{\alpha})\breve{n}.$$

Here, $\breve{n}$ is the common perpendicular between the axis of the hyperbola and the generator line. From this, we see immediately that the striction curve is the minimum diameter circle of the hyperboloid. The dual angle $\breve{\alpha}$ is determined by the twist angle, α, between the axis and the generator, and the minimum distance, d, between them, so that $\sin(\breve{\alpha}) = \sin \alpha + \varepsilon d \cos \alpha$. Hence, we can identify

$$\kappa_0 = \sin \alpha \qquad \text{and} \qquad \kappa_1 = d \cos \alpha.$$

Notice that these values are constants. Also, we see that the parameterisation chosen was not a unit parameterisation. To get a parameterisation with $\kappa_0 = 1$, we can change the parameter to $\lambda' = \lambda / \sin \alpha$ and then the first invariant, the distribution parameter, becomes

$$\kappa_1 = d \cot \alpha.$$

The line $\breve{t}$ is given by $\breve{t} = \breve{l} \times \breve{n}$. This line is orthogonal to both the generator $\breve{l}$ and the normal line $\breve{n}$. It also passes through the striction point. Hence, its twist angle with respect to the axis $\breve{a}$ is $\alpha + \pi/2$ with minimum distance still d. Thus, when we differentiate $\breve{t}$ we obtain

$$\frac{d\breve{p}}{d\lambda} = \breve{a} \times (\breve{l} \times \breve{n}) = \sin(\breve{\alpha} + \pi/2)\breve{n},$$

using the original parameterisation; see Figure 9.1. Using the unit parameterisation, we have

$$\frac{d\breve{t}}{d\lambda'} = \frac{1}{\sin \alpha} \sin(\breve{\alpha} + \pi/2)\breve{n} = (\cot \alpha - \varepsilon d)\breve{n}.$$

So, the other invariants of this surface are

$$\tau_0 = \cot \alpha \qquad \text{and} \qquad \tau_1 = -d.$$

What if $\breve{a}$ were not a line but a pitch p screw? The surface generated would be a ruled helicoid. The striction curve would be the helix traced out by the foot of the common perpendicular between the generating line and the axis. The invariants would be given by $(1 + \varepsilon p)$ times $\breve{\kappa}$ and $\breve{\tau}$. Hence, we have

$$\kappa_1 = (p + d \cot \alpha), \qquad \tau_0 = \cot \alpha, \qquad \tau_1 = (p \cot \alpha - d).$$

Notice that these invariants are constants and depend on the three independent quantities p, d and α. Hence, for a general ruled surface, one where the invariants are functions of the parameter, we could interpret the values of the invariants at a particular point as determining the ruled helicoid that most closely resembles the surface at that point.

As a final example let us look at the cylindroid. From Section 6.5.2 we can see that the lines in the cylindroid can be parameterised as

$$
\check{\mathbf{l}}(\theta) = \begin{pmatrix} \cos\theta \\ \sin\theta \\ 0 \end{pmatrix} + \varepsilon r \begin{pmatrix} -\sin\theta\cos2\theta \\ \cos\theta\cos2\theta \\ 0 \end{pmatrix}.
$$

Here we have assumed that the z-axis of our coordinates is aligned with the central axis of the cylindroid, and that the two extreme lines, $\theta = 0, \pi/2$, meet the z-axis at a distance of $+r$ and $-r$ respectively. Now differentiating with respect to the parameter θ gives

$$
\frac{d}{d\theta}\check{\mathbf{l}}(\theta) = \begin{pmatrix} -\sin\theta \\ \cos\theta \\ 0 \end{pmatrix} - \varepsilon r \begin{pmatrix} \cos\theta\cos2\theta - 2\sin\theta\sin2\theta \\ \sin\theta\cos2\theta + 2\cos\theta\sin2\theta \\ 0 \end{pmatrix}.
$$

Comparing this with $\check{\kappa}\check{\mathbf{n}}$ we see the $\kappa_0 = 1$ and so this parameterisation can be used to find the differential invariants. It is easy to see that $\kappa_1 = -2r\sin2\theta$ and that

$$
\check{\mathbf{n}} = \begin{pmatrix} -\sin\theta \\ \cos\theta \\ 0 \end{pmatrix} - \varepsilon r \begin{pmatrix} \cos\theta\cos2\theta \\ \sin\theta\cos2\theta \\ 0 \end{pmatrix}.
$$

From the results above we can find the striction point at any parameter value. By inspection,

$$
\mathbf{s} = \begin{pmatrix} 0 \\ 0 \\ r\cos2\theta \end{pmatrix},
$$

so that $\mathbf{s} \times \mathbf{n}_0 = \mathbf{n}_1$ and similarly for $\check{\mathbf{l}}$. So as we might have expected, the striction curve of the cylindroid is just its central axis. This makes the last of the three lines easy to compute, $\check{\mathbf{t}} = (0, 0, 1)^T$ and hence the other two invariants both vanish, $\tau_0 = \tau_1 = 0$.

10
A Little More Kinematics

In the previous chapter we saw how the Clifford algebra $C\ell(0,3,1)$ contains a representation of $SE(3)$, the group of rigid body motions. Here we will see that this algebra also contains representations of the points, lines and planes of Euclidean space. Moreover, the usual constructions of Euclidean geometry can be modelled by standard algebraic operations in the algebra. This provides us with a very neat setting for performing geometric computations.

These ideas are used to study the inverse kinematics of certain 6-joint serial robots. A theorem, due to Pieper and extended by Duffy, shows that if any three consecutive joints of the robot intersect at a common point or are mutually parallel, then the inverse kinematics problem is solvable. The demonstration given here is constructive, that is a more or less explicit algorithm is developed. The algorithm is illustrated with a couple of simple examples.

10.1 Clifford Algebra of Points, Lines and Planes

10.1.1 Planes

A plane can be specified by giving its unit normal vector **n** and the perpendicular distance from the origin; see Figure 10.1. As usual, the vector equation of the plane is given by

$$\mathbf{n} \cdot \mathbf{r} = d,$$

where **r** is any point on the plane. Notice that we get the same plane if the signs of **n** and d are both reversed. However, it is convenient here to consider

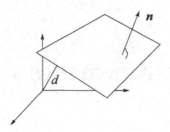

FIGURE 10.1. A Plane

oriented planes. So the two cases $\mathbf{n}$, d and $-\mathbf{n}$, $-d$, correspond to different oriented planes.

In this Clifford algebra planes can be represented by grade-1 elements with the form

$$\pi = n_x\mathbf{e}_1 + n_y\mathbf{e}_2 + n_z\mathbf{e}_3 + d\mathbf{e}.$$

These elements must satisfy the quadratic condition

$$\pi\pi^* = 1,$$

this ensures that the vector $\mathbf{n}$ has unit length. Subjecting the plane to a rigid body motion the normal vector and distance to the origin will change as follows:

$$\mathbf{n}' = R\mathbf{n}, \qquad d' = d + (R\mathbf{n}) \cdot \mathbf{t}.$$

This is most easily seen by considering the effect on the vector equation for the plane above. In the Clifford algebra this can be represented by

$$\pi' = h\pi h^*,$$

where $h = (\mathbf{g} + \frac{1}{2}\mathbf{tge})$ is a rigid body motion represented in the Clifford algebra as a rotation and a translation; see Section 9.2.

Explicitly we get

$$\pi' = (\mathbf{g} + \frac{1}{2}\mathbf{tge})(n + d\mathbf{e})(\mathbf{g}^* + \frac{1}{2}\mathbf{eg}^*\mathbf{t}) = \mathbf{gng}^* + \left(d - \frac{1}{2}(\mathbf{gng}^*\mathbf{t} + \mathbf{tgng}^*)\right)\mathbf{e},$$

where extensive use has been made of the relations of the Clifford algebra.

10.1.2 Points

In Section 9.2 points were represented by grade-1 elements. Here a different representation will be used. The advantage of this new representation is that it is consistent with the representations for the planes and lines. In particular, the group action on these points is exactly the same as for the planes and lines.

In this new representation, points are represented by grade-3 elements of the form

$$p = \mathbf{e}_1\mathbf{e}_2\mathbf{e}_3 + x\mathbf{e}_2\mathbf{e}_3\mathbf{e} + y\mathbf{e}_3\mathbf{e}_1\mathbf{e} + z\mathbf{e}_1\mathbf{e}_2\mathbf{e}.$$

The effect of a rigid body motion is given by

$$p' = hph^*.$$

Notice that these points satisfy the equation $pp^* = 1$, however, they are not the only solutions. There is another $\mathbb{R}^3$ of solutions where the coefficient of $e_1e_2e_3$ is -1 instead of $+1$. In some circumstances it might be useful to include these elements, interpreting them as 'negative' points. However, here we will only consider positive points.

In vector geometry there is a distinction between bound vectors, like the position vectors of points, and free vectors like force and angular velocity. In this Clifford algebra, free vectors can be represented by grade-3 elements with no component $e_1e_2e_3$. This is because under a rigid body motion, free vectors are unaffected by the translation, it is only rotations that transform them. So if we write

$$f = x e_2 e_3 e + y e_3 e_1 e + z e_1 e_2 e,$$

then transformation by $h = (\mathbf{g} + \frac{1}{2}\mathbf{tge})$ produces

$$f' = hfh^* = \mathbf{g}f\mathbf{g}^*.$$

Actually, angular velocities and force are better represented as parts of screws and their duals (see Chapter 12). Free vectors can be used to represent the difference between a pair of points.

10.1.3 Lines

The lines in $\mathbb{R}^3$ have been extensively studied in the previous chapter; to recap briefly, lines in $\mathbb{R}^3$ can be specified by a pair of vectors: a unit direction vector $\mathbf{v}$, and a moment vector $\mathbf{u} = \mathbf{r} \times \mathbf{v}$, where $\mathbf{r}$ is the position vector of any point on the line. In the Clifford algebra we will represent a line by a grade-2 element of the form

$$\ell = (v_x e_2 e_3 + v_y e_3 e_1 + v_z e_1 e_2) + (u_x e_1 e + u_y e_2 e + u_z e_3 e).$$

These elements must satisfy

$$\ell\ell^* = 1.$$

This relation combines the requirements that $\mathbf{v}$ is a unit vector and that $\mathbf{v}$ and $\mathbf{u}$ are orthogonal. These lines are directed lines, $-\ell$ is the same line as ℓ but with the opposite direction.

Once again the effect of a rigid body motion on these lines can be represented as

$$\ell' = h\ell h^*.$$

10.2   Euclidean Geometry

In this section we look at operations on the linear varieties defined above. Generally, the meet of a pair of linear elements is given by their exterior product and the join is given by the shuffle product of elements. Actually, we need to normalise after the product. The products vanish if there is a linear dependence between the elements, so we get incidence relations by equating the products to zero—without having to normalise.

10.2.1   Incidence

The exterior product in a Clifford algebra was introduced in Section 9.1. We begin here by looking at a point

$$p = \mathbf{e}_1\mathbf{e}_2\mathbf{e}_3 + x\mathbf{e}_2\mathbf{e}_3\mathbf{e} + y\mathbf{e}_3\mathbf{e}_1\mathbf{e} + z\mathbf{e}_1\mathbf{e}_2\mathbf{e}$$

and a plane

$$\pi = n_x\mathbf{e}_1 + n_y\mathbf{e}_2 + n_z\mathbf{e}_3 + d\mathbf{e}.$$

By direct computation we have

$$\pi \wedge p = \frac{1}{2}(\pi p - p\pi) = (xn_x + yn_y + zn_z - d)\mathbf{e}_1\mathbf{e}_2\mathbf{e}_3\mathbf{e}.$$

From the vector equation of the plane this will vanish if the point lies on the plane. So we have our first incidence relation: $\pi \wedge p = 0$ if and only if the point p lies on the plane π.

Now we introduce a line

$$\ell = v_x\mathbf{e}_2\mathbf{e}_3 + v_y\mathbf{e}_3\mathbf{e}_1 + v_z\mathbf{e}_1\mathbf{e}_2 + u_x\mathbf{e}_1\mathbf{e} + u_y\mathbf{e}_2\mathbf{e} + u_z\mathbf{e}_3\mathbf{e}$$

and look at the exterior product of a plane with a line

$$\pi \wedge \ell = \frac{1}{2}(\pi\ell + \ell\pi),$$
$$= (n_xv_x + n_yv_y + n_zv_z)\mathbf{e}_1\mathbf{e}_2\mathbf{e}_3 + (n_yu_z - n_zu_y + dv_x)\mathbf{e}_2\mathbf{e}_3\mathbf{e}$$
$$+ (n_zu_x - n_xu_z + dv_y)\mathbf{e}_3\mathbf{e}_1\mathbf{e} + (n_xu_y - n_yu_x + dv_z)\mathbf{e}_1\mathbf{e}_2\mathbf{e}.$$

Setting this equal to zero gives four conditions, $\mathbf{n} \cdot \mathbf{v} = 0$ and the vector equation $\mathbf{n} \times \mathbf{u} + d\mathbf{v} = 0$. It is not difficult to see that these equations are satisfied if and only if the line lies in the plane. Hence we have a second incidence relation $\pi \wedge \ell = 0$ if and only if the line ℓ lies in the plane π.

If we take the exterior product of a line and a point $(\ell \wedge p)$, we get zero for all lines and points. This is simply because the sum of the degrees of the elements is greater than four, so some generator will be repeated in each term. Hence we need to look elsewhere for the condition for a point to lie on a line. Suppose we write the conditions we have already found as

$$p\pi^* + \pi p^* = 0 \qquad \text{and} \qquad \ell\pi^* + \pi\ell^* = 0.$$

Here we have ignored multiplying factors and used the fact that $p^* = p$, $\ell^* = -\ell$ and $\pi^* = -\pi$. Now we can guess that the condition for a point p to lie on a line ℓ will be

$$p\ell^* + \ell p^* =$$
$$2(zv_y - yv_z + u_x)\mathbf{e}_2\mathbf{e}_3\mathbf{e} + 2(xv_z - zv_x + u_y)\mathbf{e}_3\mathbf{e}_1\mathbf{e} + 2(yv_x - xv_y + u_z)\mathbf{e}_1\mathbf{e}_2\mathbf{e}.$$

Setting this to zero is equivalent to the vector equation

$$\mathbf{p} \times \mathbf{v} = \mathbf{u},$$

where $\mathbf{p}$ is the vector given by the point p and $\mathbf{v}$ and $\mathbf{u}$ are respectively the direction and moment of the line. This equation is clearly the condition for the point to lie on the line. A little later we will see that this incidence relation can also be given by a shuffle product.

10.2.2 Meets

Most of the hard work has been done in the previous section. The intersection or meet of a pair of linear subspaces can be found by taking their exterior product and then dividing by a constant. We show that this indeed gives the intersection by verifying that the exterior product is incident on both the original linear spaces.

We begin with a pair of planes π_1 and π_2. The exterior product $\pi_1 \wedge \pi_2$ is clearly homogeneous of grade 2, but we must also show that it is a line or at least a scalar multiple of a line. That is, we must show that $(\pi_1 \wedge \pi_2)(\pi_1 \wedge \pi_2)^*$ is a scalar, certainly this quantity is not effected by Clifford conjugation. This implies that it contains only elements of grades 0, 3 or 4; in more detail we have

$$(\pi_1 \wedge \pi_2)(\pi_1 \wedge \pi_2)^* = \frac{1}{4}(\pi_1\pi_2 - \pi_2\pi_1)(\pi_2\pi_1 - \pi_1\pi_2) = \frac{1}{4}(2 - \pi_2\pi_1\pi_2\pi_1 - \pi_1\pi_2\pi_1\pi_2),$$

since $\pi_i^2 = -1$. From this we see that this expression cannot contain any elements of grade 3, remember π_i consists of elements of grade 1 only. If we look at the expression $\pi_1\pi_2\pi_1\pi_2$, we see that an element of grade 4 might result from a term like, $\mathbf{e}_i\mathbf{e}_j\mathbf{e}_k\mathbf{e}_l$ but then it would also contain the term, $\mathbf{e}_k\mathbf{e}_j\mathbf{e}_i\mathbf{e}_j$ and these terms clearly cancel. The same argument also applies to the expression $\pi_2\pi_1\pi_2\pi_1$ and hence we conclude that $\pi_1 \wedge \pi_2$ is simply proportional to a line. It is clear that this line lies on both planes since, $\pi_1 \wedge (\pi_1 \wedge \pi_2) = 0$ and $\pi_2 \wedge (\pi_1 \wedge \pi_2) = 0$ because the exterior product is associative and anti-commutative. Thus the meet of the two planes is the line given by

$$\ell = \pi_1 \wedge \pi_2 / \sqrt{(\pi_1 \wedge \pi_2)(\pi_1 \wedge \pi_2)^*}.$$

In a similar manner it is possible to show that the point where a line ℓ, and plane π, meet is given by

$$p = \ell \wedge \pi / \pm \sqrt{(\ell \wedge \pi)(\ell \wedge \pi)^*}.$$

The sign of the square root here must be chosen so that the coefficient of $\mathbf{e}_1\mathbf{e}_2\mathbf{e}_3$ is $+1$ rather than -1.

Using the two results above, the intersection of three planes π_1, π_2 and π_3 is easily seen to be

$$p = \pi_1 \wedge \pi_2 \wedge \pi_3 / \pm \sqrt{(\pi_1 \wedge \pi_2 \wedge \pi_3)(\pi_1 \wedge \pi_2 \wedge \pi_3)^*}.$$

10.2.3   Joins—The Shuffle product

For the 'join' we need another derived product, the shuffle product. This is borrowed directly from Grassmann–Cayley algebras; see Section 8.5. Here the underlying vector space is four-dimensional rather than six-dimensional. Also notice that in Section 8.5 the shuffle product was used to compute the meet of vector spaces rather than the join. The reason for this difference is that here we are representing points by degree-3 elements and planes by elements of degree 1. For more about Grassmann–Cayley algebra see [127].

In a non-degenerate algebra this is unnecessary because this operation can be represented using the exterior product and the pseudo-scalar. Here, multiplication by the pseudo-scalar $\mathbf{e}_1\mathbf{e}_2\mathbf{e}_3\mathbf{e}$, is not an isomorphism of the algebra. So we can either introduce the isomorphism explicitly, using the Hodge star operator, see [100], or introduce the shuffle product.

Let $a = a_1 \wedge a_2 \wedge \cdots \wedge a_j$ and $b = b_1 \wedge b_2 \wedge \cdots \wedge b_k$ in a general Clifford algebra, with $j + k \geq n$ the dimension of the algebra; then

$$a \vee b = \frac{1}{(n-k)!(j+k-n)!} \sum_\sigma \mathrm{sign}(\sigma)$$
$$\det(a_{\sigma(1)}, \ldots, a_{\sigma(n-k)}, b_1, \ldots, b_k)a_{\sigma(4-k+1)} \wedge \cdots \wedge a_{\sigma(j)}.$$

The sum is taken over all permutations σ of $\{1, 2, \ldots, j\}$, and in our case $n = 4$.

If we write, $a_i = a_{i1}\mathbf{e}_1 + a_{i2}\mathbf{e}_2 + \cdots + a_{in}\mathbf{e}_n$, the determinant is of a matrix with columns $a_{\sigma(1)i}, a_{\sigma(2)i}, \ldots, b_{\sigma(k)i}$. The shuffle product is extended to an entire algebra by demanding that it distribute over addition.

As an example consider $\mathbf{e}_1\mathbf{e}_2\mathbf{e}_3 \vee \mathbf{e}_2\mathbf{e}_3\mathbf{e}$. Clifford products are the same as exterior products for orthogonal generators, so

$$\mathbf{e}_1\mathbf{e}_2\mathbf{e}_3 \vee \mathbf{e}_2\mathbf{e}_3\mathbf{e} = \frac{1}{2}\det(\mathbf{e}_1, \mathbf{e}_2, \mathbf{e}_3, \mathbf{e})\mathbf{e}_2\mathbf{e}_3 - \frac{1}{2}\det(\mathbf{e}_1, \mathbf{e}_2, \mathbf{e}_3, \mathbf{e})\mathbf{e}_3\mathbf{e}_2$$
$$+ \frac{1}{2}\det(\mathbf{e}_2, \mathbf{e}_2, \mathbf{e}_3, \mathbf{e})\mathbf{e}_3\mathbf{e}_1 - \frac{1}{2}\det(\mathbf{e}_2, \mathbf{e}_2, \mathbf{e}_3, \mathbf{e})\mathbf{e}_1\mathbf{e}_3$$
$$+ \frac{1}{2}\det(\mathbf{e}_3, \mathbf{e}_2, \mathbf{e}_3, \mathbf{e})\mathbf{e}_1\mathbf{e}_2 - \frac{1}{2}\det(\mathbf{e}_3, \mathbf{e}_2, \mathbf{e}_3, \mathbf{e})\mathbf{e}_2\mathbf{e}_1,$$

where e taken as "e_4". Only the first pair of terms here are non-vanishing so, after simplification using the fact that $e_2 e_3 = -e_3 e_2$, we have the result

$$e_1 e_2 e_3 \vee e_2 e_3 e = e_2 e_3.$$

Notice that if **a** and **b** are element in the Clifford algebra with grades j and k respectively, then the their shuffle product $(\mathbf{a} \vee \mathbf{b})$ has grade $j + k - n$.

Consider a pair of points

$$p_1 = e_1 e_2 e_3 + x_1 e_2 e_3 e + y_1 e_3 e_1 e + z_1 e_1 e_2 e$$

and

$$p_2 = e_1 e_2 e_3 + x_2 e_2 e_3 e + y_2 e_3 e_1 e + z_2 e_1 e_2 e.$$

Their shuffle product is given by

$$p_1 \vee p_2 = (x_2 - x_1)e_2 e_3 + (y_2 - y_1)e_3 e_1 + (z_2 - z_1)e_1 e_2$$
$$+ (y_1 z_2 - z_1 y_2)e_1 e + (z_1 x_2 - x_1 z_2)e_2 e + (x_1 y_2 - y_1 x_2)e_3 e.$$

This is clearly proportional to the line joining the two points. So the join of two points is the line

$$\ell = p_1 \vee p_2 / \sqrt{(p_1 \vee p_2)(p_1 \vee p_2)^*}.$$

Similarly, the join of a point and a line is the plane

$$\pi = p \vee \ell / \sqrt{(p \vee \ell)(p \vee \ell)^*},$$

since

$$p \vee \ell = (xu_x + yu_y + zu_z)e + (u_x - yv_z + zv_y)e_1$$
$$+ (u_y - zv_x + xv_z)e_2 + (u_z - xv_y + yv_x)e_3.$$

To see that this is proportional to the plane in which the point p and the line ℓ lie, assume that **r** is a point on the line. Now the vector $(\mathbf{r} - \mathbf{p})$ and the vector in the direction of the line **v** both lie in the plane and so the normal vector to the plane will be given by $(\mathbf{r} - \mathbf{p}) \times \mathbf{v} = \mathbf{u} - \mathbf{p} \times \mathbf{v}$. Moreover, the perpendicular distance from the plane to the origin is proportional to $\mathbf{p} \cdot (\mathbf{u} - \mathbf{p} \times \mathbf{v}) = \mathbf{p} \cdot \mathbf{u}$, with the same constant of proportionality.

If the point lies on the line, the shuffle will be zero, so this gives another form of the incidence relation found in Section 10.2.1 above for a point to lie on a line.

From the above it is simple to see that the plane generated by three points will be given by

$$\pi = p_1 \vee p_2 \vee p_3 / \sqrt{(p_1 \vee p_2 \vee p_3)(p_1 \vee p_2 \vee p_3)^*}.$$

Again this also gives us a test for colinearity, the points will be colinear if and only if

$$p_1 \vee p_2 \vee p_3 = 0.$$

The above extends to a test for coplanarity. The shuffle product of a point p and a plane π is

$$p \vee \pi = d - x n_x - y n_y - z n_z.$$

Hence we get another version of the incidence relation given above; the point p lies on the plane π if and only if

$$p \vee \pi = 0.$$

A useful consequence of this is that we can assert that four points, p_1, p_2, p_3 and p_4 are coplanar if and only if

$$p_1 \vee p_2 \vee p_3 \vee p_4 = 0.$$

This also extends to a test for lines to be coplanar. From the above it is clear that a pair of lines ℓ_1 and ℓ_2 are coplanar if and only if

$$\ell_1 \vee \ell_2 = 0.$$

More generally, the shuffle of a pair of lines

$$\ell_1 = v_x \mathbf{e}_2 \mathbf{e}_3 + v_y \mathbf{e}_3 \mathbf{e}_1 + v_z \mathbf{e}_1 \mathbf{e}_2 + u_x \mathbf{e}_1 \mathbf{e} + u_y \mathbf{e}_2 \mathbf{e} u_z \mathbf{e}_3 \mathbf{e},$$
$$\ell_2 = w_x \mathbf{e}_2 \mathbf{e}_3 + w_y \mathbf{e}_3 \mathbf{e}_1 + w_z \mathbf{e}_1 \mathbf{e}_2 + \mu_x \mathbf{e}_1 \mathbf{e} + \mu_y \mathbf{e}_2 \mathbf{e} \mu_z \mathbf{e}_3 \mathbf{e},$$

is given by

$$\ell_1 \vee \ell_2 = v_x \mu_x + v_y \mu_y + v_z \mu_z + u_x w_x + u_y w_y + u_z w_z,$$

this is just the familiar reciprocal product of screws.

10.2.4 Perpendicularity—The Contraction

So far, using the exterior and shuffle product, we can discuss incidence, meets and joins. This is essentially projective geometry and indeed Grassmann–Cayley algebra was designed for computations in projective geometry. For Euclidean geometry we need to be able to calculate angles and distances.

In order to discuss perpendicularity neatly, we introduce yet another derived product on the Clifford algebra. This is the contraction which is defined in a very similar way to the exterior product. Consider a generator of the underlying vector space $\mathbf{x} \in V$ and an arbitrary element of the algebra $\mathbf{c}$, the contraction of $\mathbf{c}$ by $\mathbf{x}$ is defined as

$$\mathbf{x} \lrcorner \, \mathbf{c} = \frac{1}{2}\big(\mathbf{x}\mathbf{c} - \alpha(\mathbf{c})\mathbf{x}\big).$$

Compare this with the definition for the exterior product given in Section 9.1 above. Notice that for vectors $\mathbf{x}$, the Clifford product is given by

$$\mathbf{x}\mathbf{c} = \mathbf{x} \wedge \mathbf{c} + \mathbf{x} \lrcorner \, \mathbf{c}.$$

From the definition given above, it is possible to derive the relation

$$\mathbf{x} \lrcorner (\mathbf{c}_1 \wedge \mathbf{c}_2) = (\mathbf{x} \lrcorner \mathbf{c}_1) \wedge \mathbf{c}_2 + \alpha(\mathbf{c}_1) \wedge (\mathbf{x} \lrcorner \mathbf{c}_2),$$

for arbitrary elements of the algebra $\mathbf{c}_1$, $\mathbf{c}_2$, see Lounesto [69, Chap. 3.]. We extend the definition to the whole of the Clifford algebra by demanding the relation

$$(\mathbf{c}_1 \wedge \mathbf{c}_2) \lrcorner \mathbf{c}_3 = \mathbf{c}_1 \lrcorner (\mathbf{c}_2 \lrcorner \mathbf{c}_3),$$

for three arbitrary elements, and of course the property that the contraction distributes over addition.

Now we are in a position to compute the contractions of various linear elements. First consider a pair of planes

$$\pi_1 = n_x \mathbf{e}_1 + n_y \mathbf{e}_2 + n_z \mathbf{e}_3 + d\mathbf{e},$$
$$\pi_2 = m_x \mathbf{e}_1 + m_y \mathbf{e}_2 + m_z \mathbf{e}_3 + h\mathbf{e}.$$

By direct computation we have

$$\pi_1 \lrcorner \pi_2 = -(n_x m_x + n_y m_y + n_z m_z).$$

This is the negative of the cosine of the angle between the planes and clearly, the planes are mutually perpendicular if and only if $\pi_1 \lrcorner \pi_2 = 0$.

Next we look at the contraction of a line by a plane,

$$\pi \lrcorner \ell = (n_y v_z - n_z v_y)\mathbf{e}_1 + (n_z v_x - n_x v_z)\mathbf{e}_2$$
$$+ (n_x v_y - n_y v_x)\mathbf{e}_3 - (n_x u_x + n_y u_y + n_z u_z)\mathbf{e}.$$

This is almost another plane; dividing by the appropriate constant gives the plane containing the line but perpendicular to the original plane,

$$\pi^\perp = \pi \lrcorner \ell / \sqrt{(\pi \lrcorner \ell)(\pi \lrcorner \ell)^*}.$$

This plane is determined uniquely unless the original line and plane are perpendicular, in which case $\pi \lrcorner \ell = 0$.

In a similar manner the contraction of a point by a plane gives

$$\pi \lrcorner p = -n_x \mathbf{e}_2 \mathbf{e}_3 - n_y \mathbf{e}_3 \mathbf{e}_1 - n_z \mathbf{e}_1 \mathbf{e}_2$$
$$- (yn_z - zn_y)\mathbf{e}_1 \mathbf{e} - (zn_x - xn_z)\mathbf{e}_2 \mathbf{e} - (xn_y - yn_x)\mathbf{e}_3 \mathbf{e}.$$

This is a line (anti)-perpendicular to the original plane but passing through the point. Notice that we don't need to normalise here since we can assume that the normal to the plane $\mathbf{n} = (n_x, n_y, n_z)^T$ was already a unit vector. However, it might be useful to divide by -1 so that the orientations of the plane and perpendicular line agree. The perpendicular distance from the point to the plane is given by the shuffle product as we saw in the previous subsection,

$$p \vee \pi = d - xn_x - yn_y - zn_z.$$

Notice that this distance has a sign, that is we can tell which side of the plane the point lies on. The distance is positive if the point lies on the same side of the plane as the normal vector and is negative if it lies on the other side.

The contraction of a pair of lines

$$\ell_1 = v_x \mathbf{e}_2 \mathbf{e}_3 + v_y \mathbf{e}_3 \mathbf{e}_1 + v_z \mathbf{e}_1 \mathbf{e}_2 + u_x \mathbf{e}_1 \mathbf{e} + u_y \mathbf{e}_2 \mathbf{e} + u_z \mathbf{e}_3 \mathbf{e},$$
$$\ell_2 = w_x \mathbf{e}_2 \mathbf{e}_3 + w_y \mathbf{e}_3 \mathbf{e}_1 + w_z \mathbf{e}_1 \mathbf{e}_2 + \mu_x \mathbf{e}_1 \mathbf{e} + \mu_y \mathbf{e}_2 \mathbf{e} + \mu_z \mathbf{e}_3 \mathbf{e}$$

is

$$\ell_1 \lrcorner \ell_2 = -(v_x w_x + v_y w_y + v_z w_z).$$

Above we saw that the reciprocal product of screws was given by their shuffle, now we see that the Killing form is given by the contraction. This means that the lines are perpendicular if and only if $\ell_1 \lrcorner \ell_2 = 0$.

How can we find the common perpendicular to a pair of lines? That is, we seek a line ℓ, which is perpendicular to both ℓ_1 and ℓ_2 and also meets both of these lines. This can be expressed as

$$\ell \lrcorner \ell_i = 0, \qquad \ell \vee \ell_i = 0, \qquad i = 1, 2.$$

Now it is not too difficult to see that the commutator $s = [\ell_1, \ell_2] = \ell_1 \ell_2 - \ell_2 \ell_1$ satisfies these equations. The problem is that it is not generally a line; $ss^* \neq 1$. Rather s is a screw with a non-zero pitch. From above, the pitch of the screw will be $h = (s \vee s)/(-2s \lrcorner s)$. Also, if s is given by a grade-2 element of the form

$$s = v_x \mathbf{e}_2 \mathbf{e}_3 + v_y \mathbf{e}_3 \mathbf{e}_1 + v_z \mathbf{e}_1 \mathbf{e}_2 + (u_x + hv_x)\mathbf{e}_1 \mathbf{e} + (u_y + hv_y)\mathbf{e}_2 \mathbf{e} + (u_z + hv_z)\mathbf{e}_3 \mathbf{e},$$

then the contraction of the top element in the algebra with s is

$$s \lrcorner \mathbf{e}_1 \mathbf{e}_2 \mathbf{e}_3 \mathbf{e} = -(v_x \mathbf{e}_1 \mathbf{e} + v_y \mathbf{e}_2 \mathbf{e} + v_z \mathbf{e}_3 \mathbf{e}).$$

Hence, we can form the element λ:

$$\lambda = -2(s \lrcorner s)s + (s \lrcorner \mathbf{e}_1 \mathbf{e}_2 \mathbf{e}_3 \mathbf{e})(s \vee s)$$

and at last the required line is given by normalising $\ell = \lambda/\sqrt{\lambda \lambda^*}$. This procedure fails when the original lines are parallel; in such a case both, $s \lrcorner s = 0$ and $s \vee s = 0$. It is curious that this construction is so different from the others given here, but it is clear that it cannot be as simple as the others since none of the wedge, shuffle or contraction products give a result with the correct grade.

The contraction of a point p by a line ℓ is a plane,

$$\ell \lrcorner p = -v_x \mathbf{e}_1 - v_y \mathbf{e}_2 - v_z \mathbf{e}_3 - (xv_x + yv_y + zv_z)\mathbf{e}.$$

This plane is (anti)-perpendicular to the line but passes through the point. Again it is not necessary to normalise since the direction vector of the line is already a unit vector. The square of the perpendicular distance from the point to the line is given by, $(p \vee \ell)(p \vee \ell)^*$.

Finally here, we look at the distance between a pair of points p_1 and p_2. The contraction of two points doesn't tell us anything: $p_1 \lrcorner p_2 = 1$. However, from the above we can see that the square of the distance between the points is given by, $-(p_1 \vee p_2) \lrcorner (p_1 \vee p_2)$.

10.3   Pieper's Theorem

In his Ph.D. thesis, Pieper [87] showed that any 6-R robot that has three consecutive joint axes meeting at a point has solvable inverse kinematics. Later, Duffy showed [30] that this was also true when any three consecutive joints are parallel.

The exact meaning of solvability is not too important since constructive proofs were given. Clearly if non-solvability results were to be considered the precise meaning of the term 'solvable' would be very important.

The demonstration given here roughly follows the work of Pieper, but the computations using the Clifford algebra are simpler and hence the underlying geometry is much clearer.

We begin by looking at the general case, the kinematic relations for a 6-R robot with three consecutive joints parallel or intersecting.

10.3.1   Robot Kinematics

The forward kinematics of a six-joint serial manipulator can be expressed neatly in the Clifford algebra by imitating the product of exponentials formula given in Section 4.5,

$$k(\boldsymbol{\theta}) = a_1(\theta_1)a_2(\theta_2)a_3(\theta_3)a_4(\theta_4)a_5(\theta_5)a_6(\theta_6),$$

where the a_is are the exponentials of the Lie algebra elements corresponding to the robot's joints. So if ℓ_i is the line along the axis of the i-th joint, we have

$$a_i(\theta_i) = e^{\frac{\theta_i}{2}\ell_i} = \cos\frac{\theta_i}{2} + \sin\frac{\theta_i}{2}\ell_i.$$

The exponentials of these degree-2 elements in the Clifford algebra can be found in exactly the same way that the dual quaternion exponentials were computed in Section 9.3 above.

Now suppose that three consecutive joints intersect or are parallel. For the sake of illustration we assume here that it is joints $2, 3$ and 4 which have this property, but it is clear how to proceed in other cases.

1. Rearrange the kinematic equation to isolate the three coincident/parallel joints

$$a_2a_3a_4 = a_1^*ka_6^*a_5^*.$$

   To simplify notation we drop the explicit dependence on the joint angles.

2. If joints $2, 3$ and 4 are coincident, then their common point p will be preserved by a_2, a_3 and a_4, so

$$a_2a_3a_4pa_4^*a_3^*a_4^* = p = a_1^*ka_6^*a_5^*pa_5a_6k^*a_1.$$

On the other hand if the joints are parallel there will be a plane π preserved by the joints so any plane perpendicular to the parallel joint axes will do,

$$a_2 a_3 a_4 \pi a_4^* a_3^* a_4^* = \pi = a_1^* k a_6^* a_5^* \pi a_5 a_6 k^* a_1.$$

This splits the problem into two pieces, in the case of three intersecting joints we have

$$k^* a_1 p a_1^* k = a_6^* a_5^* p a_5 a_6,$$
$$a_2 a_3 a_4 = a_1^* k a_6^* a_5^* = k'$$

and for the parallel case

$$k^* a_1 \pi a_1^* k = a_6^* a_5^* \pi a_5 a_6,$$
$$a_2 a_3 a_4 = a_1^* k a_6^* a_5^* = k'.$$

3. The first of these equations only involves the joint angles θ_1, θ_5 and θ_6. Once this equation has been solved, we can evaluate $k' = a_1 k a_6^* a_5^*$ and solve the second equation for the remaining joint angles, θ_2, θ_3 and θ_4. So our first task is to solve the first of these equations. For three intersecting joints, this equation is a relation between points with the general form

$$p_\alpha = a_6^* p_\beta a_6.$$

Now the plane through p_β and perpendicular to ℓ_6 will not be affected by a rotation about ℓ_6 so we have the equation

$$\ell_6 \, \lrcorner \, p_\alpha = \ell_6 \, \lrcorner \, p_\beta.$$

We know that the normal to the plane will be anti-parallel to the direction of ℓ_6, so this gives us one equation from the coefficients of $\mathbf{e}$. To get another equation independent of θ_6, we can look at the perpendicular distances from p_α and p_β to ℓ_6,

$$(p_\alpha \vee \ell_6)(p_\alpha \vee \ell_6)^* = (p_\beta \vee \ell_6)(p_\beta \vee \ell_6)^*.$$

For three parallel joint axes, things are a little simpler. We have an equation of the form

$$\pi_\alpha = a_6^* \pi_\beta a_6,$$

to solve. The point where the plane π_β meets ℓ_6 will remain fixed. This common point is a scalar multiple of the Clifford algebra expression, $\pi_\beta \wedge \ell_6 = (\pi_\beta \ell_6 + \ell_6 \pi_\beta)/2$. This allows us to eliminate the last joint angle and write the equation as

$$\ell_6 \wedge (k^* a_1 \pi a_1^* k) = \ell_6 \wedge (a_5^* \pi a_5).$$

Observe that since $a_i = \cos(\theta_i/2) + \sin(\theta_i/2)\ell_i$, the expressions, $a_1\pi a_1^*$ and $a_5\pi a_5^*$ are linear in $\cos\theta_1$, $\sin\theta_1$ and $\cos\theta_5$, $\sin\theta_5$, respectively. There are effectively two linear equations here, the set of planes is three-dimensional and the equations tell us nothing about planes containing the line ℓ_6. We get a final pair of equations from the trigonometric identities $\cos^2\theta_1 + \sin^2\theta_1 = 1$ and $\cos^2\theta_5 + \sin^2\theta_5 = 1$. Thus, in general, we must solve a pencil of conics. There will be a maximum of four real solutions.

4. Having found θ_1 and θ_5, the angle θ_6 is simple to find using the original equation $k^* a_1 \pi a_1^* k = a_6^* a_5^* \pi a_5 a_6$. This system of equations gives essentially two linear equations in the variables $\cos\theta_6$ and $\sin\theta_6$. Hence, we obtain a unique solution for θ_6 given particular values for θ_1 and θ_5.

5. Next we must solve the second of our equations $a_2 a_3 a_4 = k'$, where $k' = a_1 k a_6^* a_5^*$. The problem has been reduced to a subgroup of the full rigid body motion group. In the case of three intersecting joints the problem has been reduced to the inverse kinematics of the spherical wrist that was studied in Section 5.1. For three parallel joints the problem has been reduced to the inverse kinematics of a planar manipulator. That is the problem has been reduced to $SO(3)$ and $SE(2)$ respectively. Notice however, that there are four possible values that k' can take corresponding to the four solutions for the angles θ_1, θ_5 and θ_6.

The relation $a_2 a_3 a_4 = k'$, is a relation between group elements, so we can eliminate a_4 by applying this group element to the 4th joint axis

$$a_2 a_3 a_4 \ell_4 a_4^* a_3^* a_2^* = a_2 a_3 \ell_4 a_3^* a_2^* = k' \ell_4 k'^*.$$

This is now a relation between lines; if the lines ℓ_2, ℓ_3 and ℓ_4 are coincident, then a_2 can be eliminated using the fact that for a pair of lines ℓ_α, ℓ_β the expression $\ell_\alpha \lrcorner \ell_\beta$ is an invariant, with respect to the group of rigid body motions. This leads us to the expression

$$a_3 \ell_4 a_3^* \lrcorner \ell_2 = k' \ell_4 k'^* \lrcorner \ell_2.$$

Again, this is linear in the variables $\cos\theta_3$ and $\sin\theta_3$. Solving this linear equation with the trigonometric identity $\cos^2\theta_3 + \sin^2\theta_3 = 1$ gives two solutions. Notice the shuffle product gives zero here because the lines are coincident.

On the other hand, if the three lines are parallel we can eliminate a_2 by setting $\ell_\alpha = \ell_2$ and $\ell_\beta = k'\ell_4 k'* = a_2 a_3 \ell_4 a_3^* = a_4^*$ then finding the expression, $(1/2)(\ell_\alpha \ell_\beta^* - \ell_\beta \ell_\alpha^*)$. Because the lines are parallel this will give $s_x e_1 e + s_y e_2 e + s_z e_3 e$ where $\mathbf{s} = (s_x, s_y, s_z)^T$ is a vector from one line to the other, perpendicular to both. The length of this vector $s^2 = s_x^2 + s_y^2 + s_z^2$ is then invariant under an overall rigid motion and will depend only on θ_3. In fact the expression we get will be the cosine rule for the triangle

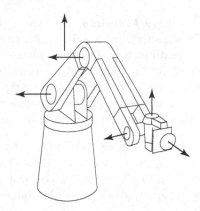

FIGURE 10.2. The T^3 Robot

formed by the three lines meeting a perpendicular plane. Hence we obtain two solutions for θ_3 corresponding to the two possible signs for $\sin\theta_3$.

6. In either case we can now retrace our steps and solve

$$a_2 a_3 \ell_4 a_3^* a_2^* = k' \ell_4 k'^*$$

to get a unique answer for θ_2.

7. Finally we use

$$a_4 = a_3^* a_2^* k'$$

to recover θ_4.

Notice that we have shown that these robots have a maximum of eight distinct solutions for their inverse kinematics. For a general 6-R robot, where no three consecutive joints intersect or are parallel, it can be shown that the inverse kinematic problem has 16 solutions; see Section 11.5.

In the following two sections a pair of examples is given in order to make the procedure more concrete.

10.3.2   The T^3 Robot

This is a large industrial robot manufactured by Cincinnati Milacron. The acronym T^3 was intended to stand for "The Tomorrow Tool". The second, third and fourth joint axes of this robot are parallel; see Figure 10.2.

We begin with a list of the joint axes in their home configuration and we choose a simple configuration with the arm stretched along the vertical z-axis,

$$\ell_1 = \mathbf{e}_1 \mathbf{e}_2,$$
$$\ell_2 = \mathbf{e}_2 \mathbf{e}_3,$$
$$\ell_3 = \mathbf{e}_2 \mathbf{e}_3 + l_1 \mathbf{e}_2 \mathbf{e},$$

$$\ell_4 = \mathbf{e}_2\mathbf{e}_3 + (l_1 + l_2)\mathbf{e}_2\mathbf{e},$$
$$\ell_5 = \mathbf{e}_3\mathbf{e}_1 - (l_1 + l_2 + l_3)\mathbf{e}_1\mathbf{e},$$
$$\ell_6 = \mathbf{e}_1\mathbf{e}_2.$$

Here the dimensions l_1, l_2 and l_3 are the constant design parameters of the robot. Notice that the real robot may have joint limits and so might not be able to reach this home position. This is not too important, since it is only a reference configuration from which to measure the joint angles.

A plane perpendicular to joints $2, 3$ and 4 is given by $\pi = \mathbf{e}_1$. The first equation we have to solve is

$$\ell_6 \wedge (k^* a_1 \pi a_1^* k) = \ell_6 \wedge (a_5^* \pi a_5).$$

Recall that $a_i = \cos(\theta_i/2) + \sin(\theta_i/2)\ell_i$, so after some computation, the right-hand side of the equation becomes

$$\ell_6 \wedge (a_5^* \pi a_5) = \ell_6 \wedge \left(\cos\theta_5 \mathbf{e}_1 + \sin\theta_5 \mathbf{e}_3 + (l_1 + l_2 + l_3)\sin\theta_5 \mathbf{e} \right),$$
$$= \sin\theta_5 \mathbf{e}_1\mathbf{e}_2\mathbf{e}_3 + (l_1 + l_2 + l_3)\sin\theta_5 \mathbf{e}_1\mathbf{e}_2\mathbf{e}.$$

The left-hand side requires more computation since we must include a general rigid motion k. Let us write this general motion as a rotation followed by a translation,

$$k = \mathbf{r} + \frac{1}{2}\mathbf{t}\mathbf{r}\mathbf{e},$$

where

$$\mathbf{r} = \cos\frac{\phi}{2} + v_x \sin\frac{\phi}{2}\mathbf{e}_2\mathbf{e}_3 + v_y \sin\frac{\phi}{2}\mathbf{e}_3\mathbf{e}_1 + v_z \sin\frac{\phi}{2}\mathbf{e}_1\mathbf{e}_2$$

and

$$\mathbf{t} = l_x\mathbf{e}_1 + l_y\mathbf{e}_2 + t_z\mathbf{e}_3.$$

It is useful at this stage to write

$$k^* a_1 \pi a_1^* k = N_x\mathbf{e}_1 + N_y\mathbf{e}_2 + N_z\mathbf{e}_3 + D\mathbf{e}.$$

where

$$N_x = (\cos\phi + v_x^2(1 - \cos\phi))\cos\theta_1 + (v_z\sin\phi + v_xv_y(1 - \cos\phi))\sin\theta_1,$$
$$N_y = (v_xv_y(1 - \cos\phi) - v_z\sin\phi)\cos\theta_1 + (\cos\phi + v_y^2(1 - \cos\phi))\sin\theta_1,$$
$$N_z = (v_y\sin\phi + v_xv_z(1 - \cos\phi))\cos\theta_1 + (v_yv_z(1 - \cos\phi) - v_x\sin\phi)\sin\theta_1,$$
and $D = N_xt_x + N_yt_y + N_zt_z.$

Now the left-hand side of the equation can be written as

$$\ell_6 \wedge (k^* a_1 \pi a_1^* k) = N_z\mathbf{e}_1\mathbf{e}_2\mathbf{e}_3 + D\mathbf{e}_1\mathbf{e}_2\mathbf{e}.$$

Comparing the coefficients of the basis elements gives us just two equations in the first and fifth joint angles,

$$N_z = \sin\theta_5, \quad \text{and} \quad D = (l_1 + l_2 + l_3)\sin\theta_5.$$

So we don't have to solve a pair of quadratic equations here, we can eliminate $\sin\theta_5$ to get a linear equation in the sine and cosine of θ_1,

$$D - (l_1 + l_2 + l_3)N_z = 0.$$

Solving this with the trigonometric identity $\cos^2\theta_1 + \sin^2\theta_1 = 1$ gives two solutions in general,

$$\cos\theta_1 = \frac{-\alpha\gamma \pm \beta\sqrt{\alpha^2 + \beta^2}}{\alpha^2 + \beta^2} \quad \text{with} \quad \sin\theta_1 = -(\alpha\cos\theta_1)/\beta.$$

The coefficients α and β are functions only of the end-effector's position and orientation,

$$\begin{aligned}
\alpha = &\left((1 - v_x^2)t_x - v_x v_y t_y - v_x v_z(t_y - l_1 - l_2 - l_2)\right)\cos\phi \\
&+ \left(v_y(t_z - l_1 - l_2 - l_3) - v_z t_y\right)\sin\phi \\
&+ v_y t_y - v_z(t_z - l_1 - l_2 - l_3)),
\end{aligned}$$

$$\begin{aligned}
\beta = &\left((1 - v_y^2)t_y - v_x v_y t_x - v_y v_z(t_z - l_1 - l_2 - l_3)\right)\cos\phi \\
&+ v_x\left(v_x t_x + \left(v_z t_x - v_x(t_z - l_1 - l_2 - l_3)\right)\right)\sin\phi \\
&+ v_y\left(v_x t_x + v_y t_y + v_z(t_z - l_1 - l_2 - l_3)\right).
\end{aligned}$$

For each of the two solutions for θ_1 we get two solutions for θ_5 given by

$$\sin\theta_5 = N_z \quad \text{and} \quad \cos\theta_5 = \pm\sqrt{1 - N_z^2}.$$

To find θ_6 we solve the linear equations

$$(k^* a_1 \pi a_1^* k) = a_6^*(a_5^* \pi a_5)a_6.$$

That is,

$$\begin{aligned}
(N_x \mathbf{e}_1 + N_y \mathbf{e}_2 + N_z \mathbf{e}_3 + D\mathbf{e}) &= a_6^*\left(\cos\theta_5\mathbf{e}_1 + \sin\theta_5\mathbf{e}_3 + (l_1 + l_2 + l_3)\sin\theta_5\mathbf{e}\right)a_6, \\
&= \cos\theta_5\cos\theta_6\mathbf{e}_1 - \cos\theta_5\sin\theta_6\mathbf{e}_2 \\
&\quad + \sin\theta_5\mathbf{e}_3 + (l_1 + l_2 + l_3)\sin\theta_5\mathbf{e}.
\end{aligned}$$

Comparing coefficients we have that

$$\cos\theta_6 = N_x/\cos\theta_5, \quad \text{and} \quad \sin\theta_6 = -N_y/\cos\theta_5.$$

From the results above we can calculate $k' = a_1 k a_6^* a_5^*$. This must be an element of the sub-group generated by a_2, a_3 and a_4 which is the group of motions in the yz-plane. Hence we can write k' as a rotation about the x-axis followed by a translation in the yz-plane,

$$k' = \cos\frac{\phi'}{2} + \sin\frac{\phi'}{2}\mathbf{e}_2\mathbf{e}_3 + \frac{1}{2}(t'_y\cos\frac{\phi'}{2} + t'_z\sin\frac{\phi'}{2})\mathbf{e}_2\mathbf{e} + \frac{1}{2}(t'_z\cos\frac{\phi'}{2} - t'_y\sin\frac{\phi'}{2})\mathbf{e}_3\mathbf{e}.$$

Now we must solve the second part of the problem,

$$a_2a_3a_4 = k'.$$

To eliminate a_2 and a_4 we examine

$$\frac{1}{2}(a_3\ell_4a_3^*\ell_2^* - \ell_2a_3\ell_4^*a_3^*) = -l_2\sin\theta_3\mathbf{e}_2\mathbf{e} + (l_1 + l_2\cos\theta_3)\mathbf{e}_3\mathbf{e}$$
$$= \frac{1}{2}(k'\ell_4k'^*\ell_2^* - \ell_2k'\ell_4^*k'^*),$$

so $s^2 = l_1^2 + l_2^2 + 2l_1l_2\cos\theta_3$. From the right-hand side of the equation we have

$$s^2 = (l_1 + l_2)^2 + 2(l_1 + l_2)(t'_z\cos\phi' - t'_y\sin\phi') + t'^{\,2}_y + t'^{\,2}_z.$$

The two solutions for θ_3 are thus

$$\cos\theta_3 = (s^2 - l_1^2 - l_1^2)/2l_1l_2, \quad\text{and}\quad \sin\theta_3 = \pm\sqrt{1 - \cos^2\theta_3}.$$

Next we can find θ_2 from

$$a_2a_3\ell_4a_3^*a_2^* = k'\ell_4k'^*.$$

The right-hand side can be written

$$k'\ell_4k'^* = \mathbf{e}_2\mathbf{e}_3 + ((l_1 + l_2)\cos\phi' + t'_z)\mathbf{e}_2\mathbf{e} + ((l_1 + l_2)\sin\phi' - t'_y)\mathbf{e}_3\mathbf{e},$$

which we can abbreviate to $k'\ell_4k'^* = \mathbf{e}_2\mathbf{e}_3 + X\mathbf{e}_2\mathbf{e} + Y\mathbf{e}_3\mathbf{e}$, with

$$X = (l_2 + l_3)\cos\phi' + t'_z,$$
$$Y = (l_2 + l_3)\sin\phi' - t'_y.$$

The left-hand side is

$$a_2a_3\ell_4a_3^*a_2^* = \mathbf{e}_2\mathbf{e}_3 + (\cos\theta_2(l_1 + l_2\cos\theta_3) - l_1\sin\theta_2\sin\theta_3)\mathbf{e}_2\mathbf{e}$$
$$+ (\sin\theta_2(l_1 + l_2\cos\theta_3) + l_1\cos\theta_2\sin\theta_3)\mathbf{e}_3\mathbf{e}.$$

So we get a unique solution for θ_2,

$$\cos\theta_2 = \big(X(l_1 + l_2\cos\theta_3) + Yl_2\sin\theta_3)\big)/(l_1^2 + l_2^2 + 2l_1l_2\cos\theta_3),$$
$$\sin\theta_2 = \big(Y(l_1 + l_2\cos\theta_3) - Xl_2\sin\theta_3)\big)/(l_1^2 + l_2^2 + 2l_1l_2\cos\theta_3).$$

Finally we must solve for θ_4, which can be found from

$$a_4 = a_3^* a_2^* g'.$$

By looking at the rotation part of this we get

$$\theta_4 = \phi' - \theta_2 - \theta_3.$$

There are several other commercially available robots with three consecutive parallel joints. In the 1990s Telequipment produced a small 'table-top' manipulator which was used as a home experiment kit for an Open University course on robotics. Again the second, third and fourth joints of this robot were parallel but the wrist was slightly different from the T^3. In SCARA robots the first three joints are parallel. These are usually 5-joint robots and were developed in Japan for assembling electronic circuit boards. The acronym stands for "Selective Compliance Robot Arm".

10.3.3   The PUMA

The PUMA robot was designed by the Unimation company for General Motors. The name is again an acronym, this time for "Programmable Universal Machine Arm". It was originally intended to replace human workers on the production line. So the design is (loosely) based on the capabilities of the human arm. Hence this design is sometimes referred to as the anthropomorphic arm. A diagram of this design is given in Figure 3.2. It can be seen that the last three joints of the robot meet at a common point, usually called the wrist-centre.

As usual we begin by listing the joint axes in their home configuration, again the home configuration has been chosen so that the arm is stretched along the vertical z-axis,

$$\ell_1 = \mathbf{e}_1\mathbf{e}_2,$$
$$\ell_2 = \mathbf{e}_2\mathbf{e}_3,$$
$$\ell_3 = \mathbf{e}_2\mathbf{e}_3 + l_2\mathbf{e}_2\mathbf{e},$$
$$\ell_4 = \mathbf{e}_1\mathbf{e}_2 - d_2\mathbf{e}_2\mathbf{e},$$
$$\ell_5 = \mathbf{e}_2\mathbf{e}_3 + (l_2 + l_3)\mathbf{e}_2\mathbf{e},$$
$$\ell_6 = \mathbf{e}_1\mathbf{e}_2 - d_2\mathbf{e}_2\mathbf{e}.$$

Again l_2, d_2 and l_3 represent design parameters. The point fixed by the last three joints is the wrist-centre

$$p = \mathbf{e}_1\mathbf{e}_2\mathbf{e}_3 + d_2\mathbf{e}_2\mathbf{e}_3\mathbf{e} + (l_2 + l_3)\mathbf{e}_1\mathbf{e}_2\mathbf{e}.$$

The two equations we have to solve are now

$$a_2 a_3 p a_3^* a_2^* = a_1^* kpk^* a_1,$$
$$a_4 a_5 a_6 = k a_1^* a_2^* a_3^* = k'.$$

Taking the first of these equations, we see that it is exactly the same problem that we looked at in Section 5.2.2 with $l_1 = 0$. As mentioned above the second problem is simply the inverse kinematics of a spherical wrist. Again, this was studied in Section 5.1. So rather than repeat the computation some more general remarks will be made.

From the first of the equations above we get the equation

$$\ell_2 \lrcorner (a_3 p a_3^*) = \ell_2 \lrcorner (a_1^* k p k^* a_1).$$

This equation is linear in the sines and cosines of θ_1 and θ_3 and does not involve θ_2. As explained above this equation relates two planes perpendicular to ℓ_2 through two points $p_\alpha = a_3 p a_3^*$ and $p_\beta = a_1^* k p k^* a_1$. However, there are other ways to find this equation. For example, we could compare the orthogonal projections of these points onto the line ℓ_2. In the formalism developed above this could be written as the meet of the line with the perpendicular planes

$$\ell_2 \wedge (\ell_2 \lrcorner p_\alpha) = \ell_2 \wedge (\ell_2 \lrcorner p_\beta).$$

However, recall from Section 10.2.1 that

$$\frac{1}{2}(\ell p - p\ell) - (zv_y - yv_z + u_x)\mathbf{e}_2\mathbf{e}_3\mathbf{e} + (xv_z - zv_x + u_y)\mathbf{e}_3\mathbf{e}_1\mathbf{e} + (yv_x - xv_y + u_z)\mathbf{e}_1\mathbf{e}_2\mathbf{e}.$$

This is a free vector whose length is the perpendicular distance from the point p to the line ℓ; its direction is normal to the plane containing ℓ and p. To get the free vector perpendicular to the line and joining the line to the point we can rotate the above vector about the line. The appropriate rotation is $\mathbf{g} = (1 - \ell)/\sqrt{2})$ and the rotation angle is $-\pi/2$ so that $\theta/2 = -\pi/4$; this gives

$$\mathbf{g}\frac{1}{2}(\ell p - p\ell)\mathbf{g}^* = \frac{1}{4}(1 - \ell)(\ell p - p\ell)(1 + \ell) = \frac{1}{2}(p + \ell p \ell),$$

remember that $\ell^2 = -1$. The perpendicular projection of p onto ℓ is therefore

$$p - \frac{1}{2}(p + \ell p \ell) = \frac{1}{2}(p - \ell p \ell).$$

So for our linear equation in the sines and cosines of θ_1 and θ_3 we can use the equation

$$p_\alpha - \ell_2 p_\alpha \ell_2 = p_\beta - \ell_2 p_\beta \ell_2.$$

Notice that the above gives us a useful way of writing the result of rotating a general point p about a line ℓ. If we write the rotation as $a(\theta) = \cos(\theta/2) + \ell \sin(\theta/2)$, the position of the point p becomes

$$a(\theta) p a(\theta)^* = \frac{1}{2}(p - \ell p \ell) + \frac{1}{2}(p + \ell p \ell)\cos(\theta) + \frac{1}{2}(\ell p - p\ell)\sin(\theta).$$

The second equation we need can be found from the length of either of the free vectors $(p + \ell p \ell)/2$ or $(\ell p - p \ell)/2$. As we saw in Section 5.2.1, after using the trigonometric identity $\cos^2\theta + \sin^2\theta = 1$, this turns into a linear equation in the sines and cosines of θ_1 and θ_3. So again we are led to solve a pair of conics.

When we have solved for the first three joint angles we must compute $k' = ka_1^* a_2^* a_3^*$. This will be a rotation about the wrist centre, so we can ignore the translational parts of the equation,

$$a_4 a_5 a_6 = k',$$

and treat it exactly as a 3-R wrist as in Section 5.1.

In Section 10.3.1 above, an algorithm was outlined which derives the inverse kinematic relations for any 6-R robot with three consecutive intersecting or parallel joints. This method may not give the most efficient derivation but this is not important since, for any robot the derivation will only be performed once. The fact that the computations are rather lengthy for hand calculation is also not a particular problem since Clifford algebra is ideally suited to automation using symbolic algebra computer programs.

Using the results given above it would be straightforward to write computer programs to find the inverse kinematics for these machines numerically; values would have to be assigned to the design parameters of course. Care should be taken, however, because the problem of singularities has not been addressed. This is not too difficult for these examples since we only need to look for divisors vanishing or the discriminant of a quadratic disappearing.

More generally the problem of singularities in 6-R robots is an important but subtle question. The control systems of robots often run into difficulties near singularities where the Jacobian becomes small. Since the columns of the robot's Jacobian are the joint screws in their current position (Section 4.5) a linear relation between them will mean that at a singularity the joint axes will lie in a line complex; see Section 6.6. However, this doesn't say anything about the nature of the particular singularity nor which designs of robots have which types of singularities. Much work has been done in this area by Adolf Karger, [62].

Can the methods outlined above be used for other types of serial robots? An obvious application would be to robots containing prismatic joints. From the above it is reasonably clear what to do, we must look for sets of consecutive joints which form sub-groups of the group of rigid body motions. If these sub-groups fix a point, a line or a plane, then we can eliminate the corresponding joint angles from the kinematic equation and hence simplify it.

11
The Study Quadric

11.1 Study's Soma

It was probably Study who first considered the possible positions of a rigid body as points in a non-Euclidian space; see Study [118]. His idea was to specify the position of the body by attaching a coordinate frame to it. He called these 'points' **soma**, which is Greek for *body*. He then used dual quaternions as coordinates for the space. As we saw in Section 9.3, using the dual quaternion representation, the elements of the group of rigid body motions can be thought of as the points of a six-dimensional projective quadric (excluding a 3-plane of 'ideal' points). If we fix a particular position of the rigid body as the home position, then all other positions of the body can be described by the unique transformation that takes the home configuration to the present one. In this way, we see that Study's somas are just the points of the six-dimensional projective quadric, the Study quadric, (not forgetting to exclude the points on the special 3-plane).

In modern terminology, the Study quadric is a particular example of a flag manifold. This is a generalisation of the Grassmann manifolds that we met in Section 6.8. The points of a Grassmannian $G(n, m)$ are n-planes in $\mathbb{R}^m$. Points in a flag manifold are **flags**, that is, nested sequences of linear subspaces. Given a sequence of increasing integers $i_1 < i_2 < \cdots < i_n$ a flag of type $(i_1, i_2, \ldots, i_n)$ is a sequence of vector spaces $V_1 \subset V_2 \subset \cdots \subset V_n$ such that the dimension of V_k is i_k for all $k = 1, \ldots, n$. We will denote the space of all flags of type $(i_1, i_2, \ldots, i_n)$ in $\mathbb{R}^m$ as $F(i_1, i_2, \ldots, i_n; m)$. See Hiller [51], for example. Thus, we have the obvious identities

$$F(1; m) = \mathbb{PR}^m \qquad \text{and} \qquad F(n; m) = G(n, m).$$

Recall from Section 6.8 that the Klein quadric of lines in $\mathbb{R}^3$ can be thought of as the Grassmannian $G(2, 4)$, excluding the 'lines at infinity'. This identification was achieved by intersecting the 2-planes in $\mathbb{R}^4$ with a 3-plane not containing the origin. Suppose we extend this idea to the type $(1, 2)$ flags in $\mathbb{R}^4$. The intersection of such a flag with the 3-plane will generally be a line containing a point. We may think of such a figure as a **pointed line**. Hence, with the usual exclusion of the infinite lines, we can identify the flag manifold $F(1, 2; 4)$ with the set of pointed lines in $\mathbb{R}^3$.

In order to represent Study's soma as flags, we need to introduce the concept of oriented flags. In most texts, no reference is made to the orientation of the vector spaces comprising a flag. This is partly because most texts only treat the case where the field of scalars is the complex numbers. However, when working with vector spaces over the real numbers we have a choice of orientations. We can specify an orientation on a vector space most easily by giving an orthonormal basis for the space. For a one-dimensional space—a line—a unit vector picks out a direction along the line. More generally, for an n-dimensional space, two orthonormal bases, $\mathcal{E} = (\mathbf{e}_1, \mathbf{e}_2, \ldots, \mathbf{e}_n)$ and $\mathcal{F} = (\mathbf{f}_1, \mathbf{f}_2, \ldots, \mathbf{f}_n)$, determine the same orientation if the determinants

$$\det(\mathcal{E}) = \mathbf{e}_1 \wedge \mathbf{e}_2 \wedge \cdots \wedge \mathbf{e}_n \qquad \text{and} \qquad \det(\mathcal{F}) = \mathbf{f}_1 \wedge \mathbf{f}_2 \wedge \cdots \wedge \mathbf{f}_n,$$

have the same sign.

Now, to define an **oriented flag**, we specify that the vector subspaces $V_1 \subset V_2 \subset \cdots \subset V_n$ must be oriented subspaces. Moreover, we require that the orientations of the subspaces must be compatible with each other. This can be achieved by giving an orthonormal basis for the largest subspace that restricts to an orthonormal basis for all the smaller subspaces. The space of oriented flags of type $(i_1, i_2, \ldots, i_n)$ in some $\mathbb{R}^m$ will be denoted $F^+(i_1, i_2, \ldots, i_n; m)$.

For the Study soma, a coordinate frame in $\mathbb{R}^3$ corresponds to an oriented flag in $\mathbb{R}^4$. We may specify a point in the flag manifold, that is, an oriented $(1, 2, 3)$, flag by a sequence of three mutually orthogonal unit vectors $\mathbf{e}_1$, $\mathbf{e}_2$ and $\mathbf{e}_3$. The line is specified by the first vector, the 2-plane by the span of the first two vectors and the 3-plane by the span of all three vectors. To see the correspondence, we stereographically project the flag onto a fixed 3-plane not containing the origin. The presence of an orientation means that this is different from simply taking the intersection of the flag with the 3-plane. If the coordinates of $\mathbb{R}^4$ are (x, y, z, w), then we may take the 3-plane to be $w = 1$ and project from the point $(0, 0, 0, -1)$. The projection is then given by

$$pr : (x, y, z, w) \longmapsto \left(\frac{2x}{1 + w}, \frac{2y}{1 + w}, \frac{2z}{1 + w}, 1 \right).$$

Notice that this (partial) mapping takes all points in $\mathbb{R}^4$ to a point on the 3-plane, except for those with $w = -1$. Moreover, if we take the 3-sphere $x^2 + y^2 + z^2 + w^2 = 1$ and remove the point, $w = -1$, then on the remaining portion of the space, the projection is bijective onto the 3-plane.

The unit vector $\mathbf{e}_1$ of the flag will project to a point in the fixed 3-plane that we may take to be the origin of the coordinate frame. The x-axis of the coordinate frame can be defined as the line joining the origin to the projection of $\mathbf{e}_2$. The direction of this axis will be taken as along the direction from the origin to $pr(\mathbf{e}_2)$. The $x - y$ plane of the frame can be taken as the plane defined by the three points $pr(\mathbf{e}_1)$, $pr(\mathbf{e}_2)$ and $pr(\mathbf{e}_3)$. Finally, the z-axis of the frame is directed along the perpendicular to the $x - y$ plane. It is given by the vector product $(pr(\mathbf{e}_2) - pr(\mathbf{e}_1)) \times (pr(\mathbf{e}_3) - pr(\mathbf{e}_1))$. The soma obtained in this way will always be a right-handed coordinate frame. Hence, we see that the space of all somas lies in the flag manifold $F^+(1, 2, 3; 4)$. The flags that do not give somas are just the ones for which $\mathbf{e}_1$ is in the $-w$ direction in $\mathbb{R}^4$. If any of the other unit vectors lie in this direction, then we can still find the coordinate frame in the 3-plane by projecting the linear spaces of the flag.

The flag manifold $F^+(1, 2, 3; 4)$ is homeomorphic to the manifold of the group $SO(4)$. This is because the group $SO(4)$ acts transitively on the flags, simply transforming the unit vectors. The isotropy group of this action is trivial: every group element, except the identity, transforms flags into different ones. So, if we choose a standard, or home, flag we may identify the flag manifold with $SO(4)$. The flag corresponding to some group element is the flag obtained by applying that element to the standard flag. Another way to see this is to identify the flag given by the mutually orthogonal unit vectors $\mathbf{e}_1$, $\mathbf{e}_2$ and $\mathbf{e}_3$ with the $SO(4)$ matrix with columns $\mathbf{e}_1$, $\mathbf{e}_2$, $\mathbf{e}_3$ and $\mathbf{e}_4 = \mathbf{e}_1 \wedge \mathbf{e}_2 \wedge \mathbf{e}_3$.

We may also picture the flag manifold $F^+(1, 2, 3; 4)$ as the set of frames, or frame bundle, of the 3-sphere S^3. Here, $\mathbf{e}_1$ defines a point on the sphere. The vectors $\mathbf{e}_2$, $\mathbf{e}_3$ and $\mathbf{e}_4$ as defined above are orthogonal tangent vectors at the point and hence give an orthonormal basis for the tangent space at the point.

Next, we identify the $SO(4)$ with the whole of the Study quadric. To do this, we look at the double covering of $SO(4)$, the group Spin(4). Recall from Section 9.1 that this group lies in the Clifford algebra $C\ell^+(0, 4, 0)$. A typical element of this space has the form

$$\tilde{h} = a_0 + a_1\mathbf{e}_2\mathbf{e}_3 + a_2\mathbf{e}_3\mathbf{e}_1 + a_3\mathbf{e}_1\mathbf{e}_2 + c_1\mathbf{e}_1\mathbf{e}_4 + c_2\mathbf{e}_2\mathbf{e}_4 + c_3\mathbf{e}_3\mathbf{e}_4 + c_0\mathbf{e}_1\mathbf{e}_2\mathbf{e}_3\mathbf{e}_4.$$

Notice that the basis element $\mathbf{e}_1\mathbf{e}_2\mathbf{e}_3\mathbf{e}_4$ commutes with all the other elements of the even subalgebra $C\ell^+(0, 4, 0)$. However, unlike the dual quaternion case we met in Section 9.3, this element squares to $+1$. Hence, we have the isomorphism

$$C\ell^+(0, 4, 0) = \mathbb{H} \otimes C\ell(1, 0, 0).$$

The algebra $C\ell(1, 0, 0)$ is sometimes called the ring of double numbers; see Yaglom [130, Chap. 1]. We may write the generator of $C\ell(1, 0, 0)$ as σ. This is subject to the condition $\sigma^2 = 1$. Returning to Spin(4), we see that this group lies in the group of units of the double quaternions. A general double quaternion has the form

$$\tilde{h} = a + c\sigma,$$

where a and c are quaternions.

The action of these double quaternions on vectors in $\mathbb{R}^4$ can be written

$$(x_0' + x_1'i\sigma + x_2'j\sigma + x_3'k\sigma) = (h_0 + h_1\sigma)(x_0 + x_1i\sigma + x_2j\sigma + x_3k\sigma)(h_0^* - h_1^*\sigma).$$

Notice that this is not the standard action of Spin(4) on $\mathbb{R}^4$ that was introduced in Section 9.1; although equivalent, the above has the advantage of being compatible with the double quaternion interpretation of $C\ell^+(0, 4, 0)$.

The usual Euclidian metric on this version of $\mathbb{R}^4$ can be written as

$$x_0^2 + x_1^2 + x_2^2 + x_3^2 = \mathbf{x}\mathbf{x}^*,$$

where we have written $\mathbf{x} = x_0 + x_1i\sigma + x_2j\sigma + x_3k\sigma$. Elements of Spin(4) must preserve this metric. Hence, we have

$$\mathbf{x}\mathbf{x}^* = (h_0 + h_1\sigma)\mathbf{x}(h_0^* - h_1^*\sigma)(h_0 - h_1\sigma)\mathbf{x}^*(h_0^* + h_1^*\sigma).$$

This will be true when the quaternions h_0 and h_1 satisfy the equations

$$h_0h_0^* + h_1h_1^* = 1 \qquad \text{and} \qquad h_0h_1^* + h_1h_0^* = 0,$$

that is, elements of the group Spin(4) can be thought of as double quaternions obeying the above conditions.

Suppose g_1, g_2 are elements of Spin(3), that is, unit quaternions. Then an element of Spin(4) is given by

$$h_0 + h_1\sigma = \frac{1}{2}(g_1 + g_2) + \frac{1}{2}(g_1 - g_2)\sigma.$$

It is a simple matter to check that this satisfies the required conditions. The above defines an accidental isomorphism, Spin(4) = Spin(3) × Spin(3).

To get to $SO(4)$ we must identify double quaternions of opposite sign. As in Section 9.3, we do this by looking at the projective space $\mathbb{PR}^7$, whose homogeneous coordinates are given by the components of h_0 and h_1. A line through the origin in $\mathbb{R}^8$ intersects the seven-dimensional sphere $h_0h_0^* + h_1h_1^* = 1$ in two antipodal points, giving the required identification. So, once again, we are left with a single condition for points to lie in the group $SO(4)$. The condition is, of course

$$h_0h_1^* + h_1h_0^* = 0 = a_0c_0 + a_1c_1 + a_2c_2 + a_3c_3.$$

This is exactly the Study quadric, but this time we do not have to make any exceptions; each element of the group corresponds to a point in the quadric and, conversely, each point in the quadric represents an element of the group $SO(4)$.

As we have already observed, not every $(1, 2, 3)$ flag in $\mathbb{R}^4$ corresponds to a soma. The space of somas sits inside the flag manifold $F^+(1, 2, 3; 4)$. The flags not corresponding to somas are simply determined by $h_0 = 0$. The introduction of the compact group $SO(4)$ is rather useful, for it allows us to make some sense out of the completion of the Study quadric. That is, we have some sort of interpretation for the 'somas at infinity'. In some ways, working with $SO(4)$ would be easier than using $SE(3)$; see McCarthy [72].

There is a standard algebraic 'trick' that allows us to go from $SO(n)$ to $SE(n-1)$. It is called the Saletan contraction; see Gilmore [38, Chap.10]. Imagine the group $SO(n)$ as the symmetry group of an $(n-1)$-dimensional, radius r sphere in $\mathbb{R}^n$. In the limit $r \to \infty$ the symmetry group jumps from $SO(n)$ to $SE(n-1)$. In the Clifford algebra calculations outlined above, it is not too difficult to see that we can achieve this transformation by writing $\sigma^2 = 1/r^2$ and then taking the limit $r \to \infty$.

11.2 Linear Subspaces

In this section, we look at linear spaces that lie entirely inside the Study quadric. We begin with the lines through the identity.

11.2.1 Lines

The identity is given by the dual quaternion $1 = (1 + \varepsilon 0)$. Now suppose

$$(a + \varepsilon c) = (a_0 + a_1 i + a_2 j + a_3 k) + \varepsilon(c_0 + c_1 i + c_2 j + c_3 k)$$

is some other point in the quadric. The projective line joining this point to the identity is given by the points $\alpha + \beta(a + \varepsilon c)$, where α and β are arbitrary. For this line to lie entirely in the quadric, we must have

$$(\alpha + \beta a_0)c_0 + \beta(a_1 c_1 + a_2 c_2 + a_3 c_3) = 0,$$

for all α and β. Since we know that $(a + \varepsilon c)$ already lies in the quadric, we must conclude that $c_0 = 0$. Along any line through the identity, there will always be a point where $a_0 = 0$, and hence the lines in the Study quadric through the identity have the form

$$\alpha + \beta(a_1 i + a_2 j + a_3 k) + \beta\varepsilon(c_1 i + c_2 j + c_3 k).$$

Now, compare these with the one-parameter subgroups of $SE(3)$. At the end of Section 9.3, we saw a general expression for the exponential of a Lie algebra element written as a dual angle $\check{\theta} = \theta + p\varepsilon d$ times a dual vector $\check{v} = v_0 + \varepsilon v_1$, where $v_0 \cdot v_1 = 0$. If we restrict our attention to pitch zero elements, that is, rotations about lines in $\mathbb{R}^3$, we get

$$e^{\theta \check{v}/2} = \cos\frac{\theta}{2} + \sin\frac{\theta}{2}v_0 + \sin\frac{\theta}{2}\varepsilon v_1.$$

Hence, these one-parameter subgroups determine projective lines in the Study quadric. The one-parameter translation subgroups also represent lines in the Study quadric, since they have the form

$$1 + \frac{\lambda}{2}\varepsilon(t_x i + t_y j + t_z k).$$

It is not difficult to see from the above that all lines through the identity have one of these two forms. So, every line in the Study quadric through the identity is either a one-parameter subgroup of rotations about some line in $\mathbb{R}^3$ or a one-parameter subgroup of translations parallel to some vector in $\mathbb{R}^3$.

11.2.2   3-planes

Next, we turn to the 3-planes in the quadric. The following should be familiar, as it is essentially the same derivation as in sections 6.3 and 6.5.1. Grouping the homogeneous coordinates of $\mathbb{PR}^7$ as a pair of four-dimensional vectors

$$(\mathbf{a}^T : \mathbf{c}^T) = (a_0 : a_1 : a_2 : a_3 : c_0 : c_1 : c_2 : c_3),$$

we note that the equations for the Study quadric can be written as $\mathbf{a} \cdot \mathbf{c} = 0$. The 3-planes in the quadric are given by

$$(I_4 - M)\mathbf{a} + (I_4 + M)\mathbf{c} = 0,$$

where $M \in O(4)$; that is, M is a 4×4 orthogonal matrix. Hence, we have two 6-dimensional families of 3-planes depending on whether $\det(M) = +1$ or $\det(M) = -1$.

This works because when we change coordinates as follows,

$$\mathbf{a} = (\mathbf{x} + \mathbf{y}) \qquad \text{and} \qquad \mathbf{c} = (\mathbf{x} - \mathbf{y}),$$

the quadric now becomes

$$\mathbf{a} \cdot \mathbf{c} = \mathbf{x} \cdot \mathbf{x} - \mathbf{y} \cdot \mathbf{y} = 0,$$

and the 3-planes become

$$(I_4 - M)\mathbf{a} + (I_4 + M)\mathbf{c} = 2\mathbf{x} - 2M\mathbf{y} = \mathbf{0}.$$

This is certainly a set of four linear equations. We can see that the planes lie in the quadric by substituting into the original equation

$$\mathbf{a} \cdot \mathbf{c} = \mathbf{y}^T M^T M \mathbf{y} - \mathbf{y}^T \mathbf{y} = 0.$$

The equation is satisfied identically since M is an orthogonal matrix. When $\det(M) = +1$ we call the plane an A-plane, and if $\det(M) = -1$ the plane is said to be a B-plane.

Now we look at the 3-planes containing the identity. For the identity element, we have $\mathbf{a}^T = (1,\, 0,\, 0,\, 0)$ and $\mathbf{c}^T = (0,\, 0,\, 0,\, 0)$, so any 3-plane containing the identity must be given by an orthogonal matrix satisfying

$$(I_4 - M) \begin{pmatrix} 1 \\ 0 \\ 0 \\ 0 \end{pmatrix} + (I_4 + M) \begin{pmatrix} 0 \\ 0 \\ 0 \\ 0 \end{pmatrix} = \begin{pmatrix} 0 \\ 0 \\ 0 \\ 0 \end{pmatrix}.$$

From this, we see that any such M must have the partitioned form

$$M = \begin{pmatrix} 1 & 0 \\ 0 & R \end{pmatrix}$$

where the 3×3 matrix R is orthogonal $R \in O(3)$. Notice that this means that all points on a 3-plane through the identity must have $c_0 = 0$. Any 3-plane can be generated by four points. If we take one point to be the identity, we may take the three other points as lying in the hyperplane $a_0 = 0$. Hence, a 3-plane through the identity in the Study quadric can be generated by the identity together with three points of the form $(0,\, a_1,\, a_2,\, a_3,\, 0,\, c_1,\, c_2,\, c_3)$ that satisfy

$$a_1 c_1 + a_2 c_2 + a_3 c_3 = 0.$$

This is simply the Klein quadric; see Section 6.3. So we can think of the three points that determine the 3-plane through the identity as screws in the Klein quadric. Furthermore, the condition that the 3-plane lies in the Study quadric implies that the space spanned by the three screws lies inside the Klein quadric. That is, each 3-plane through the identity is determined by a type II 3-system of screws. In fact, we see that each A-plane through the identity in the Study quadric corresponds to an α-plane in the Klein quadric and each B-plane through the identity corresponds to a β-plane. See Section 8.3. From the completion groups listed in Table 8.2, we can see that these α-planes are always subalgebras of $se(3)$. Now, a line through the identity and one of these points is a one-parameter subgroup of $SE(3)$, as we saw above. If we multiply points on two such lines, that is, any pair of points in the 3-plane, the result will be a point on a line joining the identity with some other element of the subalgebra. Hence we see that the A-planes through the identity are subgroups, either $SO(3)$ or $SE(2)$. The $SO(3)$ subgroups correspond to the IIA $(p = 0)$ screw systems, and the $SE(2)$ subgroups are determined by the IIC $(p = 0)$ systems. These 3-planes through the identity are distinguished by how they meet the A-plane of 'infinite somas' $a_0 = a_1 = a_2 = a_3 = 0$. This A-plane does not contain the identity. It is given by the orthogonal matrix $M = -I_4$. The A-planes corresponding to $SO(3)$ subgroups do not meet the A-plane at infinity while those corresponding to $SE(2)$ subgroups meet the infinite 3-plane in a line, see later.

Only one B-plane through the identity is a subgroup. This is the $\mathbb{R}^3$ subgroup, determined by the IID screw system. Most B-planes through the identity meet the A-plane at infinity in a single point. The B-plane corresponding to the $\mathbb{R}^3$ subgroup, is exceptional in that it meets the A-plane at infinity in a 2-plane.

Suppose that an A-plane through the identity corresponds to the subgroup of rotations about some point $\mathbf{p}$ in $\mathbb{R}^3$; call it $SO(3)_{\mathbf{p}}$. Then if $\mathbf{g} \in SE(3)$ is a group element not lying in the subgroup, the coset of elements of the form $\mathbf{g}SO(3)_{\mathbf{p}}$ gives an A-plane not containing the identity. Similarly, if we let $SE(2)_{\mathbf{n}}$ be the symmetry group of a plane in $\mathbb{R}^3$ normal to $\mathbf{n}$, then the coset of elements of the form $\mathbf{g}SE(2)_{\mathbf{n}}$ also gives an A-plane. Moreover, it is easy to see that any A-plane not through the identity must have one of these two forms.

11.2.3 Intersections of 3-planes

How do these A-planes and B-planes intersect? To investigate this, let us use the homogeneous coordinates $(\mathbf{x}, \mathbf{y})$, in which the equation for the Study quadric has the form

$$\mathbf{x}^2 - \mathbf{y}^2 = 0$$

and the A- and B-planes can be written

$$\mathbf{x} - M\mathbf{y} = 0.$$

The intersection of two of these 3-planes is given by a pair of these vector equations with different orthogonal matrices M. These are eight homogeneous linear equations, which can be neatly written in partitioned form:

$$\begin{pmatrix} I_4 & -M_1 \\ I_4 & -M_2 \end{pmatrix} \begin{pmatrix} \mathbf{x} \\ \mathbf{y} \end{pmatrix} = \mathbf{0}.$$

The solution and hence the character of the intersection of the 3-planes depends on the determinant of the coefficient matrix. If the determinant is non-zero, the intersection will be null. Further, the dimension of the intersection is given by the corank of the matrix. If we take the first four rows of the coefficient matrix from the last four, the matrix becomes

$$\begin{pmatrix} I_4 & 0 \\ -I_4 & I_4 \end{pmatrix} \begin{pmatrix} I_4 & -M_1 \\ I_4 & -M_2 \end{pmatrix} = \begin{pmatrix} I_4 & -M_1 \\ 0 & M_1 - M_2 \end{pmatrix}.$$

Thus, the determinant of the coefficient matrix can be seen to be $\det(M_1 - M_2)$. Since M_2 is non-singular, this is equivalent to $\pm \det(M_1 M_2^T - I_4)$, and this vanishes when 1 is an eigenvalue of the orthogonal matrix $M_1 M_2^T$. Indeed, the corank of the original matrix will be the multiplicity of 1 in the characteristic equation for $M_1 M_2^T$. Now, if the two 3-planes are both A-planes or both B-planes, the matrix $M_1 M_2^T$ will have determinant $+1$, that is, $M_1 M_2^T \in SO(4)$. Any matrix in $SO(4)$ can be written in the general form

$$M = R^T \begin{pmatrix} \cos\theta & -\sin\theta & 0 & 0 \\ \sin\theta & \cos\theta & 0 & 0 \\ 0 & 0 & \cos\phi & -\sin\phi \\ 0 & 0 & \sin\phi & \cos\phi \end{pmatrix} R$$

where $R \in SO(4)$. That is every element of the group is conjugate to an element of the maximal torus of the group, see Curtis [23, Chap. 7]. From this we see that in general these matrices do not have 1 as an eigenvalue so in general pairs of A-planes and pairs of B-planes do not meet. Exceptionally, if θ or ϕ vanish, an $SO(4)$ matrix will have 1 as an eigenvalue with multiplicity 2. Hence, sometimes pairs of A-plane and pairs of B-planes can meet in a line.

On the other hand, if the 3-planes are of opposite type, then $M_1 M_2^T$ has determinant -1, that is, an element in $O(4)$ not in $SO(4)$. A determinant -1 element of $O(4)$ can be written as

$$M = R^T \begin{pmatrix} -\cos\theta & \sin\theta & 0 & 0 \\ \sin\theta & \cos\theta & 0 & 0 \\ 0 & 0 & \cos\phi & -\sin\phi \\ 0 & 0 & \sin\phi & \cos\phi \end{pmatrix} R,$$

where $R \in SO(4)$. That is, the product of a reflection and a rotation. Now in general this matrix has a single unit eigenvalue, so we can infer that in general an A-plane will meet a B-plane in a single point. If $\phi = 0$, then the matrix will have three unit eigenvalues and hence, exceptionally an A-plane can meet a B-plane along a 2-plane.

As an example we will look at how the 3-planes through the identity of meet A-plane of infinite soma. The A-plane at infinity is given by the orthogonal matrix $M_1 = -I_4$ and a 3-plane through the identity is given by,

$$M_2 = \begin{pmatrix} 1 & 0 \\ 0 & R \end{pmatrix},$$

where now $R \in O(3)$. So we are looking for unit eigenvalues of $-M_2$. When $R \in SO(3)$ the 3-plane through the identity is an A-plane. A general rotation R corresponds to a subgroup of rotation about some point in space. In this case $-M_2$ has no unit eigenvalues and hence these A-planes do not meet the A-plane at infinity. The planar subgroups $SE(2)$ correspond to the case where R is a rotation of π radians. Now $-M_2$ has a pair of unit eigenvalues indicating that these A-planes meet the A-plane at infinity in a line. The B-planes are given by othogonal matrices with $R \in O(3)$ and $\det(R) = -1$. In general such a matrix R will have -1 as an eigenvalue with multiplicity one, so $-M_2$ has a single unit eigenvalue and thus these B-planes meet the A-plane at infinity in a single point. However, it is possible to have $R = -I_3$ and in this case the corresponding B-plane will meet the A-plane at infinity in a 2-plane. It is easy to see that the case $R = -I_3$ corresponds to the $\mathbb{R}^3$ subgroup.

The above calculations apply to any $2n$-dimensional quadric. The results depend on the eigenvalues of the matrices in $O(n)$. Recall that in the case of the Klein quadric α-planes and β-planes do not generally meet, but each α-plane meets every other α-plane in a point. Similarly, pairs of β-planes meet at points. This is because elements of $SO(3)$ have 1 as an eigenvalue with multiplicity 1 (except the identity), but the elements of the other connected component of

$O(3)$ do not generally have eigenvalue 1. In general, if we consider the maximal torus in $SO(n)$ we can see that when n is odd there is a single unit eigenvalue but if n is even there is no unit eigenvalue; see [23, Chap. 7].

11.2.4   Quadric Grassmannians

Finally, we look at the set of all linear spaces of some dimension lying in the quadric. To help us, we introduce the notion of a **quadric Grassmannian**; see Porteous [88, Chap.12]. Consider the space $\mathbb{R}^n$ endowed with a non-degenerate metric $Q(\mathbf{x}, \mathbf{y}) = \mathbf{x}^T Q \mathbf{y}$. A subspace is said to be **isotropic** if $Q(\mathbf{x}, \mathbf{x}) = 0$ for every $\mathbf{x}$ in the subspace. The set of isotropic k-dimensional subspaces forms a manifold called a quadric Grassmannian and written $G^Q(k, n)$. Clearly, the quadric Grassmannians are submanifolds of the ordinary Grassmannians:

$$G^Q(k, n) \subseteq G(k, n).$$

In Fulton and Harris [36, sect. 23.3] these spaces are called Lagrangian Grassmannians or Orthogonal Grassmannians, depending on the quadric.

If we think of the equation $Q(\mathbf{x}, \mathbf{x}) = 0$ as defining a projective quadric in $\mathbb{PR}^{n-1}$, then a $k - 1$-plane lying in this quadric corresponds to an isotropic k-dimensional subspace of $\mathbb{R}^n$. Hence, for the Study quadric we will be interested in the hyperbolic metric on $\mathbb{R}^8$ given by

$$\mathrm{Hy}((\mathbf{a}_1, \mathbf{c}_1), (\mathbf{a}_2, \mathbf{c}_2)) = \mathbf{a}_1 \cdot \mathbf{c}_2 + \mathbf{c}_1 \cdot \mathbf{a}_2,$$

where $\mathbf{a}_i, \mathbf{c}_i \in \mathbb{R}^4$. The 1-dimensional isotropic subspaces correspond to the points of the quadric, so we have

$$G^{\mathrm{Hy}}(1, 8) = SO(4)$$

from the remarks in Section 11.1. The comments above on the A- and B-planes in the quadric mean that we can write

$$G^{\mathrm{Hy}}(4, 8) = O(4).$$

Notice that we have three homeomorphic manifolds, the Study quadric itself, the space of A-planes, and the space of B-planes. Each of these is isomorphic to the manifold of the group $SO(4)$. This observation by Study led to the concept of **triality** in Lie groups; see Porteous [88, Chap.21].

The other quadric Grassmannians can also be described as homogeneous spaces. This, again, is most easily seen using the coordinates in which the Study quadric has the form $\mathbf{x} \cdot \mathbf{x} - \mathbf{y} \cdot \mathbf{y}$. With these coordinates, the metric Hy becomes

$$\mathrm{Hy}' = \begin{pmatrix} I_4 & 0 \\ 0 & -I_4 \end{pmatrix}.$$

The lines in the Study quadric correspond to isotropic 2-planes in this space. Suppose the two vectors $\mathbf{v}_1 = (\mathbf{x}_1, \mathbf{y}_1)^T$ and $\mathbf{v}_2 = (\mathbf{x}_2, \mathbf{y}_2)^T$ span such an isotropic 2-space. Then we must have

$$\mathrm{Hy}'(\mathbf{v}_1, \mathbf{v}_1) = 0 \qquad \text{and} \qquad \mathrm{Hy}'(\mathbf{v}_2, \mathbf{v}_2) = 0.$$

But to ensure that any linear combination of these vectors is also isotropic, we need

$$\mathrm{Hy}'(\mathbf{v}_1, \mathbf{v}_2) = 0.$$

Now, the symmetry group of the metric Hy' is $O(4,4)$ (see Section 3.1) but for our purposes here we only need the subgroup $O(4) \times O(4)$, consisting of matrices of the form

$$\begin{pmatrix} M_1 & 0 \\ 0 & M_2 \end{pmatrix}, \quad M_1, M_2 \in O(4).$$

By choosing M_1 and M_2 suitably, we can transform the vector $\mathbf{v}_1$ into the standard form $\mathbf{v}_1' = (1,0,0,0,\ 1,0,0,0)^T$. To transform the 2-plane, we must apply this matrix to $\mathbf{v}_2$ as well. Next, we produce a new generator for the 2-plane by taking a multiple of $\mathbf{v}_1'$ from $\mathbf{v}_2'$. The new generator can have the form $\mathbf{v}_2'' = (0, x_1, x_2, x_3, 0, y_1, y_2, y_3)^T$. This is possible because the relation $\mathrm{Hy}'(\mathbf{v}_1', \mathbf{v}_2') = 0$ still holds, so that making the first coordinate zero forces the fifth to vanish too. Now, we transform again, this time using a matrix with

$$M_1 = \begin{pmatrix} 1 & 0 \\ 0 & R_1 \end{pmatrix} \qquad \text{and} \qquad M_2 = \begin{pmatrix} 1 & 0 \\ 0 & R_2 \end{pmatrix},$$

where $R_1, R_2 \in O(3)$. This does not affect the first generator, but by a suitable choice of R_1 and R_2 we can reduce the second generator to $(0,1,0,0,\ 0,1,0,0)^T$.

Hence, we have shown that an isotropic 2-plane can be reduced to a standard 2-plane generated by the vectors

$$\mathbf{v}_1 = (1,\ 0,\ 0,\ 0,\ 1,\ 0,\ 0,\ 0)^T \qquad \text{and} \qquad \mathbf{v}_2 = (0,\ 1,\ 0,\ 0,\ 0,\ 1,\ 0,\ 0)^T.$$

Moreover, the group $O(4) \times O(4)$ acts transitively on the space of all isotropic 2-planes. The isotropy group of the standard form is simple to find; it consists of matrices of the form

$$\begin{pmatrix} A & 0 & 0 & 0 \\ 0 & B & 0 & 0 \\ 0 & 0 & A & 0 \\ 0 & 0 & 0 & C \end{pmatrix},$$

with $A, B, C \in O(2)$. The matrices B and C have no effect on the generators, while A moves the generators but only within the standard 2-plane. So, we can write the quadric Grassmannian $G^{\mathrm{Hy}}(2,8)$ as the quotient space

$$G^{\mathrm{Hy}}(2,8) = \frac{O(4) \times O(4)}{O(2) \times O(2) \times O(2)}.$$

By a similar argument, we also have

$$G^{\text{Hy}}(3,8) = \frac{O(4) \times O(4)}{O(3) \times O(1) \times O(1)},$$

where $O(1)$ is the discrete group $\mathbb{Z}_2$. See also Griffiths and Harris [42, sect. 6.1] for an algebraic approach to linear subspaces of quadrics.

11.3   Partial Flags and Projections

In Section 11.1, we introduced flag manifolds. In this section we take a closer look at some of these spaces that are relevant to robotics.

A flag of type $(1,2,3,\ldots,n-1)$ in $\mathbb{R}^n$ is called a **complete flag**, while flags of other types are **partial flags**. From a complete flag manifold, we have projections onto the partial flag manifolds. These projections are obtained by 'forgetting' some of the parts of the complete flag. For example, we have a projection from $F^+(1,2,3;4)$ to $F^+(1;4)$ which is given by ignoring the 2- and 3-planes of the flag and just taking the line. Clearly, this gives a smooth mapping between the spaces. It is also easy to see that the mapping is compatible with the group action; that is, the group action commutes with the mapping. Given a complete flag $V_1 \subset V_2 \subset V_3$ and a group element $\mathbf{g}$, the action of this element is given by $\mathbf{g}V_1 \subset \mathbf{g}V_2 \subset \mathbf{g}V_3$; then projecting to $F^+(1;4)$ gives $\mathbf{g}V_1$. Projecting first then applying the group element gives the same result. In our case, as we have seen, the complete flag manifold is homeomorphic to the group $SO(4)$, and once we have chosen a reference or 'home' flag we can identify the complete flags with group elements. For brevity, let's call the image of the home flag under the projection the base point. Now consider the set of group elements that leaves the base point unchanged. This set of group elements is the isotropy group of the base point. In the complete flag manifold, these elements correspond to complete flags that project to the base point. That is, the isotropy group corresponds to the pre-image of the base point or is the fibre of the mapping over the base point.

For our first example, we look at type (1) flags in $\mathbb{R}^4$. Notice that under the stereographic projection defined in Section 11.1 a type (1) flag projects to a point in $\mathbb{R}^3$. The isotropy group of a type (1) flag is $SO(3)$, and so we have the homogeneous space representation of the flag manifold $F^+(1;4) = SO(4)/SO(3) = S^3$. We saw in the last section that this isotropy group and hence the fibre over the base point, is an A-plane through the identity in $F^+(1,2,3;4)$, the Study quadric. In fact, we saw in the last section that this A-plane consists of the group of elements generated by a IIA $(p=0)$ 3-system of screws. We may call this A-plane $SO(3)_{\mathbf{o}}$.

Next, let's look at the set of group elements that takes the base point to some other partial flag. Suppose $\mathbf{g}$ is a group element that takes the base point to the target partial flag. Pre-multiplying by any element of the isotropy group of the base point, $SO(3)_\mathbf{o}$, will give more elements that take the base point to the target point. The set of all such elements is thus given by $\mathbf{g}SO(3)_\mathbf{o}$, that is, an A-plane through $\mathbf{g}$. This A-plane corresponds to the fibre of the projection over the target point. Notice that every group element must lie in one of these A-planes but cannot lie in more than one of them. Such a structure is called a fibre bundle. The base space of the bundle is $F^+(1; 4) = S^3$, the space of partial flags. The total space is $F^+(1, 2, 3; 4)$. The Study quadric and the fibres are a collection of A-planes in the quadric, each isomorphic to $SO(3)$.

Slightly more generally, the set of group elements that take a point in $\mathbb{R}^3$ to some other point in $\mathbb{R}^3$ is given by the A-plane of the form $\mathbf{g}_1 SO(3)_\mathbf{o} \mathbf{g}_2^{-1}$, where $\mathbf{g}_1$ and $\mathbf{g}_2$ are the group elements that take the base point to the target and source points, respectively. See Hunt and Parkin [57].

Our next example is $F^+(3; 4)$. Under the stereographic projection, a flag of type (3) in $\mathbb{R}^4$ becomes an oriented 2-plane in $\mathbb{R}^3$. It is not difficult to see that $F^+(3; 4)$ is homeomorphic to $F^+(1; 4)$. This is because in $\mathbb{R}^4$ an oriented 3-plane through the origin can be specified by giving its normal vector, which is just a type (1) flag. Hence, the isotropy group of the base point is again $SO(3)$. However, there is a slight subtlety here. The isotropy group consists of rotations about the normal vector to the 3-plane. But in $\mathbb{R}^3$ these symmetries must preserve a 2-plane. Hence, under the Saletan contraction this $SO(3)$ must turn into $SE(2)$. Thus, the fibre over the base point of the projection from $F^+(1, 2, 3; 4)$ to $F^+(3; 4)$ will be an A-plane through the identity, generated by a type IIC ($p = 0$) screw system. The fibres of the projection over other points in $F^+(3; 4)$ are just A-planes of this type through other group elements. Also, the set of group elements that transforms one 2-plane into another in $\mathbb{R}^3$ consists of the elements given by $\mathbf{g}_1 SE(2)_\mathbf{o} \mathbf{g}_2^{-1}$, where $SE(2)_\mathbf{o}$ is the isotropy group of the base point, and $\mathbf{g}_1$, $\mathbf{g}_2$ are any elements taking the source and target 2-planes to the base point, respectively.

Next, we look at the flag manifold $F^+(1, 2; 4)$. Under the stereographic projection, flags of this type become pointed directed lines in $\mathbb{R}^3$. The space $F^+(1, 2; 4)$ is easily seen to be isomorphic to the space $F^+(1, 3; 4)$. The correspondence is given by taking the $(1, 2)$ flag specified by the line in a $(1, 3)$ flag and the normal to 3-plane. Note that a $(1, 3)$ flag becomes a pointed oriented plane on stereographic projection to $\mathbb{R}^3$. The isotropy group of a $(1, 2)$ flag is simply the set of rotations in $\mathbb{R}^4$ that preserve the two orthogonal vectors that define the flag. For a $(1, 3)$ flag, the isotropy group consists of rotations in the 3-plane that fix the line. In either case, the isotropy group is isomorphic to $SO(2)$, and we have the homogeneous space representation

$$F^+(1, 2; 4) = F^+(1, 3; 4) = \frac{SO(4)}{SO(2)}.$$

In the last section, we saw that $SO(2)$ subgroups of the Study quadric are one-dimensional linear subspaces, that is, lines through the identity. Hence, the projection gives a line bundle, a fibre bundle whose fibres are lines. Notice that the set of group elements that takes a pointed directed line in $\mathbb{R}^3$ to another forms a line in the Study quadric.

Elements of the flag manifold $F^+(2, 3; 4)$ become lined planes when stereographically projected to $\mathbb{R}^3$. The isotropy group of a type $(2, 3)$ flag is the $SO(2)$ of rotations preserving the 2-plane in the flag. After the Saletan contraction, however, this $SO(2)$ must turn into $SE(1) = \mathbb{R}$, the group of translations that preserves the lined plane. The isotropy group is still a line in the Study quadric, but now it is a line generated by a translation, that is, a line that meets the A-plane of infinite somas.

Finally, we look at $F^+(2; 4)$. Such a flag becomes a directed line after stereographic projection to $\mathbb{R}^3$. Remember, this space is not the Klein quadric but its double cover; see Section 6.8. The isotropy group for a type (2) flag is $SO(2) \times SO(2)$. This consists of the rotations of the 2-plane in the flag and the rotations of the 2-plane orthogonal to the flag. After Saletan contraction, we obtain the symmetry group of a directed line, $SO(2) \times \mathbb{R}$. We have not met this subgroup before as a subspace of the Study quadric. In the next section it will be shown that the fibres of the projection from $F^+(1, 2, 3; 4)$ to $F^+(2; 4)$ are quadric surfaces, each the intersection of the Study quadric with a 3-plane.

These results, the identification of the flag manifolds with homogeneous spaces, are summarised in Table 11.1.

Notice that these flags can be represented as elements of the Clifford algebra we met in Chapter 10. For example, a pointed line could be represented as an element:

$$f_{12} = \frac{1}{\sqrt{2}}(p + \ell)$$

Notice that this element is not homogeneous, it consists of elements with different grades. The action of the group of rigid motions on these flags is simply,

$$f'_{12} = g f_{12} g^*.$$

Not all such elements represent pointed lines, the line must be a line, so we must have $\ell\ell^* = 1$ and, to be a point, the coefficient of $\mathbf{e}_1\mathbf{e}_2\mathbf{e}_3$ must be 1. Moreover, for the point to lie on the line the point and line must satisfy, $p\ell^* + \ell p^* = 0$. Most of these equations can be written in terms of the flag itself,

$$f_{12} f_{12}^* = 1.$$

Comparing coefficients of the various basis elements gives us all the equations except that we only have that $pp^* + \ell\ell^* = 2$. If we include the relation that the coefficient of $\mathbf{e}_1\mathbf{e}_2\mathbf{e}_3$ must be 1 then we have that $pp^* = 1$ and hence that $\ell\ell^* = 1$. Thus, we see that the space of all pointed lines form an affine algebraic variety. This is also true for all of these flag manifolds.

TABLE 11.1. The Flag Manifolds of $\mathbb{R}^4$

Flag Manifold	Homogeneous Space	Figure in $\mathbb{R}^3$
$F^+(1,2,3;4)$	$SO(4)$	Soma
$F^+(1,2;4)$	$SO(4)/SO(2)$	Pointed Line
$F^+(2,3;4)$	$SO(4)/SO(2)$	Lined Plane
$F^+(1,3;4)$	$SO(4)/SO(2)$	Pointed Plane
$F^+(1;4)$	$SO(4)/SO(3) = S^3$	Point
$F^+(2;4)$	$SO(4)/SO(2) \times SO(2)$	Directed Line
$F^+(3;4)$	$SO(4)/SO(3) = S^3$	Oriented Plane

11.4 Some Quadric Subspaces

As promised above we look at the isotropy group of a line in space. To be definite consider the z-axis. The isotropy group consists of the rotations about the z-axis together with the translations in the z direction. As dual quaternions, we may write the elements of this subgroup as

$$\left(\cos\frac{\theta}{2} + k\sin\frac{\theta}{2}\right)\left(1 + \frac{t}{2}k\varepsilon\right) = \left(\cos\frac{\theta}{2} + k\sin\frac{\theta}{2}\right) + \left(\frac{t}{2}\cos\frac{\theta}{2}k - \frac{t}{2}\sin\frac{\theta}{2}\right)\varepsilon.$$

Using $(a_0 : a_1 : a_2 : a_3 : c_0 : c_1 : c_2 : c_3)$ as homogeneous coordinates in $\mathbb{PR}^7$, we see that the subgroup above lies on the 3-plane defined by the equations $a_1 = a_2 = c_1 = c_2 = 0$. Hence, the subgroup lies in the intersection of the 3-plane with the Study quadric. Thus, the subgroup lies in a two-dimensional quadric. It is easy to see that the points in this quadric that are not elements of the subgroup are just the 'line at infinity' $a_2 = a_3 = 0$. Now, although we have done the above computations for a single line, they easily generalise to the isotropy group of any directed line in $\mathbb{R}^3$. This is because the action of the whole Euclidean group on $\mathbb{PR}^7$ is the projectivisation of a linear representation. So, the result of the group action on a 3-plane will be to move it to another 3-plane.

The methods that we have just used to find the geometry of the group $SO(2)\times$ $\mathbb{R}$ in the Study quadric can also be used to find the subspace generated by a pair of rotations. Suppose we have a pair of revolute joints $\mathbf{s}_1$ and $\mathbf{s}_2$ connected in series. The set of all possible rigid transformations that this simple machine can effect are given by the product

$$K(\theta_1, \theta_2) = e^{\theta_1 \mathbf{s}_1} e^{\theta_2 \mathbf{s}_2}.$$

Now, take s_1 to be along the z-axis and s_2 to be translated a distance l along the x-axis and rotated by an angle α about the x-axis. The exponential can be written as dual quaternions:

$$e^{\theta_1 s_1} = \cos\frac{\theta_1}{2} + k\sin\frac{\theta_1}{2}$$

$$e^{\theta_2 s_2} = \cos\frac{\theta_2}{2} - j\sin\frac{\theta_2}{2}\sin\alpha + k\sin\frac{\theta_2}{2}\cos\alpha - j\varepsilon l\sin\frac{\theta_2}{2}\cos\alpha - k\varepsilon l\sin\frac{\theta_2}{2}\sin\alpha.$$

Multiplying these exponentials gives

$$e^{\theta_1 s_1}e^{\theta_2 s_2} = (a_0 + ia_1 + ja_2 + ka_3) + (c_0 + c_1 i + c_2 j + c_3 k)\varepsilon$$

where

$$a_0 = \cos\tfrac{\theta_1}{2}\cos\tfrac{\theta_2}{2} - \sin\tfrac{\theta_1}{2}\sin\tfrac{\theta_2}{2}\cos\alpha, \quad c_0 = l\sin\tfrac{\theta_1}{2}\sin\tfrac{\theta_2}{2}\sin\alpha,$$

$$a_1 = \sin\tfrac{\theta_1}{2}\sin\tfrac{\theta_2}{2}\sin\alpha, \quad c_1 = l\sin\tfrac{\theta_1}{2}\sin\tfrac{\theta_2}{2}\cos\alpha,$$

$$a_2 = -\cos\tfrac{\theta_1}{2}\sin\tfrac{\theta_2}{2}\sin\alpha, \quad c_2 = -l\cos\tfrac{\theta_1}{2}\sin\tfrac{\theta_2}{2}\cos\alpha,$$

$$a_3 = \sin\tfrac{\theta_1}{2}\cos\tfrac{\theta_2}{2} + \cos\tfrac{\theta_1}{2}\sin\tfrac{\theta_2}{2}\cos\alpha, \quad c_3 = -l\cos\tfrac{\theta_1}{2}\sin\tfrac{\theta_2}{2}\sin\alpha.$$

By inspection, these coordinates satisfy four independent linear equations:

$$la_2 - c_3 = 0,$$
$$la_1 - c_0 = 0,$$
$$la_1\cos\alpha - c_1\sin\alpha = 0,$$
$$la_2\cos\alpha - c_2\sin\alpha = 0.$$

So we see that the space of possible rigid motions produced by a pair of revolute joints lies in a 3-plane. Hence, the two-dimensional quadric is defined by the intersection of this plane and the Study quadric itself. As usual, we appeal to the properties of the group action to show that this is true for an arbitrary pair of revolute joints.

11.5   Intersection Theory

In this section, we look at some questions from enumerative geometry. These questions have been very popular in the mechanisms literature, where problems concerning the number of assembly configurations for various types of machines have been studied. However, the methods used to tackle these difficult problems are only just beginning to catch up with the mathematics that has been developed in this area. Unfortunately, only a brief sketch of these methods can be given here. The extensive mathematics literature should be consulted for further detail.

In mathematics, enumerative problems have a long and distinguished history going back to the fundamental theorem of algebra, which counts the number of roots of a polynomial of a single variable. If we are considering real roots, then the degree of the polynomial gives an upper bound for the number of roots. In the complex case, the degree gives precisely the number of roots, provided they are counted with the appropriate multiplicity. These ideas were extended to count the solutions of systems of n homogeneous polynomial equations in n variables. The answer, given by Bézout's theorem, is that the number of complex solutions is generally given by the product of the degrees of the polynomials. The 'generally' here is to cover the case where the polynomials are not all independent, and hence there are an infinite number of solutions. One way of viewing these problems is as follows: An n variable homogeneous polynomial defines a set in $\mathbb{CP}^{n-1}$, its set of zeros. This set, or variety, will have dimension $n - 2$. The solution to a system of n such polynomials is the intersection of n such varieties. Notice that we can think of the degree of a subvariety as the number of times it intersects a generic plane of complementary dimension.

Modern intersection theory generalises this to the intersection of subvarieties in other spaces, not just projective space $\mathbb{CP}^n$; see, for example, Fulton [35]. An early result in this direction was Halphen's theorem [109, Chap. X sect. 3.1]. Halphen's theorem concerns the intersections of line congruences, which can be thought of as 2-dimensional subvarieties of the Klein quadric. To each algebraic congruence, we can associate a bidegree (m, n), where m gives the number of intersections with a general α-plane, and n is the number of intersections of the congruence with a generic β-plane. Notice that in particular an α-plane will have bidegree $(1, 0)$, while a β-plane has bidegree $(0, 1)$. Halphen's theorem states that the number of lines common to a pair of congruences, one with bidegree (m_1, n_1) and the other with (m_2, n_2), will be $m_1 m_2 + n_1 n_2$, provided, of course, that the congruences do not have a component in common.

Schubert extended this work to give results on the intersections of other line systems and eventually to subvarieties in any complex Grassmannian; see [42, Chap. 1 sect. 5]. Later progress was made by the introduction of homology theory from algebraic topology.

Homology theory was invented by Poincaré at the turn of the twentieth century. In essence, two subspaces of a topological space are **homologous** if together they form the boundary of another subspace. Actually, we also need to take account of the orientation of the subspace. So if we write ∂X for the boundary of a subspace X, then two subspaces Y_1 and Y_2 are homologous when they satisfy

$$Y_1 - Y_2 = \partial X$$

for some subspace X. The minus in the above equation denotes union of subspaces but where the second subspace has its orientation reversed.

Subspaces that are already the boundary of some subspace are homologous to the null space. Also, the boundary of a boundary subspace is always null, that is,

$$\partial^2 X = 0.$$

However, there may also be subspaces with empty boundaries but that are not themselves boundaries of any subspace, that is, spaces X for which $\partial X = 0$ but $X \neq \partial Y$ for any subspace Y. A subspace with no boundary, $\partial X = 0$, is said to be **closed** while a closed space that is itself a boundary, $X = \partial Y$, is called an **exact** subspace.

Now, the general idea behind homology theory is to classify the n-dimensional closed subspaces of a space up to homology equivalence, that is, to find all the closed subspaces modulo the exact ones. However, in order to make the algebra work properly we have to consider formal integer sums of subspaces. To interpret an expression like $2X$, where X is a subspace, we might consider the union of two subspaces, both homologous to X. Notice that, in homology, the idea of moving a subspace to general position is represented by taking a homologous subspace.

The n-dimensional homology group of a space is the set of closed n-dimensional subspaces modulo the exact ones. With a little work, it is possible to show that this set is actually a commutative group. Hence, the elements of the group are equivalence classes of subspaces; two closed subspaces belong to the same class if they differ by an exact subspace.

The reward for all this algebra is that we get a pairing between the homology classes. If we intersect two subspaces, the result is a smaller subspace in general. This intersection pairing between subspaces respects homology. That is, the homology class of the result depends only on the homology classes of the subspaces being intersected. So, we have an intersection pairing defined on the homology groups.

In a connected space, the 0-dimensional homology group is simply $\mathbb{Z}$, the group of integers. This is because the points in the space are the closed 0-dimensional subspaces, and any point is homologous to any other point, since there is always a path connecting the two points. The path forms a 1-dimensional subspace whose boundary is the pair of points. So if the result of an intersection pairing is a 0-dimensional subspace, its homology class will be an integer. This integer simply counts the number of points in the intersection, where the usual remarks about multiplicities of points and general position of the intersecting subspaces apply.

When the space is a manifold, we can compute the homology groups using a cellular decomposition of the manifold. This is a decomposition of the manifold into a collection of disjoint cells, each cell being homeomorphic to the interior of a unit ball in some $\mathbb{R}^n$, which in turn is homeomorphic to $\mathbb{R}^n$ itself. Usually, we also need to know how the cell boundary, the sphere S^{n-1}, sits in the manifold. From this information, the homology groups of the manifold are relatively easy to calculate; see for example Maunder [71, Chap. 8].

It turns out that for non-singular algebraic varieties that are complex manifolds, the homology theory and the intersection theory are exactly the same; see [42, Chap. 0 sect. 4]. The homology class of the intersection of two subvarieties corresponds to the pairing of the homology classes of the subvarieties. For zero-dimensional subvarieties, the homology class simply counts the number of points which make up the subvariety.

To see how this works, consider the projective space $\mathbb{CP}^n$. A cellular decomposition is given by a complete flag $V_1 \subset V_2 \subset \cdots \subset V_n$. In the projective space, this becomes a nested sequence of projective spaces. The difference between any consecutive pair is homeomorphic to a cell of real dimension $2k$, that is, complex dimension k. Homology in dimension $2k$ is generated by the single class α_k corresponding to the cell of dimension $2k$. Since there are cells only in every other dimension, there are no relations between the generators. See Greenberg [40, Chap. 19].

Cells of complementary complex dimension intersect in a single point, when moved to general positions. That is, $\alpha_k \cap \alpha_{n-k} = 1$, where the class of a point α_0 has been written as 1. Hence, we can find the class of any complex k-dimensional subvariety by intersecting it with a general plane of complex dimension $n - k$. If there are d intersections, then the class of the subvariety is just $d\alpha_k$. This is, of course, just the degree of the subvariety. More generally, the intersection of a pair of classes will be $\alpha_k \cap \alpha_l = \alpha_{k+l-n}$ if $k + l \geq n$ and zero otherwise.

A slightly more complicated example is given by the Cartesian product of a pair of projective spaces, $\mathbb{CP}^m \times \mathbb{CP}^n$. Notice that a subvariety of this space can be thought of as the zero set for a system of multi-homogeneous polynomials. That is, let $(x_0 : x_1 : \cdots : x_m)$ be homogeneous coordinates for $\mathbb{CP}^m$ and $(y_0 : y_1 : \cdots : y_n)$ be similar coordinates for $\mathbb{CP}^n$. Then a multi-homogeneous polynomial will be separately homogeneous in the x_i coordinates and the y_j coordinates. In $\mathbb{CP}^m \times \mathbb{CP}^n$ the homology is generated by the Cartesian products of pairs of cells of each component. Suppose we label the classes in $\mathbb{CP}^m$ by α_i and those from $\mathbb{CP}^n$ by β_j; then in dimension k the generators for the homology of the Cartesian product are $\alpha_k\beta_0, \alpha_{k-1}\beta_1, \ldots, \alpha_0\beta_k$. A single multi-homogeneous equation defines a subvariety of homology class $d_x\alpha_{m-1}\beta_n + d_y\alpha_m\beta_{n-1}$, where d_x is the equation's degree in the x variables and d_y is the degree in the y variables. A general k-dimensional subvariety will have a homology class given by an integer linear combination of the generators, $d_0\alpha_k\beta_0 + d_1\alpha_{k-1}\beta_1 + \cdots + d_k\alpha_0\beta_k$. The intersection class of a pair of generators is given by their intersections in the individual projective spaces, that is,

$$(\alpha_i\beta_j) \cap (\alpha_k\beta_l) = (\alpha_i \cap \alpha_k)(\beta_j \cap \beta_l).$$

This intersection pairing distributes over the sum of classes.

So, for example, in $\mathbb{CP}^1 \times \mathbb{CP}^1$ the equation

$$x_0 y_1^2 + x_1 y_0^2 = 0,$$

defines a subvariety of class $\alpha_0\beta_1 + 2\alpha_1\beta_0$. Intersecting this variety with another of class $2\alpha_0\beta_1 + \alpha_1\beta_0$ would, in general, result in a subvariety of class

$$(\alpha_0\beta_1 + 2\alpha_1\beta_0) \cap (2\alpha_0\beta_1 + \alpha_1\beta_0) = 5\alpha_0\beta_0.$$

Now, since $\alpha_0\beta_0$ is the class of a point, we expect five points of intersection.

This algebra has been used by Wampler to look at several enumerative problems in kinematics; see for example [124, 123]. Note, however, that this work uses cohomology rather than homology. The difference is only that cohomology is dual to homology. This extra complication will not be used here, even though it makes the algebra rather more elegant.

The most important example in robotics, however, is the homology of the Study quadric. The following cellular decomposition of the Study quadric is based on the general results given by Ehresmann [32].

Consider a tangent hyperplane to the Study quadric at a point P_0. The points on this hyperplane satisfy

$$P_0^T Q_S X = 0.$$

Call this hyperplane P_6. The intersection of P_6 with the quadric is a five-dimensional singular quadric, Q_5, say. The singular set is the single point P_0. Now the points of the Study quadric not on P_6 form a cell of 12 real dimensions. To see this, imagine linearly projecting the Study quadric from the point P_0 onto some other hyperplane P_6' not containing P_0 and meeting P_6 in a 5-dimensional plane. The points in the intersection between the Study quadric and P_6 will project to the intersection between P_6 and P_6'. The rest of the Study quadric projects onto P_6' minus this 5-plane. In this region, the projection is easily seen to be 1-to-1, and we have seen above that $\mathbb{PC}^n - \mathbb{PC}^{n-1}$ is homeomorphic to a real $2n$-dimensional cell.

Next, we look at the cells in the singular 5-dimensional quadric Q_5. Consider a line P_1 lying in the Study quadric that contains P_0. The polar 5-plane to this line is the set of points that satisfy

$$P^T Q_S X = 0,$$

for all points P on the line. Call this 5-plane P_5 and its intersection with the Study quadric Q_4. Clearly P_5 sits inside P_6, and Q_4 has a line of double points along P_1. The set of points in Q_5 not in Q_4 forms a cell with 10 real dimensions. To see this, we project from P_1 onto a 5-plane P_5' that doesn't meet P_1 and intersects P_5 in a 4-plane. The points of Q_4 are mapped into the intersection of P_5 and P_5', and the other points of Q_5 map to the rest of P_5'.

We may repeat this for a 2-plane P_2 in the Study quadric containing the line P_1. The polar plane of P_2 will be a 4-plane P_4 that will be contained in P_5. However, in this case the intersection of P_4 with the quadric Q_4 will be a degenerate quadric, consisting of an A-plane and a B-plane meeting along P_2. Let us represent this degenerate quadric as $P_A \cup P_B$. Now, as in the previous cases above, we can show that the points of Q_4 not in $P_A \cup P_B$ form a cell, this time with eight real dimensions.

In six real dimensions, we have two cells, one given by the points of P_A not in P_2 and the other by the points of P_B not in P_2. The rest of the cellular decomposition is simple, now, since we have the sequence of projective planes, P_2, P_1, P_0, each contained in the previous one. The differences between these spaces give cells with real dimensions 4, 2 and 0.

To summarise, we have defined the following subspaces of the Study quadric Q_S, linked by injective maps:

$$
\begin{array}{ccccccccc}
 & & & & P_A & & & & \\
 & & & \nearrow & & \searrow & & & \\
P_0 & \to & P_1 & \to & P_2 & & Q_4 & \to & Q_5 & \to & Q_S. \\
 & & & \searrow & & \nearrow & & & \\
 & & & & P_B & & & &
\end{array}
$$

The difference between any pair of consecutive spaces in this sequence is a cell. Moreover, all these cells are even-dimensional; hence, we do not have any relations in the homology. The homology groups of the Study quadric are therefore generated by

$$1 = [P_0], \qquad \sigma_1 = [P_1], \qquad \sigma_2 = [P_2],$$

$$\sigma_A = [P_A], \qquad \sigma_B = [P_B],$$

$$\sigma_4 = [Q_4], \qquad \sigma_5 = [Q_5], \qquad \sigma_6 = Q_S.$$

The notation [ ] here is intended to denote "the homology class of...".

Our next task is to compute the intersections of these classes. The cellular decomposition given above is really many decompositions since it depends on the choice of P_0, P_1 and P_2. To compute an intersection, we may use different representatives from the homology class, that is, cells given by different decompositions. For classes of complementary dimension, we have

$$\sigma_1 \cap \sigma_5 = 1, \qquad \sigma_2 \cap \sigma_4 = 1.$$

These can be seen by considering the intersection of a P_1 or a P_2 with a P_5 or a P_4. In each case, the result will normally be a single point. The point will always be in the Study quadric and hence in the corresponding Q_5 or Q_4, since the P_1 and P_2 lie in the Study quadric. Next, from the properties of A-planes and B-planes in the Study quadric, we have

$$\sigma_A \cap \sigma_A = \sigma_B \cap \sigma_B = 0, \qquad \sigma_A \cap \sigma_B = 1.$$

This result is analogous to Halphen's theorem.

These results give us a duality between homology classes of complementary dimensions. This is an example of a more general result known as Poincaré duality. This is extremely useful for computing the other intersections. For example, suppose we require $\sigma_5 \cap \sigma_5$. From dimensional arguments, we know that the answer must be a multiple of the class σ_4. We can fix this multiplier by

TABLE 11.2. The Intersection Pairing for the Study Quadric

$\cap$	σ_1	σ_2	σ_A	σ_B	σ_4	σ_5
σ_1	0	0	0	0	0	1
σ_2	0	0	0	0	1	σ_1
σ_A	0	0	0	1	σ_1	σ_2
σ_B	0	0	1	0	σ_1	σ_2
σ_4	0	1	σ_1	σ_1	$2\sigma_2$	$\sigma_A + \sigma_B$
σ_5	1	σ_1	σ_2	σ_2	$\sigma_A + \sigma_B$	σ_4

taking the intersection with the dual to σ_4, that is, σ_2. This triple intersection, $\sigma_5 \cap \sigma_5 \cap \sigma_2$, can be seen to be homologous to a point, so that

$$\sigma_5 \cap \sigma_5 \cap \sigma_2 = 1.$$

This is because two tangent hyperplanes and a 2-plane will normally meet at a point. So we may conclude that

$$\sigma_5 \cap \sigma_5 = \sigma_4.$$

Moreover, the associativity of the intersection means that we also have

$$\sigma_5 \cap \sigma_2 = \sigma_1,$$

since σ_1 is dual to σ_5. We can compute all the other intersections between generators with just two more triple intersections,

$$\sigma_5 \cap \sigma_4 \cap \sigma_A = 1 \quad \text{and} \quad \sigma_5 \cap \sigma_4 \cap \sigma_B = 1.$$

The results are summarised in Table 11.2.

As another example, we can find the homology class of the subspace generated by a pair of rotations. Recall from above that this subspace is the intersection of the Study quadric with a 3-plane. The intersection of such a 3-plane with a P_5 is a line that will, in general, meet the Study quadric in two points. Hence, the intersection of this subspace with a generic Q_4 is a pair of points, and we can conclude that the class of the subspace is $2\sigma_2$.

At last, we are in a position to look at how this theory can be used to compute the number of assembly configurations of some particular types of mechanisms.

11.5.1   Postures for General 6-R Robots

Consider a robot comprising six revolute joints arranged in series. If we fix the end-effector, how many ways can we arrange the links so that the relative position and orientation of consecutive joints remains the same? In other words, how many different solutions to the inverse kinematic problem are there? Each different solution to the inverse kinematic problem is called a posture, or pose, of the robot; see Section 5.1. Another way of looking at this problem is to think of the robot with its end-effector fixed as a single loop mechanism, that is, with the base connected to the end-effector with a rigid link. In this case, the different solutions are known as different assembly configurations. This is because to change from one solution to another we would have to disassemble the mechanism and reassemble it in the new configuration, since in general six revolute joints in a ring will form a rigid spatial structure.

In order to solve this problem, we imagine breaking the middle link of the chain. We now have two 3-R mechanisms. In the Study quadric, these will trace out a pair of three-dimensional subspaces. The intersection of these subspaces will generally consist of a finite number of group elements. These group elements correspond to configurations of the middle link, relative to some home configurations, for which the two halves of the robot can be reconnected. Hence, to find the number of possible reconnections, we need to find the homology class of the space generated by a 3-R chain. The subspace is 3-dimensional and thus will have a homology class with the general form $\lambda \sigma_A + \mu \sigma_B$. To find μ, the coefficient of σ_B, we can intersect the subspace with a general A-plane. Now, as we saw in Section 11.3, a general A-plane consists of the set of group elements that take one point in space to another. Hence, the intersection of the space generated by three revolute joints with an A-plane consists of the number of ways that the 3-R machine can take one point to another. But this is just the inverse kinematic problem that we solved in Section 5.2, where we found the answer to be 4. So $\mu = 4$. Now, to find λ we could intersect with a B-plane. However, there is a simpler method. Recall the projection from $SE(3)$ to the quotient $SO(3)$ obtained by just taking rotations and forgetting the translations; see Section 3.5. This map extends to a smooth map from the complex Study quadric to $\mathbb{PC}^3$. Recall that the group manifold of $SO(3)$ is the real projective space $\mathbb{PR}^3$; see Section 2.2.2. Physically, this map takes the 3-R manipulator to its spherical indicatrix, that is, the spherical mechanism given by the directions of the joints. In homology, we see that A-planes map to all of $\mathbb{PC}^3$ while the B-planes map to subspaces. In homology, the projection takes $\sigma_A \longmapsto \alpha_3$ and $\sigma_B \longmapsto 0$. Thus, we only need to find the homology class of the spherical indicatrix, which is just a 3-R wrist; see Section 5.1. In Section 5.1, we saw that, in general, for every rotation there are two possible configurations for the wrist; that is, the configuration space double covers $\mathbb{PC}^3$ almost everywhere. Hence, the homology class of the image is $2\alpha_3$ and so $\lambda = 2$.

The class of the subspace generated by three revolute joints is thus, in general, $2\sigma_A + 4\sigma_B$. The number of intersections between two such spaces is given by

$$(2\sigma_A + 4\sigma_B) \cap (2\sigma_A + 4\sigma_B) = 2 \cdot 4 + 4 \cdot 2 = 16.$$

The result is that a general 6-R robot has sixteen postures. Care must be taken with these results however. Intersections must be counted with the correct multiplicity. Some solutions may be complex, and also some solutions may lie on the A-plane of infinite somas.

The above result was first proved by more laborious means by Lee and Liang [67]. Later, Manseur and Doty found a 6-R robot with 16 real finite postures [70].

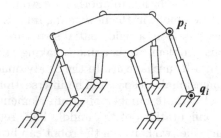

FIGURE 11.1. A General Stewart Platform

11.5.2 Conformations of the 6–3 Stewart Platform

Next, we turn to the conformations of the 6-3 Stewart platform. This mechanism is a simplified form of the general 6-6 Stewart platform.

The Stewart (or sometimes Gough–Stewart) platform is a parallel manipulator. It was developed for use in aircraft simulators but more recently has been used for 'hexapod' machine tools. The platform, on which the cockpit sits, is connected to the ground by six hydraulic rams. At either end of each ram is a passive ball and socket joint; see Figure 11.1. However, note that in general the centres of the ball and socket joints do not have to be coplanar, although in a practical machine there is usually some symmetry in the design.

Suppose we set the lengths of the hydraulic rams. In how many different ways could we assemble the mechanism? Each different assembly is called a conformation of the parallel mechanism. Notice that for a parallel mechanism like this one the inverse kinematics are straightforward. Given the position and orientation of the platform, the leg lengths are simple to calculate. It is the forward kinematic problem that is hard, that is finding the position of the platform given the leg-lengths.

As in the general 6-6 platform that we study later, the 6-3 platform has six legs consisting of a hydraulic ram with a passive spherical joint at either end. The simplification comes from the positioning of the centres of the passive spherical joints. The six spherical joints on the ground link can have arbitrary position but the joints on the platform are coincident in pairs. Two hydraulic rams which have coincident spherical joints at their ends form a triangular structure; see Figure 11.2.

When the lengths of the legs are fixed the possible motions allowed by this triangular mechanism are clearly the same as an R–S pair, that is, a revolute joint about axis joining the two spherical joints of the original mechanism in the ground link. The spherical joint of the new mechanism coincides with the coincident spherical joints of the original machine. Now we can study the conformations of this manipulator by intersecting the varieties generated by three arbitrary R–S linkages.

An R–S linkage will generate a subspace of the Study quadric of dimension 4, hence its homology class will be a multiple of σ_4. To find the coefficient we need to intersect the linkage variety with a general 2-plane lying in the Study quadric. Rather than do this directly we will find the curve given by the intersection with a general A-plane first. Suppose that the A-plane is the space of rotations about a point p in three-dimensional space. Let the centre of the spherical joint be labelled s. Now the intersections of the linkage variety with the A-plane can be pictured as the assembly configurations when the linkage is closed by adding a link to a spherical joint at p. The length of the link is not important but it should be fixed. If we can 'connect up' this closed loop mechanism, then any rotation of the link joining the two spherical joints at p and s will be in the intersection. In other words we expect the intersection to consist of a number of lines; recall that the rotations about a fixed axis form a line in the Study quadric. Now consider the possible positions of the spherical joint s; as it rotates about the revolute joint it traces out a circle in a plane π, perpendicular to the axis of the revolute joint. However, if we consider s as attached to the spherical joint at p, then s traces out a sphere. This sphere will intersect the plane π in a circle and the two circles in this plane will meet in at most two points. Hence, the intersection of the linkage variety of an R–S link meets a general A-plane in a pair of lines. The intersection of this with a general 2-plane in the A-plane is therefore a pair of points and we may conclude that the homology class of the linkage variety is $2\sigma_4$.

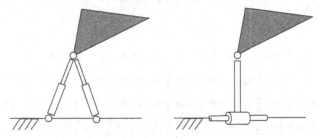

FIGURE 11.2. Triangle of Two Legs in a 6-3 Stewart Platform and Equivalent R–S Pair

Intersecting three of these varieties gives

$$(2\sigma_4)^3 = 16,$$

that is, at most 16 different conformations.

11.5.3 The Tripod Wrist

This mechanism consists of three legs joining the movable platform to the fixed base and an additional passive spherical joint joining the platform to the base. As above each leg is a hydraulic ram with a passive spherical joint at each end. To find the number of conformations of this mechanisms we can find the number of intersections between the subspaces formed by the three legs and the A-plane of rotations about the additional joint.

To proceed, we need to find the homology class of the subspace generated by a single leg. So consider a leg as an S–S mechanism. The space of possible rigid body transformations that such a link can perform will be a five-dimensional subspace of the Study quadric, since the rotation about the axis joining the centres of the spherical joints should only be counted once here.

Consider a standard S–S link for which the first joint is located at the origin and the second is located a distance l above the first along the z-axis. The subspaces generated by other S–S links will be related to this one by a rigid transformation and a change in the value of l.

Rotations about the first joint can be parameterised as

$$e^{\phi s_1} = \cos\frac{\phi}{2} + \sin\frac{\phi}{2}(iu_x + ju_y + ku_z),$$

with $\mathbf{u} = (u_x, u_y, u_z)^T$ an arbitrary unit vector. Rotations about the second joint are given by

$$e^{\psi s_2} = \cos\frac{\psi}{2} + \sin\frac{\psi}{2}(iv_x + jv_y + kv_z) + l\sin\frac{\psi}{2}\varepsilon(-iv_y + jv_x),$$

with $\mathbf{v} = (v_x, v_y, v_z)^T$ another arbitrary unit vector. The product of these dual quaternions gives a parameterisation of the subspace

$$e^{\phi s_1}e^{\psi s_2} = (a_0 + a_1 i + a_2 j + a_3 k) + (c_0 + c_1 i + c_2 j + c_3 k)\varepsilon,$$

where

$$a_0 = \cos\tfrac{\phi}{2}\cos\tfrac{\psi}{2} - \sin\tfrac{\phi}{2}\sin\tfrac{\psi}{2}(u_x v_x + u_y v_y + u_z v_z),$$
$$a_1 = \cos\tfrac{\phi}{2}\sin\tfrac{\psi}{2}v_x + \sin\tfrac{\phi}{2}\cos\tfrac{\psi}{2}u_x + \sin\tfrac{\phi}{2}\sin\tfrac{\psi}{2}(u_y v_z - u_z v_y),$$
$$a_2 = \cos\tfrac{\phi}{2}\sin\tfrac{\psi}{2}v_y + \sin\tfrac{\phi}{2}\cos\tfrac{\psi}{2}u_y + \sin\tfrac{\phi}{2}\sin\tfrac{\psi}{2}(u_z v_x - u_x v_z),$$
$$a_3 = \cos\tfrac{\phi}{2}\sin\tfrac{\psi}{2}v_z + \sin\tfrac{\phi}{2}\cos\tfrac{\psi}{2}u_z + \sin\tfrac{\phi}{2}\sin\tfrac{\psi}{2}(u_x v_y - u_y v_x),$$

$$c_0 = l\sin\tfrac{\phi}{2}\sin\tfrac{\psi}{2}(u_x v_y - u_y v_x),$$
$$c_1 = -l\cos\tfrac{\phi}{2}\sin\tfrac{\psi}{2}v_y - l\sin\tfrac{\phi}{2}\sin\tfrac{\psi}{2}v_x u_z,$$
$$c_2 = l\cos\tfrac{\phi}{2}\sin\tfrac{\psi}{2}v_x - l\sin\tfrac{\phi}{2}\sin\tfrac{\psi}{2}v_y u_z,$$
$$c_3 = l\sin\tfrac{\phi}{2}\sin\tfrac{\psi}{2}(u_x v_x + u_y v_y).$$

In addition to the quadratic relation defining the Study quadric, this variety also satisfies the quadratic relation

$$(c_0 - la_3)c_0 + (c_1 + la_2)c_1 + (c_2 - la_1)c_2 + (c_3 + la_0)c_3 = 0,$$

which is straightforward to check. A line in the Study quadric will meet such a variety in two places. Hence, the variety generated by the S–S mechanism has homology class $2\sigma_5$.

Intersecting three such subspaces will give a three-dimensional subspace in the Study quadric with homology,

$$(2\sigma_5)^3 = 8\sigma_A + 8\sigma_B.$$

Finally, we intersect this with an A-plane with homology class σ_A to give eight general conformations of the tripod wrist.

There is a little more that can be said about this situation. Notice that the intersection of the subspace generated by the S–S link with the A-plane at infinity, $a_0 = a_1 = a_2 = a_3 = 0$ is given by the two-dimensional quadric

$$c_0^2 + c_1^2 + c_2^2 + c_3^2 = 0, \quad a_0 = a_1 = a_2 = a_3 = 0.$$

This quadric does not depend on the length of the S–S link l and is easily seen to be invariant with respect to rigid transformations. So, the variety generated by any S–S link will contain this quadric, and any intersection of such varieties will contain this two-dimensional quadric. Hence, if we intersect the subspace generated by three legs with an A-plane that meets the A-plane at infinity, we will get only six finite intersections. Two of the general eight intersections will be accounted for by intersections in the A-plane at infinity. Recall, from Section 11.2 above, there are two types of A-planes. Those of the form $\mathbf{g}SO(3)_\mathbf{p}$ do not intersect the A-plane at infinity but those of the form $\mathbf{g}SE(2)_\mathbf{n}$ intersect the infinite A-plane in a line. So this does not affect the result for the tripod wrist above, but if the platform were restricted to move in a plane (rather than about a point) then we should expect only six conformations.

11.5.4   The 6-6 Stewart Platform

Unfortunately, to find the number of conformations of the general Stewart platform we cannot simply take the sixfold self-intersection of the homology class of the S–S link found above. This is because the subspace generated by each leg contains the two dimensional quadric in the A-plane at infinity, see the previous section. Hence, the intersection of these six varieties will contain a component of dimension 2 and so it will not be a complete intersection and in these circumstance the homological methods described above do not work. However this is now a classic problem in robotics which has attracted the interest of several mathematicians. The answer is known to be 40. That is, there are at most forty different conformations of the general Stewart platform.

This result has been found by many different methods, we briefly sketch here the approach of Wampler [123].

As above each leg is considered as an S–S link, with leg lengths l_i for $i = 1, \ldots, 6$. Now suppose $\mathbf{p}_i$ and $\mathbf{q}_i$ to be the centres of the spherical joints at the base and platform respectively. These points are given in some home configuration, not necessarily a configuration with the given leg lengths. For each leg we have an equation,

$$l_i^2 = (g(\mathbf{q}_i) - \mathbf{p}_i)^2, \qquad i = 1, 2, \ldots, 6,$$

where g is the rigid motion undergone by the platform from the home position to a valid configuration. Let p_i and q_i be the pure quaternions representing the points $\mathbf{p}_i$ and $\mathbf{q}_i$. The leg equations become

$$l_i^2 = (rq_i r^* + t - p_i)(rq_i r^* + t - p_i)^*, \qquad i = 1, 2, \ldots, 6.$$

Now in terms of dual quaternions the rotation r and translation t are given by

$$r = h_0, \quad \text{and} \quad t = 2h_1 h_0^* = -2h_0 h_1^*.$$

Substituting this into the leg equations, using the fact that $t^* = -t$ and homogenising by assuming $h_0 h_0^* = 1$ we get

$$(|p_i|^2 + |q_i|^2 - l_i^2)h_0 h_0^* + h_0 q_i h_0^* p_i + p_i h_0 q_i h_0^*$$
$$+ 2h_0 q_i h_1^* - 2h_1 q_i h_0^* + 2p_i h_1 h_0^* - 2h_0 h_1^* p_i + 4h_1 h_1^* = 0, \quad i = 1, \ldots, 6.$$

This is clearly a homogeneous quadratic equation in the coordinates, $(a_0 : a_1 : a_2 : a_3 : c_0 : c_1 : c_2 : c_3)$. The A-plane at infinity is given by $h_0 = 0$ and hence the 2-dimensional quadric determined by all the legs is given by the term $4h_1 h_1^* = 0$. Now in [123] Wampler subtracts one of the six quadrics given by the leg equations from the other five thus removing the $4h_1 h_1^* = 0$ term. In this way we obtain five quadric, which all contain the A-plane at infinity. Rather than use the Study quadric as the representation of $SE(3)$, Wampler uses its double cover: the affine quartic determined by the equations $h_0 h_0^* = 1$ and $h_0 h_1^* + h_1 h_0^* = 0$. A subtle change of coordinates produces a set of bihomogeneous equations in two sets of variables. The change of coordinates depends on a parameter λ and the original equations for the Stewart platform only correspond to the bihomogeneous set in the limiting case when $\lambda = 0$. Using the bi-homogeneous version of Bézout's theorem to count the intersections, Wampler shows that, independent of λ, the set of equations has 80 solutions. After identifying solutions that differ by an overall sign change we have the maximum of 40 solutions.

Counting the number of postures or conformations of various mechanisms is a pleasant pastime but for practical work it is more important to be able to find the solutions explicitly. In [59] Husty has given an algorithm for finding the 40 conformations of a Stewart platform given a set of leg lengths. In the end this algorithm involves the numerical solution of a degree 40 univariate polynomial. However, the elimination procedure which results in this degree 40 polynomial owes much to a detailed examination of the algebraic varieties described above.

12
Statics

12.1 Co-Screws

We begin here by introducing the notion of a **co-screw**. These are elements of $se^*(3)$, the dual of the Lie Algebra; see Section 7.5. Co-screws are linear functionals on the velocities, that is, functions

$$\mathcal{F} : se(3) \longrightarrow \mathbb{R} \qquad \text{satisfying} \quad \mathcal{F}(a\mathbf{s}_1 + b\mathbf{s}_2) = a\mathcal{F}(\mathbf{s}_1) + b\mathcal{F}(\mathbf{s}_2),$$

where a and b are constants. The map $\mathcal{F}(\mathbf{s})$ is usually called the evaluation map of the functional. The space of all such functionals forms a vector space with the same dimension as the original space of velocity vectors. In linear algebra, this vector space of functions is usually called the dual vector space. But to avoid confusion with the dual numbers, six-component velocity vectors will be called screws and the linear functionals co-screws. In older language, the screws would be covariant vectors and the co-screws contravariant vectors.

The reason for introducing these objects is that in modern approaches to mechanics the momentum of a system is thought of as dual to the system's velocity—a linear functional. See, for example, Arnol'd [2, sect. 37] and Abraham and Marsden [1, sect. 3.7]. The evaluation map is a pairing between the momentum and velocity that gives a scalar—the kinetic energy of the system. Several other important quantities are best thought of as co-screws. For example, the time derivative of momentum is force, and hence a generalised force on a rigid body is a co-screw. These co-screws are also called wrenches. Another example is the rows of the inverse Jacobian matrix for a six-joint robot. These are co-screws because the columns of the Jacobian are screws, as we saw in

Section 4.5. The pairing between screws and co-screws is given by the matrix product of a row by a column, as we shall see in a moment.

It is important to note that screws and co-screws are really different. Although the two vector spaces are isomorphic, they are not naturally isomorphic. That is, there is no intrinsically defined isomorphism that turns screws into co-screws. The difference is essentially in how they transform under rigid body motions.

Co-screws are elements of a 6-dimensional vector space, so like screws, we may write them as column vectors

$$\mathcal{F} = \begin{pmatrix} \mathbf{M} \\ \mathbf{p} \end{pmatrix}.$$

The evaluation map can be written as a matrix product

$$\mathcal{F}(\mathbf{s}) = (\mathbf{M}^T, \mathbf{p}^T) \begin{pmatrix} \boldsymbol{\omega} \\ \mathbf{v} \end{pmatrix},$$

which is certainly a linear map. The result of the above pairing is supposed to be a scalar, that is, invariant under any rigid motion. Now, we know that under a rigid motion a screw transforms according to the adjoint representation of $SE(3)$; see Section 4.2. The partitioned form of this transformation is given by

$$\begin{pmatrix} \boldsymbol{\omega}' \\ \mathbf{v}' \end{pmatrix} = \begin{pmatrix} R & 0 \\ TR & R \end{pmatrix} \begin{pmatrix} \boldsymbol{\omega} \\ \mathbf{v} \end{pmatrix}.$$

Thus, to keep the pairing invariant, under the same transformation a co-screw becomes

$$\begin{pmatrix} \mathbf{M}' \\ \mathbf{p}' \end{pmatrix} = \begin{pmatrix} R & TR \\ 0 & R \end{pmatrix} \begin{pmatrix} \mathbf{M} \\ \mathbf{p} \end{pmatrix}.$$

This representation of $SE(3)$ is the co-adjoint representation.

In the older literature, the distinction between screws and co-screws is not made. The invariant pairing is achieved using the reciprocal product of screws; see Section 6.4. The disadvantage of such an approach only becomes clear when we look at inertias. For the sake of consistency, we will adopt the modern view in the following.

12.2   Forces, Torques and Wrenches

Consider a force acting on a rigid body; see Figure 12.1. Suppose the force causes the body to move with an instantaneous velocity screw

$$\mathbf{s} = \begin{pmatrix} \boldsymbol{\omega} \\ \mathbf{v} \end{pmatrix}.$$

Then the point of application of the force $\mathbf{r}$ has velocity $\boldsymbol{\omega} \times \mathbf{r} + \mathbf{v}$; see Section 4.1. From elementary mechanics, the power exerted by the force is thus given by

$$\mathbf{F} \cdot (\boldsymbol{\omega} \times \mathbf{r} + \mathbf{v}) = \boldsymbol{\omega} \cdot (\mathbf{r} \times \mathbf{F}) + \mathbf{F} \cdot \mathbf{v}.$$

Since the power is independent of coordinates, we are led to consider a co-screw of the form

$$\mathcal{W} = \begin{pmatrix} \mathbf{r} \times \mathbf{F} \\ \mathbf{F} \end{pmatrix},$$

so the power is given by the pairing

$$\text{Power} = \mathcal{W}(\mathbf{s}) = \mathcal{W}^T \mathbf{s}.$$

Notice that now we don't have to have $\mathbf{r}$ as the point of application of the force; any point on the line of application of the force will do. This is because points on the line can be written as $\mathbf{r}' = \mathbf{r} + \lambda \mathbf{F}$ and hence

$$\mathbf{r}' \times \mathbf{F} = \mathbf{r} \times \mathbf{F}.$$

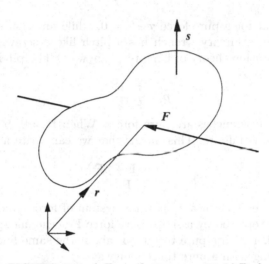

FIGURE 12.1. A Force Acting on a Rigid Body

Such co-screws are called wrenches. The first three components of a wrench are given by the moment of the force about the origin. The condition for the static equilibrium of a rigid body is that the sum of the forces and moments acting on the body are zero. In terms of our six-component wrenches, this condition simply says that the vector sum of the wrenches acting on the body must be zero.

Since pairing a wrench with a velocity screw gives the rate of work, if the pairing is zero then no work can be done. This means that a wrench cannot cause a rigid body to move in the direction of a screw if the pairing of the wrench with the screw gives zero.

A common combination of forces is given by a couple, that is, a pair of equal and opposite forces acting along parallel lines. For example,

$$\mathcal{W}_1 = \begin{pmatrix} \mathbf{r} \times \mathbf{F} \\ \mathbf{F} \end{pmatrix} \quad \text{and} \quad \mathcal{W}_2 = \begin{pmatrix} (-\mathbf{r}) \times (-\mathbf{F}) \\ (-\mathbf{F}) \end{pmatrix}.$$

If these wrenches act on a rigid body, the total wrench acting on the body will be

$$\mathcal{W}_1 + \mathcal{W}_2 = \begin{pmatrix} 2\mathbf{r} \times \mathbf{F} \\ \mathbf{0} \end{pmatrix}.$$

That is, the total force acting on the body is zero, but there is still a non-zero moment. Such a wrench is called a pure torque.

When we have several forces and torques acting on a rigid body, we can compute the total wrench acting on the body. This total wrench is not usually a pure force or a pure torque but something with the form

$$\mathcal{W} = \begin{pmatrix} \boldsymbol{\tau} \\ \mathbf{F} \end{pmatrix}.$$

We saw above that for a pure force $\boldsymbol{\tau} \cdot \mathbf{F} = 0$, while for a pure torque $\mathbf{F} = \mathbf{0}$. More generally, an arbitrary wrench has a pitch like a screw. (We knew this from the representation theory of Chapter 7 anyway.) The pitch of a wrench is given by

$$p = \frac{\boldsymbol{\tau} \cdot \mathbf{F}}{\mathbf{F} \cdot \mathbf{F}},$$

so that pitch $p = 0$ wrenches are pure forces. When $\mathbf{F} = 0$, by convention we say that the pitch is infinite. This means that we can write a general pitch p wrench as

$$\mathcal{W} = \begin{pmatrix} \mathbf{r} \times \mathbf{F} + p\mathbf{F} \\ \mathbf{F} \end{pmatrix},$$

so long as p is finite. This means that any system of forces and torques acting on a body can be replaced by a single pure force $\mathbf{F}$ acting along a line through the point $\mathbf{r}$ together with a pure torque $p\mathbf{F}$ about the same line. If p is infinite here, we are dealing with a pure torque anyway.

In the following, we look at some applications of these definitions to robotics.

12.3   Wrist Force Sensor

Our first example is a wrist force sensor. Such a sensor is often used to measure the force and torque exerted by a robot's end-effector. Usually, the sensor is placed at the robot's wrist so that it measures the wrench between the last link of the robot and its tool. Many different designs of sensor are possible; usually strain gauges are used to measure the deflection of small sections of metal. The forces on these metal rods are then inferred, and hence the wrench at the tool can be calculated. A common design of sensor is shown in Figure 12.2; strain gauges are arranged so that the eight forces shown in the figure are measured.

To find the total wrench acting on the tool, we simply sum the eight wrenches as measured by the strain gauges. With the coordinates as shown in Figure 12.2,

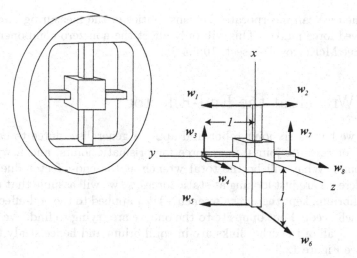

FIGURE 12.2. The Four Beam Wrist Force Sensor

the eight wrenches are

$$W_1 = \begin{pmatrix} lw_1\mathbf{k} \\ w_1\mathbf{j} \end{pmatrix}, \quad W_2 = \begin{pmatrix} -lw_2\mathbf{j} \\ w_2\mathbf{k} \end{pmatrix}, \quad W_3 = \begin{pmatrix} -lw_3\mathbf{k} \\ w_3\mathbf{i} \end{pmatrix}, \quad W_4 = \begin{pmatrix} lw_4\mathbf{i} \\ w_4\mathbf{k} \end{pmatrix},$$

$$W_5 = \begin{pmatrix} -lw_5\mathbf{k} \\ w_5\mathbf{j} \end{pmatrix}, \quad W_6 = \begin{pmatrix} lw_6\mathbf{j} \\ w_6\mathbf{k} \end{pmatrix}, \quad W_7 = \begin{pmatrix} lw_7\mathbf{k} \\ w_7\mathbf{i} \end{pmatrix}, \quad W_8 = \begin{pmatrix} -lw_8\mathbf{i} \\ w_8\mathbf{k} \end{pmatrix}$$

where we have written $\mathbf{i}$, $\mathbf{j}$ and $\mathbf{k}$ for the unit vectors in the x, y and z directions, respectively. The total wrench is thus

$$W_{tot} = \begin{pmatrix} l(w_4 - w_8) \\ l(w_6 - w_2) \\ l(w_1 - w_3 - w_5 + w_7) \\ w_3 + w_7 \\ w_1 + w_5 \\ w_2 + w_4 + w_6 + w_8 \end{pmatrix}.$$

This relation is often expressed as a matrix equation:

$$W_{tot} = \begin{pmatrix} 0 & 0 & 0 & l & 0 & 0 & 0 & -l \\ 0 & -l & 0 & 0 & 0 & -l & 0 & 0 \\ l & 0 & -l & 0 & -l & 0 & l & 0 \\ 0 & 0 & 1 & 0 & 0 & 0 & 1 & 0 \\ 1 & 0 & 0 & 0 & 1 & 0 & 0 & 0 \\ 0 & 1 & 0 & 1 & 0 & 1 & 0 & 1 \end{pmatrix} \begin{pmatrix} w_1 \\ w_2 \\ w_3 \\ w_4 \\ w_5 \\ w_6 \\ w_7 \\ w_8 \end{pmatrix}.$$

The matrix that transforms the eight w_is into the total wrench is called the resolved force matrix. If we think of the w_is as the signals, from the strain gauges then we can incorporate the sensitivities of the measuring circuits into the resolved force matrix. This will only affect the non-zero components of the matrix. See McKerrow [75, sect. 10.5.4.3].

12.4   Wrench at the End-Effector

Suppose we have a six-joint robot and apply a generalised force to each joint, a torque for revolute joints, pure force for prismatic joints, and a wrench for helical joints. What will be the total wrench at the end-effector due to these forces? Here we are just looking at static forces, so we will assume that the robot is in equilibrium, kept there by a wrench $-\mathcal{W}_{tot}$ applied to the end-effector. This final wrench is equal but opposite to the one we are trying to find. We may also assume that all of the other links are in equilibrium and hence study them one by one; see Figure 12.3.

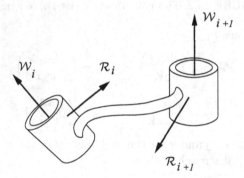

FIGURE 12.3. The Wrenches Acting on the Links of a Robot

For each link except the final one, we have a wrench equation for equilibrium:

$$\mathcal{W}_i + \mathcal{R}_i = \mathcal{W}_{i+1} + \mathcal{R}_{i+1}, \qquad i = 1, 2 \ldots 5.$$

The term $\mathcal{W}_i$ is the wrench due to the motor at the i-th joint. The power that such a wrench exerts on its joint screw is $w_i = \mathcal{W}_i(\mathbf{s}_i)$. The $\mathcal{R}_i$s are the reaction wrenches at each joint. These reaction wrenches can do no work along the joint screw, so $\mathcal{R}_i(\mathbf{s}_i) = 0$. At the last link, we have

$$\mathcal{W}_6 + \mathcal{R}_6 = \mathcal{W}_{tot}.$$

These equations can be combined to give the following system of equations:

$$\mathcal{W}_i + \mathcal{R}_i = \mathcal{W}_{tot}, \qquad i = 1, 2 \ldots 6.$$

Now, pairing each of these with the joint screws gives

$$w_i = \mathcal{W}_i(\mathbf{s}_i) = \mathcal{W}_{tot}(\mathbf{s}_i), \qquad i = 1, 2 \ldots 6$$

and hence

$$\mathcal{W}_{tot}^T J = (w_1, w_2, w_3, w_4, w_5, w_6),$$

or, when the Jacobian is non-singular,

$$\mathcal{W}_{tot} = (J^{-1})^T \begin{pmatrix} w_1 \\ w_2 \\ \vdots \\ w_6 \end{pmatrix}.$$

This gives a simple relation between the power at each joint and the wrench delivered to the end-effector. It should be emphasised that this relation holds only when the robot is in static equilibrium.

The above analysis can be of use when we study robots that are not perfectly stiff. In real robots, there is always a certain amount of compliance due to movement in the drive motors, the transmission, and possibly flexing of the links. A very simple model of this elasticity is given by a generalisation of Hook's law: $w_i = k_i \delta\theta_i$. This gives the stress at joint i in terms of the strain $\delta\theta_i$ and a constant k_i called the joint stiffness. For six joints, we can write a matrix equation

$$\begin{pmatrix} w_1 \\ w_2 \\ \vdots \\ w_6 \end{pmatrix} = K \begin{pmatrix} \delta\theta_1 \\ \delta\theta_2 \\ \vdots \\ \delta\theta_6 \end{pmatrix}$$

where $K = \mathrm{diag}(k_1, k_2, \ldots, k_6)$ is the diagonal matrix of joint stiffnesses and is called the stiffness matrix of the robot. Now, the Jacobian can be thought of as relating displacements of the end-effector to joint displacements:

$$\begin{pmatrix} \delta\theta_x \\ \delta\theta_y \\ \delta\theta_z \\ \delta x \\ \delta y \\ \delta z \end{pmatrix} = J \begin{pmatrix} \delta\theta_1 \\ \delta\theta_2 \\ \delta\theta_3 \\ \delta\theta_4 \\ \delta\theta_5 \\ \delta\theta_6 \end{pmatrix}.$$

So, combining this with the above relations, we get

$$\mathcal{W}_{tot} = (J^{-1})^T K (J^{-1}) \begin{pmatrix} \delta\theta_x \\ \delta\theta_y \\ \delta\theta_z \\ \delta x \\ \delta y \\ \delta z \end{pmatrix}.$$

This gives the wrench on the robot's end-effector necessary to cause the given displacement. Often, this will be written the other way around as

$$\begin{pmatrix} \delta\theta_x \\ \delta\theta_y \\ \delta\theta_z \\ \delta x \\ \delta y \\ \delta z \end{pmatrix} = C\mathcal{W}_{tot},$$

giving the displacement caused by the wrench applied to the end-effector. The matrix $C = JK^{-1}J^T$ is called the compliance matrix of the robot.

12.5   Gripping

Imagine trying to grasp a solid object using a multifingered hand; see Figure 12.4. If there is no friction, then the only forces that can be applied to the object will be along the surface normals to the object at the contact points. Usually, we want to hold the object in such a way that we can balance any external wrench applied to the body, such as the body's weight. A grasp with the property that any external wrench can be balanced is called a **force closed** grasp. This, perhaps, should be referred to as a force-torque closed grasp to emphasise that both forces and torques must be balanced.

We can write the normal wrenches as $w_i\mathcal{N}_i$, where w_i is the force applied to finger i, and $\mathcal{N}_i$ is the wrench representing a unit force along the normal line. The condition for a force closed grasp is that the normal wrenches must span the space of co-screws:

$$\text{span}\,(\mathcal{N}_1, \mathcal{N}_2, \ldots, \mathcal{N}_m) = se^*(3).$$

So, for a force closed grasp, we need at least six fingers. If the above relation is not satisfied, then there will be at least one external wrench that cannot be balanced. That is, there will be some wrenches $\mathcal{W}$ for which no finger forces can be found that satisfy

$$w_1\mathcal{N}_1 + w_2\mathcal{N}_2 + w_3\mathcal{N}_3 + \cdots w_m\mathcal{N}_m = \mathcal{W}.$$

This can happen, even when we have six fingers, if there is a linear dependence among the normal lines, that is, if the lines satisfy a relation

$$w_1\mathcal{N}_1 + w_2\mathcal{N}_2 + w_3\mathcal{N}_3 + \cdots w_6\mathcal{N}_6 = \mathbf{0},$$

for some, not all zero, set of force magnitudes w_i.

There are some surfaces where no force closed grasps can exist. In fact, these surfaces are precisely the Reuleaux pairs that we met in Section 3.6. To see this,

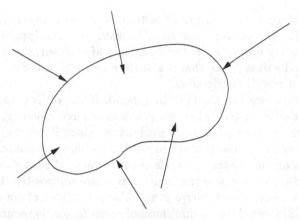

FIGURE 12.4. Several Fingers Acting on a Rigid Body

we note that the normal lines to such a surface must be linearly dependent and hence will lie in some line complex. If we assume that the surface is given as the zeros of some differentiable function $\Phi : \mathbb{R}^3 \longrightarrow \mathbb{R}$, then the normal lines can be written as

$$\mathcal{N} = \begin{pmatrix} \mathbf{r} \times \nabla\Phi \\ \nabla\Phi \end{pmatrix},$$

where ∇ is the usual gradient operator in $\mathbb{R}^3$. We don't have to worry about the normalisation here. The fact that the normal lines lie in a line complex can be expressed by pairing the normal with a constant co-screw:

$$\mathcal{N}(\mathbf{s}) = 0 = \boldsymbol{\omega} \cdot \mathbf{r} \times \nabla\Phi + \mathbf{v} \cdot \nabla\Phi.$$

This gives a partial differential equation for the function Φ and hence for the 'ungrippable' surfaces. Rather than solve this equation directly, we look instead at functions that are symmetric under a one-parameter group of rigid transformations, that is, differentiable functions $\Psi : \mathbb{R}^3 \longrightarrow \mathbb{R}$ such that $\Psi(e^{\lambda \mathbf{s}}\mathbf{r}) = \Psi(\mathbf{r})$ for all parameters λ and all points $\mathbf{r}$. Clearly, if the zero set of such a function exists, then it will be invariant with respect to the one-parameter group. The differential of the action on the function will be zero:

$$0 = \left. \frac{d\Psi(e^{\lambda \mathbf{s}}\mathbf{r})}{d\lambda} \right|_{\lambda=0} = \frac{\partial \Psi}{\partial x}\frac{dx}{d\lambda} + \frac{\partial \Psi}{\partial y}\frac{dy}{d\lambda} + \frac{\partial \Psi}{\partial z}\frac{dz}{d\lambda} = \nabla\Psi \cdot \frac{d\mathbf{r}}{d\lambda}$$

where (x, y, z) are the coordinates of the point $\mathbf{r}$. In terms of differential geometry, this operation is the Lie derivative of Ψ along a Killing field of $\mathbb{R}^3$; see Schutz [99, sect. 3.1]. The velocity of the point $\mathbf{r}$ was found in Section 4.1, so we can write the above equation as

$$\nabla\Psi \cdot (\boldsymbol{\omega} \times \mathbf{r} + \mathbf{v}) = 0.$$

Cycling the triple products, we recover the equation for ungrippable surfaces.

Hence, a surface that is invariant with respect to a one-parameter group of rigid motions will be ungrippable. Moreover, any ungrippable surface will have this symmetry property. So, these surfaces are the surface of the Reuleaux pairs, as claimed earlier. Note that if a surface has a 2- or 3-dimensional group of symmetries it is still ungrippable.

From the above, we can see that in general, if the object we are trying to hold is not a Reuleaux pair, then we can find a force closed grasp with just six fingers. This, however, is a little misleading, since it assumes that we can impose any force along the contact normal at a finger. Usually, we can only push, so assuming the contact normals point towards the interior of the body, we should study grasps where the finger forces are all positive in magnitude. Such a grasp is known as a positive grasp. Using methods of convexity theory, Mishra et al. [78] showed that a minimum of seven fingers is required for a force closed positive grasp.

There is a different way of viewing the problem of grasping a solid object with a multifingered hand. Rather than concentrating on the forces and torques, consider the possible velocities that the object can have. Ideally, when we have grasped the object, it should not be possible to move it, at least not relative to the hand. When there are no fingers, the object is free to move in any direction, so its velocity could be any screw. Placing a finger on the object's surface stops the object from moving along the contact normal, or if we are considering positive grasps, then only velocities against the direction of the contact normal are disallowed. Now, assume that the finger is just a mathematical point and the surface of the object is flat; that is, the object is a polyhedron. With these assumptions, the object will be able to rotate about the contact point and to translate tangentially to the contact normal. Alternatively, we may assume that the object is curved but that we are only considering movements up to first order. The rotational motions are called rolls, and the translations are slides. The velocity screw of the object can thus lie on a 5-system in $se(3)$, the one that is dual to translations along the contact normal. For positive grasps, the possible velocities are restricted to a half-space bounded by this 5-system. The set of possible velocities that the object can acquire will be referred to as the **space of feasible velocities**. Adding more fingers imposes more constraints on the object's velocity and restricts the space of feasible velocities. If there remains a non-zero screw that the object can acquire, then it can elude our grasp by moving with that velocity. The object is thus immobilised when we have placed fingers on its surface in such a way that the space of feasible velocities is null, that is, consists of the zero velocity only. A grasp with this property is called a **form closed** grasp. Note that the space of feasible velocities will be the intersection of the 5-systems defined by the fingers, or, for positive grasps, it will be the intersection of the half-spaces.

For point fingers and polyhedral objects, the conditions for form closure and force closure are precisely the same. To see this, notice that the condition for force closure could be thought of as requiring that the contact normals span

$se(3)$; that is, we can consider these lines as screws rather than co-screws. The duals to the contact normals are elements of $\wedge^5 se(3)$; their intersection will be null precisely when the union of the contact normals is all of $se(3)$; see Section 8.5. The same is also true for positive grasps.

If the object is curved rather than polyhedral, then the two concepts can differ. This is because at a contact point on a curved surface the feasible motions of the object may not form a linear system. The feasible velocities are rolls about the contact point as before, but the slides must be along the surface. Recall from Section 6.5.3 that the curvature axes at a point on a curved surface form a rational cubic ruled surface.

To find the possible finger motions along the surface we will write the equation for the quadric approximating the surface in the form

$$(\mathbf{q}^T, 1)Q \begin{pmatrix} \mathbf{q} \\ 1 \end{pmatrix} = 0$$

where $\mathbf{q}^T = (x, y, z)$ is the position vector of a point in space and Q is the symmetric matrix

$$Q = \begin{pmatrix} k_x & 0 & 0 & 0 \\ 0 & k_y & 0 & 0 \\ 0 & 0 & 0 & -1 \\ 0 & 0 & -1 & 0 \end{pmatrix}.$$

Now consider the following function of a single parameter t:

$$f(t) = (\mathbf{q}^T, 1)e^{\theta S^T} Q e^{\theta S} \begin{pmatrix} \mathbf{q} \\ 1 \end{pmatrix}.$$

That is, we are assuming that θ is a function of t, and we will also set $\theta(0) = 0$. This means that $f(0) = 0$ implies that the point $\mathbf{q}$ is on the quadric. The point on the quadric that we are interested in is the finger contact at the origin, so let's modify the function to look at this point:

$$f(t) = (\mathbf{0}^T, 1)e^{\theta S^T} Q e^{\theta S} \begin{pmatrix} \mathbf{0} \\ 1 \end{pmatrix}.$$

Suppose we move the finger along the screw S; we use the 4-dimensional representation here. If the path of the finger remains on the quadric, then we will have $f(t) = 0$. This is unlikely, but we can require that the finger's path should be tangential to the quadric; that is, we can ask which screws generate finger paths that have first order contact with the quadric. The condition for this is found by differentiating $f(t)$ and setting it equal to zero at the contact point; compare this with Section 5.3. We get

$$\dot{f}(0) = 0 = \dot{\theta}(\mathbf{0}^T, 1)(S^T Q + Q S) \begin{pmatrix} \mathbf{0} \\ 1 \end{pmatrix},$$

writing the screw S as

$$S = \begin{pmatrix} ad(\boldsymbol{\omega}) & \mathbf{v} \\ 0 & 0 \end{pmatrix}.$$

The condition for first order contact reduces to $v_z = 0$, reproducing our earlier assertion that the contact prohibits translations along the contact normal. However, we can go further now by demanding second order contact with the surface by differentiating once more and setting the result to zero:

$$\ddot{f}(0) = 0$$

$$= \ddot{\theta}(\mathbf{0}^T, 1)(S^T Q + QS) \begin{pmatrix} \mathbf{0} \\ 1 \end{pmatrix} + \dot{\theta}^2(\mathbf{0}^T, 1)((S^T)^2 Q + 2S^T QS + QS^2) \begin{pmatrix} \mathbf{0} \\ 1 \end{pmatrix}.$$

The first term on the right-hand side is zero already since first order contact must still hold. The second term simplifies to give a quadratic relation on the coordinates of S:

$$k_x v_x^2 + k_y v_y^2 - \omega_x v_y + \omega_y v_x = 0.$$

So, screws that generate finger motions that remain in contact with the surface to second order must satisfy these two equations, and hence they lie on a four-dimensional affine quadric in $se(3)$, or projectively a three-dimensional quadric in $\mathbb{PR}^5$. However, we are more interested in positive grasps, and the above equations determine the boundaries to spaces of feasible grasps. Also, the above equations were derived for a fixed object with a moving finger. What we really need is the equations for fixed fingers with a moving object. This only involves changing the signs of the screws. So, assuming that the finger is above the surface, that is, that finger positions satisfy

$$k_x x^2 + k_y y^2 - 2z \leq 0,$$

then the feasible region for velocities of the object that maintain contact to first order is

$$v_z \leq 0.$$

To include second order effects, we take the intersection of this region with the region defined by the inequality

$$k_x v_x^2 + k_y v_y^2 - \omega_x v_y + \omega_y v_x \leq 0.$$

For several fingers, we take the intersection of several of these feasible sets. The feasible sets for different finger positions but with the same curvature are simply related by rigid transformations. Hence, it is a simple matter to find the relations that determine the feasible set for any finger position from the results given above.

The point of this is that a grasp that is not immobile to first order may be immobile when the curvature of the surface is taken into account. If the object's surface is convex at the contact point, then the feasible space for positive grasps will contain the feasible space for first order positive grasps since translations perpendicular to contact normal will be feasible positive velocities. But at concave points a translation perpendicular to the contact normal will not be feasible since it would put the contact point inside the object. For several fingers, we get a feasible space for each, and if the intersection of these spaces is null the object is immobile. The general lesson is to place the fingers at dimples on the object's surface. These ideas have been extended to the case studied by Rimon and Burdick [94] where the fingers are also solid and curved.

Notice that, to second order, the slide motions for a single finger are rotations about the surface's curvature axes.

12.6   Friction

Finally, we cannot finish a chapter on statics without looking at the subject of friction. In reality, friction is a very complicated subject, and most mathematical models of friction are inadequate over-simplifications. However, even the simplest models can lead to some interesting geometry. Consider static friction. Suppose we have two bodies in contact at a point, say a finger touching an object. The maximum frictional force at the contact point is μ times the normal force between the bodies. The coefficient of friction μ is a number between 0 and 1. However, its exact value depends on many factors such as the cleanliness of the surfaces, the ambient temperature, and humidity. So, in practice, conservative estimates of μ are used. The possible forces exerted on the object at the point of contact are a combination of the force normal to the surface of the object together with the frictional force. Assuming that friction is the same in all directions tangential to the surface, the maximum frictional forces describe a circle in the tangent plane. The radius of the circle is μ times the magnitude of the normal force; hence, the set of forces that can be exerted on the object by a finger lies inside a cone. The axis of the cone is the normal line to the surface at the point of contact, and the semi-angle of the cone is $\arctan(\mu)$. This cone is called the **friction cone**. Notice that the axes of the forces in the friction cone all pass through the point of contact. Hence, the lines in the friction cone all lie in the α-plane of the Klein quadric corresponding to the point of contact; see Section 6.3. In fact, the interior of the friction cone corresponds to the interior of a circle in the α-plane. This can be seen by considering the friction cone in a standard coordinate system. Assume that the point of contact is the origin and the contact normal is aligned with the z-axis. The wrenches that we can apply to the body that is located above the $x - y$ plane have the form

$$
\mathcal{W} = \begin{pmatrix} 0 \\ 0 \\ 0 \\ \gamma F \cos \theta \\ \gamma F \sin \theta \\ F \end{pmatrix}, \qquad \text{where} \quad F \geq 0, \qquad 0 \leq \gamma \leq \mu.
$$

Using a rigid body motion, we can always bring a friction cone into this form. Notice that we assume that the finger can only push the object.

Suppose we have two fingers. We might suppose that any applied wrench could be balanced if we ignore the restrictions imposed by the fact that μ is finite. This is incorrect, since any two α-planes intersect; hence, the wrenches from a pair of friction cones cannot generate all of $se^*(3)$. It is not hard to see that we can never generate a torque about the line joining the two contact points. So we need at least three fingers to be able to balance any applied wrench on the body, more if we take account of the finite nature of μ and the fact that the normal force can only push. The linear algebra problem of finding the finger forces and frictional forces given an applied wrench is underdetermined: there are only six equations in the nine unknown force magnitudes. Hence, many solutions are possible—the system is statically indeterminate. However, if there is a solution that satisfies all the constraints imposed by the friction cones, then the object will not move.

Thus, we are led to consider the set of positive linear combinations of forces from the friction cones. Here we are thinking of the friction cone as the set of wrenches in $se^*(3)$. The friction cone is a closed convex cone in wrench space. There is a substantial body of theory concerned with closed convex cones, originally developed for convex programming. In this context, a cone is defined as a subset C of a vector space V that satisfies

$$
\mathbf{0} \in C
$$

and

$$
\text{if} \quad \mathbf{v} \in C \quad \text{then} \quad \lambda \mathbf{v} \in C, \quad \text{for all} \quad \lambda > 0,
$$

see Nering [79, Chap. VI sect. 2]. The notions of closure and convexity have their usual meaning. Now, it is possible to show that the intersection of two closed convex cones is another closed convex cone. Also, the positive linear combination of a pair of closed convex cones forms a closed convex cone. We will denote this operation as $C_1 + C_2$ for two cones C_1 and C_2. For any closed convex cone, we can find its dual, which is also a closed convex cone but in the dual space V^*. The definition is as follows:

$$
C^* = \{ \mathbf{f} \in V^* : \mathbf{f}(\mathbf{v}) \geq 0, \quad \text{for all} \quad \mathbf{v} \in C \}.
$$

Notice that for our standard friction cone the dual friction cone consists of the screws of the form

$$\mathbf{s} = \begin{pmatrix} 0 \\ 0 \\ 0 \\ \lambda l \cos\phi \\ \lambda l \sin\phi \\ l \end{pmatrix}, \qquad \text{where} \quad l \geq 0, \qquad 0 \leq \lambda \leq \frac{1}{\mu}.$$

If one cone is contained in another, $C_1 \subseteq C_2$, then on dualising we have $C_2^* \subseteq C_1^*$. It is also possible to show that a version of de Morgan's Law holds; that is, we have

$$(C_1 + C_2)^* = C_1^* \cap C_2^*.$$

Hence, rather than considering the positive linear combinations of the friction cones it may be easier to study the intersections of their duals. In particular, any applied wrench can be balanced if the intersection of the dual friction cones is null.

As a simple application, consider the following well-known result concerning two fingered, planar, positive grasps. Such a grasp is force closed if and only if each contact point lies inside the friction cone of the opposite finger or in the negative of the friction cone of the opposite finger; see [80]. Another way of putting the condition is that the line joining the contact points must lie in the friction cones, and the friction cones must be directed towards each other or away from each other. This is now simple to prove. Consider forces along the line connecting the contact points. This line must lie in both friction cones in order that forces in both directions along it can be balanced. To show that the conditions are also sufficient to guarantee force closure, we look at the duals to the friction cones and to the forces along the line. The dual to a force along a line is simply a half-space. In this case, the boundary of the half-space will be two-dimensional since $se^*(2)$ is only three-dimensional. Since the force along the line lies in the friction cone in wrench space, the dual of the friction cone lies in the half-space. The intersection of the two half-spaces corresponding to forces in both directions along the line is simply the boundary 2-plane. This must contain the intersection of the dual friction cones, which is thus null.

Notice that in two dimensions the boundary of a friction cone is just a pair of lines. Hence, in the planar case all grasping problems can be reduced to linear programming problems. This is not possible in three dimensions. However, a common technique here is to use linear programming by approximating the friction cones by polyhedral cones, that is, cones with flat faces.

13
Dynamics

13.1 Momentum and Inertia

For a rigid body, the velocity is given by a screw $\mathbf{s}$, and the momentum is given by a dual vector, or co-screw, $\mathcal{M}$. The pairing between the velocity and momentum gives the kinetic energy of the body

$$\mathcal{M}^T \mathbf{s} = \mathbf{j} \cdot \omega + \mathbf{p} \cdot \mathbf{v} = 2E_K,$$

where

$$\mathbf{s} = \begin{pmatrix} \omega \\ \mathbf{v} \end{pmatrix} \quad \text{and} \quad \mathcal{M} = \begin{pmatrix} \mathbf{j} \\ \mathbf{p} \end{pmatrix},$$

with $\mathbf{j}$ and $\mathbf{p}$ the usual three-dimensional angular and linear momenta.

Traditionally, in rigid body mechanics angular and linear momentum are combined into a screw, and the kinetic energy is given by the reciprocal product of the velocity and momentum; see Section 6.4. The fact that this works is yet another accidental property of three-dimensional space. The view taken here is more general, and further, it will make inertias easier to deal with.

Having separated the notions of velocity and momentum, we now turn to inertias. In line with modern classical mechanics, we think of inertias as operators (tensors) that convert velocities to momenta; see Arnol'd [2, Appendix 2 sect. C]. The inertia operator provides a linear isomorphism

$$N : se(3) \longrightarrow se^*(3).$$

The operator N can be represented by a 6×6 symmetric matrix.

So, the momentum co-screw of a rigid body with velocity $\mathbf{s}$ is given by $\mathcal{M} = N\mathbf{s}$. There is no unique choice for N. However, once we are given a rigid body, N is precisely determined by the equations of elementary mechanics

$$\mathbf{j} = \mathbf{I}\boldsymbol{\omega} + m(\mathbf{c} \times \mathbf{v}), \qquad \mathbf{p} = m\mathbf{v} + m(\boldsymbol{\omega} \times \mathbf{c}).$$

Here m is the body's mass, $\mathbf{c}$ the position of its centre of mass, and $\mathbf{I}$ its 3×3 inertia tensor. The partitioned form of N is thus

$$N = \begin{pmatrix} \mathbf{I} & mC \\ mC^T & mI_3 \end{pmatrix}.$$

As usual, I_3 is the 3×3 identity and $C = \mathrm{ad}(\mathbf{c})$.

The kinetic energy of the body is then given by

$$E_K = \frac{1}{2}\mathcal{M}(\mathbf{s}) = \frac{1}{2}N\mathbf{s}(\mathbf{s}) = \frac{1}{2}\mathbf{s}^T N\mathbf{s}.$$

The kinetic energy is a scalar, independent of the body's location and orientation. Moreover, we know that the velocity screws transform according to the adjoint representation of $SE(3)$, and hence we can infer the transformation properties of the inertia matrix. We have

$$E_K = \frac{1}{2}\mathbf{s}^T N\mathbf{s} = \frac{1}{2}\mathbf{s}'^T N'\mathbf{s}' = \frac{1}{2}\mathbf{s}^T H^T N' H\mathbf{s},$$

so in the new position

$$N' = (H^T)^{-1} N H^{-1},$$

where H is the matrix representing the move in the adjoint representation. This equation is a generalisation of both the tensor properties of the 3×3 inertia matrix and the parallel axis theorem. For instance, suppose we have a rigid body positioned such that its centre of mass coincides with the origin. Its inertia matrix will have the form

$$N = \begin{pmatrix} \mathbf{I} & 0 \\ 0 & mI_3 \end{pmatrix}.$$

A pure translation is represented by a matrix

$$H = \begin{pmatrix} I_3 & 0 \\ T & I_3 \end{pmatrix}.$$

Hence, after a translation the inertia matrix of our body will be

$$N' = \begin{pmatrix} I_3 & T \\ 0 & I_3 \end{pmatrix}\begin{pmatrix} \mathbf{I} & 0 \\ 0 & mI_3 \end{pmatrix}\begin{pmatrix} I_3 & 0 \\ -T & I_3 \end{pmatrix} = \begin{pmatrix} \mathbf{I} - mT^2 & mT \\ -mT & mI_3 \end{pmatrix}.$$

This is simply the parallel axis theorem; see Woodhouse [129, sect. 3.1] for example. The above relation is more general, so perhaps it should be called the "Skew axis theorem".

The inertia matrix N also determines a bilinear symmetric form on the Lie algebra $se(3)$. This form is positive definite since the kinetic energy of a body is never negative. The form is not a group invariant, as we saw above. However, since the matrix is positive definite we should be able to diagonalise it by a suitable change of coordinates. Clearly, from the comments in the previous paragraph, we must place the origin of our new coordinates at the centre of mass of the body. Finally, using the standard theory of the 3×3 inertia matrix, we rotate the axes into coincidence with the principal axes of $\mathbf{I}$. These principal axes are the eigenvectors of $\mathbf{I}$. When the rotation is complete, our matrix N will be diagonal.

In his famous treatise [6], Ball introduced what he called **conjugate screws of inertia**. With our present notation, we can interpret this idea as a pair of screws that satisfy

$$\mathbf{s}_1^T N \mathbf{s}_2 = 0.$$

The physical idea behind this is that if we give the body an impulse so that it begins to move with velocity $\mathbf{s}_1$, then no work can be done on screw $\mathbf{s}_2$. This relation between screws is symmetrical since transposing gives

$$\mathbf{s}_1^T N \mathbf{s}_2 = \mathbf{s}_2^T N \mathbf{s}_1.$$

Using Gramm–Schmidt orthogonalisation, we can always find six linearly independent, mutually conjugate screws. For a single rigid body under the influence of an arbitrary system of forces and torques, this is very useful since it enables us to find coordinates in which the equations of motion decouple. Notice that the condition for a pair of screws to be conjugate can be thought of as a quadric in $se(3) \oplus se(3)$. By changing coordinates to $\mathbf{s}_a = \mathbf{s}_1 + \mathbf{s}_2$ and $\mathbf{s}_b = \mathbf{s}_1 - \mathbf{s}_2$, we have planes of solutions

$$\mathbf{s}_b = M \mathbf{s}_a.$$

where M is a matrix satisfying $M^T N M = N$, that is, a matrix that preserves the inertia matrix. The set of all such matrices M forms a group conjugate to $O(6)$; see Section 3.1. In terms of the original screws, we have that conjugate screws of inertia satisfy

$$(I_6 - M)\mathbf{s}_1 - (I_6 + M)\mathbf{s}_2 = \mathbf{0}.$$

The coadjoint action of the group $SE(3)$ induces a representation of the Lie algebra $se(3)$ on its dual space $se^*(3)$. If we write the action of the group as $\mathcal{M}' = e^{-t\, ad^T(\mathbf{s})} \mathcal{M}$, then we find the action of the Lie algebra by differentiating with respect to t, and then setting $t = 0$. Hence, the coadjoint representation of the Lie algebra is given by the product $-ad^T(\mathbf{s})\mathcal{M}$. This action of a screw on a co-screw will be written as $\{\mathbf{s}, \mathcal{M}\} = -ad^T(\mathbf{s})\mathcal{M}$. In terms of column vectors, it is straightforward to show that

$$\{\mathbf{s}, \mathcal{M}\} = \left\{ \begin{pmatrix} \boldsymbol{\omega} \\ \mathbf{v} \end{pmatrix}, \begin{pmatrix} \mathbf{j} \\ \mathbf{p} \end{pmatrix} \right\} = \begin{pmatrix} \boldsymbol{\omega} \times \mathbf{j} + \mathbf{v} \times \mathbf{p} \\ \boldsymbol{\omega} \times \mathbf{p} \end{pmatrix}.$$

Notice that if we have a pairing between a screw and a co-screw $\mathcal{M}(s_1)$ and we move both about another screw s_2, then the result of the pairing should be unchanged:

$$\frac{d}{dt}\mathcal{M}(s_1) = 0 = \{s_2, \mathcal{M}\}(s_1) + \mathcal{M}([s_2, s_1]),$$

remembering that $\mathrm{ad}(s_2)s_1 = [s_2, s_1]$; see Section 4.3. Hence, for any pair of screws s_1, s_2 and any co-screw $\mathcal{M}$,

$$\{s_2, \mathcal{M}\}(s_1) = \mathcal{M}([s_1, s_2]).$$

Notice that we have the relation

$$\{s_1, Q_0 s_2\} = Q_0[s_1, s_2]$$

where Q_0 is the Klein quadric, thought of here as an element of the representation $ad^* \otimes ad^*$, that is, a linear operator that transforms screws to co-screws.

If a screw s and a co-screw $\mathcal{M}$ have the same line of action, then $\{s, \mathcal{M}\} = 0$. To see this, we write the screw and co-screw as

$$s = \begin{pmatrix} v \\ r \times v + pv \end{pmatrix} \quad \text{and} \quad \mathcal{M} = \begin{pmatrix} r \times v + qv \\ v \end{pmatrix}.$$

The co-adjoint action then gives

$$\{s, \mathcal{M}\} = \begin{pmatrix} v \times (r \times v + pv) + (r \times v + qv) \times v \\ v \times v \end{pmatrix} = 0.$$

The converse is unfortunately not true, except in the case where the pitches of the screw and co-screw are finite. When there are no infinite pitch screws or co-screws, then the vanishing of the co-adjoint action of the screw on a co-screw implies that they have the same line of action.

The analogue of the principal axes of a 3×3 inertia matrix in the case of 6×6 inertias is the **principal screws of inertia**. These are defined as screws s for which s and Ns have the same pitch and line of action. This means that we must have

$$Ns = \lambda Q_0 s$$

where, as above, Q_0 is the Klein quadric. This is now an eigenvalue problem. For solutions to exist, the constant λ must satisfy

$$\det(N - \lambda Q_0) = 0.$$

In general, there will be six different solutions λ_i and, corresponding to these, six linearly independent principal screws of inertia, s_i. Now suppose we have two different principal screws of inertia, s_i and s_j. Pairing these against the inertia matrix gives

$$s_j^T N s_i = \lambda_i s_j^T Q_0 s_i.$$

The pairing is symmetric, and hence we also have the same relation, with λ_j replacing λ_i. Subtracting these two relations gives

$$0 = (\lambda_i - \lambda_j)\mathbf{s}_j^T Q_0 \mathbf{s}_i,$$

and since the two eigenvalues are different we must have $\mathbf{s}_j^T Q_0 \mathbf{s}_i = 0$. That is, principal screws of inertia are mutually reciprocal. Moreover, rearranging the above relation, we can get the equation

$$0 = \left(\frac{1}{\lambda_i} - \frac{1}{\lambda_j}\right) \mathbf{s}_j^T N \mathbf{s}_i,$$

so principal screws of inertia are also mutually conjugate. Of course, the converse does not necessarily hold.

To study principal screws of inertia a little more closely, we will look at the diagonal form of the inertia matrix. We can then generalise our results by invoking group invariance. Hence, we assume that the inertia matrix can be written

$$N = \begin{pmatrix} d_1 & 0 & 0 & 0 & 0 & 0 \\ 0 & d_2 & 0 & 0 & 0 & 0 \\ 0 & 0 & d_3 & 0 & 0 & 0 \\ 0 & 0 & 0 & m & 0 & 0 \\ 0 & 0 & 0 & 0 & m & 0 \\ 0 & 0 & 0 & 0 & 0 & m \end{pmatrix}.$$

Now we have that

$$\det(N - \lambda Q_0) = (md_1 - \lambda^2)(md_2 - \lambda^2)(md_3 - \lambda^2).$$

Thus, the six eigenvalues are $\lambda_i^{\pm} = \pm\sqrt{md_i}$. The corresponding eigenvectors are

$$\mathbf{s}_i^{\pm} = \begin{pmatrix} \mathbf{x}_i \\ \pm\sqrt{d_i/m} \ \mathbf{x}_i \end{pmatrix}.$$

Here $\mathbf{x}_i$ is a 3-vector in the direction of the i-th principal axes of the 3×3 inertia matrix.

So, we can see that generally when the principal values d_i of the 3×3 inertia matrix are all different, the principal screws of inertia have lines of action that pass through the body's centre of mass. Their direction is the same as the direction of the principal axes of the 3×3 inertia matrix. The pitches of the principal screws are $\pm\sqrt{d_i/m}$, where d_i is a principal value of the 3×3 inertia matrix and m is the body's mass.

We can also write down the commutation properties of the principal screws of inertia:

$$[\mathbf{s}_i^+, \mathbf{s}_i^-] = 0$$

and

$$[s_i^\rho, s_j^\sigma] = \frac{1}{2\sqrt{d_k}}(\sqrt{d_k} + \rho\sqrt{d_i} + \sigma\sqrt{d_j})s_k^+ + \frac{1}{2\sqrt{d_k}}(\sqrt{d_k} - \rho\sqrt{d_i} - \sigma\sqrt{d_j})s_k^-$$

where $\rho = \pm 1$ and $\sigma = \pm 1$; also, $\mathbf{x}_i$, $\mathbf{x}_j$ and $\mathbf{x}_k$ are a right handed set of 3-vectors. Again, these relations are most easily computed for the case where the inertia matrix is diagonal but also apply to the case of a general inertia matrix.

Finally, note that the principal screws of inertia also satisfy $\{s, Ns\} = 0$.

13.2   Robot Equations of Motion

In this section, we will develop the equations of motion for a general robot with six joints connected in series. We begin by looking at a single rigid body.

13.2.1   Equations for a Single Body

The equations of motion for a single rigid body are given by Newton's second law: the rate of change of momentum is equal to the applied force. That is,

$$\frac{d}{dt}\mathcal{M} = \mathcal{W},$$

where $\mathcal{W}$ is the applied wrench and $\mathcal{M}$ the momentum co-screw. If the body is moving with a velocity screw $\dot{\mathbf{q}}$, then the momentum co-screw can be written as $\mathcal{M} = N\dot{\mathbf{q}}$, as we saw above.

To make further progress, we need to know how to find the time derivatives of screws and co-screws. Consider the situation where $\mathbf{s}$ is a screw attached to a moving rigid body. Think of $\mathbf{s}$ as the screw associated with a joint on the body. As a function of time, the position of $\mathbf{s}$ will be given by

$$\mathbf{s}(t) = e^{\text{ad}(\mathbf{z})}\mathbf{s}(0),$$

where $\mathbf{z}$ is some function of t. Differentiating, we get

$$\frac{d}{dt}e^{\text{ad}(\mathbf{z})}\mathbf{s}(0) = \text{ad}(\mathbf{z}_d)e^{\text{ad}(\mathbf{z})}\mathbf{s}(0) = \text{ad}(\mathbf{z}_d)\mathbf{s}(t) = [\mathbf{z}_d, \mathbf{s}(t)],$$

see Section 4.5.2. The screw $\mathbf{z}_d$ is the instantaneous velocity screw of the motion.

For co-screws, we must use the coadjoint representation

$$\mathcal{M}(t) = e^{-ad^T(\mathbf{z})}\mathcal{M}(0).$$

Hence, when we differentiate we get

$$\frac{d}{dt}e^{-ad^T(\mathbf{z})}\mathcal{M}(0) = -\text{ad}^T(\mathbf{z}_d)e^{-ad^T(\mathbf{z})}\mathcal{M}(0) = \{\mathbf{z}_d, \mathcal{M}(t)\}.$$

It is also a simple matter to find the time derivative of the 6×6 inertia matrix. The action of the group is given by

$$N(t) = e^{-\operatorname{ad}^T(\mathbf{z})} N(0) e^{-\operatorname{ad}(\mathbf{z})}.$$

Hence, we have

$$\frac{d}{dt} N(t) = -\operatorname{ad}^T(\mathbf{z}_d) N(t) - N(t) \operatorname{ad}(\mathbf{z}_d).$$

Now, we apply the above results to a single rigid body. The instantaneous velocity screw of the body has appeared in two forms above that can be identified, $\mathbf{z}_d = \dot{\mathbf{q}}$, so the rate of change of momentum is

$$\frac{d}{dt} N\dot{\mathbf{q}} = \mathcal{W}.$$

Evaluating the derivative of the inertia matrix, we get

$$\frac{d}{dt} N\dot{\mathbf{q}} = N\ddot{\mathbf{q}} - (\operatorname{ad}^T(\dot{\mathbf{q}})N + N\operatorname{ad}(\dot{\mathbf{q}}))\dot{\mathbf{q}} = N\ddot{\mathbf{q}} + \{\dot{\mathbf{q}}, N\dot{\mathbf{q}}\}.$$

The last term of the derivative contains $[\dot{\mathbf{q}}, \dot{\mathbf{q}}]$ and hence disappears. For brevity, the dependance of N on t will not be written explicitly from now on.

Finally, we can write the equations of motion of a single rigid body in the form

$$N\ddot{\mathbf{q}} + \{\dot{\mathbf{q}}, N\dot{\mathbf{q}}\} = \mathcal{W}.$$

This is not quite the screw combination of Newton's and Euler's equations. Certainly, the above contains Newton's equations as the description of the velocity of the body's centre of mass. However, we have implicitly used a single inertial frame of reference in the above. Hence, unlike Euler's equations, the inertia matrix is not constant in the above but changes as the body moves. If we really wanted to derive the equations of motion for a single rigid body, we could fix coordinates in the body and use the Coriolis theorem to find derivatives with respect to this frame; see Woodhouse [129, sects. 1.2. and 3.2].

13.2.2   Serial Robots

Our purpose, however, is to derive the equations of motion for a robot with six serially connected links. Since each link is a rigid body, we have an equation like the one above for each. We can say a little more about the wrench acting on each link. We have three kinds of wrench to consider. To begin with, each link is subject to gravity. Hence, on each link there is a wrench due to gravity, $\mathcal{G}_i$, that is a pure force acting along a line through the link's centre of mass. Next, we have the wrenches due to the motors $\mathcal{T}_i$. These have the same pitch and axis as the joint screws of the robot, so they are pure torques for revolute joints. Each link, except the last, contains two joints and hence is acted on by two motor wrenches. Finally, we have the reaction wrenches at the joints $\mathcal{R}_i$. As usual, a reaction wrench can do no work on its joint screw and so pairing a reaction wrench with its joint screw annuls it. This is almost exactly the same situation that we had for static forces in Section 12.4. Numbering the links and joints from the base up to the end-effector, we get six screw equations of motion:

$$N_1 \ddot{\mathbf{q}}_1 + \{\dot{\mathbf{q}}_1, N_1 \dot{\mathbf{q}}_1\} = \mathcal{G}_1 + \mathcal{T}_1 - \mathcal{T}_2 + \mathcal{R}_1 - \mathcal{R}_2$$
$$N_2 \ddot{\mathbf{q}}_2 + \{\dot{\mathbf{q}}_2, N_2 \dot{\mathbf{q}}_2\} = \mathcal{G}_2 + \mathcal{T}_2 - \mathcal{T}_3 + \mathcal{R}_2 - \mathcal{R}_3$$
$$\vdots$$
$$N_5 \ddot{\mathbf{q}}_5 + \{\dot{\mathbf{q}}_5, N_5 \dot{\mathbf{q}}_5\} = \mathcal{G}_5 + \mathcal{T}_5 - \mathcal{T}_6 + \mathcal{R}_5 - \mathcal{R}_6$$
$$N_6 \ddot{\mathbf{q}}_6 + \{\dot{\mathbf{q}}_6, N_6 \dot{\mathbf{q}}_6\} = \mathcal{G}_6 + \mathcal{T}_6 + \mathcal{R}_6.$$

If we add the last equation here to the one above it, we can eliminate the $\mathcal{R}_6$ term. This equation can be used to get rid of the $\mathcal{R}_5$ term in the previous equation. We can continue this process to obtain the six screw equations

$$\mathcal{T}_i + \mathcal{R}_i = \sum_{j=i}^{6} (N_j \ddot{\mathbf{q}}_j + \{\dot{\mathbf{q}}_j, N_j \dot{\mathbf{q}}_j\} - \mathcal{G}_j) \qquad i = 1, 2, \ldots 6.$$

To get rid of the reactions, we can pair with the joint screws and produce six scalar equations. We will write the result of the pairing $\mathcal{T}_i(\mathbf{s}_i)$ as τ_i. This is the magnitude of the generalised force delivered by the motor, a torque if the joint is revolute. So, the equations of motion are

$$\tau_i = \sum_{j=i}^{6} (\ddot{\mathbf{q}}_j^T N_j \mathbf{s}_i + \dot{\mathbf{q}}_j^T N_j [\mathbf{s}_i, \dot{\mathbf{q}}_j] - \mathcal{G}_j^T \mathbf{s}_i) \qquad i = 1, 2, \ldots 6.$$

Notice that we have made liberal use of the fact that $\mathbf{q}^T N \mathbf{s} = \mathbf{s}^T N \mathbf{q}$ to tidy up the above equations.

If we take the direction of gravity as being in the $-z$ direction, then the gravity wrenches will all have the form

$$\mathcal{G}_i = \begin{pmatrix} -mg\mathbf{c}_i \times \mathbf{k} \\ -mg\mathbf{k} \end{pmatrix}$$

where g is acceleration due to gravity and $\mathbf{k}$ the unit vector in the z direction. Notice that we can tidy up our equations a little by introducing a gravity screw

$$\mathbf{g} = \begin{pmatrix} \mathbf{0} \\ -g\mathbf{k} \end{pmatrix}.$$

Now, we have

$$\mathcal{G}_i = N_i \mathbf{g}$$

and hence we can absorb the gravity terms into the acceleration part of the equations of motion:

$$\tau_i = \sum_{j=i}^{6} \left((\ddot{\mathbf{q}}_j - \mathbf{g})^T N_j \mathbf{s}_i + \dot{\mathbf{q}}_j^T N_j [\mathbf{s}_i, \dot{\mathbf{q}}_j] \right) \qquad i = 1, 2, \ldots 6.$$

This is a very neat way of writing the dynamics of a robot, but it is not terribly useful in practice. It is more usual to write the dynamic equations in terms of the joint rates and their accelerations. To do this, recall from Section 4.5 that for a six-joint robot the velocity screw of the i-th link is given by

$$\dot{\mathbf{q}}_i = \dot{\theta}_1 \mathbf{s}_1 + \dot{\theta}_2 \mathbf{s}_2 + \cdots + \dot{\theta}_i \mathbf{s}_i = \sum_{j=1}^{i} \dot{\theta}_j \mathbf{s}_j.$$

Hence, the second derivative has the form

$$\ddot{\mathbf{q}}_i = \ddot{\theta}_1 \mathbf{s}_1 + \ddot{\theta}_2 \mathbf{s}_2 + \cdots + \ddot{\theta}_i \mathbf{s}_i + \dot{\theta}_1 \frac{d}{dt}\mathbf{s}_1 + \dot{\theta}_2 \frac{d}{dt}\mathbf{s}_2 + \cdots + \dot{\theta}_i \frac{d}{dt}\mathbf{s}_i,$$

$$= \ddot{\theta}_1 \mathbf{s}_1 + \ddot{\theta}_2 \mathbf{s}_2 + \cdots + \ddot{\theta}_i \mathbf{s}_i + \dot{\theta}_1 [\dot{\mathbf{q}}_1, \mathbf{s}_1] + \dot{\theta}_2 [\dot{\mathbf{q}}_2, \mathbf{s}_2] + \cdots + \dot{\theta}_i [\dot{\mathbf{q}}_i, \mathbf{s}_i],$$

$$= \sum_{j=1}^{i} \ddot{\theta}_j \mathbf{s}_j + \sum_{1 \le k < l \le i} \dot{\theta}_k \dot{\theta}_l [\mathbf{s}_k, \mathbf{s}_l].$$

Substituting this into the equations of motion that we found above, we get

$$\tau_i = \sum_{j=i}^{6} \left(\sum_{k=1}^{j} \ddot{\theta}_k \mathbf{s}_i^T N_j \mathbf{s}_k + \sum_{1 \le k < l \le j} \dot{\theta}_k \dot{\theta}_l \mathbf{s}_i^T N_j [\mathbf{s}_k, \mathbf{s}_l] + \sum_{k,l=1}^{j} \dot{\theta}_k \dot{\theta}_l \mathbf{s}_k^T N_j [\mathbf{s}_i, \mathbf{s}_l] + \mathcal{G}_j^T \mathbf{s}_i \right).$$

Very often the dynamics of a robot will be summarised by an equation of the form

$$\tau_i = A_{ij} \ddot{\theta}_j + B_{ijk} \dot{\theta}_j \dot{\theta}_k + C_i,$$

where summation is intended over repeated indices. From the work above, we can identify the parameters in this expression. The 6×6 matrix A is called the generalised inertia matrix of the robot. Its elements are

$$A_{ij} = \begin{cases} \mathbf{s}_i^T (N_i + \cdots + N_6) \mathbf{s}_j & \text{if } i \ge j, \\ \mathbf{s}_j^T (N_j + \cdots + N_6) \mathbf{s}_i & \text{if } i < j. \end{cases}$$

The "B" terms are sometimes called the Coriolis terms. There are many possible choices we could make since the above only gives the sums over j and k. However, we will see a little later that a good choice is to take

$$B_{ijk} = \frac{1}{2}(\mathbf{s}_j^T (N_x + \cdots + N_6)[\mathbf{s}_i, \mathbf{s}_k] + \mathbf{s}_k^T (N_x + \cdots + N_6)[\mathbf{s}_i, \mathbf{s}_j]$$

$$- \mathbf{s}_i^T (N_x + \cdots + N_6)[\mathbf{s}_k, \mathbf{s}_j])$$

when $j \leq k$, and

$$B_{ijk} = \frac{1}{2}(\mathbf{s}_j^T(N_x + \cdots + N_6)[\mathbf{s}_i, \mathbf{s}_k] + \mathbf{s}_k^T(N_x + \cdots + N_6)[\mathbf{s}_i, \mathbf{s}_j]$$
$$- \mathbf{s}_i^T(N_x + \cdots + N_6)[\mathbf{s}_j, \mathbf{s}_k])$$

when $k < j$, where in both cases $x = \max(i, j, k)$.

Finally, the gravity terms are

$$C_i = \mathbf{s}_i^T(N_i + \cdots + N_6)\mathbf{g}.$$

13.2.3 Change in Payload

An immediate application of these formulas is to how the dynamics of the robot changes when it picks up an object. The equations of motion of the robot plus payload can be summarised as

$$\tau_i = A'_{ij}\ddot{\theta}_j + B'_{ijk}\dot{\theta}_j\dot{\theta}_k + C'_i.$$

Now suppose that the inertia matrix of the payload is N_p. Then, since inertias are additive and also since only the inertia of the last link is affected, we may write the coefficients in the equations of motion for the robot plus payload as $A' = A + A^p$, $B' = B + B^p$ and $C' = C + C^p$, where, A, B and C are the coefficients in the equations of motion of the robot without a payload. The changes in the coefficients due to the payload are easily computed from the formulas above:

$$A^p_{ij} = \mathbf{s}_i^T N_p \mathbf{s}_j$$

and when $j \leq k$:

$$B^p_{ijk} = \frac{1}{2}(\mathbf{s}_j^T N_p[\mathbf{s}_i, \mathbf{s}_k] + \mathbf{s}_k^T N_p[\mathbf{s}_i, \mathbf{s}_j] - \mathbf{s}_i^T N_p[\mathbf{s}_k, \mathbf{s}_j]),$$

but when $j > k$ we have

$$B^p_{ijk} = \frac{1}{2}(\mathbf{s}_j^T N_p[\mathbf{s}_i, \mathbf{s}_k] + \mathbf{s}_k^T N_p[\mathbf{s}_i, \mathbf{s}_j] - \mathbf{s}_i^T N_p[\mathbf{s}_j, \mathbf{s}_k]).$$

The gravity terms are given by

$$C^p_i = \mathbf{s}_i^T N_p \mathbf{g}.$$

13.3 Recursive Formulation

In the control of robots, it is often necessary to compute the torque required to make the robot follow some specified path. That is, given the path in terms of

$\theta_i(t)$, $\dot{\theta}_i(t)$ and $\ddot{\theta}_i(t)$ we have to compute values for τ_i. This forms the basis of the computed torque method of control. The idea is to use an explicit model of the arm's rigid body dynamics to compute the torques necessary to have the end-effector follow a desired trajectory. These torques are are then fed-forward to the robot's joint servos. This effectively linearises the dynamics of the system so that feedback loops can be put around the system to provide disturbance rejection, and to handle errors due to model mismatch and unmodelled effects such as friction. This gives a feed-forward/feedback controller where the feed-forward encodes explicit knowledge of the system's dynamics while the feedback deals with uncertainty.

Although the results of the last section are rather neat, it would be completely crazy to use them to perform calculations. The equations given in the last section are computationally inefficient because both the screws s_i and the inertias N_i are dependent on position.

A computationally more efficient method of calculating the torques is given by a recursive method, first developed by Hollerbach [53]. We can give a reasonably clean description of this method using the notation introduced above. Note that this version of the recursive algorithm was first given in [103], but see also Park and Bobrow [83].

There are two sources of improvement in the algorithm we describe. First, notice that a pairing between a wrench and a screw $\mathcal{W}^T s$ is coordinate independent—we get the same results whichever coordinate frame we use. If we use coordinate frames at rest with respect to the robot's links, then we only need to know the inertia of the links in their 'home' positions. This will involve us in performing many coordinate transformations. The second source of improvement is that these transformations can be computed recursively. This seems to be the only place in robotics where it is advantageous to work with several different coordinate frames at once. However, we could take a complementary point of view, one more in line with our active approach, and think of transforming quantities back to the home position. Alternatively, we could think of the whole process as a purely algebraic one.

Now we saw above that the equations of motion for the robot can be written as

$$\tau_i = \sum_{j=i}^{6} \left((\ddot{\mathbf{q}}_j - \mathbf{g})^T N_j \mathbf{s}_i + \dot{\mathbf{q}}_j^T N_j [\mathbf{s}_i, \dot{\mathbf{q}}_j] \right) \qquad i = 1, 2, \ldots, 6,$$

so if we write

$$\mathcal{Q}_j = N_j(\ddot{\mathbf{q}}_j - \mathbf{g}) + \{\dot{\mathbf{q}}_j, N_j \dot{\mathbf{q}}_j\},$$

then we can express the equations of motion as

$$\tau_i = \sum_{j=i}^{6} \mathcal{Q}_j^T \mathbf{s}_i \qquad i = 1, 2, \ldots, 6.$$

In fact, we can do better. Since the wrenches $\mathcal{Q}_i$ have the same form for every joint, we can put

$$\mathcal{P}_i = \sum_{j=i}^{6} \mathcal{Q}_j = \mathcal{Q}_i + \mathcal{P}_{i+1}$$

and then

$$\tau_i = \mathcal{P}_i^T \mathbf{s}_i \qquad i = 1, 2, \ldots, 6.$$

Physically, we could think of $\mathcal{Q}_i$ as the wrench due to link i, while $\mathcal{P}_i$ contains the total wrench acting on the i-th link. By the remarks above, if we transform these wrenches back to the home position we can pair them with the joint screws in their home positions to give the same result as before:

$$\tau_i = (\mathcal{P}_i^0)^T \mathbf{s}_i^0 \qquad i = 1, 2, \ldots, 6,$$

where the superscript 0 denotes the quantity in the home position.

So, let us fix a home position for the robot and label all quantities in this po-sition with a superscript 0. The home positions of the inertias, for example, will be N_i^0. In some subsequent position, with joint coordinates $\boldsymbol{\theta} = (\theta_1, \theta_2, \ldots, \theta_6)$, the joint screws will be

$$\mathbf{s}_i(\boldsymbol{\theta}) = H_1(\theta_1)H_2(\theta_2) \cdots H_i(\theta_i)\mathbf{s}_i^0,$$

where the 'H' matrices are given by $H_j(\theta_j) = \mathrm{Ad}(e^{\theta_j \mathbf{s}_j})$, that is, the adjoint representation of $SE(3)$. Actually, the last transformation is unnecessary here since $\mathbf{s}_i$ is unchanged by H_i. However, the above relation is true for any screw attached to the i-th link. For brevity, we will drop the explicit dependence of quantities on $\boldsymbol{\theta}$. The inertia matrices are therefore given by

$$N_i = (H_1^T)^{-1}(H_2^T)^{-1} \cdots (H_i^T)^{-1} N_i^0 (H_i)^{-1} \cdots (H_2)^{-1}(H_1)^{-1}.$$

The wrench due to each link is then

$$\mathcal{Q}_j^0 = N_j^0(\ddot{\mathbf{q}}_j^0 - \mathbf{g}_j^0) + \{\dot{\mathbf{q}}_j^0, N_j^0 \dot{\mathbf{q}}_j^0\}.$$

Notice that we must transform the gravity screw along with everything else. We have

$$\mathbf{g} = H_1 H_2 \cdots H_j \mathbf{g}_j^0.$$

The link velocities also transform in the same way, so that

$$\dot{\mathbf{q}}_j = H_1 H_2 \cdots H_j \dot{\mathbf{q}}_j^0.$$

But we also have that

$$\dot{\mathbf{q}}_j = \dot{\theta}_1 \mathbf{s}_1 + \dot{\theta}_2 \mathbf{s}_2 + \cdots + \dot{\theta}_j \mathbf{s}_j.$$

In terms of the joint screws in their home positions, we have

$$\dot{\mathbf{q}}_j^0 = (\dot{\theta}_1 H_j^{-1} \cdots H_2^{-1} \mathbf{s}_1^0) + (\dot{\theta}_2 H_j^{-1} \cdots H_3^{-1} \mathbf{s}_2^0) + \cdots + \dot{\theta}_j \mathbf{s}_j^0.$$

These equations are not quite so fearsome as they seem, since we can compute the link velocities recursively. To begin with, we have

$$\dot{\mathbf{q}}_1^0 = \dot{\theta}_1 \mathbf{s}_1^0.$$

Then subsequently we have

$$\dot{\mathbf{q}}_j^0 = \dot{\theta}_j \mathbf{s}_j^0 + H_j^{-1} \dot{\mathbf{q}}_{j-1}^0.$$

This now requires just five matrix operations to get all six link velocities.

We can do much the same for the link 'accelerations'. First, we have

$$\ddot{\mathbf{q}}_j = \ddot{\theta}_1 \mathbf{s}_1 + \ddot{\theta}_2 \mathbf{s}_2 + \cdots + \ddot{\theta}_j \mathbf{s}_j + \dot{\theta}_2 [\dot{\mathbf{q}}_2, \mathbf{s}_2] + \dot{\theta}_3 [\dot{\mathbf{q}}_3, \mathbf{s}_3] + \cdots + \dot{\theta}_j [\dot{\mathbf{q}}_j, \mathbf{s}_j],$$

then, in terms of the home position quantities

$$\ddot{\mathbf{q}}_j^0 = (\ddot{\theta}_1 H_j^{-1} \cdots H_2^{-1} \mathbf{s}_1^0) + (\ddot{\theta}_2 H_j^{-1} \cdots H_3^{-1} \mathbf{s}_2^0) + \cdots + (\ddot{\theta}_j \mathbf{s}_j^0)$$
$$+ (\dot{\theta}_2 H_j^{-1} \cdots H_3^{-1} [\dot{\mathbf{q}}_2^0, \mathbf{s}_2^0]) + (\dot{\theta}_3 H_j^{-1} \cdots H_4^{-1} [\dot{\mathbf{q}}_3^0, \mathbf{s}_3^0]) + \cdots + (\dot{\theta}_j [\dot{\mathbf{q}}_j^0, \mathbf{s}_j^0]).$$

Recursively, we begin with

$$\ddot{\mathbf{q}}_1^0 = \ddot{\theta}_1 \mathbf{s}_1^0$$

and then subsequently

$$\ddot{\mathbf{q}}_j^0 = \ddot{\theta}_j \mathbf{s}_j^0 + \dot{\theta}_j [\dot{\mathbf{q}}_j^0, \mathbf{s}_j^0] + H_j^{-1} \ddot{\mathbf{q}}_{j-1}^0.$$

This becomes even more economical if we include the gravity terms; that is, we first calculate

$$(\ddot{\mathbf{q}}_1^0 - \mathbf{g}_1^0) = \ddot{\theta}_1 \mathbf{s}_1^0 - H_1^{-1} \mathbf{g}$$

and then recursively find the higher terms:

$$(\ddot{\mathbf{q}}_j^0 - \mathbf{g}_j^0) = \ddot{\theta}_j \mathbf{s}_j^0 + \dot{\theta}_j [\dot{\mathbf{q}}_j^0, \mathbf{s}_j^0] + H_j^{-1} (\ddot{\mathbf{q}}_{j-1}^0 - \mathbf{g}_{j-1}^0).$$

Now that we have all the velocities and accelerations, it is a simple matter to find the wrenches due to each link, the $\mathcal{Q}_i^0$ terms. But to find the $\mathcal{P}_i^0$, terms we need to do some more transformations. Remember that

$$\mathcal{P}_i = \mathcal{Q}_i + \mathcal{P}_{i+1}.$$

Wrenches transform according to the coadjoint representation of $SE(3)$, and so

$$\mathcal{P}_i = (H_1^T)^{-1} (H_2^T)^{-1} \cdots (H_i^T)^{-1} \mathcal{P}_i^0.$$

Thus

$$\mathcal{P}_i^0 = \mathcal{Q}_i^0 + (H_{i+1}^T)^{-1} \mathcal{P}_{i+1}^0.$$

This gives another recursion scheme, this time beginning with

$$\mathcal{P}_6^0 = \mathcal{Q}_6^0$$

and then using the relation above to compute the wrenches $\mathcal{P}_i^0$ down to $i = 1$. The final step then is to pair these wrenches with the joint screws in their home position.

The algorithm can be summarised as follows:

Inputs:
  Current joint angle θ_i, rate $\dot{\theta}_i$ and acceleration $\ddot{\theta}_i$,
  joint screw in home position s_i^0,
  link inertia matrices in home position N_i^0

Outputs:
  torque required at each joint τ_i

Method:

$$H_i \leftarrow e^{\theta_i \, \text{ad}(s_i^0)} \qquad\qquad i = 1, 2, \ldots, 6$$

$$\dot{\mathbf{q}}_1^0 \leftarrow \dot{\theta}_1 s_1^0$$

$$\dot{\mathbf{q}}_i^0 \leftarrow \dot{\theta}_i s_i^0 + H_i^{-1} \dot{\mathbf{q}}_{i-1}^0 \qquad\qquad i = 2, 3, \ldots, 6$$

$$(\ddot{\mathbf{q}}_1^0 - \mathbf{g}_1^0) \leftarrow \ddot{\theta}_1 s_1^0 - H_1^{-1} \mathbf{g}$$

$$(\ddot{\mathbf{q}}_i^0 - \mathbf{g}_i^0) \leftarrow \ddot{\theta}_i s_i^0 + \dot{\theta}_i [\dot{\mathbf{q}}_i^0, s_i^0] + H_i^{-1}(\ddot{\mathbf{q}}_{i-1}^0 - \mathbf{g}_{i-1}^0) \qquad i = 2, 3, \ldots, 6$$

$$\mathcal{Q}_i^0 \leftarrow N_i^0(\ddot{\mathbf{q}}_i^0 - \mathbf{g}_i^0) + \{\dot{\mathbf{q}}_i^0, N_i^0 \dot{\mathbf{q}}_i^0\} \qquad\qquad i = 1, 2, \ldots, 6$$

$$\mathcal{P}_6^0 \leftarrow \mathcal{Q}_6^0$$

$$\mathcal{P}_i^0 \leftarrow \mathcal{Q}_i^0 + (H_{i+1}^T)^{-1} \mathcal{P}_{i+1}^0 \qquad\qquad i = 5, 4, \ldots, 1$$

$$\tau_i \leftarrow (\mathcal{P}_i^0)^T s_i^0 \qquad\qquad i = 1, 2, \ldots, 6$$

As it stands, this algorithm could be used in a computer algebra system to generate the equations of motion for a general robot. However, for control applications, where speed is crucial, we should avoid 6×6 matrices at all costs. Hence, rather than compute terms like $N_i^0 \dot{\mathbf{q}}_i^0$ using 6×6 matrix multiplication, we should at least decompose the inertias into 3×3 blocks and use the vector product instead of multiplying by anti-symmetric 3×3 matrices. The problem of computing the torques from the robot's equations of motion is an important one that has been extensively studied. Many refinements on the above are possible; see, for example, Featherstone [33] and the references therein.

13.4   Lagrangian Dynamics of Robots

In this section, we rederive the equations of motion for a six-joint robot but using the Lagrangian approach to dynamics. The Lagrangian function of the robot is given by $L = E_K - E_P$, where the kinetic energy is given by

$$E_K = \frac{1}{2} \sum_{i=1}^{6} \dot{\mathbf{q}}_i^T N_i \dot{\mathbf{q}}_i = \frac{1}{2} \sum_{i,j=1}^{6} A_{ij} \dot{\theta}_i \dot{\theta}_j.$$

The most natural way to express the potential energy E_P would be to use the four-dimensional representation of the group $SE(3)$. Let us write $\tilde{\mathbf{c}}_i$ for the four-dimensional vector

$$\tilde{\mathbf{c}}_i = \begin{pmatrix} m_i \mathbf{c}_i \\ m_i \end{pmatrix}$$

where m_i is the mass of the i-th link and $\mathbf{c}_i$ is the position of its centre of mass. We will also write the four-vector defining the direction in which gravity acts by

$$\tilde{\mathbf{g}} = \begin{pmatrix} -g\mathbf{k} \\ 0 \end{pmatrix}.$$

Then the potential energy of the whole six-joint robot will be the sum

$$E_P = \sum_{i=1}^{6} \tilde{\mathbf{g}}^T \tilde{\mathbf{c}}_i.$$

13.4.1 Euler–Lagrange Equations

The equations of motion are now given by the Euler–Lagrange equations

$$\frac{d}{dt}\left(\frac{\partial L}{\partial \dot{\theta}_i}\right) - \frac{\partial L}{\partial \theta_i} = \tau_i.$$

So we see that we must evaluate the derivatives of the kinetic and potential energies given above. We will begin with the easiest cases. Since only the kinetic energy depends explicitly on $\dot{\theta}_i$ and moreover the matrix A_{ij} is symmetric, we get

$$\frac{\partial L}{\partial \dot{\theta}_i} = A_{ij} \dot{\theta}_j.$$

Alternatively, we could write this result in terms of screws and inertias as

$$\frac{\partial L}{\partial \dot{\theta}_i} = \sum_{j=i}^{6} \mathbf{s}_i^T N_j \dot{\mathbf{q}}_j.$$

The time derivative of this last form is more easily computed since we have the results from Section 13.2:

$$\frac{d}{dt} N_j \dot{\mathbf{q}}_j = N_j \ddot{\mathbf{q}}_j + \{\dot{\mathbf{q}}_j, N_j \dot{\mathbf{q}}_j\} \qquad \text{and} \qquad \frac{d}{dt} \mathbf{s}_i = [\dot{\mathbf{q}}_i, \mathbf{s}_i].$$

Putting this together, we get

$$\frac{d}{dt}\left(\frac{\partial L}{\partial \dot{\theta}_i}\right) = \sum_{j=i}^{6}(\mathbf{s}_i^T N_j \ddot{\mathbf{q}}_j + \dot{\mathbf{q}}_j^T N_j[\mathbf{s}_i, \dot{\mathbf{q}}_j] - \dot{\mathbf{q}}_j^T N_j[\mathbf{s}_i, \dot{\mathbf{q}}_i]).$$

The term involving the partial derivatives of L with respect to the joint variables can be split into two pieces. The first is

$$\frac{\partial E_K}{\partial \theta_i} = \frac{1}{2}\frac{\partial}{\partial \theta_i}\left(\sum_{j=1}^{6} \dot{\mathbf{q}}_j^T N_j \dot{\mathbf{q}}_j\right).$$

To evaluate this, we need to find the partial derivatives of the velocity screws and inertias with respect to the joint variables. The j-th joint screw is given by

$$\mathbf{s}_j = e^{\theta_1 \text{ ad}(\mathbf{s}_1)}e^{\theta_2 \text{ ad}(\mathbf{s}_2)}\cdots e^{\theta_j \text{ ad}(\mathbf{s}_j)}\mathbf{s}_j^0,$$

so it is easy to see that

$$\frac{\partial \mathbf{s}_j}{\partial \theta_i} = \begin{cases} [\mathbf{s}_i, \mathbf{s}_j], & \text{if } i \le j, \\ 0 & \text{otherwise,} \end{cases}$$

since $\text{ad}(\mathbf{s}_i)\mathbf{s}_j = [\mathbf{s}_i, \mathbf{s}_j]$. Now, the velocity screws are given by

$$\dot{\mathbf{q}}_j = \dot{\theta}_1 \mathbf{s}_1 + \dot{\theta}_2 \mathbf{s}_2 + \cdots + \dot{\theta}_j \mathbf{s}_j$$

and hence the derivative with respect to θ_i is

$$\frac{\partial \dot{\mathbf{q}}_j}{\partial \theta_i} = \begin{cases} [\mathbf{s}_i, \dot{\mathbf{q}}_j - \dot{\mathbf{q}}_i], & \text{if } j \le i, \\ 0 & \text{otherwise.} \end{cases}$$

The inertia matrix transforms according to the symmetric square of the coadjoint representation as we have seen in sections 13.1 and 13.2. So, the derivative is given by

$$\frac{\partial N_j}{\partial \theta_i} = \begin{cases} -\text{ad}(\mathbf{s}_i)^T N_j + N_j \text{ ad}(\mathbf{s}_i), & \text{if } j \le i, \\ 0 & \text{otherwise.} \end{cases}$$

Assembling the various parts of the derivative of E_K and simplifying using $\text{ad}(\mathbf{s}_i)\mathbf{s}_j = [\mathbf{s}_i, \mathbf{s}_j]$, we obtain the result

$$\frac{\partial E_K}{\partial \theta_i} = -\dot{\mathbf{q}}_j^T N_j[\mathbf{s}_i, \dot{\mathbf{q}}_i].$$

The final term we need is straightforward:

$$\frac{\partial E_P}{\partial \theta_i} = \sum_{j=1}^{6} \tilde{\mathbf{g}}^T \frac{\partial \tilde{\mathbf{c}}_j}{\partial \theta_i} = \sum_{j=i}^{6} -m_j g\mathbf{k} \cdot (\boldsymbol{\omega}_i \times \mathbf{c}_j + \mathbf{v}_i).$$

In Section 13.2 above, we wrote this as

$$\frac{\partial E_P}{\partial \theta_i} = -\sum_{j=i}^{6} \mathbf{s}_i^T N_j \mathbf{g} = -C_i.$$

Assembling these results, we regain the equations of motion

$$\tau_i = \sum_{j=i}^{6} \left((\ddot{\mathbf{q}}_j - \mathbf{g})^T N_j \mathbf{s}_i + \dot{\mathbf{q}}_j^T N_j [\mathbf{s}_i, \dot{\mathbf{q}}_j] \right) \qquad i = 1, 2, \ldots, 6.$$

13.4.2   Derivatives of the Generalised Inertia Matrix

We could also have worked with the Lagrangian in terms of the joint variables
and their derivatives, but the computations would have been more involved.
However, we can make at least one observation. From the above, we can see
that

$$B_{ijk} \dot{\theta}_j \dot{\theta}_k = \frac{dA_{ij}}{dt} \dot{\theta}_j - \frac{1}{2} \frac{\partial A_{jk}}{\partial \theta_i} \dot{\theta}_j \dot{\theta}_k.$$

Using the chain rule to develop the time derivative of A_{ij}, we can write

$$B_{ijk} \dot{\theta}_j \dot{\theta}_k = \frac{\partial A_{ij}}{\partial \theta_k} \dot{\theta}_j \dot{\theta}_k - \frac{1}{2} \frac{\partial A_{jk}}{\partial \theta_i} \dot{\theta}_j \dot{\theta}_k.$$

This can be written more symmetrically as

$$B_{ijk} \dot{\theta}_j \dot{\theta}_k = \frac{1}{2} \left(\frac{\partial A_{ij}}{\partial \theta_k} + \frac{\partial A_{ik}}{\partial \theta_j} - \frac{\partial A_{jk}}{\partial \theta_i} \right) \dot{\theta}_j \dot{\theta}_k.$$

Hence, as expected from more general theory (see Whittaker [128, p. 39]) we
have

$$B_{ijk} = \frac{1}{2} \left(\frac{\partial A_{ij}}{\partial \theta_k} + \frac{\partial A_{ik}}{\partial \theta_j} - \frac{\partial A_{jk}}{\partial \theta_i} \right).$$

This is consistent with formulas given for the Coriolis terms in Section 13.2,
since we can compute the partial derivatives. There are essentially three cases
to consider. First, when $i \leq j \leq k$ we have

$$\frac{\partial A_{jk}}{\partial \theta_i} = 0.$$

Then, when $j < i \leq k$ we get

$$\frac{\partial A_{jk}}{\partial \theta_i} = \mathbf{s}_k^T (N_k + \cdots + N_6)[\mathbf{s}_j, \mathbf{s}_i].$$

Finally, when $j \leq k < i$ the result is

$$\frac{\partial A_{jk}}{\partial \theta_i} = \mathbf{s}_k^T (N_i + \cdots + N_6)[\mathbf{s}_j, \mathbf{s}_i] + \mathbf{s}_j^T (N_i + \cdots + N_6)[\mathbf{s}_k, \mathbf{s}_i].$$

No sum on k is intended in the above. The results for the cases when $k \leq j$ are given by interchanging j and k in these relations.

These calculations are evaluated as follows. For $j \leq k$ the element of the generalised inertia matrix is,

$$A_{jk} = \mathbf{s}_k^T (N_j + \cdots + N_6)\mathbf{s}_j,$$

as found above. In the case that $i \leq j \leq k$ all of these terms are functions of θ_i and so

$$\frac{\partial A_{jk}}{\partial \theta_i} = \mathbf{s}_k^T \operatorname{ad}(\mathbf{s}_i)^T (N_j + \cdots + N_6)\mathbf{s}_j$$
$$- \mathbf{s}_k^T \operatorname{ad}(\mathbf{s}_i)^T (N_j + \cdots + N_6)\mathbf{s}_j - \mathbf{s}_k^T (N_j + \cdots + N_6)\operatorname{ad}(\mathbf{s}_i)\mathbf{s}_j$$
$$+ \mathbf{s}_k^T (N_j + \cdots + N_6)\operatorname{ad}(\mathbf{s}_i)\mathbf{s}_j = 0.$$

This could have been expected since all the links and joint above the i-th one simply rotate about the i-th joint axis (if it is a revolute joint). So A_{jk} will be a scalar in this case. The other results are calculated in a similar fashion but now some of the terms will be independent of θ_i.

13.4.3 Small Oscillations

An application of the Lagrangian theory is to the stability of points of equilibrium and the study of small oscillations about stable equilibrium points. Suppose we try to hold the end-effector of the robot steady at some position in its work space. When the end-effector is stationary, we must have $\ddot{\theta}_i = \dot{\theta}_i = 0$ for each joint. However we will still have to supply torques at the joints $\tau_i = C_i$ to balance the force of gravity. We can now look at small movements about this equilibrium point. To do this, imagine that the joint variables can be written as $\theta_i + \phi_i$, where the θ_is are constant and the ϕ_is are small enough for their products to be negligible. The linearised equations of motion about the point defined by the θ_is is thus

$$A_{ij}\ddot{\phi}_j + Q_{ij}\phi_j = 0 \qquad i = 1, 2, \ldots, 6,$$

where Q_{ij} is the Hessian of the potential energy function:

$$Q_{ij} = \frac{\partial C_i}{\partial \theta_j} = \frac{\partial^2 E_P}{\partial \theta_i \partial \theta_j}.$$

We can evaluate this Hessian for a general six-joint robot using the results found above:

$$Q_{ij} = \begin{cases} -g \sum_{l=i}^{6} \mathbf{k} \cdot \boldsymbol{\omega}_j \times (\boldsymbol{\omega}_i \times m_l \mathbf{c}_l + \mathbf{v}_i) & \text{if } i \geq j, \\ -g \sum_{l=j}^{6} \mathbf{k} \cdot \boldsymbol{\omega}_i \times (\boldsymbol{\omega}_j \times m_l \mathbf{c}_l + \mathbf{v}_j) & \text{if } i < j, \end{cases}$$

or in terms of the inertia matrices:

$$Q_{ij} = \begin{cases} \mathbf{s}_i^T(N_i + \cdots + N_6)[\mathbf{g}, \mathbf{s}_j], & \text{if } i \geq j, \\ \\ \mathbf{s}_j^T(N_j + \cdots + N_6)[\mathbf{g}, \mathbf{s}_i], & \text{if } i < j. \end{cases}$$

The linearised equations are, of course, linear equations in the variables ϕ_i, and so explicit solutions can be found if we know the values of the constant matrices A_{ij} and Q_{ij}. Texts on classical mechanics give the solutions to such equations as sinusoidal functions

$$\phi_i(t) = a_i \cos(\lambda t - \delta) \qquad i = 1, 2, \ldots, 6,$$

where the possible frequencies λ are the roots of the equation

$$\det(Q_{ij} - \lambda^2 A_{ij}) = 0;$$

see Goldstein [39, Chap. 6], for example. This is a consequence of the normal form of a pair of symmetric matrices that we found in Section 3.4. The matrix A for the kinetic energy is always positive definite; hence, in the normal form the equations decouple and become six independent second order equations.

The frequencies given by the above equation are the normal mode frequencies, and the corresponding solutions $\phi_i(t)$ are the normal mode solutions. The stability of the solution is determined by the sign of λ^2. The normal mode is stable if and only if λ^2 is positive. For 'imaginary frequencies' the normal modes contain diverging exponentials.

A general solution to the problem is an admixture of these normal mode solutions. However, once started in a normal mode the system will continue to move according to this solution, since there are no dissipative mechanisms in our model. If we start the system from rest, the phase δ will be zero. The velocity of the robot's end-effector in such a stable normal mode is given by

$$\dot{\mathbf{q}}_6 = \dot{\phi}_1 \mathbf{s}_1 + \dot{\phi}_2 \mathbf{s}_2 + \cdots + \dot{\phi}_6 \mathbf{s}_6 = K \sin(\lambda t)\mathbf{h},$$

where the constant K depends on the initial position of the end-effector. The screw $\mathbf{h}$ is given by

$$\mathbf{h} = a_1 \mathbf{s}_1 + a_2 \mathbf{s}_2 + \cdots + a_6 \mathbf{s}_6.$$

For a stable equilibrium position where there are six different normal mode frequencies, there will be six screws like this. Ball [6] termed these screws **harmonic screws**. So, the normal mode solutions for the end-effector are simply oscillations about a constant screw.

If the normal mode frequencies are all different, then the normal mode solutions have nice orthogonality properties. Suppose ϕ_i and ψ_i are normal mode solutions with frequencies λ and μ, respectively. Then, we have

$$Q_{ij}\phi_j = \lambda^2 A_{ij}\phi_j \qquad \text{and} \qquad Q_{ij}\psi_j = \mu^2 A_{ij}\psi_j.$$

Taking the scalar product of these equations with the other normal mode and using the fact that both A and Q are symmetric matrices, we can subtract the equations we get to yield

$$0 = (\lambda^2 - \mu^2) A_{ij} \phi_i \psi_j$$

and since the eigenvalues are different, we must have

$$A_{ij} \phi_i \psi_j = 0.$$

Returning to the original equations, we also have that

$$Q_{ij} \phi_i \psi_j = 0.$$

The relations can be extended to the harmonic screws. For two different harmonic screws $\mathbf{h}_a$ and $\mathbf{h}_b$ we have

$$\mathbf{h}_a^T (J^T)^{-1} A J^{-1} \mathbf{h}_b = 0, \quad \text{and} \quad \mathbf{h}_a^T (J^T)^{-1} Q J^{-1} \mathbf{h}_b = 0,$$

where J is the robot's Jacobian matrix.

13.5   Hamiltonian Dynamics of Robots

In this short section, we take a brief look at how Hamiltonian mechanics can be used to describe the dynamics of robots. This is an area that has not received much attention to date, but see Section 13.6.3 later. For a more detailed account of Hamiltonian mechanics, refer to Goldstein [39, Chap. 9].

The generalised momentum conjugate to the joint variable θ_i is given by

$$\pi_i = \frac{\partial L}{\partial \dot{\theta}_i} = A_{ij} \dot{\theta}_j.$$

The Hamiltonian function h of the robot is then given by the Legendre transform

$$h = \pi_i \dot{\theta}_i - L = \frac{1}{2} A_{ij} \dot{\theta}_i \dot{\theta}_j + E_P = \frac{1}{2} A_{ij}^{-1} \pi_i \pi_j + E_P.$$

From the general theory of Hamiltonian mechanics, the equations of motion are then

$$\dot{\theta}_i = \frac{\partial h}{\partial \pi_i},$$

$$i = 1, 2, \ldots, 6.$$

$$\dot{\pi}_i = -\frac{\partial h}{\partial \theta_i} + \tau_i,$$

Care must be taken here to avoid the 'second fundamental confusion of calculus'; see Woodhouse [129, Chap. 4]. The partial derivatives $\partial h / \partial \theta_i$ must be calculated with the π_js held constant, so, unlike Lagrangian mechanics, the terms $\partial \dot{\theta}_i / \partial \theta_j$ are non-zero. For a six-joint robot, we get

$$\dot{\theta}_i = A_{ij}^{-1} \pi_j,$$

$$i = 1, 2, \ldots, 6,$$

$$\dot{\pi}_i = \frac{1}{2} \frac{\partial A_{jk}}{\partial \theta_i} \dot{\theta}_j \dot{\theta}_k - C_i + \tau_i,$$

or, if we use the definition and symmetry properties of the "B" terms

$$\dot{\theta}_i = A_{ij}^{-1} \pi_j,$$

$$i = 1, 2, \ldots, 6.$$

$$\dot{\pi}_i = B_{jik} \dot{\theta}_j \dot{\theta}_k - C_i + \tau_i,$$

These are now twice as many first order equations, but in twice as many variables, as the Lagrangian equations of motion. The partial derivatives of the inertia matrix A_{jk} were computed above in Section 13.4.

Hamilton's equations can be written very neatly if put

$$\eta = (\theta_1, \theta_2, \ldots, \theta_6, \pi_1, \pi_2, \ldots, \pi_6),$$

that is, in phase space, the space whose coordinates are θ_i and π_i. Now, when there are no external forces Hamilton's equations become

$$\dot{\eta}_i = E_{ij} \frac{\partial h}{\partial \eta_j}$$

where the matrix E is the 12×12 anti-symmetric matrix

$$E = \begin{pmatrix} 0 & I_6 \\ -I_6 & 0 \end{pmatrix}.$$

A smooth change of coordinates in phase space is called a **canonical transformation** if it preserves the form of Hamilton's equations. Suppose the new coordinates are

$$\xi = (\theta_1', \theta_2', \ldots, \theta_6', \pi_1', \pi_2', \ldots, \pi_6')$$

and are given by $\xi = f(\eta)$. In our previous notation, this would be written

$$\theta_i' = f_i(\theta_1, \ldots, \pi_6),$$
$$\pi_i' = f_{6+i}(\theta_1, \ldots, \pi_6).$$

Now, a tangent vector in the new coordinates is given by

$$\dot{\xi} = J \dot{\eta}$$

where J is the Jacobian matrix of the transformation

$$J_{ij} = \frac{\partial f_i}{\partial \eta_j}.$$

Substituting Hamilton's equations into the above, we get

$$\dot{\xi}_i = J_{ij} E_{jk} \frac{\partial h}{\partial \eta_k} = J_{ij} E_{jk} J_{lk} \frac{\partial h}{\partial \xi_l}.$$

These equations are in the same form as the original equations, provided $JEJ^T = E$. So, for a smooth coordinate change to be a canonical transformation its Jacobian matrix must be a symplectic matrix at each point of phase space. Actually, in modern texts the definition of the canonical transformation is taken the other way around. That is, a canonical transformation is defined as a symplectic transformation of phase space; see Arnol'd [2, Chap. 9] for example.

Next, we introduce the Poisson bracket $[-, -]$ a bilinear, anti-symmetric operation on pairs of functions defined on phase space. The definition of the Poisson bracket is

$$[f, g] = \sum_{i=1}^{6} \left(\frac{\partial f}{\partial \theta_i} \frac{\partial g}{\partial \pi_i} - \frac{\partial f}{\partial \pi_i} \frac{\partial g}{\partial \theta_i} \right).$$

It is a simple matter to check that this operation satisfies the Jacobi identity, and hence it defines a Lie algebra structure on these functions. However, we have no guarantee that this Lie algebra is finite-dimensional.

The Poisson bracket allows us to write the equations of motion as

$$\dot{\theta}_i = [\theta_i, h], \qquad i = 1, 2, \ldots, 6.$$
$$\dot{\pi}_i = [\pi_i, h] + \tau_i,$$

Without doing any more partial differentiation, we can see from the above that

$$[\theta_i, h] = A_{ij}^{-1} \pi_j,$$

$$i = 1, 2, \ldots, 6.$$

$$[\pi_i, h] = B_{jik} \dot{\theta}_j - C_i + \tau_i,$$

The time derivative of any function on phase space is given by

$$\frac{df}{dt} = [f, h] + \frac{\partial f}{\partial \pi_i} \tau_i.$$

In the case where there are no external forces, we have just; $df/dt = [f, h]$. A function on phase space that is invariant with respect to the evolution in time of the system is called an **invariant of the motion**. Clearly, such invariants are characterised by the property $[f, h] = 0$; that is, they must commute with the Hamiltonian function. However, we will not pursue this further, except to note that recently there has been much interest in the subject of numerical methods for integrating Hamilton's equations. Symplectic integration techniques have been introduced that automatically preserve the invarants of motion. It is claimed that these methods give a better qualitative picture of the solutions than do traditional numerical methods. The key feature of these methods is that the solution at the next time step is given by a canonical transformation. A review of these methods can be found in [98] . More recently these ideas have been extended to deal with differential equations on Lie groups and manifolds in general, see [60]

13.6   Simplification of the Equations of Motion

In this section, we look at how to design a robot so that its equations of motion are simplified.

In this area it is traditional to ignore gravity. The absence of the C_i terms in the equations of motion could be interpreted as assuming the robot is statically balanced. To do this, we design the robot so that the centre of mass of the links above and including the i-th one lie on the i-th joint axis. On the other hand we might interpret the absence of gravity by simply assuming that the robot is working in outer space.

In the following we look at three slightly different design problems. Two-joint robots are considered to keep things simple. There are many other design problems that could be studied. Here we keep to examples that have been discussed to some extent in the literature.

Of course the careful designs that will simplify the robot's dynamics will be disturbed as soon as the machine picks up a payload. However, this does not necessarily render the study of these problems irrelevant. The complications introduced by a payload can be easily computed; see Section 13.2 above. So, it may be possible to design a control system that can adapt to changes in payload.

13.6.1   Decoupling by Design

The material in this section follows quite closely the ideas in Asada and Youcef-Toumi [4]. The dynamics of a two-joint manipulator, in the absence of gravity can be summarised as

$$A_{ij}\ddot{\theta}_j + B_{ijk}\dot{\theta}_j\dot{\theta}_k = \tau_i, \qquad i, j = 1, 2,$$

where summation over repeated indices is assumed. In order to decouple these equations, we first need to design the robot so that the manipulator inertia matrix A_{ij} is diagonal. For the two-link example, we can write the generalised inertia matrix as

$$A = \begin{pmatrix} \mathbf{s}_1^T(N_1 + N_2)\mathbf{s}_1 & \mathbf{s}_1^T N_2 \mathbf{s}_2 \\ \mathbf{s}_2^T N_2 \mathbf{s}_1 & \mathbf{s}_2^T N_2 \mathbf{s}_2 \end{pmatrix}.$$

So, to diagonalise the matrix, we must have

$$\mathbf{s}_2^T N_2 \mathbf{s}_1 = 0.$$

That is, the joint screws must be conjugate with respect to N_2. These quantities depend on position, so just making the off-diagonal terms vanish at a single position is not very useful. What we need to ensure is that the term is constant. To do this, we look at how the term varies with joint angle. Assuming that $\mathbf{s}_1$ and $\mathbf{s}_2$ are now fixed, we have

$$A_{21}(\theta_1, \theta_2) = \mathbf{s}_2^T N_2 (e^{-\theta_2 \, \mathrm{ad}(\mathbf{s}_2)} \mathbf{s}_1).$$

Now, if the second joint, $\mathbf{s}_2$, is a revolute joint, the formula for the exponential given in Section 4.4.3 simplifies a little and the equation becomes

$$A_{21}(\theta_1, \theta_2) = \mathbf{s}_2^T N_2 \Big(I_6 - \sin\theta_2 \, \mathrm{ad}(\mathbf{s}_2) + (1 - \cos\theta_2)\, \mathrm{ad}(\mathbf{s}_2)^2 \Big) \mathbf{s}_1$$
$$= \mathbf{s}_2^T N_2 \mathbf{s}_1 - \sin\theta_2 \mathbf{s}_2^T N_2 [\mathbf{s}_2, \mathbf{s}_1] + (1 - \cos\theta_2) \mathbf{s}_2^T N_2 [\mathbf{s}_2, [\mathbf{s}_2, \mathbf{s}_1]].$$

To make this term zero for all positions of the machine and hence diagonalise the inertia matrix, we must have

$$\mathbf{s}_2^T N_2 \mathbf{s}_1 = 0, \qquad \mathbf{s}_2^T N_2 [\mathbf{s}_2, \mathbf{s}_1] = 0, \qquad \text{and} \qquad \mathbf{s}_2^T N_2 [\mathbf{s}_2, [\mathbf{s}_2, \mathbf{s}_1]] = 0.$$

Now suppose we write $\mathbf{s} = Q_0 N_2 \mathbf{s}_2$, so that

$$\{\mathbf{s}_2, N_2 \mathbf{s}_2\} = \{\mathbf{s}_2, Q_0 \mathbf{s}\} = Q_0[\mathbf{s}_2, \mathbf{s}],$$

see Section 13.1 above. The three equations to be satisfied for a diagonal inertia matrix thus become

$$\mathbf{s}^T Q_0 \mathbf{s}_1 = 0, \qquad [\mathbf{s}_2, \mathbf{s}]^T Q_0 \mathbf{s}_1 = 0, \qquad [\mathbf{s}_2, [\mathbf{s}_2, \mathbf{s}]]^T Q_0 \mathbf{s}_1 = 0.$$

Hence, we see that $\mathbf{s}_1$ must lie in the screw system reciprocal to the one generated by $\mathbf{s}$, $[\mathbf{s}_2, \mathbf{s}]$ and $[\mathbf{s}_2, [\mathbf{s}_2, \mathbf{s}]]$. Assuming that $\mathbf{s}_2$ is a line we can write it as $\mathbf{s}_2^T = (\boldsymbol{\omega}^T, \mathbf{r} \times \boldsymbol{\omega})$, where $\mathbf{r}$ is a point on the line. Now we have,

$$\mathbf{s} = \begin{pmatrix} 0 & I_3 \\ I_3 & 0 \end{pmatrix} \begin{pmatrix} I & mC \\ mC^T & mI_3 \end{pmatrix} \begin{pmatrix} \boldsymbol{\omega} \\ \mathbf{r} \times \boldsymbol{\omega} \end{pmatrix} = \begin{pmatrix} m(\mathbf{r} - \mathbf{c}) \times \boldsymbol{\omega} \\ I\boldsymbol{\omega} + m\mathbf{c} \times (\mathbf{r} \times \boldsymbol{\omega}) \end{pmatrix},$$

where the general form of an inertia matrix has been used, see Section 13.1 above. So we can easily see that $\mathbf{s}$ and $\mathbf{s}_2$ lie on orthogonal axes, that is, $\mathbf{s}^T Q_\infty \mathbf{s}_2 = 0$. From the example given in Section 8.4, we can see that this gives a IIB 3-system. The pitch of the system is given by

$$p = \frac{\mathbf{s}^T Q_0 \mathbf{s}}{\mathbf{s}^T Q_\infty \mathbf{s}}.$$

If we require that the first joint, s_1, is also a revolute joint, then the reciprocal of this 3-system and hence the system itself must have pitch zero. Hence, we have a condition that must be satisfied for the inertia matrix to diagonalise:

$$s^T Q_0 s = s_2^T N_2 Q_0 N_2 s_2 = 0.$$

Notice that this condition determines a quadratic line complex; see Griffiths and Harris [42, sect. 6.2] . By taking coordinates aligned with the principal axes of inertia, the equations simplify to

$$(d_1 - d_3)y(\omega_z \omega_x) + (d_2 - d_1)z(\omega_x \omega_y) + (d_3 - d_2)x(\omega_y \omega_z) = 0$$

where d_1, d_2, d_3 are the eigenvalues of the 3×3 inertia matrix, $\omega^T = (\omega_x, \omega_y, \omega_z)$ is the direction of the line s_2, and $r^T = (x, y, z)$ is any point on the line. Through any point r, there is a conic of lines in the complex given by

$$(\omega_x, \omega_y, \omega_z) \begin{pmatrix} 0 & (d_2 - d_1)z & (d_1 - d_3)y \\ (d_2 - d_1)z & 0 & (d_3 - d_2)x \\ (d_1 - d_3)y & (d_3 - d_2)x & 0 \end{pmatrix} \begin{pmatrix} \omega_x \\ \omega_y \\ \omega_z \end{pmatrix} = 0.$$

The surface of points for which the conic is degenerate is an object of much classical study. It is called the Kummer surface of the complex; see [42, sect. 6.2]. In our case, the Kummer surface is very simple. It is given by the vanishing of the determinant of the symmetric form in the above equation:

$$(d_2 - d_1)(d_1 - d_3)(d_3 - d_2)xyz = 0.$$

So, when the principal values of the inertia matrix are all different, the Kummer surface is given by the three planes perpendicular to the principal axes of inertia.

Returning to the design problem, the first joint, s_1, must lie in the 3-system generated by s, $[s_2, s]$ and $[s_2, [s_2, s]]$, since this system is self-reciprocal. As we have seen, this is a IIB system with $p = 0$; hence it is a β-plane in the Klein quadric and therefore represents a plane of lines in $\mathbb{R}^3$. It is easy to see that this plane is perpendicular to the line s_2 and contains s. Notice that this implies that the lines s_1 and s_2 must be perpendicular.

Unfortunately, if we succeed in diagonalising the inertia matrix, we may not have decoupled the equations of motion. This is because there may still be 'B' terms that couple $\dot{\theta}_1$ and $\dot{\theta}_2$. On examining these terms, however, all that is required is to have

$$s_1^T N_2 [s_2, s_1] = 0,$$

for all positions of the mechanism. Notice that if we do this, then since the 'B' terms are essentially the partial derivatives of the elements of the inertia matrix, the 'A' terms, we will also have succeeded in linearising the equations of motion. That is, the elements of the inertia matrix will all be constant.

For three or more joints, we can easily see that it is not possible to have a diagonal inertia matrix. This is because we have seen above that the joint axes would have to be mutually perpendicular, but this can only be true for specific configurations of the robot.

13.6.2 Ignorable Coordinates

The material in this section is loosely based on the work of Stokes and Brockett, see [117].

Recall from Section 13.5 that the momentum conjugate to a joint angle satisfies the differential equation

$$\dot{\pi}_i = \frac{1}{2} \frac{\partial A_{jk}}{\partial \theta_i} \dot{\theta}_j \dot{\theta}_k + \tau_i,$$

where, as usual, we have assumed there are no gravity terms. Now if the generalised inertia matrix A_{jk}, is independent of the i-th joint angle θ_i, we say that θ_i is an **ignorable** or **cyclic** coordinate. Clearly in these circumstances, the equation of motion for the conjugate momentum is very simple. Moreover, if all the coordinates are ignorable the generalised inertia matrix will be constant and the equations of motion will be linear.

In Section 13.4 above, formulas were derived for the partial derivatives of the generalised inertia matrix. From these results it is clear that θ_1, the first joint angle, is always ignorable. To linearise the dynamics of a two joint robot we need to have

$$\frac{\partial A_{11}}{\partial \theta_2} = -\mathbf{s}_1^T \big(\mathrm{ad}(\mathbf{s}_2)^T N_2 + N_2 \,\mathrm{ad}(\mathbf{s}_2) \big) \mathbf{s}_1 = -2\mathbf{s}_1^T N_2 [\mathbf{s}_2, \, \mathbf{s}_1] = 0$$

and

$$\frac{\partial A_{12}}{\partial \theta_2} = -\mathbf{s}_1^T \big(\mathrm{ad}(\mathbf{s}_2)^T N_2 + N_2 \,\mathrm{ad}(\mathbf{s}_2) \big) \mathbf{s}_2 = -\mathbf{s}_2^T N_2 [\mathbf{s}_2, \, \mathbf{s}_1] = 0.$$

Notice that it is always true that

$$\mathbf{s}_2^T \big(\mathrm{ad}(\mathbf{s}_2)^T N_2 + N_2 \,\mathrm{ad}(\mathbf{s}_2) \big) \mathbf{s}_2 = 0.$$

By taking appropriate multiples of these equations and summing we see that, for arbitrary α and β we have,

$$\big(\alpha \mathbf{s}_1 + \beta \mathbf{s}_2\big)^T \big(\mathrm{ad}(\mathbf{s}_2)^T N_2 + N_s \,\mathrm{ad}(\mathbf{s}_2) \big) \big(\alpha \mathbf{s}_1 + \beta \mathbf{s}_2\big) = 0.$$

So any element of the screw system generated by $\mathbf{s}_1$ and $\mathbf{s}_2$ must lie in the quadric given by the symmetric matrix $Q = \big(\mathrm{ad}(\mathbf{s}_2)^T N_2 + N_2 \,\mathrm{ad}(\mathbf{s}_2) \big)$. Now if we assume that both joints are revolute and so $\mathbf{s}_1$ and $\mathbf{s}_2$ are lines, then there are just three possible 2-systems; see Section 8.4.2.

The above condition, that the screw system lies in the quadric, must be true for all configurations of the robot. We can ensure this by rotating the screw system about $\mathbf{s}_2$ and demanding that the result still lies in the quadric Q. Notice, that it might be more natural to think of rotating the quadric with the screw system fixed, however this is clearly equivalent to doing things the other way around. We treat each case in turn, the first is the most general case where the lines are skew and form a IA $(p_a \neq p_b)$ 2-system. Rotating about s_2 does not affect s_2 but the line s_1 will sweep out a regulus of a cylindrical hyperboloid, see Figure 9.1. Hence we generate a cone with vertex s_2 in $\mathbb{PR}^5$. This cone must lie in Q for the rotation angle θ_2 to be ignorable. Now choose coordinates so that the two lines are

$$s_1 = \begin{pmatrix} \cos\phi \mathbf{i} - \sin\phi \mathbf{j} \\ l\sin\phi \mathbf{i} + l\cos\phi \mathbf{j} \end{pmatrix} \quad \text{and} \quad s_2 = \begin{pmatrix} \mathbf{i} \\ \mathbf{0} \end{pmatrix}.$$

This is slightly different from Section 8.4.2 but l is still the perpendicular distance between the lines and ϕ the angle between them. Now the regulus is represented in $\mathbb{PR}^5$ by a conic curve parameterised as

$$s_1(\theta_2) = \begin{pmatrix} \cos\phi \\ -\cos\theta_2 \sin\phi \\ -\sin\theta_2 \sin\phi \\ l\sin\phi \\ l\cos\theta_2 \cos\phi \\ l\sin\theta_2 \cos\phi \end{pmatrix}.$$

The equations that must be satisfied are given by

$$s_1(\theta_2)^T Q s_1(\theta_2) = 2s_1(\theta_2)^T N_2[s_2, s_1(\theta_2)] = 0$$

and

$$s_2^T Q s_1(\theta_2) = 2s_2^T N_2[s_2, s_1(\theta_2)] = 0.$$

If we write the inertia matrix N_2 using variables to be found,

$$N_2 = \begin{pmatrix} I_{11} & I_{12} & I_{13} & 0 & -mc_z & mc_y \\ I_{12} & I_{22} & I_{23} & mc_z & 0 & -mc_x \\ I_{13} & I_{23} & I_{33} & -mc_y & mc_x & 0 \\ 0 & mc_z & -mc_y & m & 0 & 0 \\ -mc_z & 0 & mc_x & 0 & m & 0 \\ mc_y & -mc_x & 0 & 0 & 0 & m \end{pmatrix}$$

and then substitute this into the two equations above, we get two equations in these variables and the joint angle θ_2. These equations must hold whatever the joint angle and so we can equate to zero the coefficients of $\cos\theta_2$, $\sin\theta_2$, $\cos^2\theta_2$ and $\sin^2\theta_2$ separately. This leads to six linear equations in the unknown entries of N_2,

$$I_{23}\sin^2\phi = 0,$$
$$(I_{33} - I_{22})\sin^2\phi = 0,$$
$$-I_{13}\cos\phi \sin\phi + mlc_y = 0,$$
$$I_{12}\cos\phi \sin\phi + mlc_z = 0,$$
$$-I_{13}\sin\phi + mlc_y \cos\phi = 0,$$
$$I_{12}\sin\phi + mlc_z \cos\phi = 0.$$

These equations are very simple to solve and lead to the result

$$
N_2 = \begin{pmatrix}
I_{11} & 0 & 0 & 0 & 0 & 0 \\
0 & I_{22} & 0 & 0 & 0 & -mc_x \\
0 & 0 & I_{22} & 0 & mc_x & 0 \\
0 & 0 & 0 & m & 0 & 0 \\
0 & 0 & mc_x & 0 & m & 0 \\
0 & -mc_x & 0 & 0 & 0 & m
\end{pmatrix},
$$

where I_{11}, I_{22}, c_x and the total mass m are arbitrary. This means that the inertia matrix of the second link has s_2 as an axis of rotational symmetry. This can be expressed as

$$
\mathrm{ad}(s_2)^T N_2 + N_2 \, \mathrm{ad}(s_2) = 0,
$$

which is easily seen to satisfy our original conditions for θ_2 to be ignorable. This will always be a solution whatever the screw system defined by s_1 and s_2. Moreover, if we have an n-joint serial robot, then we can always make the last joint angle ignorable by arranging the inertia of the final link to be symmetrical about the last joint axis.

For the other screw systems there may also be other solutions. So we look at these now, beginning with the case where the joint axes meet at a point and so determine a IIA $(p = 0)$ 2-system of screws. Rotating this about s_2 gives all the lines through the common point and this is an α-plane in the Klein quadric. For θ_2 to be ignorable this α-plane must also lie in Q. Using the six equations above with $l = 0$ we see that the only constraints we obtain are that $I_{12} = I_{13} = I_{23} = 0$ and $I_{22} = I_{33}$. That is, the inertia matrix must have an axis of rotational symmetry, and that axis must be parallel to the second joint axis s_2, but now it is not necessary that the centre of mass of the link lies on s_2. Notice that in this case the algebra generated by s_1 and s_2 using iterated commutators, is a proper subalgebra of $se(3)$, it is in fact $so(3)$, see Table 8.1. Moreover, the condition on the inertia found above can be expressed as

$$
s_i^T \big(\mathrm{ad}(s_2)^T N_2 + N_2 \, \mathrm{ad}(s_2) \big) s_j = 0, \qquad \text{for all} \quad s_i, s_j \in so(3).
$$

The other sub-case is where the joint axes are parallel. The 2-system generated by s_1 and s_2 consists of all lines in the plane determined s_1 and s_2 and parallel to them. This is a IIB $(p = 0)$ 2-system. Rotating this about s_2 gives all the lines in space parallel to s_2. Again this is an α-plane in the Klein quadric, this time meeting the α-plane of infinite lines. If we set $\sin \phi = 0$ (and $\cos \phi = \pm 1$) in the six equations above we get the conditions $c_y = c_z = 0$. Hence θ_2 is ignorable in this case if the centre of mass of link 2 lies on the axis of joint 2 with no restriction on the 3×3 inertia matrix. Again the algebra generated by the two joint screw is a proper subalgebra of $se(3)$, this time $se(2)$, and again the constraints found on the inertia can be expressed as

$$
s_i^T \big(\mathrm{ad}(s_2)^T N_2 + N_2 \, \mathrm{ad}(s_2) \big) s_j = 0, \qquad \text{for all} \quad s_i, s_j \in se(2).
$$

For a three-joint robot the generalised inertia matrix would be

$$A = \begin{pmatrix} \mathbf{s}_1^T(N_1 + N_2 + N_3)\mathbf{s}_1 & \mathbf{s}_1^T(N_2 + N_3)\mathbf{s}_2 & \mathbf{s}_1^T N_3 \mathbf{s}_3 \\ \mathbf{s}_2^T(N_2 + N_3)\mathbf{s}_1 & \mathbf{s}_2^T(N_2 + N_3)\mathbf{s}_2 & \mathbf{s}_2^T N_3 \mathbf{s}_3 \\ \mathbf{s}_3^T N_3 \mathbf{s}_1 & \mathbf{s}_3^T N_3 \mathbf{s}_2 & \mathbf{s}_3^T N_3 \mathbf{s}_3 \end{pmatrix}.$$

For the third joint angle to be ignorable the following equations must hold:

$$\mathbf{s}_i^T\big(\operatorname{ad}(\mathbf{s}_3)^T N_3 + N_3 \operatorname{ad}(\mathbf{s}_3)\big)\mathbf{s}_j, \qquad i, j = 1, \, 2, \, 3.$$

That is, the screw system spanned by $\mathbf{s}_1$, $\mathbf{s}_2$ and $\mathbf{s}_3$ must lie in the quadric $\big(\operatorname{ad}(\mathbf{s}_3)^T N_3 + N_3 \operatorname{ad}(\mathbf{s}_3)\big)$. These relations must be independent of both θ_3 and θ_2. The same solutions as in the 2-joint case above will also work here; we could have, $\big(\operatorname{ad}(\mathbf{s}_3)^T N_3 + N_3 \operatorname{ad}(\mathbf{s}_3)\big) = 0$ implying that N_3 was rotationally symmetric about $\mathbf{s}_3$. Alternatively, the three-joint could meet at a point, and hence generate a IIA ($p = 0$) 3-system, and in this case the inertia of the third link only needs to satisfy

$$\mathbf{s}_i^T\big(\operatorname{ad}(\mathbf{s}_3)^T N_3 + N_3 \operatorname{ad}(\mathbf{s}_3)\big)\mathbf{s}_j = 0, \qquad \text{for all} \quad \mathbf{s}_i, \, \mathbf{s}_j \in so(3).$$

Similarly when the three-joint axes are parallel they generate a IIC ($p = 0$) 3-system and the inertia only needs to satisfy

$$\mathbf{s}_i^T\big(\operatorname{ad}(\mathbf{s}_3)^T N_3 + N_3 \operatorname{ad}(\mathbf{s}_3)\big)\mathbf{s}_j = 0, \qquad \text{for all} \quad \mathbf{s}_i, \, \mathbf{s}_j \in se(2).$$

Next, to make the second joint angle ignorable and hence the generalised inertia matrix constant, we must ensure that the following relations are satisfied,

$$\mathbf{s}_i^T\big(\operatorname{ad}(\mathbf{s}_2)^T(N_2 + N_3) + (N_2 + N_3)\operatorname{ad}(\mathbf{s}_2)\big)\mathbf{s}_j = 0, \qquad j = 1, \, 2$$

and

$$[\mathbf{s}_1, \, \mathbf{s}_2]^T N_3 \mathbf{s}_3 = 0.$$

The first set of equations here is exactly the same as the relations for a 2-joint robot to have constant generalised inertia matrix, except that the inertia matrix of the second link is now the sum $N_2 + N_3$ of inertia matrices above it. To make the last equation valid for all values of θ_2 all that is required is that:

$$[[\mathbf{s}_1, \, \mathbf{s}_2], \, \mathbf{s}_2]^T N_3 \mathbf{s}_3 = 0,$$

in addition to the original equation. To investigate this suppose we have designed the robot so that the third joint coincides with the rotational symmetry axis of the third link. The equations above are reminiscent of the problem in the previous section. So let us write $\mathbf{s} = Q_0 N_3 \mathbf{s}_3$ and, after a little rearrangement, the pair of equations above become

$$\mathbf{s}_1^T Q_0 [\mathbf{s}_2, \, \mathbf{s}] = 0, \qquad \text{and} \qquad \mathbf{s}_1^T Q_0 [\mathbf{s}_2, \, [\mathbf{s}_2, \, \mathbf{s}]] = 0.$$

Choosing coordinates so that

$$s_2 = \begin{pmatrix} \cos\phi\mathbf{i} - \sin\phi\mathbf{j} \\ l\sin\phi\mathbf{i} + l\cos\phi\mathbf{j} \end{pmatrix} \quad \text{and} \quad s_3 = \begin{pmatrix} \mathbf{i} \\ \mathbf{0} \end{pmatrix}$$

we can compute the Lie bracket factors

$$[s_2, s] = \begin{pmatrix} \mathbf{0} \\ I_{11}\sin\phi\mathbf{k} \end{pmatrix} \quad \text{and} \quad [s_2, [s_2, s]] = \begin{pmatrix} \mathbf{0} \\ -I_{11}\sin\phi(\sin\phi\mathbf{i} + \cos\phi\mathbf{j}) \end{pmatrix}.$$

It is easy to see that these two screws form a 2-system of type IIC. To satisfy the equations the first joint axis, s_1 must lie in the reciprocal $\overline{\text{IIC}}$ system. The lines in this system are all parallel to s_2. So finally we will have to design the inertia of the second link of the robot so that the centre of mass of the composite of the second and third links lies on the second joint axis.

13.6.3   Decoupling by Coordinate Transformation

The ideas in this section come from Spong [114], who in turn credits Koditschek [66]. We attempt to simplify the equations of motion by a coordinate change. Consider the joint space of a robot as a metric space with the generalised inertia matrix A_{ij} as metric. When there is no gravity and the geodesics of this metric are solutions to the equations of motion,

$$A_{ij}\ddot{\theta}_j + B_{ijk}\dot{\theta}_j\dot{\theta}_k = 0;$$

see Section 13.2. If we include the driving forces, the equations are just

$$A_{ij}\ddot{\theta}_j + B_{ijk}\dot{\theta}_j\dot{\theta}_k = \tau_i.$$

In Hamiltonian form, the equations of motion are

$$\dot{\theta}_i = A_{ij}^{-1}\pi_j,$$

$$i = 1, 2, \ldots, 6;$$

$$\dot{\pi}_i = B_{jik}\dot{\theta}_j\dot{\theta}_k + \tau_i,$$

see Section 13.5.

Now consider a smooth transformation of joint space. The new coordinates $\phi^T = (\phi_1, \phi_2, \ldots, \phi_6)$ will be functions of the original joint variables $\boldsymbol{\theta}$. Such a transformation is called a point transformation in the mechanics literature. It is known that all point transformations are canonical; see Goldstein [39, sect. 9.2]. We can write the Jacobian of the transformation as

$$K_{ij} = \frac{\partial\phi_i}{\partial\theta_j},$$

so that
$$\dot{\phi} = K\dot{\theta}.$$

Next, we assume that the generalised inertia matrix A_{ij} satisfies
$$A = K^T K.$$

Thinking of this matrix as a metric again, we may use the transformation to pull back the metric and hence turn the transformation into an isometry. In the new coordinates, the kinetic energy of the robot is given by

$$E_K = \frac{1}{2} A_{ij} \dot{\theta}_i \dot{\theta}_j = \frac{1}{2} \dot{\phi}_i \dot{\phi}_i.$$

Since there is no potential energy, this expression is also the Lagrangian function and Hamiltonian function $E_K = L = h$. Thus, the momentum conjugate to the ϕ_i coordinate is given by

$$\sigma_i = \frac{\partial L}{\partial \dot{\phi}_i} = \dot{\phi}_i.$$

On the other hand, the momentum conjugate to the coordinate θ_i is given by

$$\pi_i = A_{ij} \dot{\theta}_j = K_{ji} \dot{\phi}_i = K_{ji} \sigma_i,$$

that is, $\boldsymbol{\pi} = K^T \boldsymbol{\sigma}$. Differentiating this with respect to time and using the equations of motion, we get

$$\dot{\pi}_i = B_{jik} \dot{\theta}_j \dot{\theta}_k + \tau_i = \dot{K}_{ji} \sigma_j + K_{ji} \dot{\sigma}_j.$$

Now, by substituting for the partial derivatives of A in the formula

$$B_{ijk} = \frac{1}{2} \left(\frac{\partial A_{ij}}{\partial \theta_k} + \frac{\partial A_{ik}}{\partial \theta_j} - \frac{\partial A_{jk}}{\partial \theta_i} \right)$$

we obtain the relation

$$B_{ijk} = K_{li} \frac{\partial K_{lj}}{\partial \theta_k}$$

and thus

$$B_{jik} \dot{\theta}_j \dot{\theta}_k = \dot{K}_{ji} \sigma_j$$

and hence

$$\tau_i = K_{ji} \dot{\sigma}_j.$$

The equations of motion in the new coordinates are thus

$$\dot{\phi}_i = \sigma_i,$$
$$\dot{\sigma}_i = (K^{-1})_{ji} \tau_j, \qquad i = 1, 2, \ldots, 6,$$

or in Lagrangian form

$$\ddot{\phi} = (K^T)^{-1} \boldsymbol{\tau}.$$

Notice that in these coordinates the equations of motion are decoupled and linearised.

However, the question still remains, Does such a transformation exist? Remember, it must satisfy

$$A_{ij} = \frac{\partial \phi_k}{\partial \theta_i} \frac{\partial \phi_k}{\partial \theta_j}.$$

Notice that in the new coordinates the metric is simply the usual Euclidean metric. This metric is flat, that is, it has zero curvature tensor R^i_{jkl}. The vanishing of this tensor is independent of coordinates, and hence a necessary condition for the linearising transformation to exist is that the curvature tensor of the metric A_{ij} must vanish. It is known that this condition is also sufficient; see Spong [114].

We can look in a little more detail at the general 2-joint example; compare this with Section 13.6.1. If the joints are revolute, then the configuration space is a torus. It is well known that it is topologically possible for a torus to have a flat metric; see O'Neill [81, Chap. VII sect. 2] for example. This is also true for other joints.

The curvature tensor is in general hard to compute. In this simple case, we are lucky since we may work with the Gaussian curvature of the surface; see Bishop and Crittenden [11, sect. 9.1]. This may be found using the method of moving frames, [81, Chap. VII sect. 2]. First, we need a pair of 1-forms that are orthonormal with respect to the metric

$$\mu_1 = \frac{A_{22}}{\sqrt{A_{22}\Delta}} d\theta_1 - \frac{A_{12}}{\sqrt{A_{22}\Delta}} d\theta_2 \quad \text{and} \quad \mu_2 = \frac{1}{\sqrt{A_{22}}} d\theta_2$$

where A_{ij} are elements of the metric A and $\Delta = \det(A)$ is the determinant of the metric. Next, we find the connection form ω by taking the exterior derivative and using the equations

$$d\mu_1 = \omega \wedge \mu_2, \qquad d\mu_2 = -\omega \wedge \mu_1.$$

Now, all the elements of A are independent of θ_1, and A_{22} is also independent of θ_2, so we have

$$d\mu_1 = \frac{\sqrt{A_{22}}}{2} \Delta^{-3/2} \frac{\partial \Delta}{\partial \theta_2} d\theta_1 \wedge d\theta_2, \qquad d\mu_2 = 0.$$

This gives us

$$\omega = \frac{\sqrt{A_{22}}}{2} \Delta^{-1} \frac{\partial \Delta}{\partial \theta_2} \mu_1 = \frac{1}{2} \Delta^{-3/2} \frac{\partial \Delta}{\partial \theta_2} (A_{22} d\theta_1 - A_{12} d\theta_2).$$

Finally, the exterior derivative of the connection form gives the Gaussian curvature K:

$$d\omega = -K(\mu_1 \wedge \mu_2).$$

After a little computation, the Gaussian curvature is

$$K = \frac{A_{22}}{4}\left(\frac{2\Delta\Delta'' - 3(\Delta')^2}{\Delta^2}\right),$$

where we have written Δ' and Δ'' for $\partial\Delta/\partial\theta_2$ and $\partial^2\Delta/\partial\theta_2^2$, respectively. From Section 13.6.1, we have that the determinant is given by

$$\Delta = A_{11}A_{22} - A_{12}^2$$
$$= (s_1^T N_1 s_1)(s_2^T N_2 s_2) + (s_1^T N_2 s_1)(s_2^T N_2 s_2) - (s_1^T N_2 s_2)(s_1^T N_2 s_2)$$

at the home position. The derivatives at this point can also be found:

$$\Delta' = 2(s_1^T N_2[s_1, s_2])(s_2^T N_2 s_2) - 2(s_2^T N_2[s_1, s_2])(s_1^T N_2 s_2)$$

and

$$\Delta'' = 4(s_1^T N_2[s_2, [s_2, s_1]])(s_2^T N_2 s_2) + 4([s_2, s_1]^T N_2[s_2, s_1])(s_2^T N_2 s_2)$$
$$- 2(s_2^T N_2[s_2, [s_2, s_1]])(s_1^T N_2 s_2) - 2(s_2^T N_2[s_1, s_2])(s_2^T N_2[s_1, s_2]).$$

As usual, although we have done the computations for the home configuration, there is nothing special about this point, and hence the formulas apply at any position. However, the vanishing of the Gaussian curvature at some point in the configuration space does not imply that the metric is flat. For that, the curvature must disappear at every point of the manifold. It will be appreciated from the above that the explicit dependence of the curvature on the joint angle θ_2 is complicated and hence is probably best studied for particular cases.

Notice that the curvature is certainly zero when the design has been chosen to decouple the equations of motion as in Section 13.6.1. There may be other zero-curvature solutions.

14
Constrained Dynamics

In this chapter we extend the ideas of the previous chapter to more complicated manipulator designs. Specifically we look at the dynamics of tree and star-structured mechanisms. Then we look at robots with constrained end-effectors; we only consider time invariant holonomic constraints here. This type of constraint can occur when the end-effector of the robot interacts with the environment. When these ideas are combined with the dynamics of tree and star-structured robots, it is possible to derive the dynamics of some kinematic loops and parallel robots. Finally, we look at some examples.

Most of the ideas here first appeared in [106] and [107].

14.1 Trees and Stars

The definitions and notation that are introduced here are based loosely on the graph theory definitions for rooted trees, see for example [10, Chap. 9].

Tree and star mechanisms are characterised by having several terminal links and no loops. We call these terminal links **leaf-links**. Leaf-links have a single joint. All other links of the mechanism are **internal links** and these have at least two, possibly many, joints.

A tree mechanism is grounded, that is, one of the links is connected to a fixed base by a single joint. Star mechanisms are not grounded; this is the only distinction between the two. Possible examples of stars are multi-leg walking machines and multi-body space-craft.

In any analysis we always distinguish one link as the **root link**. In a tree, the root link will always be the link connected to ground. For a star mechanism the choice of the root link is arbitrary, but a natural choice is usually self-evident.

The distance of each link from the root determines its level. The root link is at level 1; all links attached to the root are level 2; any link connected to a link at level 2, but not the root, is at level 3; and so on. The link immediately below a given link is its **parent** and all links immediately above it are its **children**.

For each link there is exactly one joint connecting it to its parent link. This includes the root link in a tree, where the ground is at level 0. For link i, say, the i-th joint connects the link to a link at a lower level. Notice that such a machine always has the same number of links as joints.

The links **above** a given link are those at levels higher than the link in question but connected to it by a sequence of joints and links, that is by the descendants of the link. Leaf-links are childless. If link j is above link i or equal to it, we write $i \preceq j$. Technically, the tree structure defines a partial order on the set of links.

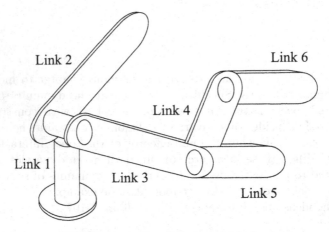

FIGURE 14.1. A Tree Structured Mechanism

For any link in the mechanism there is a unique sequence of links and joints connecting the given link with the root. From link i, for example, we can pass through joint i to the link immediately below i and continue in this way, going down one level at a time, until we reach the root link at level 1. The sequence of links (or joints) is the **ancestry** of the link. For example, in Figure 14.1 the ancestry of link 6 is $(1, 3, 4, 6)$. Observe that we have included the end-points in this path specification.

14.1.1   Dynamics of Tree and Star Structures

With these definitions we are able to give the equations of motion for a tree-structured mechanism. For each link i we get an equation of the form,

$$N_i \ddot{\mathbf{q}}_i + \{\dot{\mathbf{q}}_i, N_i \dot{\mathbf{q}}_i\} = \mathcal{T}_i + \mathcal{R}_i + \mathcal{G}_i - \sum_j (\mathcal{T}_j + \mathcal{R}_j).$$

As in Section 13.2 above, the wrenches at joint i consist of torques $\mathcal{T}_i$ and reactions $\mathcal{R}_i$. The velocity screw of the link is $\dot{\mathbf{q}}_i$ and its acceleration is $\ddot{\mathbf{q}}_i$. The sum is over all other joints attached to the link. These equations will be referred to as the link equations of the mechanism. To isolate the wrenches at joint i we must follow the tree structure, adding the link equations cumulatively. That is, we add up the equations corresponding to link i and all the link equations for links above i,

$$\sum_{i \preceq j} \left(N_j \ddot{\mathbf{q}}_j + \{\dot{\mathbf{q}}_j, N_j \dot{\mathbf{q}}_j\} - \mathcal{G}_j \right) = \mathcal{T}_i + \mathcal{R}_i \qquad \text{all } i.$$

Then we pair this equation with the i-th joint screw $\mathbf{s}_i$ to give the amplitude of the torque:

$$\sum_{i \preceq j} \left(\ddot{\mathbf{q}}_j^T N_j \mathbf{s}_i + \dot{\mathbf{q}}_j^T N_j [\mathbf{s}_i, \dot{\mathbf{q}}_j] - \mathcal{G}_j^T \mathbf{s}_i \right) = \tau_i \qquad \text{all } i.$$

These equations can be called the joint equations for the mechanism.

Star mechanisms can be treated in almost the same way. Because the root of a star mechanism is not grounded, there is no reaction wrench from the root joint and no driving force or torque either. If we label the links so that the root link is 1, then we have the same joint equations as before,

$$\sum_{i \preceq j} \left(\ddot{\mathbf{q}}_j^T N_j \mathbf{s}_i + \dot{\mathbf{q}}_j^T N_j [\mathbf{s}_i, \dot{\mathbf{q}}_j] - \mathcal{G}_j^T \mathbf{s}_i \right) = \tau_i \qquad \text{all } i \neq 1.$$

Then, for the root link, we add all the link equations to get,

$$\sum_{1 \preceq j} \left(N_j \ddot{\mathbf{q}}_j + \{\dot{\mathbf{q}}_j, N_j \dot{\mathbf{q}}_j\} - \mathcal{G}_j \right) = 0.$$

We can pair this six-component screw equation with any six linearly independent screws to produce six scalar equations, if needed.

Many of the mechanisms we want to consider have multi-degree of freedom joints. Usually these joints are passive, that is they are not driven. For example, the passive ball-and-socket joints of a Stewart platform that connect the actuated leg-rods to the platform-body and base-body. Any such spherical joint introduces three variables into the equations of motion. Hence we need three equations for each such joint to make the system determinate.

The joint can be replaced by an instantaneously equivalent system of screws. All of the screws in this system will be dual to any possible reaction wrench, that is they will be unable to perform any work. Then we pair the appropriate equation with the basis elements from this system of joint screws. For a three-degree-of-freedom joint this gives three joint equations,

$$\sum_{i \preceq j} \left(\ddot{\mathbf{q}}_j^T N_j \mathbf{s}_{i1} + \dot{\mathbf{q}}_j^T N_j [\mathbf{s}_{i1}, \dot{\mathbf{q}}_j] - \mathcal{G}_j^T \mathbf{s}_{i1} \right) = 0,$$

$$\sum_{i \preceq j} \left(\ddot{\mathbf{q}}_j^T N_j \mathbf{s}_{i2} + \dot{\mathbf{q}}_j^T N_j [\mathbf{s}_{i2}, \dot{\mathbf{q}}_j] - \mathcal{G}_j^T \mathbf{s}_{i2} \right) = 0,$$

$$\sum_{i \preceq j} \left(\ddot{\mathbf{q}}_j^T N_j \mathbf{s}_{i3} + \dot{\mathbf{q}}_j^T N_j [\mathbf{s}_{i3}, \dot{\mathbf{q}}_j] - \mathcal{G}_j^T \mathbf{s}_{i3} \right) = 0,$$

where $\mathbf{s}_{i1}$, $\mathbf{s}_{i2}$ and $\mathbf{s}_{i3}$ form a basis for the freedom screws of the joint. The right-hand sides are zero because the joint is passive.

14.1.2   Link Velocities and Accelerations

Consider a link i having ancestry $(1, i_1, i_2, \ldots, i)$. If the mechanism is a tree, the kinematics of the link are given by a product of exponentials formula

$$K_i = e^{\theta_1 \mathbf{s}_1} e^{\theta_{i_1} \mathbf{s}_{i_1}} e^{\theta_{i_2} \mathbf{s}_{i_2}} \cdots e^{\theta_i \mathbf{s}_i},$$

where the θ_j are joint variables and $\mathbf{s}_j$ are the joint screws in the home position of the mechanism, see Section 4.5. The choice of home is arbitrary but once chosen it is identified as the configuration where all the joint variables $\theta_j = 0$. The transformation K_i takes link i from its home position to that determined by the joint variables $\theta_1, \theta_{i_1}, \theta_{i_2}, \ldots, \theta_i$. This implicitly assumes that all the joints are one-degree-of-freedom joints, however, as mentioned in the previous section, we often need to consider multi-degree-of-freedom joints such as spherical joints. For such a joint the exponential term will be

$$e^{\phi_{j1} \mathbf{s}_{j1} + \phi_{j2} \mathbf{s}_{j2} + \phi_{j3} \mathbf{s}_{j3}},$$

where the ϕ_{jk}s are local coordinates that determine the position of the joint and the $\mathbf{s}_{jk}$s are a system of basis screws that span the system of freedoms determined by the joint. Alternatively, we could use different local coordinates, say Euler angles, and we would get a product of three exponential terms

$$e^{\phi_j \mathbf{s}_{j1}} e^{\theta_j \mathbf{s}_{j2}} e^{\psi_j \mathbf{s}_{j3}}.$$

In a star mechanism there is no root joint so the formula

$$K_i = e^{\theta_{i_1} \mathbf{s}_{i_1}} e^{\theta_{i_2} \mathbf{s}_{i_2}} \cdots e^{\theta_i \mathbf{s}_i},$$

gives the transformation relative to the root link. Because a star mechanism is not grounded, we could consider it a tree with a six-degree-of-freedom root joint.

The velocity of any link is found by differentiating the forward kinematics. For a link i in a tree we have

$$\dot{\mathbf{q}}_i = \dot{\theta}_1 \mathbf{s}_1 + \dot{\theta}_{i_1} \mathbf{s}_{i_1} + \dot{\theta}_{i_2} \mathbf{s}_{i_2} + \cdots + \dot{\theta}_i \mathbf{s}_i.$$

In this formula the joint screws $\mathbf{s}_j$, must be interpreted as indicating the current position of the linkage. In a star-mechanism we have

$$\dot{\mathbf{q}}_i = \dot{\mathbf{q}}_1 + \dot{\theta}_{i_1} \mathbf{s}_{i_1} + \dot{\theta}_{i_2} \mathbf{s}_{i_2} + \cdots + \dot{\theta}_i \mathbf{s}_i,$$

where $\dot{\mathbf{q}}_1$ is the velocity of the root link.

Relations for the link accelerations, that is the time derivatives of the link velocities, are also similar to those for serial robots. For a tree we sum along the ancestry of the link, that is,

$$\ddot{\mathbf{q}}_i = \sum_{1 \preceq j \preceq i} \left(\ddot{\theta}_j \mathbf{s}_j + \dot{\theta}_j [\dot{\mathbf{q}}_j, \mathbf{s}_j] \right).$$

For a star we perform the same summation but need to add in the acceleration of the root link,

$$\ddot{\mathbf{q}}_i = \ddot{\mathbf{q}}_1 + \sum_{1 \prec j \preceq i} \left(\ddot{\theta}_j \mathbf{s}_j + \dot{\theta}_j [\dot{\mathbf{q}}_j, \mathbf{s}_j] \right).$$

14.1.3   Recursive Dynamics for Trees and Stars

The torque or force at each joint of a tree can be computed by adapting the standard recursive procedure used for serially articulated robot, see Section 13.3. For a serial robot the recursion is done in two passes: one from the base to the tip to compute the velocities and accelerations, followed by a second pass, from tip back to base, to compute the joint torques or forces. For tree mechanisms, we can similarly recurse up then down the tree structure. For example, to compute the velocities and accelerations we start at the root link and work our way up the tree. It would probably be easier to use a 'breadth first' order to traverse the tree; that is, we complete the calculations for each level before moving to the next level. This would help to minimise the storage requirements.

For each link, working our way up the levels in the tree, the following procedure must be computed.

Inputs:

Current joint angle θ_i, rate $\dot{\theta}_i$ and acceleration $\ddot{\theta}_i$
parent link velocity $\dot{\mathbf{q}}_{i-1}^0$ and acceleration $\ddot{\mathbf{q}}_{i-1}^0$
joint screw in home position $\mathbf{s}_i^0$
link inertia matrix in home position N_i^0

Outputs:

>   Link velocity $\dot{\mathbf{q}}_i^0$ and acceleration $\ddot{\mathbf{q}}_i^0$
>   wrench due to the link $\mathcal{Q}_i^0$

Method:

$$\dot{\mathbf{q}}_i^0 \leftarrow \dot{\theta}_i \mathbf{s}_i^0 + \mathrm{Ad}(-\theta_i \mathbf{s}_i^0)\dot{\mathbf{q}}_{i-1}^0$$

$$\ddot{\mathbf{q}}_i^0 \leftarrow \ddot{\theta}_i \mathbf{s}_i^0 + \dot{\theta}_i [\dot{\mathbf{q}}_i^0,\, \mathbf{s}_i^0] + \mathrm{Ad}(-\theta_i \mathbf{s}_i^0)\ddot{\mathbf{q}}_{i-1}^0$$

$$\mathcal{Q}_i^0 \leftarrow N_i^0 \ddot{\mathbf{q}}_i^0 + \dot{\mathbf{q}}_i^0,\, N_i^0 \dot{\mathbf{q}}_i^0$$

The superscript '0' reminds us that all the quantities are to be taken in their home positions. For clarity, gravity has not been included in the above, but it is simple to include it by subtracting the gravitational acceleration $\mathbf{g}$ from $\ddot{\mathbf{q}}_i^0$ and $\ddot{\mathbf{q}}_{i-1}^0$. Note that the starting point for the recursion is

$$\dot{\mathbf{q}}_1^0 = \dot{\theta}_1 \mathbf{s}_1^0, \qquad \text{and} \qquad \ddot{\mathbf{q}}_1^0 = \ddot{\theta}_1 \mathbf{s}_1^0.$$

The second stage of the algorithm recurses from the leaves back towards the root. Again this can be done level by level, performing the following procedure at each link:

Inputs:

>   Current joint angles of child links θ_{i+1}
>   joint screw in home position $\mathbf{s}_i^0$
>   wrench due to the link $\mathcal{Q}_i^0$
>   wrenches acting on the child links $\mathcal{P}_{i+1}^0$

Outputs:

>   wrench acting on link $\mathcal{P}_i^0$
>   Joint torque τ_i

Method:

$$\mathcal{P}_i^0 \leftarrow \mathcal{Q}_i^0 + \sum \mathrm{Ad}(-\theta_i \mathbf{s}_{i+1}^0)^T \mathcal{P}_{i+1}^0$$

$$\tau_i \leftarrow \left(\mathcal{P}_i^0\right)^T \mathbf{s}_i^0$$

The summation is over all child links. The starting conditions are, for all leaf-links

$$\mathcal{P}_l^0 = \mathcal{Q}_l^0.$$

The same procedure can be used for star mechanisms though we must start and stop the recursions before reaching the root joint and we must know the velocity and acceleration of the root link.

14.2 Serial Robots with End-Effector Constraints

14.2.1 Holonomic Constraints

The main idea here is that the forces and torques maintaining a system of constraints form a system of wrenches dual to the motion screws allowed by the constraints.

As usual, if we agree on a home position for a rigid body, then all subsequent positions and orientations can be specified by elements of the group of rigid body motions $SE(3)$. So each point in the group manifold corresponds to a possible configuration of the body.

The constraints that we are interested in are time invariant holonomic constraints, sometimes called scleronomic constraints. Such constraints can be represented by vanishing of a number of functions defined on the group

$$\phi_i(g) = 0, \qquad i = 1, \ldots, n.$$

The common solutions to these constraint equations will, in general, define a submanifold in the group. Exceptionally, the Jacobian of the map defined by the functions may drop rank and singularities will occur on the constraint subspace. However, in the simple case we consider below this does not appear to happen.

In Section 4.1 we saw that tangent vectors to the group manifold represent velocities. To put these velocities into the fixed coordinate system we must transform them back to the identity element, that is we turn them into Lie algebra elements. The tangent vectors to the submanifold represent the infinitesimal freedoms or virtual displacements consistent with the constraints. At a point in the constraint submanifold the space of screws tangent to the submanifold will be called the **screw system of freedoms**. In general, as we move around the constraint submanifold this system of freedoms will vary. However, there are important cases where this screw system is fixed. Suppose that the constraint submanifold is actually a subgroup. This can happen if the end-effector of the robot is constrained to lie on a lower Reuleaux pair, or in a closed loop mechanism where the last joint can be thought of as a constraint on an open kinematic chain. In such cases it is clear that the system of freedoms is simply the subalgebra corresponding to the subgroup.

The constraints are maintained by reaction wrenches. That is, if we try to move the rigid body in a constrained direction, reaction forces and torques will appear that stop the constraint-violating motion. By the principle of virtual work the constraint wrenches do no work on the infinitesimal freedoms; see [39, Chap. 1]. So the constraint wrenches form a linear system which we will call the **system of constraint wrenches**.

If the system of freedom screws has a basis $\mathbf{z}_1, \mathbf{z}_2, \ldots, \mathbf{z}_n$, then the system of constraint wrenches will have a basis $\mathcal{W}_{n+1}, \ldots, \mathcal{W}_6$ satisfying

$$\mathcal{W}_i^T \mathbf{z}_j = 0, \quad 1 \le i \le n, \ n+1 \le j \le 6.$$

For example, suppose that a single point in the rigid body is fixed, or perhaps the body is constrained to lie on a sphere. We can take the fixed point or the centre of the sphere as the origin and then a basis for the freedom screws could be

$$z_1 = \begin{pmatrix} i \\ 0 \end{pmatrix}, \quad z_2 = \begin{pmatrix} j \\ 0 \end{pmatrix}, \quad z_3 = \begin{pmatrix} k \\ 0 \end{pmatrix}$$

where, as usual i, j, k are the unit vectors in the x, y and z directions. Clearly the above basis generates the subalgebra of rotations about the origin. The dual system of constraint wrenches has a basis,

$$\mathcal{W}_4 = \begin{pmatrix} 0 \\ i \end{pmatrix}, \quad \mathcal{W}_5 = \begin{pmatrix} 0 \\ j \end{pmatrix}, \quad \mathcal{W}_6 = \begin{pmatrix} 0 \\ k \end{pmatrix}.$$

So in this case the constraint wrenches are any pure force.

As another example, suppose the rigid body is constrained so that one point on the body lies in a plane and a line in the body remains perpendicular to the plane. The motivation behind this example is hand-writing. Imagine a robot holding a pen, the point of the pen must stay on the paper and the pen must remain upright. These constraints form a subgroup, the group of planar rigid motions. If we assume the plane is the xy-plane, then a basis for the screws of freedom is

$$z_1 = \begin{pmatrix} k \\ 0 \end{pmatrix}, \quad z_2 = \begin{pmatrix} 0 \\ i \end{pmatrix}, \quad z_3 = \begin{pmatrix} 0 \\ j \end{pmatrix}$$

and the basis for the constraint wrenches could be

$$\mathcal{W}_4 = \begin{pmatrix} 0 \\ k \end{pmatrix}, \quad \mathcal{W}_5 = \begin{pmatrix} i \\ 0 \end{pmatrix}, \quad \mathcal{W}_6 = \begin{pmatrix} j \\ 0 \end{pmatrix}.$$

Notice that, in both the examples above the constraint submanifold is a subgroup. In fact in both cases it is a 3-plane in the Study quadric. As a final example, we look at a system of constraints that do not give a subgroup.

Suppose the robot's end-effector is required to follow a quadric surface, that is a surface defined by the equation

$$k_x x^2 + k_y y^2 - 2z = 0.$$

We want a point on the end-effector to stay on this surface and a line in the end-effector to coincide with the normal to the surface. This example is not as artificial as it appears at first sight. Many robot tasks require the end effector to follow a curved surface in this way and almost any smooth surface can be locally approximated by such a quadric, see [81, Chap. V sec.3]. The constants k_x and k_y give the principle curvatures of the surface and we can always place the surface at a more general position and orientation using a rigid body transformation.

Now a point on the quadratic surface can be found by a rigid-body motion of the origin,

$$(h_0 + \varepsilon h_1)(1 + \varepsilon 0)(h_0^* - \varepsilon h_1^*) = (1 + \varepsilon(xi + yj + zk)).$$

Using the parameterisation for dual quaternions introduced in Section 9.3 this gives

$$x = 2(a_0 b_1 - a_1 b_0 + a_2 b_3 - a_3 b_2),$$
$$y = 2(a_0 b_2 - a_1 b_3 - a_2 b_0 + a_3 b_1),$$
$$z = 2(a_0 b_3 + a_1 b_2 - a_2 b_1 - a_3 b_0).$$

Substituting into the original equation for the surface gives an equation for the set of motions that keep the point on the surface, we have

$$2k_x(a_0 b_1 - a_1 b_0 + a_2 b_3 - a_3 b_2)^2 + 2k_y(a_0 b_2 - a_1 b_3 - a_2 b_0 + a_3 b_1)^2$$
$$- (a_0 b_3 + a_1 b_2 - a_2 b_1 - a_3 b_0)(a_0^2 + a_1^2 + a_2^2 + a_3^2) = 0.$$

Notice that we have multiplied the z-term by $a_0^2 + a_1^2 + a_2^2 + a_3^2$ to keep the equation homogeneous. For the group of unit dual quaternions this makes no difference since $a_0^2 + a_1^2 + a_2^2 + a_3^2 = 1$, so when we project to the Study quadric, the only points affected are those on the infinite 3-plane $a_0 = a_1 = a_2 = a_3 = 0$.

In order that the end-effector be aligned with the normal to the surface we need the normal direction at any point. This is given by the gradient

$$\nabla(k_x x^2 + k_y y^2 - 2z) = \begin{pmatrix} 2k_x x \\ 2k_y y \\ -2 \end{pmatrix}.$$

As written, this is not a unit vector. At the origin the normal vector lies on the z-axis, so after a rigid transformation the vector will be proportional to

$$h_0 k h_0^* = 2(a_0 a_2 + a_1 a_3)i + 2(a_2 a_3 - a_0 a_1)j + (a_0^2 - a_1^2 - a_2^2 + a_3^2)k.$$

Comparing the two vectors gives the equations

$$k_x(a_0^2 - a_1^2 - a_2^2 + a_3^2)(a_0 b_1 - a_1 b_0 + a_2 b_3 - a_3 b_2) + 2(a_0 a_2 + a_1 a_3)(a_0^2 + a_1^2 + a_2^2 + a_3^2) = 0,$$

$$k_y(a_0^2 - a_1^2 - a_2^2 + a_3^2)(a_0 b_2 - a_1 b_3 - a_2 b_0 + a_3 b_1) + 2(a_2 a_3 - a_0 a_1)(a_0^2 + a_1^2 + a_2^2 + a_3^2) = 0.$$

In summary, we have three quartic equations. The intersection of these varieties with the Study quadric defines the three-dimensional space of possible motions that the robot's end-effector can undergo. Notice that the 3-plane $a_0 = a_1 = a_2 = a_3 = 0$ lies in all the equations but, of course, these points do not correspond to physical motions.

Now the freedom screws are not constant but change from point to point in the constraint subspace. At the identity the freedom screws have a basis

$$\mathbf{z}_1(0,0) = \begin{pmatrix} 0 \\ 0 \\ 1 \\ 0 \\ 0 \\ 0 \end{pmatrix}, \qquad \mathbf{z}_2(0,0) = \begin{pmatrix} 0 \\ 0 \\ 0 \\ 1 \\ 0 \\ 0 \end{pmatrix}, \qquad \mathbf{z}_3(0,0) = \begin{pmatrix} 0 \\ 0 \\ 0 \\ 0 \\ 1 \\ 0 \end{pmatrix}.$$

This is the same as our second example above. This is because the tangent plane to the quadric surface at the origin is simply the xy-plane. More generally, at a point on the three-dimensional subspace corresponding to the point (x, y) on the quadric surface, we have

$$
\mathbf{z}_1(x, y) = \begin{pmatrix} -k_x x \\ -k_y y \\ 1 \\ y(1 + k_y z) \\ -x(1 + k_x z) \\ (k_x - k_y)xy \end{pmatrix}, \quad
\mathbf{z}_2(x, y) = \begin{pmatrix} 0 \\ 0 \\ 0 \\ 1 \\ 0 \\ k_x x \end{pmatrix}, \quad
\mathbf{z}_3(x, y) = \begin{pmatrix} 0 \\ 0 \\ 0 \\ k_x k_y xy \\ -(1 + k_x^2 x^2) \\ -k_y y \end{pmatrix},
$$

where z is given by the equation of the surface, $z = (k_x x^2 + k_y y^2)/2$. The first of these generates a rotation about a line in the direction of the normal vector through the point (x, y, z). The other screws generate translations normal to this line. This may not be the most convenient basis to use in all circumstances. Notice that, although the constraint subspace is three-dimensional, the tangent space depends on only two parameters, x and y.

A basis for the space of constraint wrenches is

$$
\mathcal{W}_4(x, y) = \begin{pmatrix} y(1 + k_y z) \\ -x(1 + k_x z) \\ (k_x - k_y)xy \\ -k_x x \\ -k_y y \\ 1 \end{pmatrix}, \quad
\mathcal{W}_5(x, y) = \begin{pmatrix} 1 \\ 0 \\ k_x x \\ 0 \\ 0 \\ 0 \end{pmatrix}, \quad
\mathcal{W}_6(x, y) = \begin{pmatrix} k_x k_y xy \\ -(1 + k_x^2 x^2) \\ -k_y y \\ 0 \\ 0 \\ 0 \end{pmatrix}.
$$

Next we look at the equations of motion for a single rigid body subject to holonomic constraints.

14.2.2   Constrained Dynamics of a Rigid Body

Consider a single rigid body subjected to holonomic constraint. With the constraint specified it is possible, in principle at least, to reduce the problem to a subspace of the Study quadric and hence eliminate the wrenches of constraint. Finding coordinates for such a subspace in all but the simplest situations is a formidable problem. So we take an approach that is close to the standard theory of Lagrange multipliers, see [39].

We begin with the equations of motion for a rigid body subject to external forces, see Section 13.2. The external forces are of two types, the constraint wrenches and other external forces, such as gravity. Hence we can write

$$
N\ddot{\mathbf{q}} + \{\dot{\mathbf{q}}, N\dot{\mathbf{q}}\} = \mathcal{F} + \sum_{j=n+1}^{6} \lambda_j \mathcal{W}_j.
$$

Here, N is the inertia matrix of the body, $\dot{\mathbf{q}}$ its velocity screw and $\mathcal{F}$ is the set of external wrenches. The wrenches $\mathcal{W}_i$ form a basis for the system of constraint wrenches with the λ_j multipliers corresponding to the amplitudes of the basis wrenches.

The multipliers are to be considered at a set of $6 - n$ extra variables, hence we need an additional $6 - n$ equations to specify the system completely. These extra equations are given by the principle of virtual work

$$\mathcal{W}_i^T \dot{\mathbf{q}} = 0, \qquad i = n+1, \ldots, 6,$$

that is, the motion of the body does no work on the constraint wrenches. To put it another way, the velocity screw of the body lies in the space of freedom screws and so is consistent with the constraints.

If the constraints are holonomic, then it should be possible to write the equations of motion in terms of coordinates on the constraint subspace. That is, we should be able to eliminate the λ_is. To do this we pair the equation of motion with the freedom screws $\mathbf{z}_i$ to get the n equations

$$\ddot{\mathbf{q}}^T N \mathbf{z}_i + \dot{\mathbf{q}}^T N [\mathbf{z}_i, \dot{\mathbf{q}}] = \mathcal{F}^T \mathbf{z}_i, \qquad i = 1, \ldots, n.$$

In some circumstances we may want to find the constraint wrenches. To do this we need to introduce the idea of dual bases of screws and wrenches. Let $\mathbf{z}_1, \ldots, \mathbf{z}_6$ be a basis of screws generating all of $se(3)$, and let $\mathcal{W}_1, \ldots, \mathcal{W}_6$ be a basis for the dual to the Lie algebra. Further, suppose that

$$\mathcal{W}_i^T \mathbf{z}_j = \begin{cases} 1, & \text{if } i = j, \\ 0, & \text{if } i \neq j. \end{cases}$$

We call such a system of screws and wrenches a **dual basis**. Such a dual basis can be found quite simply using a slight modification of the standard Gramm–Schmidt method of finding orthonormal bases for vector spaces with positive definite metrics.

For our purposes we want, $\mathbf{z}_1, \ldots, \mathbf{z}_n$ to be the basis for the freedom screws and $\mathcal{W}_{n+1}, \ldots, \mathcal{W}_6$ to be the basis for the constraint wrenches. This does not uniquely specify the dual basis since, for example, we can add any linear combination of the freedom screws to $\mathbf{z}_{n+1}, \ldots, \mathbf{z}_6$ without affecting the duality relation above. However, if we pair the wrench equation for the body with any of the screws $\mathbf{z}_{n+1}, \ldots, \mathbf{z}_6$, we will get

$$\lambda_j = \ddot{\mathbf{q}}^T N \mathbf{z}_j + \dot{\mathbf{q}}^T N [\mathbf{z}_j, \dot{\mathbf{q}}] - \mathcal{F}^T \mathbf{z}_j, \qquad j = n+1, \ldots, 6.$$

These coefficients are independent of the choice of dual basis. That is, if we add any linear combination of the freedom screws to our screws $\mathbf{z}_{n+1}, \ldots, \mathbf{z}_6$, we get the same results.

14.2.3 Constrained Serial Robots

We can apply the theory above to find the equations of motion for a constrained robot. For definiteness we will assume that the robot has six joints. For each link we get an equation similar to that given above,

$$N_6\ddot{q}_6 + \{\dot{q}_6, N_6\dot{q}_6\} = T_6 + R_6 + G_6 + \sum_{j=n+1}^{6} \lambda_j W_j,$$

$$N_5\ddot{q}_5 + \{\dot{q}_5, N_5\dot{q}_5\} = T_5 + R_5 + G_5 - T_6 - R_6,$$

$$\vdots$$

$$N_1\ddot{q}_1 + \{\dot{q}_1, N_1\dot{q}_1\} = T_1 + R_1 + G_1 - T_2 - R_2.$$

As before we manipulate these equations to eliminate the reaction wrenches at the joints R_i by adding the equations cumulatively, starting from the top, and pairing with the appropriate joint screw s_i.

This gives

$$\sum_{j=i}^{6} \left(\ddot{q}_j^T N_j s_i + \dot{q}_j^T N_j [s_i, \dot{q}_j] - G_j^T s_i \right) = \tau_i + \sum_{k=n+1}^{6} \lambda_k W_k^T s_i, \qquad i = 1, 2, \ldots, 6.$$

The extra $6 - n$ equations needed to make the system determinate are supplied by requiring that the end-effector satisfy the constraints

$$W_k^T \dot{q}_6 = 0, \qquad k = n+1, \ldots, 6.$$

To eliminate the λ_ks we write the screws of freedom in terms of the joint screws as

$$z_i = \sum_{j=1}^{6} \alpha_{ij} s_j, \qquad i = 1, 2, \ldots, n.$$

The coefficients α_{ij} are functions of position on the constraint space. They can be found from the kinematics of the robot. Where the robot is singular a further constraint on end-effector motion is acquired. This may or may not lie in the constraint space, but in either case the formalism extends in the obvious way to account for it.

Since the constraint wrenches are dual to the freedom screws we can multiply the equations of motion by the coefficients α_{ij} to produce n equations without the λ_js, that is,

$$\sum_{i=1}^{6} \alpha_{ki} \left(\sum_{j=i}^{6} \left(\ddot{q}_j^T N_j s_i + \dot{q}_j^T N_j [s_i, \dot{q}_j] - G_j^T s_i \right) \right) = \sum_{i=1}^{6} \alpha_{ki} \tau_i, \quad k = 1, 2, \ldots, n.$$

After re-ordering the sum and relabelling some of the indices this simplifies slightly to

$$\sum_{j=1}^{6} \sum_{k=1}^{j} \left(\ddot{q}_j^T N_j (\alpha_{ik} s_k) + \dot{q}_j^T N_j [(\alpha_{ik} s_k), \dot{q}_j] - G_j^T (\alpha_{ik} s_k) \right) = \sum_{j=1}^{6} \alpha_{ij} \tau_j,$$

$$i = 1, 2, \ldots, n.$$

This represents n equations in the six variables $\theta_1, \theta_2, \ldots, \theta_6$. Hence we need $6 - n$ more equations to make the system determinate. As before, we can use the constraint equations

$$\mathcal{W}_k^T \dot{\mathbf{q}}_6 = \mathcal{W}_k^T (\dot{\theta}_1 \mathbf{s}_1 + \dot{\theta}_2 \mathbf{s}_2 + \cdots + \dot{\theta}_6 \mathbf{s}_6) = 0, \qquad k = n+1, \ldots, 6.$$

Finally the magnitudes of the constraint wrenches can be found by a slight extension to the method for a single body above. Given a choice of screws $\mathbf{z}_{n+1}, \ldots, \mathbf{z}_6$ we can extend the definition of α_{ij} so that

$$\mathbf{z}_i = \sum_{j=1}^{6} \alpha_{ij} \mathbf{s}_j \qquad i = n+1, \ldots, 6.$$

Now if we multiply the original equations of motion by α_{ij}, summing over j will pick out the λ_i term when $i > n$,

$$\lambda_i = \sum_{j=1}^{6} \left(\sum_{k=1}^{j} (\ddot{\mathbf{q}}_j^T N_j (\alpha_{ik} \mathbf{s}_k) + \dot{\mathbf{q}}_j^T N_j [(\alpha_{ik} \mathbf{s}_k), \dot{\mathbf{q}}_j] - \mathcal{G}_j^T (\alpha_{ik} \mathbf{s}_k)) - \alpha_{ij} \tau_j \right),$$

$$i = n+1, \ldots, 6.$$

In Section 15.4 below, these ideas will be extended to investigate force/position control of serial robots.

14.3   Constrained Trees and Stars

Next we look at the equations of motion of mechanisms derived from tree or star-structured mechanisms by applying appropriate holonomic constraints to the leaf-links. Often, a parallel mechanism can be considered as either a constrained tree or a constrained star. The choice of which view to take will usually be determined by the problem.

To carry through the plan we need analogues of screws of freedom and constraint wrenches for tree and star-structured mechanisms.

14.3.1   Systems of Freedom

Each constrained leaf-link has a system of screws of freedom associated with it, that is a linear space of possible velocities that the leaf-link can assume. However, the systems of freedom at different links may not be independent. As we saw in Section 14.1.2, the possible velocities of a leaf-link are given by the span of the joint screws of all joints in the ancestry of the link. Because two or more leaf-links cannot have velocities that imply different velocities for a common ancestor, it makes sense to consider a possible freedom of the mechanism as a consistent allocation of velocities to the links.

In Figure 14.1 we gave a simple example of a tree-structured mechanism. The velocities of the mechanism can be recorded using nested parenthesis notation, see [64, Section 2.3.2]. A general velocity has the form

$$\mathbf{v} = (\dot{\theta}_1(\dot{\theta}_2, \ \dot{\theta}_3(\dot{\theta}_4(\dot{\theta}_6), \dot{\theta}_5))).$$

In this notation the velocities of child links are written in parentheses after the velocity of the parent. Notice that these velocities add vectorially and can be multiplied by scalars. From these properties it is not difficult to show that they form a vector space. We can think of this vector space as a tangent space to the configuration space of the mechanism. Hence the dimension of the vector space is the same as that of the configuration space and gives the total number of (instantaneous) degrees of freedom of the tree.

If the constraints we impose are holonomic, they define a subspace of the configuration space and a linear subspace of the velocity vectors. To find this vector subspace, that is, to find a set of basis vectors for it, it is necessary to specify the system of constraint wrenches at each leaf-link. The constraint wrenches are workless so we get a linear equation for each constraint wrench $\mathcal{W}_k$ acting on a leaf-link l, that is,

$$\mathcal{W}_k^T \dot{\mathbf{q}}_l = 0 = \sum_{1 \le j \le l} \dot{\theta}_j \mathcal{W}_k^T \mathbf{s}_j.$$

The sum is over the ancestry of the leaf-link. This will give a system of linear homogeneous equations in the velocities $\dot{\theta}_j$. The solutions

$$\mathbf{v} = (\dot{\theta}_1, (\dot{\theta}_2, \ldots) \ldots) = (\alpha_1, (\alpha_2, \ldots) \ldots),$$

give the subspace of freedoms.

We could investigate the cotangent space to the configuration space but we are more interested in the constraint wrenches at the leaf-links. That is, we prefer to isolate a particular constraint wrench and produce a single equation that gives its magnitude. We will see how to do this at the end of the next section.

14.3.2   Parallel Mechanisms

A parallel mechanism here will be a linkage that can be formed by constraining the leaf-links of a tree or star. Such linkages usually have multiple loops, but this is not the most general kind of mechanism that one could invent. To analyse the dynamics of these parallel machines we proceed by including, for each constrained link, the reaction wrenches produced by the constraints. So at leaf-link l we have an equation of motion

$$N_l \ddot{\mathbf{q}}_l + \{\dot{\mathbf{q}}_l, N_l \dot{\mathbf{q}}_l\} = \mathcal{T}_l + \mathcal{R}_l + \mathcal{G}_l + \sum_j \lambda_j \mathcal{W}_j.$$

Here the $\mathcal{W}_j$s are a basis for the constraint wrenches at the leaf-link and the λ_j are amplitudes. Following the procedure in Section 14.1.1, the reaction wrenches at the joints can be eliminated by adding the link equations cumulatively according to the tree structure and then pairing with the joint screw. This produces the joint equations

$$\sum_{i \preceq j} \left(\ddot{\mathbf{q}}_j^T N_j \mathbf{s}_i + \dot{\mathbf{q}}_j^T N_j [\mathbf{s}_i, \dot{\mathbf{q}}_j] - \mathcal{G}_j^T \mathbf{s}_i \right) = \tau_i + \sum_{i \preceq k} \lambda_k \mathcal{W}_k^T \mathbf{s}_i, \qquad \text{all } i.$$

The sum on the right-hand side here is over all constraints on all leaf-links above link i.

Our task now is to eliminate the constraint amplitudes λ_k. Suppose that we have a basis for a system of freedoms, with a typical basis vector given by

$$\mathbf{v}_a = (\alpha_{a1}, (\alpha_{a2}, \ldots) \ldots).$$

The constraint terms can be eliminated from the equations of motion by adding joint equations according to freedom vectors

$$\sum_i \alpha_{ai} \left(\sum_{i \preceq j} (\ddot{\mathbf{q}}_j^T N_j \mathbf{s}_i + \dot{\mathbf{q}}_j^T N_j [\mathbf{s}_i, \dot{\mathbf{q}}_j] - \mathcal{G}_j^T \mathbf{z}_i) \right) = \sum_i \alpha_{ai} \tau_i, \qquad \text{all } a.$$

This works because a freedom vector will produce velocities at the leaf-links that annul any system of constraint wrenches. Recall that

$$\sum_{1 \preceq j \preceq l} \alpha_{aj} \mathcal{W}_k^T \mathbf{s}_j = 0.$$

This procedure gives us the same number of equations as there are freedoms, one for each basis element. To make the system determinate extra equations are needed and these are found from the constraint equations,

$$\mathcal{W}_k^T \dot{\mathbf{q}}_l = 0 = \sum_{1 \preceq j \preceq l} \dot{\theta}_j \mathcal{W}_k^T \mathbf{s}_j,$$

where $\mathcal{W}_k$ are the constraint wrenches acting on the leaf-link l. The total number of equations is now equal to the total number of unconstrained freedoms, that is, the dimension of the configuration space of the original tree-structured mechanism.

To find the magnitude of the constraint wrench $\mathcal{W}_k$ acting on leaf-link l we need a vector of the same type as the freedom vectors,

$$\mathbf{v}_k = (\alpha_{k1}, (\alpha_{k2}, \ldots) \ldots),$$

which satisfies the following system of linear equations:

$$\sum_{1 \preceq j \preceq l} \alpha_{kj} \mathcal{W}_k^T \mathbf{s}_j = 1, \qquad \text{and} \qquad \sum_{1 \preceq j \preceq t} \alpha_{kj} \mathcal{W}_m^T \mathbf{s}_j = 0, \quad \text{if } m \neq k,$$

where the index t here refers to a leaf-link, not necessarily l. These equations will not have unique solutions since we can always add an element of the kernel, that is, a freedom of the constrained mechanism. However, if we take any solution and combine it with the joint equations we get

$$\sum_i \alpha_{ki} \left(\sum_{i \preceq j} (\ddot{\mathbf{q}}_j^T N_j \mathbf{s}_i + \dot{\mathbf{q}}_j^T N_j [\mathbf{s}_i, \dot{\mathbf{q}}_j]) - \mathcal{G}_j^T \mathbf{s}_i) - \tau_i \right) = \lambda_k.$$

Finally here, the **shaking wrench** of a mechanism is the force and moment imposed on the environment by the mechanism as it moves. For star-structured mechanisms, it is simple to see that this is given by adding together all the link equations

$$\sum_{1 \preceq j} (N_j \ddot{\mathbf{q}}_j + \{\dot{\mathbf{q}}_j, N_j \dot{\mathbf{q}}_j\} - \mathcal{G}_j) = \mathcal{W}_{TOTAL}.$$

14.4   Dynamics of Planar 4-Bars

As mentioned in Section 5.4 the planar 4-bar mechanism is ubiquitous in applications. Hence its dynamics has been closely studied by mechanical engineers, see for example [110, §13-9]. However, work in this area usually concentrates on the problem of deriving models that are computationally efficient.

Here we model the 4-bar as a star structure with two leaf links each constrained to circle fixed points. The equations of motion will be derived and so will relations for the shaking forces and moments.

Since this is a planar problem we don't need to use the full six-dimensional screw theory. We can assume that the plane of the mechanism is the xy-plane and restrict our representation of velocities and wrenches to be 3-component vectors. So here a generalised velocity and a wrench will take the form,

$$\mathbf{s} = \begin{pmatrix} \omega_z \\ v_x \\ v_y \end{pmatrix}, \qquad \mathcal{W} = \begin{pmatrix} \tau_z \\ F_x \\ F_y \end{pmatrix}.$$

Note that this is slightly different from the notation given in Section 5.3, however the difference is not significant. Using the above ordering of the components means that we can simply take the middle components from the six-dimensional spatial theory.

A general inertia matrix will be a 3×3 matrix of the form

$$N = \begin{pmatrix} mk^2 & -mc_y & mc_x \\ -mc_y & m & 0 \\ mc_x & 0 & m \end{pmatrix}$$

where m is the mass of the body, k is the radius of gyration, c_x and c_y are the x and y-components of the body's centre of mass. Note that mk^2 is equivalent to I_{33} in the spatial case.

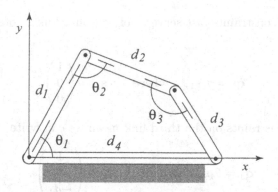

FIGURE 14.2. A Four-Bar Mechanism

Figure 14.2 shows the four-bar in a general position. For simplicity we will assume that the linkage operates in a horizontal plane and hence ignore gravity. We will also assume that the mechanism is being driven by a motor at the first joint, this motor provides a torque given by

$$\mathcal{T} = \begin{pmatrix} \tau \\ 0 \\ 0 \end{pmatrix}.$$

We have three link equations, for the first link or bar,

$$\ddot{\mathbf{q}}_1^T N_1 \mathbf{s}_2 + \dot{\mathbf{q}}_1^T N_1 [\mathbf{s}_2, \dot{\mathbf{q}}_1] = \mathcal{T}^T \mathbf{s}_2 + \mathcal{R}_1^T \mathbf{s}_2$$

where $\dot{\mathbf{q}}_i$ is the velocity screw of the ith link, N_i is the inertia matrix of the ith link and $\mathbf{s}_i$ is the ith joint screw. The reactions at the joints are written $\mathcal{R}_i$. For the third link we have the equation

$$\ddot{\mathbf{q}}_3^T N_3 \mathbf{s}_3 + \dot{\mathbf{q}}_3^T N_3 [\mathbf{s}_3, \dot{\mathbf{q}}_3] = \mathcal{R}_4^T \mathbf{s}_3.$$

The second link is the root link and we sum the equations for all the links to get rid of the reaction wrenches at joints 2 and 3. This gives the equation

$$\sum_{i=1}^{3} \left(N_i \ddot{\mathbf{q}}_i + \{\dot{\mathbf{q}}_i, N_i \dot{\mathbf{q}}_i\} \right) = \mathcal{T} + \mathcal{R}_1 + \mathcal{R}_4.$$

Next we look at the constraint equations at joint 1 we have

$$\mathcal{W}_1^T \dot{\mathbf{q}}_1 = 0, \qquad \text{and} \qquad \mathcal{W}_2^T \dot{\mathbf{q}}_1 = 0,$$

where the constraint wrenches are

$$\mathcal{W}_1 = \begin{pmatrix} 0 \\ 1 \\ 0 \end{pmatrix}, \qquad \text{and} \qquad \mathcal{W}_2 = \begin{pmatrix} 0 \\ 0 \\ 1 \end{pmatrix}.$$

Of course these constraints just serve to ensure that link 1 rotates about joint 1, so we can write

$$\dot{\mathbf{q}}_1 = \dot{\theta}_1 \mathbf{s}_1, \quad \text{where} \quad \mathbf{s}_1 = \begin{pmatrix} 1 \\ 0 \\ 0 \end{pmatrix}.$$

Likewise, the constraints on the third link mean we can write

$$\dot{\mathbf{q}}_3 = \dot{\theta}_4 \mathbf{s}_4, \quad \text{where} \quad \mathbf{s}_4 = \begin{pmatrix} 1 \\ 0 \\ -d_4 \end{pmatrix}.$$

Remember that the second and third components here correspond to the x and y-components of the moment of the joint axis. Now another way of expressing these constraints is by looking at the velocity of the second link, which can be found from the first and last pair of joints, giving us the equations

$$\dot{\mathbf{q}}_2 = \dot{\theta}_1 \mathbf{s}_1 + \dot{\theta}_2 \mathbf{s}_2 = -\dot{\theta}_3 \mathbf{s}_3 - \dot{\theta}_4 \mathbf{s}_4.$$

This now gives us the loop equation

$$\dot{\theta}_1 \mathbf{s}_1 + \dot{\theta}_2 \mathbf{s}_2 + \dot{\theta}_3 \mathbf{s}_3 + \dot{\theta}_4 \mathbf{s}_4 = \mathbf{0}.$$

Given expressions for the joint screws, we can solve the above equation for the possible velocities. That is, we must solve $\alpha_1 \mathbf{s}_1 + \alpha_2 \mathbf{s}_2 + \alpha_3 \mathbf{s}_3 + \alpha_4 \mathbf{s}_4 = \mathbf{0}$, with

$$\mathbf{s}_1 = \begin{pmatrix} 1 \\ 0 \\ 0 \end{pmatrix} \quad \mathbf{s}_2 = \begin{pmatrix} 1 \\ d_1 \sin\theta_1 \\ -d_1 \cos\theta_1 \end{pmatrix}$$

$$\mathbf{s}_3 = \begin{pmatrix} 1 \\ d_1 \sin\theta_1 + d_2 \sin(\theta_1 + \theta_2) \\ -d_1 \cos\theta_1 - d_2 \cos(\theta_1 + \theta_2) \end{pmatrix} \quad \mathbf{s}_4 = \begin{pmatrix} 1 \\ 0 \\ -d_4 \end{pmatrix}.$$

The solution is easily found to be

$$\alpha_1 = d_1 d_4 \sin\theta_2 - d_2 d_4 \sin(\theta_1 + \theta_2),$$
$$\alpha_2 = d_1 d_4 \sin\theta_1 + d_2 d_4 \sin(\theta_1 + \theta_2),$$
$$\alpha_3 = -d_1 d_4 \sin\theta_1,$$
$$\alpha_4 = -d_1 d_2 \sin\theta_2.$$

Since the equations are homogeneous, any constant multiple of these results will also be a solution.

These results can be used to eliminate the reaction wrenches from the link equations. We multiply the equations for the first and third links by $-\alpha_2$ and α_3

respectively. The equation for the root link must be paired with $\alpha_1 s_1 + \alpha_2 s_2 = -\alpha_3 s_3 - \alpha_4 s_4$. Adding these equations we get the left-hand-side,

$$
\begin{aligned}
(\ddot{\mathbf{q}}_1^T N_1 \alpha_1 s_1 &+ \dot{\mathbf{q}}_1^T N_1 [\alpha_1 s_1, \, \dot{\mathbf{q}}_1]) \\
&+ (\ddot{\mathbf{q}}_2^T N_2 (\alpha_1 s_1 + \alpha_2 s_2) + \dot{\mathbf{q}}_2^T N_2 [\alpha_1 s_1 + \alpha_2 s_2, \, \dot{\mathbf{q}}_2]) \\
&- (\ddot{\mathbf{q}}_3^T N_3 \alpha_4 s_4 + \dot{\mathbf{q}}_3^T N_3 [\alpha_4 s_4, \, \dot{\mathbf{q}}_3]).
\end{aligned}
$$

Notice here that since $\dot{\mathbf{q}}_1 \propto s_1$, $\dot{\mathbf{q}}_3 \propto s_4$ and $\dot{\mathbf{q}}_2 \propto \alpha_1 s_1 + \alpha_2 s_2$, the Lie bracket terms all vanish. On the right-hand-side we get

$$
\alpha_1 \mathcal{T}^T s_1 + \alpha_1 \mathcal{R}_1^T s_1 - \alpha_4 \mathcal{R}_4^T s_4.
$$

The last two terms here vanish because the reaction wrench can do no work on the joint screw. The first term is easily seen to be $\alpha_1 \tau$, a multiple of the motor torque. Putting this together we get

$$
\ddot{\mathbf{q}}_1^T N_1 \alpha_1 s_1 + \ddot{\mathbf{q}}_2^T N_2 (\alpha_1 s_1 + \alpha_2 s_2) - \ddot{\mathbf{q}}_3^T N_3 \alpha_4 s_4 = \alpha_1 \tau.
$$

This equation, together with the loop equation $\dot{\theta}_1 s_1 + \dot{\theta}_2 s_2 + \dot{\theta}_3 s_3 + \dot{\theta}_4 s_4 = \mathbf{0}$, determine the dynamics of the 4-bar. Notice that the loop equations could be rewritten using the α_is we found above. Clearly $\dot{\theta}_i \propto \alpha_i$, hence we could split the loop equation into scalar equations of the form $\dot{\theta}_4 \alpha_1 = \dot{\theta}_1 \alpha_4$. In this way it would be possible to eliminate θ_4 from our equations; notice they are already independent of θ_3. However, this will not be pursued here except to state that given a particular 4-bar mechanism, that is, its design parameters, masses and inertias, we could expand the expressions given above and solve the dynamics numerically in a relatively straightforward manner.

For high speed mechanisms it is important to minimise the shaking forces and shaking moment. In the literature many methods for balancing mechanisms have been reported, see for example [110, Chap. 15]. To investigate the shaking wrench we must go back to the link equation for the root link

$$
\sum_{i=1}^{3} (N_i \ddot{\mathbf{q}}_i + \{\dot{\mathbf{q}}_i, \, N_i \dot{\mathbf{q}}_i\}) = \mathcal{T} + \mathcal{R}_1 + \mathcal{R}_4.
$$

The right-hand side of this equation gives the wrench acting on the ground link or frame of the mechanism, this is the shaking wrench. To isolate the shaking moment we can pair the equation above with the screw which picks out the first component of any wrench, this is of course just s_1. So after a little rearrangement we can write

$$
\begin{aligned}
\ddot{\theta}_1 s_1^T N_1 s_1 &+ \ddot{\theta}_1 s_1^T N_2 s_1 + \ddot{\theta}_2 s_1^T N_2 s_2 + \ddot{\theta}_4 s_1^T N_3 s_4 \\
&+ 2\dot{\theta}_1 \dot{\theta}_2 s_1^T N_2 [s_1, \, s_2] + \dot{\theta}_2^2 s_1^T N_2 [s_1, \, s_2] + \dot{\theta}_4^2 s_4^T N_3 [s_1, \, s_4] = \tau + \mathcal{R}_4^T s_1.
\end{aligned}
$$

Clearly we could do the same for the two components of the shaking force, by pairing the equation with the screws $(0, 1, 0)^T$ and $(0, 0, 1)^T$. To make any progress we must incorporate the constraints but this problem will not be developed further here.

14.5 Biped Walking

Over the past few years walking machines have attracted a lot of interest in the robotics community. Several impressive machines have been built. However, controlling these machines, especially bipeds, remains a challenging problem. In this brief section we simply look at deriving the equations of motion for such a device, focusing on the problem of detecting when a foot breaks contact with the ground.

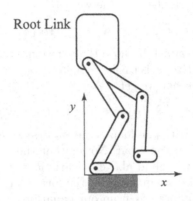

FIGURE 14.3. A Planar Biped Robot

Most studies of bipedal walking concentrate on planar devices for simplicity. We do the same here in order to explain how the methods developed above apply to this problem. Our discussion focusses on the planar walking machine shown in Figure 14.3. Extending these ideas to spatial walking mechanisms is straightforward though with more degrees-of-freedom, more variables and equations are to be expected.

Though we can think of a walking machine as a serial robot with one foot as the ground and the other as the end-effector, it is more symmetrical and more convenient to take the torso as the root link of a star-structured device so that both feet are leaf-links. A feature of these devices is the discontinuity that occurs when contact between ground and foot is broken. The robot has different equations of motion depending on whether one, two or no feet are in contact with the ground. To determine when the equations of motion change we need to monitor the magnitude of the (one-sided) constraint wrench acting on the feet. When the vertical component of this wrench disappears the foot is about to break contact with the ground.

In the previous section we saw how to restrict the spatial theory we have developed to planar motion. Recall that for motion in the xy-plane, screws and wrenches can be considered three-dimensional vectors

$$\mathbf{s} = \begin{pmatrix} \omega_z \\ v_x \\ v_y \end{pmatrix}, \qquad \mathcal{W} = \begin{pmatrix} M_z \\ F_x \\ F_y \end{pmatrix}$$

and so forth.

Now consider the biped shown in Figure 14.3. This machine has seven links. Two of the links are feet and in a practical implementation frictional forces will act whenever the feet are in contact with the ground. To keep the analysis simple we assume that, if a foot contacts the ground, it is fixed in such a way that the two-link-leg is constrained to circle the ankle joint. In other words we ignore the frictional forces or rather we take the frictional forces as granted.

Suppose that the position of the hip joints are $(x_h, y_h) = (x_{Rh}, y_{Rh}) = (x_{Lh}, y_{Lh})$ and right and left knee and ankle joints are at the points (x_{Rk}, y_{Rk}), (x_{Lk}, y_{Lk}), (x_{Ra}, y_{Ra}) and (x_{La}, y_{La}). So the joint screws are

$$s_h = \begin{pmatrix} 1 \\ y_h \\ -x_h \end{pmatrix}$$

and so forth. We get four joint equations. One for each lower leg,

$$\left(\ddot{\mathbf{q}}_{Rk}^T N_{Rk} \mathbf{s}_{Rk} + \dot{\mathbf{q}}_{Rk}^T N_{Rk}[\mathbf{s}_{Rk}, \dot{\mathbf{q}}_{Rk}] - \mathcal{G}_{Rk}^T \mathbf{s}_{Rk}\right)$$
$$= \tau_{Rk} + \lambda_{Rx} \mathcal{W}_{Rx}^T \mathbf{s}_{Rk} + \lambda_{Ry} \mathcal{W}_{Ry}^T \mathbf{s}_{Rk}$$
$$\left(\ddot{\mathbf{q}}_{Lk}^T N_{Lk} \mathbf{s}_{Lk} + \dot{\mathbf{q}}_{Lk}^T N_{Lk}[\mathbf{s}_{Lk}, \dot{\mathbf{q}}_{Lk}] - \mathcal{G}_{Lk}^T \mathbf{s}_{Lk}\right)$$
$$= \tau_{Lk} + \lambda_{Lx} \mathcal{W}_{Lx}^T \mathbf{s}_{Lk} + \lambda_{Ly} \mathcal{W}_{Ly}^T \mathbf{s}_{Lk}$$

and one for each thigh,

$$\left(\ddot{\mathbf{q}}_{Rk}^T N_{Rk} \mathbf{s}_h + \dot{\mathbf{q}}_{Rk}^T N_{Rk}[\mathbf{s}_h, \dot{\mathbf{q}}_{Rk}] - \mathcal{G}_{Rk}^T \mathbf{s}_h\right)$$
$$+ \left(\ddot{\mathbf{q}}_{Rh}^T N_{Rh} \mathbf{s}_h + \dot{\mathbf{q}}_{Rh}^T N_{Rh}[\mathbf{s}_h, \dot{\mathbf{q}}_{Rh}] - \mathcal{G}_{Rh}^T \mathbf{s}_h\right)$$
$$= \tau_{Rh} + \lambda_{Rx} \mathcal{W}_{Rx}^T \mathbf{s}_{Rh} + \lambda_{Ry} \mathcal{W}_{Ry}^T \mathbf{s}_{Rh}$$
$$\left(\ddot{\mathbf{q}}_{Lk}^T N_{Lk} \mathbf{s}_h + \dot{\mathbf{q}}_{Lk}^T N_{Lk}[\mathbf{s}_h, \dot{\mathbf{q}}_{Lk}] - \mathcal{G}_{Lk}^T \mathbf{s}_h\right)$$
$$+ \left(\ddot{\mathbf{q}}_{Lh}^T N_{Lh} \mathbf{s}_h + \dot{\mathbf{q}}_{Lh}^T N_{Lh}[\mathbf{s}_h, \dot{\mathbf{q}}_{Lh}] - \mathcal{G}_{Lh}^T \mathbf{s}_h\right)$$
$$= \tau_{Lh} + \lambda_{Lx} \mathcal{W}_{Lx}^T \mathbf{s}_{Lh} + \lambda_{Ly} \mathcal{W}_{Ly}^T \mathbf{s}_{Lh}.$$

The constraint wrenches here are forces on the ankle joints. For the right ankle we have

$$\mathcal{W}_{Ry} = \begin{pmatrix} x_{Ra} \\ 0 \\ 1 \end{pmatrix} \quad \text{and} \quad \mathcal{W}_{Rx} = \begin{pmatrix} -y_{Ra} \\ 1 \\ 0 \end{pmatrix}$$

and similarly for the left ankle. The equation for the root link, the torso, is

$$N_T \ddot{\mathbf{q}}_T + \{N_T \dot{\mathbf{q}}_T, \dot{\mathbf{q}}_T\} - \mathcal{G}_T + N_{Rh} \ddot{\mathbf{q}}_{Rh} + \{N_{Rh} \dot{\mathbf{q}}_{Rh}, \dot{\mathbf{q}}_{Rh}\} - \mathcal{G}_{Rh}$$
$$+ N_{Lh} \ddot{\mathbf{q}}_{Lh} + \{N_{Lh} \dot{\mathbf{q}}_{Lh}, \dot{\mathbf{q}}_{Lh}\} - \mathcal{G}_{Lh} + N_{Rk} \ddot{\mathbf{q}}_{Rk} + \{N_{Rk} \dot{\mathbf{q}}_{Rk}, \dot{\mathbf{q}}_{Rk}\} - \mathcal{G}_{Rk}$$
$$+ N_{Lk} \ddot{\mathbf{q}}_{Lk} + \{N_{Lk} \dot{\mathbf{q}}_{Lk}, \dot{\mathbf{q}}_{Lk}\} - \mathcal{G}_{Lk} = \mathcal{W}_{Rx} + \mathcal{W}_{Ry} + \mathcal{W}_{Lx} + \mathcal{W}_{Ly}.$$

Here the subscript T refers to the torso. The constraint equations are simply

$$\mathcal{W}_{Rx}^T \dot{\mathbf{q}}_{Rk} = 0, \quad \mathcal{W}_{Ry}^T \dot{\mathbf{q}}_{Rk} = 0, \quad \mathcal{W}_{Lx}^T \dot{\mathbf{q}}_{Lk} = 0, \quad \mathcal{W}_{Ly}^T \dot{\mathbf{q}}_{Lk} = 0.$$

We could now derive the equations of motion by eliminating the magnitudes of the constraint forces. The process is straightforward but not very illuminating. Instead, we will show how to derive an equation for the vertical constraint force on the right foot. (Of course, the force on the left foot can be found in the same way.) To do this we must first find a system of velocities dual to a vertical force at the right foot. This can be done by solving the following equations for the joint rates:

$$\mathcal{W}_{Rx}^T \dot{\mathbf{q}}_{Rk} = 0, \quad \mathcal{W}_{Ry}^T \dot{\mathbf{q}}_{Rk} = 1, \quad \mathcal{W}_{Lx}^T \dot{\mathbf{q}}_{Lk} = 0, \quad \mathcal{W}_{Ly}^T \dot{\mathbf{q}}_{Lk} = 0.$$

The system is overdetermined, so to keep the computations simple we find the solution such that the torso and left leg are stationary. We are left with the pair of equations

$$\alpha_{Rh} \left(\mathcal{W}_{Rx}^T \mathbf{s}_h \right) + \alpha_{Rk} \left(\mathcal{W}_{Rx}^T \mathbf{s}_{Rk} \right) = 0,$$
$$\alpha_{Rh} \left(\mathcal{W}_{Ry}^T \mathbf{s}_h \right) + \alpha_{Rk} \left(\mathcal{W}_{Ry}^T \mathbf{s}_{Rk} \right) = 1.$$

These simplify to

$$\alpha_{Rh}(x_{Ra} - x_h) + \alpha_{Rk}(x_{Ra} - x_{Rk}) = 0,$$
$$\alpha_{Rh}(y_{Ra} - y_h) + \alpha_{Rk}(y_{Ra} - y_{Rk}) = 1.$$

Their solution is

$$\alpha_{Rh} = (x_{Ra} - x_{Rk})/\Delta, \qquad \alpha_{Rk} = (x_h - x_{Ra})/\Delta$$

where

$$\Delta = (x_{Ra} - x_{Rk})(y_{Ra} - y_h) + (x_h - x_{Ra})(y_{Ra} - y_{Rk}).$$

The equation for the magnitude of the vertical force on the right foot is given by

$$\lambda_{Ry} = (x_{Ra} - x_{Rk})\left(\ddot{\mathbf{q}}_{Rk}^T N_{Rk} \mathbf{s}_h + \dot{\mathbf{q}}_{Rk}^T N_{Rk}[\mathbf{s}_h, \dot{\mathbf{q}}_{Rk}] - \mathcal{G}_{Rk}^T \mathbf{s}_h \right.$$
$$+ \ddot{\mathbf{q}}_{Rh}^T N_{Rh} \mathbf{s}_h + \dot{\mathbf{q}}_{Rh}^T N_{Rh}[\mathbf{s}_h, \dot{\mathbf{q}}_{Rh}] - \mathcal{G}_{Rh}^T \mathbf{s}_h - \tau_{Rh} \Big)/\Delta$$
$$+ (x_h - x_{Ra})\left(\ddot{\mathbf{q}}_{Rk}^T N_{Rk} \mathbf{s}_{Rk} + \dot{\mathbf{q}}_{Rk}^T N_{Rk}[\mathbf{s}_{Rk}, \dot{\mathbf{q}}_{Rk}] - \mathcal{G}_{Rk}^T \mathbf{s}_{Rk} - \tau_{Rk} \right)/\Delta.$$

We could use a similar derivation to find an equation for the horizontal force on the foot and hence, given a model of the friction at the foot, we could monitor whether the foot was about to slip.

It is reasonably clear how to extend this to the case where the foot is on an incline. Apart from taking account of the weight of the foot all we need to do is to use a basis for the constraint wrenches parallel and perpendicular to the slope.

14.6   The Stewart Platform

The next example sketches out an algorithm for computed torque, or rather computed force control of Stewart platform devices. That is, we give an algorithm to compute the forces that the legs must apply so that the platform-body undergoes some desired trajectory. See also the brief discussion at the beginning of section 13.3. Fully-in-parallel actuated devices are being increasingly used for precision tasks such as high precision machining, so being able to compute the inverse dynamics offers the prospect of better overall accuracy.

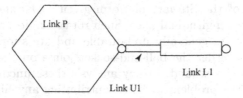

FIGURE 14.4. Platform and One Leg of a Stewart Platform

The Stewart platform consists of a platform-body together with six legs, see Figure 11.1 for a diagram of the general Stewart platform. Each leg is a linear actuator, in practice either a hydraulic ram or a lead screw, which is connected to the platform-body by a passive ball-and-socket joint. A second ball-and-socket joint on each leg connects it to the ground. When the six legs are ungrounded we can consider the mechanism as a star; we choose the platform-body to be the root link and label it P. The grounding of each leg via a ball-and-socket joint appropriately constrains the star. The dynamic equations for such a device contain a large number of variables and a correspondingly large number of equations of motion, this is because of the elaborate connection of platform to ground via the six doubled-ended leg-rods. Giving symmetry to the design can reduce complexity to manageable levels; for instance, pairs of ball-and-socket joints in the platform and base can be coalesced to give an octahedral device. If only pairs of joints on the platform are coalesced we get the 6—3 Stewart platform as in Section 11.5.2. However, the expressions remain complicated requiring the solution of systems of simultaneous linear equations as part of the algorithm. For this reason it is infeasible to write out closed form solutions in full so we limit ourselves here to a sketch of the necessary computations.

We will label the upper part of the hydraulic joints $U1, U2, \ldots, U6$ and the lower parts, that is the parts nearest the ground, $L1, L2, \ldots, L6$. Also the centres of the passive joints will be located at $(x_{U1}, y_{U1}, z_{U1}), \ldots, (x_{U6}, y_{U6}, z_{U6})$ on the platform and $(x_{L1}, y_{L1}, z_{L1}), \ldots, (x_{L6}, y_{L6}, z_{L6})$ on the ground. Figure 14.4 shows the platform and a single leg to illustrate the labelling used here.

To find the equations of motion we consider the lower passive ball-socket joints as constraints. The system of constraint wrenches at the first of these will be given by

$$
W_{1x} = \begin{pmatrix} 0 \\ z_{L1} \\ -y_{L1} \\ 1 \\ 0 \\ 0 \end{pmatrix}, \quad
W_{1y} = \begin{pmatrix} -z_{L1} \\ 0 \\ x_{L1} \\ 0 \\ 1 \\ 0 \end{pmatrix}, \quad
W_{1z} = \begin{pmatrix} y_{L1} \\ -x_{L1} \\ 0 \\ 0 \\ 0 \\ 1 \end{pmatrix}.
$$

Similar expressions give constraints for each of the other five such ball-socket joints.

The description of the Stewart platform given so far includes rotations of the legs about their longitudinal axes. Such rotations do not affect the rest of the mechanism but are generally undesirable and are suppressed in practical machines by keying either the ball-and-socket joints on the platform-body or those connected to the ground. In any analysis these unconstrained rotations add complexity to the problem without contributing anything. Here we ignore them by assuming the passive joints on the platform-body have only two degrees of freedom. We can do this by using Euler angles (or a similar scheme) as coordinates for the configuration of each ball-and-socket joint and then ignoring the final coordinate corresponding to the unwanted rotation.

The joint screws for a typical passive joint are simple enough to write down once we have chosen positions for them in the home configuration. For example we can choose the first joint screw s_{U11}, to be aligned along the axis of the prismatic joint in its home position. Of course the axis of this screw must pass through the centre of the passive joint. Let ϕ_1 denote the joint variable measured from this home position. The second joint screw on the first leg, s_{U12}, would have a joint variable θ_1 and the joint screw would be perpendicular to the first, again with its axis passing through the centre of the ball-and-socket joint. Notice that the home configuration of the mechanism refers to the star mechanism so does not necessarily have to satisfy the constraints.

For the prismatic joint in each leg we have a joint screw

$$
s_{L1} = \begin{pmatrix} 0 \\ 0 \\ 0 \\ (x_{U1} - x_{L1})/M \\ (y_{U1} - y_{L1})/M \\ (z_{U1} - z_{L1})/M \end{pmatrix}
$$

where $M^2 = (x_{U1} - x_{L1})^2 + (y_{U1} - y_{L1})^2 + (z_{U1} - z_{L1})^2$.

For each leg we get three joint equations, one for the prismatic leg actuator and two for the (passive) freedoms of the two ball-and-socket joints connecting the leg to the platform body. For the prismatic joint actuating the leg we have

$$
(\ddot{q}_{L1}^T N_{L1} s_{L1} + \dot{q}_{L1}^T N_{L1}[s_{L1}, \dot{q}_{L1}] - \mathcal{G}_{L1}^T s_{L1})
$$
$$
= F_1 + \lambda_{1x} W_{1x}^T s_{L1} + \lambda_{1y} W_{1y}^T s_{L1} + \lambda_{1z} W_{1z}^T s_{L1}
$$

where F_1 is the magnitude of the hydraulic ram force. The two equations for the freedoms of the ball-and-socket joint connecting to the platform-body are

$$\left(\ddot{\mathbf{q}}_{U1}^T N_{U1}\mathbf{s}_{U11} + \dot{\mathbf{q}}_{U1}^T N_{U1}[\mathbf{s}_{U11}, \dot{\mathbf{q}}_{U1}] - \mathcal{G}_{U1}^T\mathbf{s}_{U11}\right)$$
$$+ \left(\ddot{\mathbf{q}}_{L1}^T N_{L1}\mathbf{s}_{U11} + \dot{\mathbf{q}}_{L1}^T N_{L1}[\mathbf{s}_{U11}, \dot{\mathbf{q}}_{L1}] - \mathcal{G}_{L1}^T\mathbf{s}_{U11}\right)$$
$$= \lambda_{1x}\mathcal{W}_{1x}^T\mathbf{s}_{U11} + \lambda_{1y}\mathcal{W}_{1y}^T\mathbf{s}_{U11} + \lambda_{1z}\mathcal{W}_{1z}^T\mathbf{s}_{U11},$$
$$\left(\ddot{\mathbf{q}}_{U1}^T N_{U1}\mathbf{s}_{U12} + \dot{\mathbf{q}}_{U1}^T N_{U1}[\mathbf{s}_{U12}, \dot{\mathbf{q}}_{U1}] - \mathcal{G}_{U1}^T\mathbf{s}_{U12}\right)$$
$$+ \left(\ddot{\mathbf{q}}_{L1}^T N_{L1}\mathbf{s}_{U12} + \dot{\mathbf{q}}_{L1}^T N_{L1}[\mathbf{s}_{U12}, \dot{\mathbf{q}}_{L1}] - \mathcal{G}_{L1}^T\mathbf{s}_{U12}\right)$$
$$= \lambda_{1x}\mathcal{W}_{1x}^T\mathbf{s}_{U12} + \lambda_{1y}\mathcal{W}_{1y}^T\mathbf{s}_{U12} + \lambda_{1z}\mathcal{W}_{1z}^T\mathbf{s}_{U12}.$$

Notice that these two equations contain no driving forces or torques; this is because the joint is passive. For the platform-body we have the equation

$$N_P\ddot{\mathbf{q}}_P + \{N_P\dot{\mathbf{q}}_P, \dot{\mathbf{q}}_P\} - \mathcal{G}_P$$
$$+ \sum_{i=1}^{6}\left(N_{Ui}\ddot{\mathbf{q}}_{Ui} + \{N_{Ui}\dot{\mathbf{q}}_{Ui}, \dot{\mathbf{q}}_{Ui}\} - \mathcal{G}_{Ui} + N_{Li}\ddot{\mathbf{q}}_{Li} + \{N_{Li}\dot{\mathbf{q}}_{Li}, \dot{\mathbf{q}}_{Li}\} - \mathcal{G}_{Li}\right)$$
$$= \sum_{i=1}^{6}\lambda_{ix}\mathcal{W}_{ix} + \lambda_{iy}\mathcal{W}_{iy} + \lambda_{iz}\mathcal{W}_{iz},$$

where the subscript P refers to the platform link.

Next, we use the constraint equations to eliminate constraint wrenches at the leaf-links. Suppose that the velocity of the root link is given by

$$\dot{\mathbf{q}}_P = \dot{\omega}_{Px}\mathbf{s}_{P\omega x} + \dot{\omega}_{Py}\mathbf{s}_{P\omega y} + \dot{\omega}_{Pz}\mathbf{s}_{P\omega z} + \dot{x}_P\mathbf{s}_{Px} + \dot{y}_P\mathbf{s}_{Py} + \dot{z}_P\mathbf{s}_{Pz}.$$

That is, the screws $\mathbf{s}_{P\omega x}$ and so on, form a basis for the space of all freedoms. Now, for each of the six legs we have three constraint equations,

$$\alpha_{P\omega x}\mathcal{W}_{1x}^T\mathbf{s}_{P\omega x} + \cdots$$
$$\cdots + \alpha_{Pz}\mathcal{W}_{1x}^T\mathbf{s}_{Pz} + \alpha_{U11}\mathcal{W}_{1x}^T\mathbf{s}_{U11} + \alpha_{U12}\mathcal{W}_{1x}^T\mathbf{s}_{U12} + \alpha_{L1}\mathcal{W}_{1x}^T\mathbf{s}_{L1} = 0,$$

$$\alpha_{P\omega x}\mathcal{W}_{1y}^T\mathbf{s}_{P\omega x} + \cdots$$
$$\cdots + \alpha_{Pz}\mathcal{W}_{1y}^T\mathbf{s}_{Pz} + \alpha_{U11}\mathcal{W}_{1y}^T\mathbf{s}_{U11} + \alpha_{U12}\mathcal{W}_{1y}^T\mathbf{s}_{U12} + \alpha_{L1}\mathcal{W}_{1y}^T\mathbf{s}_{L1} = 0,$$

$$\alpha_{P\omega x}\mathcal{W}_{1z}^T\mathbf{s}_{P\omega x} + \cdots$$
$$\cdots + \alpha_{Pz}\mathcal{W}_{1z}^T\mathbf{s}_{Pz} + \alpha_{U11}\mathcal{W}_{1z}^T\mathbf{s}_{U11} + \alpha_{U12}\mathcal{W}_{1z}^T\mathbf{s}_{U12} + \alpha_{L1}\mathcal{W}_{1z}^T\mathbf{s}_{L1} = 0.$$

This gives, in total, $(6 \times 3 =)$ 18 homogeneous linear equations in the 24 unknown αs.

There are two ways to proceed from here. The most direct approach involves setting $\alpha_{L1} = 1$ and $\alpha_{Li} = 0$ for all other legs. These six assignments reduce the number of unknowns from 24 to 18 and then the solution of the 18 linear equations gives the coefficients to multiply the joint equations by. Because we have set $\alpha_{Li} = 0$, where $i \neq 1$, F_1 appears in the resulting equation but none of the other driving forces. Thus we need to repeat the procedure, setting each $\alpha_{Li} = 1$ in turn while the other leaf-joint coefficients are set equal to zero. The result is six equations having the general form

$$\sum \alpha_{ik} \sum (\ddot{\mathbf{q}}_j^T N_j \mathbf{s}_i + \dot{\mathbf{q}}_j^T N_j [\mathbf{s}_i, \dot{\mathbf{q}}_j] - \mathcal{G}_j^T \mathbf{s}_i) = F_k.$$

Unfortunately this is not the end of the story since we still need to eliminate ϕs, θs and ls to get equations that give the leg forces in terms of the motion of the platform alone. To do this we need to go back to the constraint equations. We will see how to do this in a moment, but let us go back a little further and consider a less direct method of producing these equations which may be computationally more efficient.

Rather than choose the coefficients α_{Li} in the constraint equations we could choose to solve them in terms of the velocity of the platform. That is we choose the coefficients $\alpha_{P\omega x}, \ldots, \alpha_{Pz}$ to be 1 or 0 in turn. There are two reasons for this. First the equations we must solve reduce to six sets of 3×3 equations. Second we need this solution again when we eliminate the ϕs, θs and l variables.

Combining these solutions with the joint equations gives equations of motion of the general form

$$\sum \alpha_{ik} \sum (\ddot{\mathbf{q}}_j^T N_j \mathbf{s}_i + \dot{\mathbf{q}}_j^T N_j [\mathbf{s}_i, \dot{\mathbf{q}}_j] - \mathcal{G}_j^T \mathbf{s}_i)$$
$$= \alpha_{11} F_1 + \alpha_{12} F_2 + \alpha_{13} F_3 + \alpha_{14} F_4 + \alpha_{15} F_5 + \alpha_{16} F_6.$$

So to find the forces that must be applied by each leg we need to invert a 6×6 matrix of coefficients α_{ij}. But before we do that we need to find the joint rates $\dot{\phi}_i$, $\dot{\theta}_i$, $\dot{l}_i$ and their accelerations $\ddot{\phi}_i$, $\ddot{\theta}_i$, $\ddot{l}_i$ in terms of the velocity of the platform. This is a kinematic problem which can be solved using the constraint equations given above. In principle, we have solved it already. We simply need to set, for example,

$$\dot{\phi}_1 = \dot{\omega}_{Px} \alpha_{U111} + \dot{\omega}_{Py} \alpha_{U112} + \dot{\omega}_{Pz} \alpha_{U113} + \dot{x}_P \alpha_{U114} + \dot{y}_P \alpha_{U115} + \dot{z}_P \alpha_{U116}$$

where α_{U111} is the solution for α_{U11} when $\alpha_{P\omega x} = 1$ and the others are set to zero, α_{U112} is the solution for α_{U11} when $\alpha_{P\omega y} = 1$ and the others are set to zero, and so forth.

To find the joint accelerations we need to differentiate the constraint equations. Here the constraint wrenches are constants so we simply need the derivative of the leaf-link velocities; see Section 14.1.2. Again we get six groups of three equations, each having the form

$$\mathcal{W}^T \ddot{\mathbf{q}}_P + \ddot{\phi}_i \mathcal{W}^T \mathbf{s}_{Ui1} \ddot{\theta}_i \mathcal{W}^T \mathbf{s}_{Ui2} + \ddot{l}_i \mathcal{W}^T \mathbf{s}_{Li}$$
$$= \dot{\phi}_i \mathcal{W}^T [\dot{\mathbf{q}}_P, \mathbf{s}_{Ui1}] + \cdots + \dot{\theta}_i \dot{l}_i \mathcal{W}^T [\mathbf{s}_{Ui2}, \mathbf{s}_{Li}].$$

If we move the $\mathcal{W}^T \ddot{\mathbf{q}}_P$ terms to the right-hand sides of these equations then the left-hand sides are just as before. The right-hand sides consist of the desired acceleration of the platform together with terms that can be computed from what we already have.

Finally we can sketch the complete algorithm.

Inputs:

> Desired velocity and acceleration of the platform,
> Current position of all joints,
> joint screw in home position,
> link inertia matrices in home position.

Outputs:

> Forces F_i at each leg.

Method:

1. Given the desired velocity and acceleration of the platform, use the constraint equations to compute

    (i) the α coefficients,

    (ii) the joint rates $\dot{\phi}_i$, $\dot{\theta}_i$ and $\dot{l}_i$,

    (iii) the joint accelerations $\ddot{\phi}_i$, $\ddot{\theta}_i$ and $\ddot{l}_i$.

2. Use the recursive algorithm given in Section 14.1.2 to compute the left-hand sides of the joint equations.

3. Don't forget to include the six equations for the platform.

4. Combine the joint equations using the αs.

5. Solve for the forces at the legs.

Clearly, given a particular machine, its dimensions, weights and inertias, it would be possible to turn the above sketch into a working computer program.

15
Differential Geometry

This final chapter is slightly different in character from the preceding ones. The aim is to present some less elementary examples. The examples are loosely related by their use of some concepts from differential geometry, hence the title. We will begin by looking at some differential geometry on the manifold of the group $SE(3)$.

15.1   Metrics, Connections and Geodesics

Let G be a connected Lie group. Elements of the Lie algebra can be thought of as left-invariant vector fields. The integral curves of these vector fields, at least the ones that pass through the identity element, are the one-parameter subgroups of G. Let $\{X_1, X_2, \ldots, X_n\}$ be a basis for the Lie algebra of G. In general, the integral curves corresponding to this basis do not form a coordinate basis. Indeed

$$\partial_{X_i} X_j = \partial_i X_j = [X_i, X_j] = C_{ij}^k X_k.$$

For a coordinate basis, we must have $\partial_i X_j = 0$ when $i \neq j$; see for example Schutz [99, sect. 2.15]. In the following, we will write ∂_i for the derivative with respect to the i-th basis element.

Here C_{ij}^k are the structure constants for the algebra that we met in Section 4.3. As usual, summing over repeated indices is understood. Recall that the structure constants are anti-symmetric, so that

$$C_{ij}^k = -C_{ji}^k.$$

Let Q be a invariant metric on the group, that is, a metric that is both left- and right-invariant. With respect to the Lie algebra basis given above, the components of the metric are $Q_{ij} = Q(X_i, X_j)$. As the metric is symmetric, so are the components

$$Q_{ij} = Q_{ji}.$$

The invariance of the metric can be expressed as the invariance of its components:

$$\partial_i Q_{jk} = 0$$

and hence

$$\partial_i Q_{jk} = Q([X_i, X_j], X_k) + Q(X_j, [X_i, X_k]) = 0.$$

The linearity of the metrics allows us to compare the coefficients of the basis elements after we have expanded the commutators:

$$C_{ij}^l Q_{lk} + C_{ik}^l Q_{lj} = 0.$$

Now, any metric defines a unique symmetric connection. Since we do not have a coordinate basis, we cannot simply require that the Christoffel symbols of the connection be symmetric. We must go back to the original definition for a symmetric connection:

$$\nabla_i X_j - \nabla_j X_i = [X_i, X_j];$$

see Schutz [99, sect. 6.5]. That is, the connection defines a covariant derivative ∇, which is given by

$$\nabla_i X_j = \partial_i X_j + \Gamma_{ij}^k X_k$$

where Γ is the Christoffel symbol of the connection.

The symmetry requirement now yields

$$(\Gamma_{ij}^k - \Gamma_{ji}^k) X_k = -[X_i, X_j]$$

and so

$$\Gamma_{ij}^k - \Gamma_{ji}^k = -C_{ij}^k.$$

To link such a connection to a metric, we require that the metric must be compatible with the connection. This means

$$\nabla_i Q_{jk} = Q(\nabla_i X_j, X_k) + Q(X_j, \nabla_i X_k) = 0.$$

Since we already have $\partial_i Q_{jk} = 0$, this simplifies to

$$\Gamma_{ij}^l Q_{lk} + \Gamma_{ik}^l Q_{lj} = 0.$$

Cycling the indices gives

$$\Gamma_{jk}^l Q_{li} + \Gamma_{ji}^l Q_{lk} = 0$$

and finally

$$\Gamma^l_{ki}Q_{lj} + \Gamma^l_{kj}Q_{li} = 0.$$

If we add the first two of these equations and subtract the third, we obtain

$$(\Gamma^l_{ij} + \Gamma^l_{ji})Q_{lk} + (\Gamma^l_{ik} - \Gamma^l_{ki})Q_{lj} + (\Gamma^l_{jk} - \Gamma^l_{kj})Q_{li} = 0.$$

Using the results above, we get

$$(\Gamma^l_{ij} + \Gamma^l_{ji})Q_{lk} = C^l_{ik}Q_{lj} + C^l_{jk}Q_{li} = 0.$$

If the metric is non-degenerate, then multiplying by the inverse of the metric gives

$$\Gamma^l_{ij} + \Gamma^l_{ji} = 0.$$

Notice that in these coordinates a symmetric connection has an anti-symmetric Christoffel symbol. If we add this to the result for the anti-symmetric part of the Christoffel symbol, we have

$$\Gamma^k_{ij} = -\frac{1}{2}C^k_{ij}.$$

Thus, the covariant derivative can be written

$$\nabla_i X_j = \frac{1}{2}[X_i, X_j].$$

See also Milnor [77, sect. 21].

The above ideas can be applied to geodesic curves. A geodesic curve is a curve that has minimal or at least stationary length as measured by the metric. The length of a curve is given by the integral

$$\text{length}^2(\gamma(t)) = \int Q(\dot\gamma, \dot\gamma)\, dt.$$

The Euler–Lagrange equation for this variational problem can be shown to be the geodesic equation

$$\nabla_{\mathbf{X}}\mathbf{X} = 0$$

where $\mathbf{X} = \dot\gamma$ is the tangent vector to the curve; see Schutz [99, p. 218] for example. Now, if $\mathbf{X}$ is a constant linear combination of basis elements, that is, an element of the Lie algebra, it is easy to see that the geodesic equation will be identically satisfied. Hence, the one-parameter subgroups of the group are geodesics through the identity. The standard uniqueness theorems for solutions of differential equations ensure that these are the only geodesics through this point. So, the geodesics through the identity have the form

$$\gamma : t \longmapsto e^{t\mathbf{X}}$$

where $\mathbf{X}$ is a constant Lie algebra element and the parameter of the curve is t. Geodesics through other points will just be translates of these curves, that is, curves of the form

$$\gamma : t \longmapsto ge^{t\mathbf{X}},$$

where g represents translation by a constant group element. Alternative proofs of this result can be found in Helgason [48, p. 224], Sternberg [116, p. 232], Milnor [77, p. 113], and many other differential geometry texts.

Now, we look at how the above applies to the group of rigid body motions $SE(3)$. The two symmetric, invariant bilinear forms on $se(3)$, that is, the Killing form Q_∞ and the reciprocal product Q_0, can be thought of as invariant metrics on the group. See Section 7.5. The metrics act on vector fields, but, as already mentioned, we may think of the Lie algebra elements as left-invariant vector fields on the group. Suppose we write a general invariant metric as

$$Q_p = \alpha Q_0 + \beta Q_\infty,$$

where $p = -\beta/2\alpha$. Then for a pitch p screw $\mathbf{s}$ we have

$$Q_p(\mathbf{s}, \mathbf{s}) = 0.$$

We may conclude that for any of the invariant metrics on $SE(3)$ the geodesic curves will be the same: the one-parameter subgroups and their translates. That is, the group of finite screws of constant pitch about some line will be a geodesic for all possible invariant metrics. The difference between the different metrics is that although they have the same geodesics, the 'length' of such a curve is different under each metric. Since the length of tangent vectors along a geodesic is constant, we see that under the pitch p metric, the geodesics generated by pitch p screws are null.

Now we extend the above to robots. Consider a six-joint robot arm. The forward kinematics of such a machine can be thought of as a map from the joint space of the robot to $SE(3)$; see Section 3.7. We will write this map as

$$\rho : (\theta_1, \theta_2, \theta_3, \theta_4, \theta_5, \theta_6) \longmapsto e^{\theta_1 \mathbf{s}_1} e^{\theta_2 \mathbf{s}_2} e^{\theta_3 \mathbf{s}_3} e^{\theta_4 \mathbf{s}_4} e^{\theta_5 \mathbf{s}_5} e^{\theta_6 \mathbf{s}_6}$$

where θ_i is the i-th joint variable and $\mathbf{s}_i$ the i-th joint screw for the robot in its home position. We can use this map ρ to pull back the metrics to joint space and express the connection and geodesic equations in terms of the joint variables, the coordinates in joint space. That is, we may evaluate the Christoffel symbols for the metrics on joint space in terms of the robot's joint screws.

The pullback to joint space of the metric Q_p is simply given by

$$Q_p^* = J^T Q_p J$$

where J is the Jacobian of the forward kinematic mapping ρ; see Section 4.5.

To calculate the covariant derivative for vector fields on joint space, we map them to $SE(3)$ and use the result $\nabla_i X_j = \frac{1}{2}[X_i, X_j]$ from above. So, two vector fields pushed forward to the group will have the form

$$\mathbf{U} = J\dot{\boldsymbol{\theta}} = \dot{\theta}_1 \mathbf{s}_1 + \dot{\theta}_2 \mathbf{s}_2 + \dot{\theta}_3 \mathbf{s}_3 + \dot{\theta}_4 \mathbf{s}_4 + \dot{\theta}_5 \mathbf{s}_5 + \dot{\theta}_6 \mathbf{s}_6,$$

$$\mathbf{V} = J\dot{\boldsymbol{\phi}} = \dot{\phi}_1 \mathbf{s}_1 + \dot{\phi}_2 \mathbf{s}_2 + \dot{\phi}_3 \mathbf{s}_3 + \dot{\phi}_4 \mathbf{s}_4 + \dot{\phi}_5 \mathbf{s}_5 + \dot{\phi}_6 \mathbf{s}_6.$$

Here, the joint screws $\mathbf{s}_i$ are functions of the joint variables. As the robot moves, the joint screws are transformed according to

$$\mathbf{s}_i(\boldsymbol{\theta}) =$$
$$e^{\theta_1 ad(\mathbf{s}_1)} e^{\theta_2 ad(\mathbf{s}_2)} \ldots e^{\theta_{i-1} ad(\mathbf{s}_{i-1})} \mathbf{s}_i(0) e^{-\theta_{i-1} ad(\mathbf{s}_{i-1})} \ldots e^{-\theta_2 ad(\mathbf{s}_2)} e^{-\theta_1 ad(\mathbf{s}_1)}.$$

The time derivative of one of these screws is thus

$$\frac{d}{dt}\mathbf{s}_i(\boldsymbol{\theta}) = \sum_{j=1}^{i} \dot{\theta}_j [\mathbf{s}_j(\boldsymbol{\theta}), \mathbf{s}_i(\boldsymbol{\theta})];$$

see Section 13.2. As usual, we drop the explicit dependence of the joint screws on the joint angles. Now the covariant derivative can be calculated:

$$\nabla_{\mathbf{U}} \mathbf{V} = \partial_{\mathbf{U}} \mathbf{V} - \frac{1}{2}[\mathbf{U}, \mathbf{V}]$$

$$= \sum_{i=1}^{6} \dot{\theta}_i \frac{\partial}{\partial \theta_i} \mathbf{V} - \frac{1}{2}[\mathbf{U}, \mathbf{V}]$$

$$= \sum_{i=1}^{6} \ddot{\phi}_i \mathbf{s}_i + \sum_{1 \leq i < k \leq 6} \dot{\theta}_i \dot{\phi}_k [\mathbf{s}_i, \mathbf{s}_k] - \frac{1}{2} \sum_{i,k=1}^{6} \dot{\theta}_i \dot{\phi}_k [\mathbf{s}_i, \mathbf{s}_k]$$

$$= \sum_{i=1}^{6} \ddot{\phi}_i \mathbf{s}_i + \frac{1}{2} \sum_{1 \leq i < k \leq 6} \dot{\theta}_i \dot{\phi}_k [\mathbf{s}_i, \mathbf{s}_k] - \frac{1}{2} \sum_{1 \leq k < i \leq 6} \dot{\theta}_i \dot{\phi}_k [\mathbf{s}_i, \mathbf{s}_k].$$

This general result tells us how to find the covariant derivative of vector fields and hence how to parallel translate those vectors.

The geodesic equations in joint space are thus

$$\nabla_{\mathbf{U}} \mathbf{U} = \sum_{i=1}^{6} \ddot{\theta}_i \mathbf{s}_i + \sum_{1 \leq i < j \leq 6} \dot{\theta}_i \dot{\theta}_k [\mathbf{s}_i, \mathbf{s}_j] = 0.$$

The equations can be tidied up away from singularities. When the Jacobian is non-singular, we can find six co-screws $\mathcal{W}_i$ that satisfy

$$\mathcal{W}_i(\mathbf{s}_j) = \begin{cases} 1, & \text{if } i = j, \\ 0, & \text{if } i \neq j; \end{cases}$$

see Section 12.1. These are just the rows of the inverse Jacobian J^{-1}, Section 6.7. The geodesic equations can be rewritten as

$$\ddot{\theta}_i + \sum_{1 \leq j < k \leq 6} \mathcal{W}_i([\mathbf{s}_j, \mathbf{s}_k]) \dot{\theta}_j \dot{\theta}_k = 0 \qquad i = 1, 2 \ldots, 6.$$

From here it is easy to find the Christoffel symbol of the symmetric connection on joint space induced by a invariant metric on the group. Remember from above that the Christoffel symbol must be anti-symmetric in this case:

$$\Gamma^i_{jk} = \begin{cases} \frac{1}{2}\mathcal{W}_i([\mathbf{s}_j, \mathbf{s}_k]) & \text{if } j \le k, \\ -\frac{1}{2}\mathcal{W}_i([\mathbf{s}_j, \mathbf{s}_k]) & \text{if } j \ge k. \end{cases}$$

Our final result here is to pull back the Maurier–Cartan form to joint space. The Maurier–Cartan form is a Lie algebra valued, differential 1-form on a Lie group. It is uniquely specified by the property that, when evaluated on a left-invariant vector field, it produces the corresponding element of the Lie algebra. Suppose the Lie algebra has a basis $\{\mathbf{X}_1, \mathbf{X}_2, \ldots, \mathbf{X}_n\}$ and a dual basis of 1-forms given by $\{\boldsymbol{\Omega}_1, \boldsymbol{\Omega}_2, \ldots, \boldsymbol{\Omega}_n\}$, that is,

$$\boldsymbol{\Omega}_i(\mathbf{X}_j) = \begin{cases} 1, & \text{if } i = j, \\ 0, & \text{if } i \ne j. \end{cases}$$

Now the Maurier–Cartan form can be written

$$\boldsymbol{\Omega} = \sum_{i=1}^{n} \mathbf{X}_i \otimes \boldsymbol{\Omega}_i,$$

so that for any left-invariant vector field $\mathbf{A} = a_1\mathbf{X}_1 + a_2\mathbf{X}_2 + \cdots + a_n\mathbf{X}_n$ we have $\boldsymbol{\Omega}(\mathbf{A}) = \mathbf{A}$. The Maurier–Cartan form satisfies the structure equations

$$d\boldsymbol{\Omega} = \frac{1}{2}[\boldsymbol{\Omega}, \boldsymbol{\Omega}]$$

where the exterior derivative is defined on basis elements as $d(\mathbf{X}_i \otimes \boldsymbol{\Omega}_i) = \mathbf{X}_i \otimes d\boldsymbol{\Omega}_i$, and the bracket operation is given by

$$[\mathbf{X}_i \otimes \boldsymbol{\Omega}_i, \mathbf{X}_j \otimes \boldsymbol{\Omega}_j] = [\mathbf{X}_i, \mathbf{X}_j] \otimes \boldsymbol{\Omega}_i \wedge \boldsymbol{\Omega}_j;$$

see Griffiths [41] for example.

In order to pull back the Maurier–Cartan form to joint space, we consider first the pushforward of the coordinate fields $\partial/\partial\theta_i$. From the Jacobian of the forward kinematic map that we found in Section 4.5, we have

$$\rho_* \frac{\partial}{\partial\theta_i} = \mathbf{s}_i.$$

This vector field is not left-invariant. However, we can write it in terms of a basis for the Lie algebra as

$$\mathbf{s}_i = a_{i1}\mathbf{X}_1 + a_{i2}\mathbf{X}_2 + a_{i3}\mathbf{X}_3 + a_{i4}\mathbf{X}_4 + a_{i5}\mathbf{X}_5 + a_{i6}\mathbf{X}_6, \qquad i = 1, \ldots, 6,$$

where the coefficients are now functions of position. Notice that these coefficients can be found by evaluating the dual 1-forms on the fields $\boldsymbol{\Omega}_i(\mathbf{s}_j) = a_{ji}$. So now the pullback of the dual 1-forms is given by

$$\rho^*\boldsymbol{\Omega}_i = a_{1i}\,d\theta_1 + a_{2i}\,d\theta_2 + a_{3i}\,d\theta_3 + a_{4i}\,d\theta_4 + a_{5i}\,d\theta_5 + a_{6i}\,d\theta_6, \qquad i = 1, \ldots, 6.$$

The pullback of the Maurier–Cartan form is thus

$$\rho^*\Omega = \sum_{i,j=1}^{6} \mathbf{X}_i \otimes (a_{ji}\, d\theta_j) = \sum_{i,j=1}^{6} (a_{ji}\mathbf{X}_i) \otimes d\theta_j,$$
$$= \mathbf{s}_1 \otimes d\theta_1 + \mathbf{s}_2 \otimes d\theta_2 + \mathbf{s}_3 \otimes d\theta_3 + \mathbf{s}_4 \otimes d\theta_4 + \mathbf{s}_5 \otimes d\theta_5 + \mathbf{s}_6 \otimes d\theta_6.$$

The pullback operation is a functor and this ensures that the form still satisfies the structure equations. The exterior derivative of the form is given by

$$d(\rho^*\Omega) = \rho^*(d\Omega) = \sum_{1 \le i < j \le 6} [\mathbf{s}_i, \mathbf{s}_j] \otimes d\theta_i \wedge d\theta_j$$

and thus so is the bracket $1/2[\rho^*\Omega, \rho^*\Omega]$.

15.2   Mobility of Overconstrained Mechanisms

Mobility is usually described in terms of degrees of freedom. This can be made precise by defining the **mobility** of a mechanism as the dimension of its configuration space. The configuration space of a mechanism is the set whose elements are distinct configurations of the mechanism. This space can be given the structure of a topological space; in fact, it can usually be made into an algebraic variety. We can talk about the dimension at a point in the configuration space. For regular points, this is quite easy, since by definition a regular point has a neighbourhood homeomorphic $\mathbb{R}^n$, so the dimension at the point is just n. This **local dimension** can be defined for all points, but the details are unnecessary here; see Hartshorne [45, p. 5]. In kinematics, it is well known that there are some mechanisms whose configuration space has components of different dimensions. For example, some overconstrained $6 - R$'s have configuration spaces consisting of a curve plus an isolated point. It is usual to define the dimension of a topological space as the maximum of the local dimensions of its points. To distinguish this from the idea of local dimension, it will be referred to as **global dimension**. It should be clear now that the mobility of a mechanism is the global dimension of its configuration space. In the following, we will mainly be concerned with the local dimension of configurations. If we can show that a mechanism has a configuration with local dimension 1, say, then the mechanism as a whole must have mobility at least 1.

Note that the concept of local dimension is different from infinitesimal mobility. It can be deduced that the infinitesimal mobility of a configuration is the dimension of the tangent space to the corresponding point in configuration space. For a mechanism to have mobility 1, it is a necessary but not sufficient condition that it have a configuration with infinitesimal mobility 1. The inverse function theorem tells us that at regular points the local dimension and infinitesimal mobility are the same; see Section 1.4. However, it is easy to conceive of a mechanism that is infinitesimally mobile in every configuration that it can adopt and yet is still not mobile. For example, consider a planar 4-bar with the sum of its three shortest sides equal to the length of the longest side. Its configuration space is a single point; hence, its global dimension is 0. But this configuration has infinitesimal mobility 2.

Lastly, it is common in kinematics to use algebraic geometry. This is possible when there are no helical joints. Then, either implicitly or explicitly, variables are taken to be complex. In general, we may expect the complex dimension of the variety defined by a mechanism to be the same as the global dimension of the configuration space. Unfortunately, this is not always the case. It is possible for complex projective varieties to have no real (i.e., physical) points. Hence, the mobility of a mechanism is not the dimension of its algebraic variety but the dimension of the set of real points in that variety.

In the following, we will restrict our attention to single loop mechanisms. So, consider a mechanism consisting of n rigid links connected in a loop by the same number of one-degree-of-freedom joints. Now suppose that we break the ground link into two pieces. Keeping one of the pieces fixed, the other piece becomes the last link of an n-joint serial robot. If we hold the free part of the ground link in its original, unbroken position and orientation, then the possible movements of the joints will be expressed by the **loop equation**

$$e^{\theta_1 \mathbf{s}_1} e^{\theta_2 \mathbf{s}_2} e^{\theta_3 \mathbf{s}_3} \cdots e^{\theta_n \mathbf{s}_n} = I.$$

This is the forward kinematic map ρ set equal to the identity, I. The configuration space of the mechanism is given by the pre-image of I under this map, that is, the set $\rho^{-1}(I)$. Differentiating the loop equation, we get $J\dot{\boldsymbol{\theta}} = \mathbf{0}$, which can be written more fully as

$$\sum_{i=1}^{n} \dot{\theta}_i \mathbf{s}_i = 0.$$

These are six differential equations in n variables; the initial conditions will be $\theta_1(0) = \theta_2(0) = \cdots = \theta_n(0) = 0$. The solution of this initial value problem will give a parameterisation of the mechanism's configuration space in the neighbourhood of the point $\boldsymbol{\theta} = \mathbf{0}$. Hence, in general, the local mobility of this point will be given by

$$m = n - 6,$$

which is usually known as the **Kutzbach** or **Grübler mobility formula**. Note that this is usually given for a more general class of mechanisms with arbitrarily many loops; see Hunt [54, p. 33]. Since the formula is only generally true, there are many exceptions.

The above differential equations will have only trivial solutions, and thus the mechanism will have local mobility 0, unless there is a linear dependency among the joint screws, that is, unless the dimension of the screw system $\Delta_1 = \mathrm{Span}(\mathbf{s}_1, \mathbf{s}_2, \ldots \mathbf{s}_n)$ is less than n. This is, of course, satisfied if $n > 6$, a situation that also occurs in redundant robots. More usually, in kinematics we are interested in the case where $n < 6$, that is, in overconstrained mechanisms.

In particular, it is important to know when an overconstrained mechanism is mobile. Unfortunately, if $\dim(\Delta_1) < n$, then there is no guarantee that the local mobility will be greater than 0. All we can say is that the infinitesimal mobility is greater than 0.

We can use the Campbell–Baker–Hausdorff theorem to sharpen the above result a little. Recall from Section 4.8 that the product of exponentials can be written

$$e^{\theta_1 s_1} e^{\theta_2 s_2} e^{\theta_3 s_3} \cdots e^{\theta_n s_n} = e^{f(s_1, s_2, \ldots s_n)},$$

where the function f is a function of the Lie algebra elements $s_1, s_2, \ldots, s_n$ and all their iterated commutators. So, the image of the forward kinematic map ρ always lies in the completion group of Δ_1; see Section 8.1. Generally, the local mobility will be

$$m = n - \dim(\Delta_\infty)$$

where Δ_∞ is the Lie algebra of the completion group.

In Hunt [54, p. 378] Table 13.1 lists "screw systems that guarantee 'full-cycle' mobility". Upon inspection, these screw systems turn out to be exactly those systems that are subalgebras, that is, the ones where $\Delta = \Delta_\infty$. We may understand this if we define a screw system that 'guarantees' full-cycle mobility to be one where the kinematic map ρ gives a smooth map of the joint space onto a totally geodesic submanifold of $SE(3)$ for some neighbourhood of $\theta = 0$. We also require that the dimension of the screw system be larger than the dimension of the submanifold. Recall that a totally geodesic submanifold is one where the geodesics in the manifold that are tangent to any point in the submanifold actually lie in the submanifold. To understand this, consider a cylinder in $\mathbb{R}^3$. The geodesics in $\mathbb{R}^3$ are straight lines, so at a point on the cylinder only one tangent line lies in the cylinder, the one parallel to the axis of the cylinder. Hence, a cylinder is not a totally geodesic submanifold of $\mathbb{R}^3$, but it is easy to see that a 2-plane would be.

The idea behind the definition is as follows: Suppose that we break the link between the last and next to last joints. Now, the image of the kinematic map on the first $n - 1$ joints traces out a submanifold in $SE(3)$. On the other hand, the last joint traces out a geodesic curve in $SE(3)$. If the mechanism is mobile, then this geodesic must lie in the submanifold generated by the other joints. Now suppose we make a slight error when we reconnect the broken link; that is, suppose we move the last joint slightly. If the screw system 'guarantees mobility', then we expect this perturbed mechanism to have the same mobility as the original. If the submanifold generated by the first $n - 1$ joints is totally geodesic, then so long as the new last joint is linearly dependent on the first $n - 1$ joints, the geodesic it generates will lie in the submanifold. Hence, the mechanism will still be mobile.

If the screw system Δ generated by $\{s_1, s_2, \ldots, s_{n-1}\}$ is a subalgebra of $se(3)$, then it follows from the Campbell–Baker–Hausdorff theorem that the image of ρ will be a totally geodesic submanifold.

Conversely, if the image of ρ is totally geodesic, then we can show that Δ must be a subalgebra. For a totally geodesic submanifold, the covariant derivative of any tangent vector field along any other tangent field gives a vector field that is also tangent to the submanifold; see [48, Chap. I sect. 14]. So let

$$\mathbf{U} = \dot{\theta}^1 \mathbf{s}_1 + \dot{\theta}_2 \mathbf{s}_2 + \cdots + \dot{\theta}_{n-1} \mathbf{s}_{n-1},$$
$$\mathbf{V} = \dot{\phi}_1 \mathbf{s}_1 + \dot{\phi}_2 \mathbf{s}_2 + \cdots + \dot{\phi}_{n-1} \mathbf{s}_{n-1}$$

be a pair of arbitrary vectors tangent to the submanifold. The covariant derivative $\nabla_{\mathbf{U}} \mathbf{V}$ is given by

$$\nabla_{\mathbf{U}} \mathbf{V} = \sum_{i=1}^{n-1} \ddot{\phi}_i \mathbf{s}_i + \frac{1}{2} \sum_{1 \leq i < k \leq n-1} (\dot{\theta}_i \dot{\phi}_k + \dot{\theta}_k \dot{\phi}_i)[\mathbf{s}_i, \mathbf{s}_k];$$

see Section 15.1. Now for this to be tangent to the submanifold we must certainly have that the tangent vector at the identity lies in the screw system Δ_1. Since this must be true for arbitrary $\boldsymbol{\theta}$ and $\boldsymbol{\phi}$, we must have that

$$[\mathbf{s}_i, \mathbf{s}_k] \in \Delta_1, \qquad 1 \leq i < k \leq n - 1.$$

This includes all possible commutators in Δ_1; hence, we have that Δ_1 must be a subalgebra.

Table 15.1 shows the screw systems that 'guarantee' full-cycle mobility, as in Hunt's Table 13.1, but also showing the corresponding completion groups. Note that the list of connected subgroups, and hence of subalgebras for $se(3)$, was derived in Section 3.5.

In Hervé [49] mechanisms were categorised as banal, extraordinary, or paradoxical. Banal mechanisms are mechanisms that satisfy the mobility formula. The extraordinary mechanisms come in several varieties, or families, each satisfying a mobility formula depending on the family, for example, planar or spherical mechanisms. The paradoxical mechanisms do not fit into any family and generally disobey the mobility formulas. The Bennett and Bricard mechanisms are examples of paradoxical mechanisms. Notice that if the image of the kinematic mapping ρ is not a totally geodesic submanifold of $SE(3)$, it may still be possible for it to contain a geodesic through the identity. In this case, it will be possible to close the loop by adding a joint corresponding to the same Lie algebra element as the geodesic and thus producing a single loop mechanism with full-cycle mobility.

Finally, note that these ideas are also relevant to robots. Suppose we hold the end-effector of a robot fixed. For a six-joint robot this turns the robot into a single loop mechanism. If this mechanism has infinitesimal mobility, then the configuration of the robot is singular. This is because when the joint screws are linearly dependent the robot's Jacobian will be singular. Singularities that have mobility greater than 1 are more difficult for the robot's control system to deal with. For these singularities we can hold the robot's end-effector fixed but still move the rest of the arm.

TABLE 15.1. Screw Systems That 'Guarantee' Full-Cycle Mobility

Completion Group	Gibson–Hunt Type	Normal Form
1-systems		
$\mathbb{R}$	IIB	$\mathbf{s}_1 = (0,0,0,1,0,0)^T$
H_p	IA, $\quad (p \neq 0)$	$\mathbf{s}_1 = (1,0,0,p,0,0)^T$
$SO(2)$	IA, $\quad (p = 0)$	$\mathbf{s}_1 = (1,0,0,0,0,0)^T$
2-systems		
$\mathbb{R}^2$	IIC	$\mathbf{s}_1 = (0,0,0,1,0,0)^T$ $\mathbf{s}_2 = (0,0,0,0,1,0)^T$
$SO(2) \times \mathbb{R}$	IB^0	$\mathbf{s}_1 = (1,0,0,0,0,0)^T$ $\mathbf{s}_2 = (0,0,0,1,0,0)^T$
3-systems		
$\mathbb{R}^3$	IID	$\mathbf{s}_1 = (0,0,0,1,0,0)^T$ $\mathbf{s}_2 = (0,0,0,0,1,0)^T$ $\mathbf{s}_3 = (0,0,0,0,0,1)^T$
$SO(3)$	IIA, $\quad (p = 0)$	$\mathbf{s}_1 = (1,0,0,0,0,0)^T$ $\mathbf{s}_2 = (0,1,0,0,0,0)^T$ $\mathbf{s}_3 = (0,0,1,0,0,0)^T$
$SE(2)$	IIC, $\quad (p = 0)$	$\mathbf{s}_1 = (1,0,0,0,0,0)^T$ $\mathbf{s}_2 = (0,0,0,0,1,0)^T$ $\mathbf{s}_3 = (0,0,0,0,0,1)^T$
$H_p \ltimes \mathbb{R}^2$	IIC, $\quad (p \neq 0)$	$\mathbf{s}_1 = (1,0,0,p,0,0)^T$ $\mathbf{s}_2 = (0,0,0,0,1,0)^T$ $\mathbf{s}_3 = (0,0,0,0,0,1)^T$
4-systems		
$SE(2) \times \mathbb{R}$	$\overline{\text{IIC}}$	$\mathbf{s}_1 = (1,0,0,0,0,0)^T$ $\mathbf{s}_2 = (0,0,0,1,0,0)^T$ $\mathbf{s}_3 = (0,0,0,0,1,0)^T$ $\mathbf{s}_4 = (0,0,0,0,0,1)^T$

15.3 Controlling Robots Along Helical Trajectories

As we saw in Section 13.2, the equations of motion for a robot are given by

$$\tau_i = A_{ij}\ddot{\theta}_j + B_{ijk}\dot{\theta}_j\dot{\theta}_k + C_i.$$

The generalised inertia matrix of the robot, A, contains all the information about the inertia of the links. The Coriolis terms, B terms, contain the interactions between links. The gravity terms, C, are related to the weights of the links. Finally, the τs are the generalised forces applied to the joints; for revolute joints these will be torques.

In order to control the robot, we must choose torques at the joints to drive the machine along the desired path. Conventionally, the desired path of the end-effector will be given. From this, the corresponding path in joint space can be computed. Typically this involves finding points along the path of the end-effector and then using the robot's inverse kinematics at each point. Once we have the path in joint space, or at least a discrete approximation to it, we can compute the required torques using the equations of motion above.

For a geodesic path, we can do better than this. We can dispense with the inverse kinematics completely. In Section 15.1 above, we derived the equations for paths in joint space corresponding to the geodesics in the group. Now suppose we can choose a control law that makes the closed loop dynamics of the system into these geodesic equations. That is, we choose the joint torques to be functions of the joint angles and their rates. If we succeed, the end-effector of the robot will follow a geodesic path—straight line, circle, or helix, depending on the initial conditions. Straight line and circular paths are frequently used paths in robotics; helical paths are less common but have been recommended by several authors; see [56, 97], for example.

This goal can be achieved. We set the joint torques to

$$\tau_i = (B_{ijk} - A_{il}\Gamma^l_{jk})\dot{\theta}_j\dot{\theta}_k + C_i \qquad i = 1, 2 \dots, 6.$$

Note that since this is a feedback law we assume that the values for the joint angles and their velocities are measured values. The closed loop equations of motion are now

$$A_{il}\ddot{\theta}_l + A_{il}\Gamma^l_{jk}\dot{\theta}_j\dot{\theta}_k = 0.$$

The generalised inertia matrix A is always invertible, so this is just the equation for geodesics, as promised. Hence, the solutions will be one-parameter subgroups, that is, helical motions of the robot's end-effector. The pitch of the screw motion will depend on the initial conditions, essentially the linear and angular velocities of the robot's last link at time $t = 0$.

There is a standard way of studying the stability of geodesic equations; see Schutz [99, sect. 6.9] or Milnor [77, sect. 19]. The geodesic equations can be linearised about a particular geodesic. This gives the equations of geodesic deviation

$$\nabla_{\mathbf{X}}\nabla_{\mathbf{X}}\mathbf{Z} + \mathbf{R}(\mathbf{X}, \mathbf{Z})\mathbf{X} = 0.$$

The vector field $\mathbf{X}$ is tangent to the particular geodesic, while $\mathbf{Z}$ is a field on the geodesic that expresses the difference between the geodesic and nearby geodesics. Such a field is called a Jacobi field. The term $\mathbf{R}$ is the Riemann curvature tensor of the metric. It is defined by

$$\mathbf{R}(\mathbf{U}, \mathbf{V})\mathbf{W} = -\nabla_{\mathbf{U}}\nabla_{\mathbf{V}}\mathbf{W} + \nabla_{\mathbf{V}}\nabla_{\mathbf{U}}\mathbf{W} + \nabla_{[\mathbf{U},\mathbf{V}]}\mathbf{W}.$$

(Note that it is also common to give $\mathbf{R}$ the opposite sign; the conventions in [77] will be followed here.)

The stability of this equation depends on the sign of the sectional curvature

$$K(\mathbf{X}, \mathbf{Z}) = Q(\mathbf{R}(\mathbf{X}, \mathbf{Z})\mathbf{X}, \mathbf{Z}).$$

According to Milnor [77, sect. 21], on a Lie group with invariant metric Q we have

$$\mathbf{R}(\mathbf{X}, \mathbf{Y})\mathbf{Z} = \frac{1}{4}[[\mathbf{X}, \mathbf{Y}], \mathbf{Z}]$$

and

$$Q(\mathbf{R}(\mathbf{X}, \mathbf{Y})\mathbf{Z}, \mathbf{W}) = \frac{1}{4}Q([\mathbf{X}, \mathbf{Y}], [\mathbf{Z}, \mathbf{W}])$$

so that the sectional curvature on a group is given by

$$K(\mathbf{X}, \mathbf{Z}) = Q(\mathbf{R}(\mathbf{X}, \mathbf{Z})\mathbf{X}, \mathbf{Z}) = \frac{1}{4}Q([\mathbf{X}, \mathbf{Z}], [\mathbf{X}, \mathbf{Z}]).$$

If the sectional curvature is positive, then the geodesic will be stable in the sense that nearby geodesics will stay nearby. The geodesic equations can be thought of as a Hamiltonian system; see Arnol'd [2, Appendix 1 sect. H]; thus, no energy loss is possible, and we cannot expect nearby solutions to converge.

If $\mathbf{X}$ and $\mathbf{Z}$ commute, then the sectional curvature will be zero. Hence, geodesics will not be stable against perturbations in such directions.

It is possible to modify the proposed control law so that the closed loop dynamics are absolutely stable. Consider the following equations in the group:

$$\ddot{\mathbf{q}}_6 = \lambda(\mathbf{s}_d - \dot{\mathbf{q}}_6).$$

Certainly, $\dot{\mathbf{q}}_6 = \mathbf{s}_d$ is a solution to this equation for any value of the constant λ. Moreover, nearby solutions converge to this solution. To see this, assume that $\dot{\mathbf{q}}_6$ is given by a linear combination of constant screws that form a basis of the Lie algebra:

$$\dot{\mathbf{q}}_6 = a_d\mathbf{s}_d + a_1\mathbf{z}_1 + a_2\mathbf{z}_2 + a_3\mathbf{z}_3 + a_4\mathbf{z}_4 + a_5\mathbf{z}_5$$

where the a_is are non-constant coefficients. This equation now reduces to the six equations for the coefficients

$$\dot{a}_d = \lambda(1 - a_d) \quad \text{and} \quad \dot{a}_i = -\lambda a_i \qquad i = 1, 2 \ldots, 5,$$

the solutions of which are

$$a_d(t) = 1 \quad \text{and} \quad a_i(t) = e^{-\lambda t} a_i(0) \qquad i = 1, 2 \ldots, 5.$$

If we choose λ to be positive, then it is clear that extraneous solutions will decay away.

Just as with the geodesic equations, this equation can be pulled back to joint space to give equations in the joint variables:

$$\ddot{\theta}_i + \sum_{j<k} \mathcal{W}_i([\mathbf{s}_j, \mathbf{s}_k])\dot{\theta}_j\dot{\theta}_k = \lambda \mathcal{W}_i\left(\mathbf{s}_d - \sum_{j=1}^{6} \dot{\theta}_j \mathbf{s}_j\right) \qquad i = 1, 2 \ldots, 6.$$

Recall that

$$\ddot{\mathbf{q}}_6 = \sum_{j=1}^{6} \ddot{\theta}_j \mathbf{s}_j + \sum_{1 \leq k < l \leq 6} \dot{\theta}_k \dot{\theta}_l [\mathbf{s}_k, \mathbf{s}_l];$$

see Section 13.2.

Again, it is possible to choose a control law that makes the closed loop dynamics of the robot look like the above equations:

$$\tau_i = (B_{ijk} - A_{il}\Gamma^l_{jk})\dot{\theta}_j\dot{\theta}_k + C_i + \lambda A_{il}\mathcal{W}_l(\mathbf{s}_d - \dot{\theta}_j\mathbf{s}_j) \qquad i = 1, 2 \ldots, 6.$$

Notice that our simple model of the robot's dynamics takes no account of friction at the joints, the dynamics of the motors, nor any flexibility that may be present in the links or joints. It may be argued that the dynamics of the model is too simple, especially since the motor dynamics are not included. This is, however, a common first step in modelling robot dynamics. It can also be justified by the common practice of controlling each motor separately and then having a supervisory level of control for coordinating the motions of the joints.

We can, however, introduce a simple model of friction into the dynamics as follows. Assume that the frictional forces on the joints are proportional to the joint rates. The equations of motion for the robot, including this type of friction, are just

$$A_{ij}\ddot{\theta}_j + B_{ijk}\dot{\theta}_j\dot{\theta}_k + C_i + \mu_i\dot{\theta}_i = \tau_i,$$

where the μs are the friction coefficients, so no sum over i is intended here. The same technique as above can be used in this case also. We set the joint torques to

$$\tau_i = (B_{ijk} - A_{il}\Gamma^l_{jk})\dot{\theta}_j\dot{\theta}_k + C_i + \mu_i\dot{\theta}_i + \lambda A_{il}\mathcal{W}_l(\mathbf{s}_d - \dot{\theta}_j\mathbf{s}_j) \qquad i = 1, 2, \ldots, 6.$$

In many situations, we would like the robot's end-effector to accelerate from rest or decelerate along a geodesic. In other words, we would like the end-effector to follow a path given by

$$\ddot{\mathbf{q}}_6 = e^{\gamma t^2 \mathbf{s}_d/2},$$

where the constant γ depends on the rate of acceleration along the geodesic. For this situation, we consider the equation

$$\ddot{\mathbf{q}}_6 = \gamma \mathbf{s}_d + \lambda(\gamma t \mathbf{s}_d - \dot{\mathbf{q}}_6).$$

Clearly, this has solution

$$\dot{\mathbf{q}}_6 = \gamma t \mathbf{s}_d,$$

as required. As above, we can see that this is stable when λ is positive. Hence, to make the closed loop dynamics of the robot match this equation, we set the joint torques to

$$\tau_i = (B_{ijk} - A_{il}\Gamma^l_{jk})\dot{\theta}_j\dot{\theta}_k + C_i + A_{il}W_l(\gamma(1 - \lambda t)\mathbf{s}_d - \lambda\dot{\theta}_j\mathbf{s}_j) \quad i = 1, 2, \ldots, 6.$$

Notice that all of our results above depend on the exact cancellation between the modelled and actual dynamics. That is, the quantities A, B, C and so on must be known with a high degree of precision. This will clearly be a problem when the robot picks up an unknown payload; see the end of Section 13.2. The ideas in this section first appeared in [108].

15.4 Hybrid Control

15.4.1 What is Hybrid Control?

In many tasks it is necessary to control the force exerted by the robot's end-effector in some directions whilst at the same time controlling its position in other directions. A typical example is writing with a pen; an even pressure must be applied normal to the surface of the paper while moving along some curve in the plane of the paper. Another example might be robot inspection of welds. Here the robot must move a probe around a weld, typically around a pipe, the probe must remain in contact with the weld and usually its orientation with respect to the weld must be maintained. There are many other examples and many more applications.

In [91], Raibert and Craig introduced the idea of hybrid force/position control of robot manipulators to solve this type of problem. In essence their method uses a 'splitting matrix' S and its complement $I - S$. The matrix S is used to project the positional error of the robot into the subspace of controlled positions. While the matrix $I - S$ projects the force/torque error into the subspace of controlled force/torque vectors or wrenches. The position and force are then controlled separately; see Figure 15.1, this diagram is based on one given in Spong and Vidyasagar [115, p.256].

This method soon attracted criticism; see [68], for example. The problem is that the splitting matrix is not well behaved under coordinate transformations. Lipkin and Duffy gave an example where the control action changes if the origin

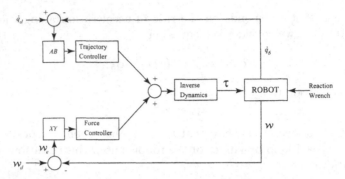

FIGURE 15.1. Hybrid Control Architecture

of the coordinates is moved. The problem stems from the (implicit) identification of the space of force/torque vectors with the space of infinitesimal motions of the robot's end-effector. In the above we have been careful to distinguish screws and wrenches; see Section 12.1.

In spite of the above comments, hybrid control has frequently been implemented with some success; see [91] for instance. This suggests that there is nothing intrinsically wrong with the method and it should be possible to find a simple coordinate-free description of it. Below we replace S and $I - S$ with simple geometrical objects with well understood transformation properties.

A related concept, that these methods easily extend to, is shared control. Here the control task is split in two, the end-effector must be constrained in some directions but in the remaining directions it should be free to move. This is mainly used in the field of teleoperation where a human operator will guide the robot in the free directions but the movement is limited by the constraints.

The ideas contained in this section first appeared in [105] and [104].

15.4.2   Constraints

When the end-effector of a robot makes contact with its environment, constraint forces materialise. It is these forces we wish to understand and control.

In the robotics literature the terms natural and artificial constraints are prevalent; see for example [3, Chap. 7] . Natural constraints are imposed by physical objects, the end-effector of the robot cannot pass through a solid object. The artificial constraints are notional and must be imposed by the control system, for example suppose we wanted to move the robot so that the end-effector stays on the surface of an imaginary sphere.

Clearly, in the shared control problem the constraints are all artificial. In the hybrid control problem the constraints are mainly natural however, we may also want to impose some artificial constraints. For example, in the pen writing example the natural constraint is that the point of the pen must stay on the paper. The artificial constraint that we may want to impose here is that the pen should remain normal to the plane of the paper. Notice also that the natural constraints tend to be one-sided. If we try to lift the pen off the paper, we generate no constraint force. We only produce a constraint force when we try to push the pen through the paper. Hence, with a natural constraint we may want to supply the other side of the constraint artificially. The upshot of this discussion is that we need to consider both natural and artificial constraints.

The constraints that we are interested in are scleronomic constraints, that is, time-invariant holonomic constraints. Recall from Section 14.2.1 that such constraints can be represented by vanishing of a number of functions defined on the group

$$\phi_i(g) = 0, \qquad i = 1, \ldots, n.$$

The tangent vectors to the constraint submanifold represent the infinitesimal freedoms or virtual displacements consistent with the constraints. By the principle of virtual work the constraint wrenches do no work on the infinitesimal freedoms; see [39, Chap.1]. Hence, the system of constraint wrenches can be identified as the system dual to the freedom screws.

The implication for hybrid control is as follows. We may maintain the constraint by supplying the appropriate constraint forces to the robot. At the same time we may control the freedom screws. There will be no interference between these control actions since the constraint wrenches are dual to the freedom screws, that is they do no work on them. The net effect will be to move the end effector over the constraint surface.

15.4.3 Projection Operators

Next we turn to the problem of dealing with error signals produced by the robot's sensors.

Suppose that the system of freedom screws is S with a basis $\mathbf{z}_1, \ldots, \mathbf{z}_n$ and the system of constraint wrenches is W with a basis, $\mathcal{W}_{n+1}, \ldots, \mathcal{W}_6$. We must project the error signals onto W and S. To do this we construct a dual basis for wrench space and screw space. That is, we have to find two complementary subspaces W' and S' that satisfy

$$W \oplus W' = se(3)^*, \quad S \oplus S' = se(3),$$

so W' is linearly independent from W and S' is linearly independent from S. Moreover we require that

$$\mathcal{W}_i^T \mathbf{s}_j = \begin{cases} 1, & \text{if } i = j, \\ 0, & \text{if } i \neq j. \end{cases}$$

Of course this is the same as the dual basis of screws and wrenches discussed in Section 14.2.2. In particular applications it is often simple to find the bases for W' and S' by inspection.

The result, however, will not be unique; we can always add elements of W to the basis elements of W' without changing anything. Similarly we may add elements of S to the basis elements of S' without effect, nevertheless the lack of uniqueness at this stage will not effect the projection operator constructed below.

We use the elements of S' to project the force errors onto W and the elements of W' to project the positional errors onto S. This is perhaps best understood by considering an example. We will use what seems to have become the standard example in the subject: the peg in a hole. The peg is partially inserted and we will assume that it is aligned along the z-axis. Hence the freedom screws represent a rotation about the z-axis and a translation along the z-axis,

$$\mathbf{z}_1 = \begin{pmatrix} \mathbf{k} \\ 0 \end{pmatrix}, \quad \mathbf{z}_2 = \begin{pmatrix} 0 \\ \mathbf{k} \end{pmatrix}.$$

Hence the constraint wrenches are

$$\mathcal{W}_3 = \begin{pmatrix} \mathbf{i} \\ 0 \end{pmatrix}, \quad \mathcal{W}_4 = \begin{pmatrix} \mathbf{j} \\ 0 \end{pmatrix}, \quad \mathcal{W}_5 = \begin{pmatrix} 0 \\ \mathbf{i} \end{pmatrix}, \quad \mathcal{W}_6 = \begin{pmatrix} 0 \\ \mathbf{j} \end{pmatrix}.$$

It is not hard to see that S' can have the basis

$$\mathbf{z}_3 = \begin{pmatrix} \mathbf{i} \\ 0 \end{pmatrix}, \quad \mathbf{z}_4 = \begin{pmatrix} \mathbf{j} \\ 0 \end{pmatrix}, \quad \mathbf{z}_5 = \begin{pmatrix} 0 \\ \mathbf{i} \end{pmatrix}, \quad \mathbf{z}_6 = \begin{pmatrix} 0 \\ \mathbf{j} \end{pmatrix}.$$

The wrench system W' can have the basis

$$\mathcal{W}_1 = \begin{pmatrix} \mathbf{k} \\ 0 \end{pmatrix} \quad \mathcal{W}_2 = \begin{pmatrix} 0 \\ \mathbf{k} \end{pmatrix}.$$

In general we have $0 < n < 6$ freedom screws. Now suppose that the sensors reveal a positional error given by a screw $\mathbf{s}_e$, then the error in the direction of $\mathbf{z}_1$ is given by

$$\delta_1 = \mathcal{W}_1^T \mathbf{s}_e,$$

the error in the direction of $\mathbf{z}_2$ is

$$\delta_2 = \mathcal{W}_2^T \mathbf{s}_e$$

and so forth. For the force control, suppose that comparison of the desired wrench with that measured by the force sensor gives an error $\mathcal{W}_e$, then in the direction of the first constraint wrench we have

$$\delta_{n+1} = \mathcal{W}_e^T \mathbf{z}_{n+1}$$

and so forth.

We may put the basis elements of W' together to form a matrix B. The rows of B are $\mathcal{W}_1^T, \mathcal{W}_2^T \dots \mathcal{W}_n^T$. So the errors are given by the matrix product:

$$B\mathbf{s}_e = \begin{pmatrix} \delta_1 \\ \vdots \\ \delta_n \end{pmatrix}.$$

The projection is then achieved by multiplying each basis element $\mathbf{z}_i$ by the appropriate δ_i and summing. Again this can be represented by a matrix product,

$$\mathbf{s}_p = AB\mathbf{s}_e,$$

where A is the matrix whose columns are the basis screws, $\mathbf{z}_1, \ldots, \mathbf{z}_n$. The projection operator can now be seen to be the 6×6 matrix AB. For the peg-in-hole example we have

$$B = \begin{pmatrix} 0 & 0 & 1 & 0 & 0 & 0 \\ 0 & 0 & 0 & 0 & 0 & 1 \end{pmatrix}$$

and

$$A = \begin{pmatrix} 0 & 0 \\ 0 & 0 \\ 1 & 0 \\ 0 & 0 \\ 0 & 0 \\ 0 & 1 \end{pmatrix}.$$

Hence the projection operator is given by:

$$AB = \begin{pmatrix} 0 & 0 & 0 & 0 & 0 & 0 \\ 0 & 0 & 0 & 0 & 0 & 0 \\ 0 & 0 & 1 & 0 & 0 & 0 \\ 0 & 0 & 0 & 0 & 0 & 0 \\ 0 & 0 & 0 & 0 & 0 & 0 \\ 0 & 0 & 0 & 0 & 0 & 1 \end{pmatrix}.$$

This is the group-invariant version of the selection matrix of Raibert and Craig. We can do the same sort of thing for the projection onto W. We have

$$W_p = XYW_e,$$

where Y has rows $\mathbf{z}_{n+1}^T, \ldots, \mathbf{z}_6^T$ and X has columns $W_{n+1}, \ldots, W_6$. For our peg-in-hole example we get

$$Y = \begin{pmatrix} 1 & 0 & 0 & 0 & 0 & 0 \\ 0 & 1 & 0 & 0 & 0 & 0 \\ 0 & 0 & 0 & 1 & 0 & 0 \\ 0 & 0 & 0 & 0 & 1 & 0 \end{pmatrix} \quad \text{and} \quad X = \begin{pmatrix} 1 & 0 & 0 & 0 \\ 0 & 1 & 0 & 0 \\ 0 & 0 & 0 & 0 \\ 0 & 0 & 1 & 0 \\ 0 & 0 & 0 & 1 \\ 0 & 0 & 0 & 0 \end{pmatrix}.$$

The projection onto W is thus given by the operator

$$XY = \begin{pmatrix} 1 & 0 & 0 & 0 & 0 & 0 \\ 0 & 1 & 0 & 0 & 0 & 0 \\ 0 & 0 & 0 & 0 & 0 & 0 \\ 0 & 0 & 0 & 1 & 0 & 0 \\ 0 & 0 & 0 & 0 & 1 & 0 \\ 0 & 0 & 0 & 0 & 0 & 0 \end{pmatrix}.$$

The matrices are exactly the same as those given in standard expositions of hybrid control so our method is equivalent to using a splitting matrix, at least in this coordinate system. The real advantage of our efforts above is that now we know how these projection operators will behave under rigid transformations. Consider passive transformations, that is, coordinate changes; under such a rigid transformation screws and wrenches change according to

$$\tilde{\mathbf{s}} = H^{-1}\mathbf{s} \quad \text{and} \quad \widetilde{\mathcal{W}} = H^T\mathcal{W},$$

where $H = \begin{pmatrix} R & 0 \\ TR & R \end{pmatrix}$. From this we can see that the matrices we defined above, built up from screws and wrenches, have the transformation properties,

$$\tilde{A} = H^{-1}A, \quad \tilde{B} = BH, \quad \tilde{X} = H^T X, \quad \tilde{Y} = YH^{-T},$$

where we have abbreviated $(H^{-1})^T$ to H^{-T}. The projection operators thus transform according to

$$\widetilde{AB} = H^{-1}(AB)H, \quad \widetilde{XY} = H^T(XY)H^{-T}.$$

To see how this works in practice consider the peg-in-hole example once more. Now suppose we change coordinates by displacing the origin a distance r units in the x-direction. For such a transformation the matrix H is given by

$$H = \begin{pmatrix} 1 & 0 & 0 & 0 & 0 & 0 \\ 0 & 1 & 0 & 0 & 0 & 0 \\ 0 & 0 & 1 & 0 & 0 & 0 \\ 0 & 0 & 0 & 1 & 0 & 0 \\ 0 & 0 & -r & 0 & 1 & 0 \\ 0 & r & 0 & 0 & 0 & 1 \end{pmatrix}.$$

In these new coordinates the projection operators are now given by

$$\widetilde{AB} = H^{-1}(AB)H = \begin{pmatrix} 0 & 0 & 0 & 0 & 0 & 0 \\ 0 & 0 & 0 & 0 & 0 & 0 \\ 0 & 0 & 1 & 0 & 0 & 0 \\ 0 & 0 & 0 & 0 & 0 & 0 \\ 0 & 0 & r & 0 & 0 & 0 \\ 0 & r & 0 & 0 & 0 & 1 \end{pmatrix}$$

and

$$\widetilde{XY} = H^T(XY)H^{-T} = \begin{pmatrix} 1 & 0 & 0 & 0 & 0 & 0 \\ 0 & 1 & 0 & 0 & 0 & r \\ 0 & 0 & 0 & 0 & r & 0 \\ 0 & 0 & 0 & 1 & 0 & 0 \\ 0 & 0 & 0 & 0 & 1 & 0 \\ 0 & 0 & 0 & 0 & 0 & 0 \end{pmatrix}.$$

Notice that these matrices are no longer diagonal. Also since they act on different vector spaces it makes no sense to add them, so they cannot be said to sum to the identity. Hence, they cannot be interpreted as splitting matrices in the sense defined by Raibert and Craig. Remember that in the present coordinates the freedom screws are

$$\tilde{\mathbf{z}}_1 = \begin{pmatrix} \mathbf{k} \\ r\mathbf{j} \end{pmatrix}, \quad \tilde{\mathbf{z}}_2 = \begin{pmatrix} \mathbf{0} \\ \mathbf{k} \end{pmatrix}$$

and the constraint wrenches are

$$\widetilde{\mathcal{W}}_3 = \begin{pmatrix} \mathbf{i} \\ \mathbf{0} \end{pmatrix}, \quad \widetilde{\mathcal{W}}_4 = \begin{pmatrix} \mathbf{j} \\ \mathbf{0} \end{pmatrix}, \quad \widetilde{\mathcal{W}}_5 = \begin{pmatrix} \mathbf{0} \\ \mathbf{i} \end{pmatrix}, \quad \widetilde{\mathcal{W}}_6 = \begin{pmatrix} -r\mathbf{k} \\ \mathbf{j} \end{pmatrix}.$$

Notice for example, that an instantaneous rotation about the new y-axis is projected onto a translation in the z-direction.

15.4.4   The Second Fundamental Form

The control scheme outlined above only takes into account the linear structure of the constraint manifold. Second order effects are important here. Suppose that a point on the robot's end-effector is constrained to lie on a surface in space. Now imagine the end-effector point at a concave point on the surface; one possible unconstrained motion of the robot would be to move so that the point on the end-effector moves along a tangent to the surface. Such a motion would penetrate the surface so we would expect a constraint wrench directed towards the surface's centre of curvature. Compare this with the material on gripping curved objects at the end of Section 12.5.

The idea behind this section is to reduce the dynamics to the constraint space. In order to do this we need to introduce 'fictitious forces'. This is familiar from particle dynamics, for example suppose a particle is constrained to circle a fixed point, say a bead on a circular wire, then centrifugal and coriolis forces must be included. From the viewpoint of differential geometry these fictitious forces correspond to the second fundamental form of the constraint manifold.

Think of the end-effector of a robot as a single rigid body subject to constraints. The equations of motion for the end-effector can be written

$$N_6 \dot{V} + \{V, N_6 V\} = \mathcal{W},$$

where N_6 is the 6×6 inertia matrix of the end-effector, V its velocity screw and $\mathcal{W}$ the total wrench acting on the body; see Chapter 13. The kinetic energy of the rigid body is given by, $E_K = (1/2)V^T N_6 V$. In the absence of external forces and potentials, the orbits of this dynamical system are given by the geodesics of the metric defined by the kinetic energy. So, the equations of motion can be rewritten in terms of a covariant derivative as

$$N_6 \nabla_V V = \mathcal{W},$$

with ∇ the covariant derivative of a metric connection; see [2, Appendix 1 & 2]. The metric will be the one given by N_6, that is, the kinetic energy of the end-effector.

Now if we restrict this equation to a subspace, for example, a constraint subspace then we must introduce the **second fundamental form** or **shape operator** of the subspace; see [82, Chap. 4],

$$N_6 \nabla_V V = N_6 \tilde{\nabla}_V V + N_6 \, II(V, V).$$

Here $\tilde{\nabla}$ is the covariant derivative restricted to the subspace. The vector field $V = \dot{\mathbf{q}}_6$, the velocity of the end-effector, must be tangent to the constraint subspace, that is the velocity must be consistent with the constraints. The term $II(V, V)$ is the second fundamental form of the constraint subspace.

In terms of geodesics we have the following interpretation of the second fundamental form: As we saw in Section 15.2, a geodesic curve in the constraint space may not be a geodesic in the ambient space. However, the difference between the two curves is given by the second fundamental form. The second fundamental form tells us about extrinsic curvature, that is, the geometry of how the subspace lies in the ambient space. In this case it concerns how the constraint space lies in the group of rigid body motions. The metric in this case will be N_6. If the constraint subspace is a subgroup, then it is flat, that is has zero curvature, with respect to a bi-invariant metric. However, since the N_6 metric will not be bi-invariant, only left-invariant, even subgroups will have non-zero curvature.

Now if we pre-multiply the second fundamental form by the metric we get the wrench needed to maintain the constraint. In the example given above concerning a robot contacting a concave surface, this gives us the force towards the centre of curvature. This force is given by the expression $N_6 \, II(V, V)$.

The metric N_6 gives us a notion of orthogonality which simplifies the projection operators considered above. It is always possible to find a basis for the Lie algebra, $\mathbf{z}_1, \ldots, \mathbf{z}_6$ such that

$$\mathbf{z}_i^T N_6 \, \mathbf{z}_j = \begin{cases} 1, & \text{if } i = j, \\ 0, & \text{if } i \neq j, \end{cases}$$

where $\mathbf{z}_1, \ldots, \mathbf{z}_n$ spans the space of freedom screw, that is, the tangent space to the constraint subspace. Notice, the constraint space is assumed to be n-dimensional. The constraint wrenches then have a basis, $\mathcal{W}_j = N_6 \mathbf{z}_j$ for $j = n+1, \ldots, 6$. This definition guarantees that the constraint wrenches and freedom screws satisfy

$$\mathcal{W}_i^T \mathbf{z}_j = \begin{cases} 1, & \text{if } i = j, \\ 0, & \text{if } i \neq j. \end{cases}$$

Now the projection operators can be written as follows; the one onto the space of freedom screws is $AA^T N_6$, where as before A is the $6 \times n$ matrix whose columns are the screws $\mathbf{z}_1, \mathbf{z}_2, \ldots, \mathbf{z}_n$. The projection onto the space of constraint wrenches is given by $XX^T N_6^{-1} = N_6 Y^T Y$, where X has columns $\mathcal{W}_{n+1}, \ldots, \mathcal{W}_6$ or in terms of screws, Y has rows $\mathbf{z}_{n+1}^T, \ldots, \mathbf{z}_6^T$.

By definition, the second fundamental form is normal to the constraint subspace, and the term $\widetilde{\nabla}_V V$ is tangent to it. So we can project onto the space of constraint wrenches to get

$$XX^T N_6^{-1}(N_6 \nabla_V V) = N_6 \, II(V, V).$$

If we write the constraint wrench as

$$N_6 \, II(V, V) = \lambda_{n+1} \mathcal{W}_{n+1} + \lambda_{n+2} \mathcal{W}_{n+2} + \cdots + \lambda_6 \mathcal{W}_6,$$

then a short calculation, using the equations of motion for a single rigid body with no external wrenches, reveals

$$\lambda_j = \mathbf{s}_j^T N_6 \dot{V} + V^T N_6 [\mathbf{s}_j, V], \qquad j = n + 1, \ldots, 6.$$

Now consider a six-joint serial robot with joint screws $\mathbf{s}_i$. Using the results of Chapter 13 above, we get

$$\sum_{j=i}^{6} (\ddot{\mathbf{q}}_j^T N_j \mathbf{s}_i + \dot{\mathbf{q}}_j^T N_j [\mathbf{s}_i, \dot{\mathbf{q}}_j] - \mathcal{G}_j^T \mathbf{s}_i) = \tau_i + \sum_{k=n+1}^{6} \lambda_k \mathcal{W}_k^T \mathbf{s}_i \qquad i = 1, 2, \ldots, 6,$$

where $\mathcal{G}_i$ is the wrench due to gravity on the i-th link, N_i is the inertia matrix of the i-th link and $\dot{\mathbf{q}}_i = \dot{\theta}_1 \mathbf{s}_1 + \cdots + \dot{\theta}_i \mathbf{s}_i$ is the velocity screw of the i-th link. The term $\sum_{k=n+1}^{6} \lambda_k \mathcal{W}_k$ is the constraint wrench acting on the end-effector as above. The λ_i terms are given as above by

$$\lambda_j = \mathbf{z}_j^T N_6 \ddot{\mathbf{q}}_6 + \dot{\mathbf{q}}_6^T N_6 [\mathbf{z}_j, \dot{\mathbf{q}}_6], \qquad j = n + 1, \ldots, 6.$$

An equivalent form of the dynamics of a robot with end-effector constraints was also given by McClamroch [74] and Yoshikawa [131].

Finally we are in a position to see how these ideas affect hybrid control. We saw above that if a body is in contact with a system of constraints, then a motion that is tangential to the constraint space can still generate reaction forces. Hence to isolate the positional control on the constraint space from the force control normal to it, we must require the desired motion to be one that produces no reaction. That is, we must filter the desired motion by requiring that the velocity is tangent to the constraint space and the derivative of the velocity produces no reaction wrench. Now in order that the λ_js all vanish we must have

$$\mathcal{W}_j^T \ddot{\mathbf{q}}_6 = \dot{\mathbf{q}}_6^T N_6 [\dot{\mathbf{q}}_6, \mathbf{z}_j] \qquad j = n + 1, \ldots, 6.$$

For a simple control method we can produce an error velocity $\dot{\mathbf{q}}_e$, from the desired velocity $\dot{\mathbf{q}}_d$ and the measured velocity $\dot{\mathbf{q}}_6$,

$$\dot{\mathbf{q}}_e = AA^T N_6(\dot{\mathbf{q}}_d - \dot{\mathbf{q}}_6).$$

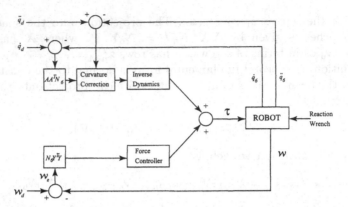

FIGURE 15.2. Hybrid Control Architecture with Compensation for Constraint-Space Curvature

The derivative of the velocity will be similar but acquires a correction term to ensure that the motion follows the constraint space,

$$\ddot{\mathbf{q}}_e = AA^T N_6(\ddot{\mathbf{q}}_d - \ddot{\mathbf{q}}_6) + AA^T\{\dot{\mathbf{q}}_e, N_6\dot{\mathbf{q}}_e\}.$$

Notice that if the motion of the robot's end-effector produces no reaction wrench, then there is no difference between physical or artificial constraints.

The desired wrench must be normal to the constraint subspace. This can be achieved using the projection operator

$$XX^T N_6^{-1} = N_6 Y^T Y.$$

The torques demanded of the joints can be split into a contribution from the motion controller τ_i^m and a contribution from the force controller τ_i^f. The ith joint will be supplied with the torque $\tau_i = \tau_i^m + \tau_i^f$.

For the force controller we have

$$\tau_i^f = \sum_{j=k+1}^{6} \mathcal{W}_d^T \mathbf{z}_j \mathcal{W}_j^T \mathbf{s}_i.$$

That is, τ_i^f is the contribution to the joint torque needed to exert the desired wrench $\mathcal{W}_d$. This control method is illustrated in Figure 15.2.

In practice, all the control methods sketched here depend on exact cancellation of terms in the equations of motion. So there will always be some mismatch between the dynamical model and the real machine. Also the output of a control system is usually a voltage applied to the motors rather than a torque or force. Hence real control systems must take account of the electo-mechanical properties of the robot's motors. Despite these complications hybrid control has been applied successfully to real robots.

References

[1] R. Abraham and J.E. Marsden. *Foundations of Mechanics.* Benjamin Cummings, Reading, MA, second edition, 1984.

[2] V.I. Arnol'd. *Geometrical Methods of Classical Mechanics*, volume 60 of *Graduate Texts in Mathematics.* Springer-Verlag, New York, 1978.

[3] H. Asada and J-J. E. Slotine. *Robot Analysis and Control.* John Wiley and Sons, New York, 1986.

[4] H. Asada and K. Youcef-Toumi. *Direct Drive Robots: Theory and Practice.* MIT Press, Cambridge MA, 1986.

[5] L. Auslander. *Differential Geometry.* Harper & Row, New York, 1967.

[6] R.S. Ball. *The Theory of Screws.* Cambridge University Press, Cambridge, 1900.

[7] P.G. Bamberg and S. Sternberg. *A Course in Mathematics for Students of Physics*, volume 1. Cambridge University Press, Cambridge, 1988.

[8] M. Berger and B. Gostiaux. *Differential Geometry: Manifolds, Curves and Surfaces.* Springer-Verlag, New York, 1988.

[9] H.A. Bernstein and A.V. Phillips. Fiber bundles and quantum theory. *Scientific American*, 245:94–109, 1981.

[10] N.L. Biggs. *Discrete Mathematics.* Oxford University Press, Oxford, 1985.

[11] R.L. Bishop and R.J. Crittenden. *Geometry of Manifolds*. Academic Press, New York, 1964.

[12] W. Blaschke. *Kinematic und Quaternionen*. VEB Verlag, Berlin, 1960.

[13] W.M. Boothby. *An Introduction to Differentiable Manifolds and Riemannian Geometry*. Academic Press, Orlando, FL, second edition, 1986.

[14] O. Bottema and B. Roth. *Theoretical Kinematics*. North-Holland Publishing, Amsterdam, 1979. Reprinted by Dover, New York in 1990.

[15] Th. Bröcker and K. Jänich. *Introduction to Differentiable Topology*. Cambridge University Press, Cambridge, 1982.

[16] R.W. Brockett. *Robotic Manipulators and Products of Exponential Formulas*, volume 58 of *Lecture Notes in Computer Science*, pages 120–129. Springer-Verlag, New York, 1984. Proceedings of conference held in Beer Sheva 1983.

[17] F. Bullo and R.M. Murray. Proportional derivative (PD) control on the Euclidean group. In *Proceedings of the 1995 European Control Conference, Rome, Italy*, 1995.

[18] J.W. Burdick. A classification of 3R regional manipulator singularities and geometries. *Mechanism and Machine theory*, 30:71–89, 1995.

[19] E. Celledoni and B. Owren. Lie group methods for rigid body dynamics and time integration on manifolds. http://www.math.ntnu.no/num/synode, 1999.

[20] W.K. Clifford. Preliminary sketch of biquaternions. *Proceedings of the London Mathematical Society*, iv(64/65):381–395, 1873.

[21] P.M. Cohn. *Algebra*, volume 1. John Wiley and Sons, London, 1974.

[22] A.J. Coleman. The greatest mathematical paper of all time. *Mathematical Intelligencer*, 11(3):29–38, 1989.

[23] M.L. Curtis. *Matrix Groups*. Springer-Verlag, New York, 1979.

[24] D. Downing, A. Samuel and K. Hunt. Identification of the special configuration of the octahedral manipulator using the pure condition. *International Journal of Robotics Research*, 21:147–159, 2002.

[25] M.G. Darboux. De l'emploi des fonctions elliptiques dans la théorie du quadrilatère plan. *Bull. des Sciences Mathématiques*, pages 109–128, 1879.

[26] J.A. Dieudonné and J.B. Carrell. *Invariant Theory, Old and New*. Academic Press, New York, 1971.

[27] F.M. Dimentberg. *The screw calculus and its applications in mechanics.* Izd. Nauka, Moscow, 1965. Translated into English, Foriegn Technology Division, Wright-Patterson AFB, Ohio, 1968.

[28] P.S. Donelan and C.G. Gibson. First-order invariants of euclidean motions. *Acta Appl. Math.*, 24:233–251, 1991.

[29] P.S. Donelan and C.G. Gibson. On the hierarchy of screw systems. *Acta Appl. Math.*, 32:267–296, 1993.

[30] J. Duffy. *Analysis of Mechanisms and Robot Manipulators.* Edward Arnold, London, 1980.

[31] W.L. Edge. *The Theory of Ruled Surfaces.* Cambridge University Press, Cambridge, 1931.

[32] C. Ehresmann. Sur la topologie de certaines espaces homogènes. *Ann. of Math.*, 396–443, 1934.

[33] R. Featherstone. *Robot Dynamics Algorithms.* Kluwer, Boston, MA, 1991.

[34] A. Fijany and A.K. Bejczy. Efficient jacobian inversion for the control of simple robot manipulators. In *Proceedings of the 1988 IEEE International Conference on Robotics and Automation, Philadelphia, PA.*, 1988.

[35] W. Fulton. *Intersection Theory.* Springer-Verlag, Berlin, 1984.

[36] W. Fulton and J. Harris. *Representation Theory*, volume 129 of *Graduate Texts in Mathematics.* Springer-Verlag, New York, 1991.

[37] C.G. Gibson and K.H. Hunt. Geometry of screw systems. *Mechanism and Machine Theory*, 25:1–27, 1990.

[38] R. Gilmore. *Lie Groups, Lie Algebras, and some of their Applications.* John Wiley and Sons, New York, 1974.

[39] H. Goldstein. *Classical Mechanics.* Addison-Wesley, Reading, MA, second edition, 1980.

[40] M. Greenberg. *Lectures on Algebraic Topology.* Benjamin, New York, 1966.

[41] P. Griffiths. On Cartan's method of Lie groups and moving frames as applied to uniqueness and existence questions in differential geometry. *Duke Math. Journal*, 41:775–814, 1974.

[42] P. Griffiths and J. Harris. *Principles of Algebraic Geometry.* John Wiley and Sons, New York, 1976.

[43] H.W. Guggenheimer. *Differential Geometry*. McGraw-Hill, New York, 1963.

[44] J. Harris. *Algebraic Geometry, A First Course*, volume 133 of *Graduate Texts in Mathematics*. Springer-Verlag, New York, 1992.

[45] R. Hartshorne. *Algebraic Geometry*, volume 52 of *Graduate Texts in Mathematics*. Springer-Verlag, New York, 1977.

[46] F. Hausdorff. Die symbolische exponential formel in den gruppen theorie. *Berichte de Sächicen Akademie de Wissenschaften (Math Phys Klasse)*, 58:19–48, 1906.

[47] T. Hawkins. The birth of Lie's theory of groups. *Mathematical Intelligencer*, 16(2):6–17, 1994.

[48] S. Helgason. *Differential Geometry, Lie Groups, and Symmetric Spaces*, volume 80 of *Pure and Applied Mathematics*. Academic Press, New York, 1978.

[49] J.M. Hervé. Analyse structurelle de méchanismes par groupe des déplacement. *Mechanism and Machine Theory*, 13:437–450, 1978.

[50] D. Hestenes and G. Sobczyk. *Clifford Algebra to Geometric Calculus: A Unified Language for Mathematics and Physics*. D. Reidel, Dordrecht, 1984.

[51] H. Hiller. *Geometry of Coxeter Groups*, volume 54 of *Research Notes in Mathematics*. Pitman, London, 1982.

[52] W.V.D. Hodge and D. Pedoe. *Methods of Algebraic Geometry*, volume 2. Cambridge University Press, Cambridge, 1952.

[53] J.M. Hollerbach. A recursive Lagrangian formulation of manipulator dynamics and a comparative study of dynamics formulation complexity. *IEEE Trans. on Systems, Man and Cybernetics*, SMC-10(11):730–736, 1980.

[54] K.H. Hunt. *Kinematic Geometry of Mechanisms*. Clarendon Press, Oxford, 1978.

[55] K.H. Hunt. Robot kinematics—a compact analytic inverse solution for velocities. Presented at A.S.M.E. 19th Mechanisms Conference, Columbus OH as paper 86-DET-127, October 1986.

[56] K.H. Hunt. Manipulating a body through a finite screw displacement. In *Proceedings of the 7th World Congress, Seville*, volume 1, pages 187–191, 1987.

[57] K.H. Hunt and I.A. Parkin. Finite displacements of points, planes, and lines via screw theory. *Mechanism and Machine Theory*, 30(2):177–192, 1995.

[58] D. Husemöller. *Fibre Bundles*, volume 20 of *Graduate Texts in Mathematics*. Springer-Verlag, New York, third edition, 1993.

[59] M.L. Husty. An algorithm for solving the direct kinematics of Stewart-Gough-type platforms. Technical report, McGill University, June 1994. Centre for Intelligent Machines, cim-94-1.

[60] A. Iserles and S.P. Norsett. On the solution of linear differential equations in lie groups. *Phil. Trans. Royal Soc. A*, 357:983–1019, 1999.

[61] K. Kanatani. *Geometric Computation For Machine Vision*. Clarendon Press, Oxford, 1993.

[62] A. Karger. Singularity analysis of serial robot-manipulators. *A.S.M.E. J. Mech. Design,*, 118:520–525, 1996.

[63] F. Klein. Notiz betreffend dem Zusammenhang der Linengeometrie mit der Mechanik starrer Körper. *Math. Ann.*, 4:403–415, 1871.

[64] D.E. Knuth. *The Art of Computer Programming*, volume 1. Addison Wesley, Reading MA, third edition, 1997.

[65] S. Kobayashi and K. Nomizu. *Foundations of Differential Geometry*. John Wiley and Sons, New York, 1969.

[66] D. Koditschek. Robot kinematics and coordinate transformations. In *Proceedings of the 24th Conference on Decision and Control, Ft. Lauderdale, Fl*, pages 1–4. IEEE, 1985.

[67] H.Y. Lee and C.G. Liang. Displacement analysis of the general 7-link 7-R mechanism. *Mechanisms and Machine Theory*, 23(3):219–226, 1988.

[68] H. Lipkin and J. Duffy. Hybrid twist and wrench control for a robotic manipulator. *Trans. ASME J. Mechanism, Transmission Automation Design*, 110:138–144, 1988.

[69] P. Lounesto. *Clifford Algebras and Spinors*. Number 286 in London Mathemetical Society Lecture Note Series. Cambridge University Press, Cambridge, second edition, 2001.

[70] R. Manseur and K.L. Doty. A robot manipulator with 16 real inverse kinematic solution sets. *International Journal of Robotics Research*, 8(5):75–79, 1989.

[71] C.R.F. Maunder. *Algebraic Topology*. Cambridge University Press, Cambridge, 1980.

[72] J.M. McCarthy. The generalization of line trajectories in spatial kinematics to trajectories of great circles on a hypersphere. *A.S.M.E. Journal of Mechanisms, Transmissions and Automation in Design*, 108:60–64, 1986.

[73] J.M. McCarthy. *An Introduction to Theoretical Kinematics*. MIT Press, Cambridge, MA, 1990.

[74] N.H. McClamroch and D. Wang. Feedback stabilization and tracking of constrained robots. *IEEE Trans. Automatic Control*, 33(5):419–426, 1988.

[75] P.J. McKerrow. *Introduction to Robotics*. Addison-Wesley, Sydney, 1991.

[76] W. Miller. *Lie Theory and Special Functions*. Academic Press, New York, 1968.

[77] J. Milnor. *Morse Theory*, volume 51 of *Annals of Mathematics Studies*. Princeton University Press, Princeton, NJ, 1969.

[78] B. Mishra, J.T. Schwartz, and M. Sharir. On the existence and synthesis of multifingered positive grips. *Algorithmica*, 2:541–558, 1987.

[79] E.D. Nering. *Linear Algebra and Matrix Theory*. John Wiley, New York, second edition, 1970.

[80] V.D. Nguyen. Constructing force closure grasps. *International Journal of Robotics Research*, 7(3):3–16, 1988.

[81] B. O'Neill. *Elementary Differential Geometry*. Academic Press, New York, 1966.

[82] B. O'Neill. *Semi-Riemannian geometry: with applictions to relativity*. Academic Press, New York, 1983.

[83] F.C. Park and J.E. Bobrow. A recursive algorithm for robot dynamics. In *Proceedings of the 1994 IEEE International Conference on Robotics and Automation, San Diego*, volume 2, pages 1535–1540. IEEE, 1994.

[84] R.P. Paul and C.N. Stevenson. Kinematics of robot wrists. *International Journal of Robotics Research*, 2(1):31–38, 1983.

[85] J. Phillips. *Freedom in Machinery: Introducing Screw Theory*, volume 1. Cambridge University Press, Cambridge, 1984.

[86] J. Phillips. *Freedom in Machinery: Screw Theory Exemplified*, volume 2. Cambridge University Press, Cambridge, 1990.

[87] D.L. Pieper. *The Kinematics of Manipulators under Computer Control*. PhD thesis, Stanford University, 1968.

[88] I.R. Porteous. *Topological Geometry*. Cambridge University Press, Cambridge, second edition, 1981.

[89] I.R. Porteous. *Geometric Differentiation*. Cambridge University Press, Cambridge, second edition, 2001.

[90] M. Postnikov. *Lie Groups and Lie Algebras*. Mir, Moscow, 1986.

[91] M.H. Raibert and J.J. Craig. Hybrid position/force control of manipulators. *A.S.M.E. Journal of Dyn. Sys. Measurement Control*, 102:126–133, 1981.

[92] M. Reid. *Undergraduate Algebraic Geometry*, volume 12 of *London Mathematical Society Student Texts*. Cambridge University Press, Cambridge, 1990.

[93] F. Reuleaux. *Theortische Kinematic: Grunzüge einer Theorie des Maschinwesens*. Vieweg, Braunschweig, 1875. Reprinted as "Kinematics of Machinery" by Dover, New York, 1963.

[94] E. Rimon and J. Burdick. Mobility of bodies in contact–I: A new 2nd order mobility index for multiple-finger grasps. In *Proceedings of the 1994 IEEE International Conference on Robotics and Automation, San Diego*, volume 3, pages 2329–2335, 1994.

[95] J.J. Rooney. On the principle of transference. In *Proceedings of the Fourth World Congress on the Theory of Machines and Mechanisms*, pages 1089–1094, London, 1975. Institute of Mechanical Engineers.

[96] G. Salmon. *Conic Sections*. Longmans, Green and Co., London, sixth edition, 1879.

[97] A.E. Samuel, P.R. McAree, and K.H. Hunt. Unifying screw geometry and matrix transformations. *International Journal of Robotics Research*, 10(5):454–472, 1991.

[98] J.M. Sanz-Sena and M.P. Calvo. *Numerical Hamiltonian Problems*, volume 7 of *Applied Mathematics and Mathematical Computation*. Chapman and Hall, London, 1994.

[99] B.F. Schutz. *Geometrical Methods of Mathematical Physics*. Cambridge University Press, Cambridge, 1980.

[100] J M. Selig. Clifford algebra of points, lines and planes. *Robotica*, 18:545–556, 2000.

[101] J M. Selig and J.J. Rooney. Reuleaux pairs and surfaces that cannot be gripped. *International Journal of Robotics Research*, 8(5):79–86, 1989.

[102] J.M. Selig. A note on the principle of transference. Presented at A.S.M.E. 19th Mechanisms Conference, Columbus OH as paper 86-DET-174, October 1986.

[103] J.M. Selig. *Introductory Robotics*. Prentice Hall, London, 1992.

[104] J.M. Selig. Curvature in force/position control. In *Proceedings of the 1998 IEEE International Conference on Robotics and Automation, Leuven, Belgium*, volume 2, pages 1761–1766, 1998.

[105] J.M. Selig and P.R McAree. A simple approach to invariant hybrid control. In *Proceedings of the 1996 IEEE International Conference on Robotics and Automation, Minneapolis*, volume 3, pages 2238–2245, 1996.

[106] J.M. Selig and P.R McAree. Constrained robot dynamics I: Serial robots with end-effector constraints. *Journal of Robotic Systems*, 16(9):471–486, 1999.

[107] J.M. Selig and P.R McAree. Constrained robot dynamics II: Parallel machines. *Journal of Robotic Systems*, 16(9):487–498, 1999.

[108] J.M. Selig and A.I. Ovseevitch. Manipulating robots along helical trajectories. *Robotica*, 14(2):261–267, 1996.

[109] J.G. Semple and L. Roth. *Introduction to Algebraic Geometry*. Clarendon Press, Oxford, 1949. Reprinted in 1985 with index.

[110] J.E. Shigley and J.J. Uicker. *Theory of Machines and Mechanisms*. McGraw Hill, New York, 1980.

[111] G.E. Shilov. *Linear Algebra*. Prentice Hall, Englewood Cliffs, NJ, 1971.

[112] D.R. Smith. *The Design of Solvable 6R Manipulators*. PhD thesis, Georgia Institute of Technology, 1990.

[113] G. Sobczyk. The generalized spectral decomposition of a linear operator. *College Mathematics Journal*, pages 27–38, 1997.

[114] M.W. Spong. Remarks on robot dynamics: Canonical transformations and Riemannian geometry. In *Proceedings of the 1992 IEEE International Conference on Robotics and Automation, Nice, France*, volume 1, pages 554–559, 1992.

[115] M.W. Spong and M. Vidyasagar. *Robot Dynamics and Control*. John Wiley and Sons, New York, 1989.

[116] S. Sternberg. *Lectures on Differential Geometry*. Prentice Hall, Englewood Cliffs, NJ, 1964.

[117] A. Stokes and R. Brockett. Dynamics of kinematic chains. *International Journal of Robotics Research*, 15(4):393–405, 1996.

[118] E. Study. von den Bewegungen und Umlegungen. *Math. Ann.*, 39:441–566, 1891.

[119] J.D. Talman. *Special Functions: A Group Theoretic Approach*. Benjamin, New York, 1968.

[120] W.P. Thurston and J.R. Weeks. The mathematics of three-dimensional manifolds. *Scientific American*, 251(1):94–106, 1984.

[121] N.J. Vilenkin. *Special Functions and the Theory of Group Representations*, volume 22 of *Translations of Mathematical Monographs*. American Mathematical Society, Providence, RI, 1968.

[122] J.K. Waldron. A method of studying joint geometries. *Mechanism and Machine Theory*, 7:347–355, 1972.

[123] C.W. Wampler. Forward displacement analysis of general six-in-parallel SPS (Stewart) platform manipulators using soma coordinates. *Mechanism and Machine Theory*, 31(3):331–337, 1996.

[124] C.W. Wampler and A.P. Morgan. Solving the kinematics of general 6R manipulators using polynomial continuation. In K. Warwick, editor, *Robotics: Applied Mathematical and Computational Aspects*, volume 41 of *The Institute of Mathematics and its Applications Conference Series*, pages 57–69, Oxford, 1993. Oxford University Press.

[125] P. Wenger. Classification of 3R positioning manipulators. *A.S.M.E J. Mech. Design*, 120(2):327–332, 1998.

[126] H. Weyl. *The Classical Groups*. Princeton University Press, Princeton, NJ, 1946.

[127] N. White. Grassmann-Cayley algebra and robotics. *J. Intell. Robot Syst.*, 11:97–10, 1994.

[128] E.T. Whittaker. *Analytical Dynamics of Particles and Rigid Bodies*. Cambridge University Press, Cambridge, fourth edition, 1937.

[129] N.M.J. Woodhouse. *Introduction to Analytical Dynamics*. Clarendon Press, Oxford, 1987.

[130] I.M. Yaglom. *Complex Numbers in Geometry*. Academic Press, New York, 1968.

[131] T. Yoshikawa. Dynamic hybrid position/force control of robot manipulators—description of hand constraints and calculation of joint driving force. *IEEE J. Robotics and Automation*, RA-3(5):386–392, 1987.

Index

A-matrices, 47, 71
A-planes, 246, 248, 250, 252, 253,
    260, 263
Abraham, R, 271
acceleration, 100, 103, 104, 109,
    166, 294, 295, 299, 363
accidental isomorphisms, 20, 133,
    204, 244
accidental property, 56, 59, 287
actuators, 1
adjoint representation, 54–57, 79,
    157, *see also under*
    individual groups
    of a Lie algebra, 58, 141
affine variety, 5, 14, 123, 282
aircraft simulators, 264
algebraic geometry, 4–7, 113, 356
algebraic set, 7
    irreducible, 5
algebraic topology, 257
α-planes, 118, 119, 176, 182, 247,
    249, 257, 283
alternating tensor, 54, 160
angular momentum, 287

angular velocity, 53, 74–76, 100,
    108, 109, 360
anti-commutators, 199
anti-involution, 201
anti-symmetric matrices, 34, 39,
    52, 53, 58, 63, 118, 131,
    132, 136, 137, 159, 300,
    307
anti-symmetric powers, 146, 153,
    164, 193, 206, 214
anti-symmetric product, 136, 145,
    147, 164, 173, 184, 205
Arnol'd, V.I., 271, 287, 308, 361,
    370
artificial intelligence, 3
Artobolevskii, I.I, 3
Asada, H., 309, 364
assembly configurations, 256, 262,
    263
associative algebras, 198, 205
associativity, 12, 14, 22
Auslander, L., 9

B-planes, 246, 248, 250, 260, 263
ball and socket joint, 44, 264

Ball's point, 104–105, 108, 110
Ball, R.S., v, 2, 51, 105, 159, 163, 289, 305
Bamberg, P.G., 133
Bejczy, A.K, 133
Berger, M., 9
Bernstein, H.A., 77
β-planes, 118, 119, 176, 182, 247, 249, 257, 311
Bézout's theorem, 7, 107, 257, 268
bidegree, 257
bifurcation surface, 94, 96
Biggs, N.L., 321
bijective maps, 20, 201, 243
bilinear form, 18
    anti-symmetric, 17, 39
    degenerate, 80, 202
    invariant, 80, 121, 144, 156, 157, 165, 352
    positive definite, 15, 144, 289
    symmetric, 33, 80, 120, 122, 146, 199, 202, 289, 311, 352
biped walking, 340–342
Bishop, R.L., 9, 318
Blaschke, W., 197
Bobrow, J.E., 297
Bolshevik revolution, 159
Bottema, O., 110
Bröcker, Th., 51
Brockett, R.W., 71, 312
Bullo, F., 73, 77
Burdick, J., 97, 283
Burmester points, 106–108

Calvo, M.P., 309
Campbell–Baker–Hausdorff formula, 81–83, 357
    for $se(2)$, 99, 209
    for $so(3)$, 81
canonical transformations, 17, 307–309, 316
Carrell, J.B., 147
Cartan, E., vi, 8, 11

Cayley, A., 11, 147
Cayley-Hamilton theorem, 65
Celledoni, E., 76
cellular decomposition, 258
centre of curvature, 101, 105, 369
centre of mass, 288, 289, 291, 293, 301, 309
centre of rotation, 98, 100, 208
Chasles's theorem, 25
Chebychev, P.L., 2
Christoffel symbols, 350, 352, 354
Cincinnati Milacron, 234
circle, 13, 16, 90, 93, 101, 104, 107, 110, 165, 219, 283, 360
classical groups, 15
Clifford algebra
    products, 199
Clifford algebras, 197–220, 243, 245
    anti-involution, 201
    even subalgebra, 200, 206, 210
    main involution, 202, 204, 207
    odd subspace, 200
    units, 202
Clifford, W.K., 2, 11, 163, 198
closed subspace, 258
co-screws, 195, 271–273, 278, 281, 287, 289, 292, 353
coadjoint representation, 154
codimension, 6
coefficient of friction, 283
Cohn, P.M., 33, 34, 38, 39
cohomology, 260
Coleman, A.J., 11
collineation, 4
commutation relations, 59, 60, 80, 148, 154, 157, 208, 291
commutative algebra, 4
commutative groups, 12, 14, 22, 258
commutators, 57–61, 78, 141, 166, 206, 208, 209, 212, 350, 357, 358
complete flags, 252, 259
complete intersection, 6, 267

complex numbers, 142, 148, 199,
        242, 356
    conjugate, 13, 18, 107, 201
    unit modulus, 12, 15, 18–20,
        207
complex plane, 15
compliance, 277
compliance matrix, 278
computed torque control, 297, 343
computer algebra system, 70, 76,
        240, 300
computer controlled machine tools,
        3
computer graphics, 37
computer scientists, 3, 197
cone, 283
configuration space, 318, 355
conformal group, 133
congruence, 23, 33, 34, 38, 199
conical double point, 96
conjugacy class, 43, 194
conjugate, 33, 43, 55
conjugate momentum, 306, 317
conjugate screws of inertia, 289,
        291, 310
conjugation, 26, 32, 35, 41, 42, 47,
        54, 57, 98, 167, 168, 170,
        201, 208
connection, 318, 352
    symmetric, 350, 354
consistent equations, 27, 105, 115,
        180
continuous groups, 11
contractible loop, 78
contraction, 228–230
contravariant vectors, 271
control systems, 72, 296, 300, 309,
        358, 360–372
control theory, 38, 46
convex programming, 284
convex sets, 284
convexity theory, 280
coodinate basis, 349
coordinate patch, 7, 20

corank, 248
Coriolis terms, 295, 303, 307, 311,
        360
Coriolis theorem, 293
coset space, 34
cosets, 34, 248
couple, 273
covariant, 186
covariant derivative, 350–353, 358
covariant vectors, 271
covariants, 139
Craig, J.J., 363, 367
Crittenden, R.J., 9, 318
crystallographic groups, 41
cubic of stationary curvature,
        105–106, 108, 110
Curtis, M.L., 62, 203, 249, 250
curvature, 101, 102, 105, 113
    principal axes, 129
curves, 4, 44, 45, 53, 101–103
    asymptote, 106
    conic, 91, 93, 104, 125, 126,
        180, 311
    cubic, 4, 105, 107
    elliptic, 108
    evolute, 105
    geodesic, 351, 357
    helix, 111, 219, 360
    in the Klein quadric, 124, 127
cuspidal robot, 97
cyclic coordinate, 312
cylinder, 44, 45, 132, 357
cylindroid, 126–128, 173, 175, 220

Darboux, G., 2, 108
de Morgan's law
    for cones, 285
    for screw systems, 195
decomposable representation, 143
degree, 4–6, 14, 117, 122, 146, 147,
        200, 201, 206, 208, 257,
        259
Descartes, R., 4
design, 309, 311

design parameters, 88, 94, 95, 235, 238, 240, 339

determinant, 9, 14, 16, 18, 21, 24, 34, 35, 52, 62, 72, 73, 78, 89, 103, 105, 106, 109, 118, 124, 131, 136, 137, 190, 242, 248, 311, 318

Devol, G., 3

diagonal matrix, 33, 36, 38, 39, 172, 178, 199, 277, 289, 291, 292, 309, 310

Dieudonné, J.A., 147

diffeomorphism, 21

differentiable map, 8, 12, 19, 21, 32, 34, 36, 48, 72, 81

differential equations, 8, 37, 44, 61, 147, 279, 351, 356

differential forms, 8, 318, 354

differential geometry, 7–9, 52, 105, 113, 123, 127, 198, 214, 279, 349–372

differential invariants, 217, 219

differential operator, 8, 147, 153

dihedral group D_2, 171

dimension, 5, 9, 14, 16, 19, 20
    of a cell, 259
    of a Clifford algebra, 200, 210
    of a configuration space, 355
    of a Grassmannian, 138
    of a group, 52
    of a Lie algebra, 52, 208
    of a linear space, 250
    of a representation, 22, 24, 146, 147, 158
    of a screw system, 164, 356
    of a subgroup, 41, 44
    of a topological space, 355
    of a variety, 257
    of a vector space, 22, 36, 58, 142, 145, 163, 241, 271
    of an algebraic set, 4

Dimentberg, F.M., 3, 197

dimples, 283

direct product, 138

direct sum, 143, 144

directed lines, 123, 138, 214, 215, 254

discriminant, 4, 86, 87, 91

distribution parameter, 217, 219

Donelan, P.S., 164, 173, 183

Doty, K.L., 264

double covering, 19, 77, 123, 138, 141, 150, 203, 204, 206, 207, 210, 211, 243, 254

double numbers, 243

double quaternions, 244

Downing, D., 194

dual angle, 214, 216, 219, 245

dual numbers, 158–161, 200, 210, 212, 213, 217

dual of a cone, 284

dual of a representation, 142, 146, 154

dual of a vector space, 8, 142, 194, 271, 281, 289, 354

dual quaternions, 3, 210–214, 241, 243, 245, 255, 256, 266
    conjugate, 211

dual scalar product, 213, 217

dual vector product, 212, 213, 217

dual vectors, 212, 214, 245, 287

Duffy, J., viii, 221, 231, 363

dynamics, 3, 40, 73, 287–319

Edge, W.L., 123

Ehresmann, C., 260

eigenvalues, 27, 33, 38–40, 62, 143, 145, 148, 150, 151, 155, 248, 290, 291, 306, 311

eigenvectors, 27, 86, 145, 148, 149, 151, 289, 291

elasticity, 277

elbow manipulator, 92

electrical engineers, 3

embedding, 9

end-effector, 46, 47, 71, 72, 78, 85, 262, 274, 276, 277, 304, 305, 358, 360, 362, 364, 369, 371

endomorphisms, 140, 146
Engleberger, J., 3
enumerative geometry, 256
equations of motion, 289, 306
  decoupled, 309–319
  for a robot, 292–297, 300, 301,
    303, 309, 360
equivalence classes, 34, 51, 79, 258
equivalence relation, 34, 51, 79
Euclidean group $SE(2)$, 42, 45, 98,
    108, 171, 176, 247
  C-B-H formula, 99, 209
  Clifford algebra, 207–209
  exponential map, 99, 102, 208
  Lie algebra, 208
Euclidean group $SE(3)$, 22–29, 46,
    52, 71, 119, 122, 148, 172,
    175–177, 211, 245, 263,
    352
  adjoint rep., 60, 68, 153, 154,
    164, 180, 213, 272, 288,
    298
  Clifford algebra, 198, 210–214
  coadjoint rep., 154, 155, 184,
    195, 272, 289, 292, 299
  exponential map, 66, 90, 213,
    245, 256, 310
  group manifold, 349
  Lie algebra, 60, 113, 122, 126,
    128, 153, 156, 163, 212,
    357
  one-parameter subgroups, 245,
    247, 279, 280, 352, 360
  representations, 71, 140, 144,
    153–161, 183, 214, 255,
    301, 302
  subgroups, 41–44, 167, 168,
    176, 247, 357, 358
Euclidean groups $SE(n)$, 24, 207,
    245
  Clifford algebra, 206–207, 210
Euler's equations of motion, 293
Euler's relation, 129
Euler, L., 7

Euler–Lagrange equations, 301,
    351
Euler-Savary equation, 101–103
evaluation map, 142, 154, 157, 195,
    271
exact subspace, 258
exponential map, 61–71, 77, 85, 90,
    128, 141, 168, 357, see
    also under individual
    groups
exponential of a matrix, 61, 81
exterior derivative, 318, 354, 355
exterior product, 205, 225, 226, 228
extrinsic geometry, 9, 113

faithful representation, 140, 154,
    157
feasible region, 280, 282
Featherstone, R., 300
fibre bundles, 20, 253
Fijany, A., 133
fingers, 93, 193, 278, 280, 281, 283,
    285
finite screw, see screw motion
fixed axode, 128
fixed centrode, 102, 108
flag manifolds, 241, 243, 245,
    252–254
flags, 241
force closed grasps, 278, 280, 285
forces, 271–274, 276, 278, 280, 289,
    292, 293, 304, 307, 308,
    316
form closed grasps, 280
forward kinematics, 46, 48, 72, 78,
    85, 231, 352, 354, 357
4-bar mechanism, 2, 108–111,
    336–339, 356
Frenet–Serret formulas, 217
Freudenstein, F., 3
friction, 283–285, 362
friction cone, 283–285
Fulton, W., 139, 151, 203, 250, 257

Galois, E., 11

Gauss, C.F., 7
Gaussian curvature, 130, 318
general linear group $GL(4)$, 132
general linear groups, 164
general linear groups $GL(n)$, 14,
        18, 19, 21, 22, 31, 33, 36,
        39, 147
    complex, 37, 40
General Motors, 3, 238
generalised inertia matrix, 295,
        301, 303, 305, 307,
        309–311, 316, 360
geodesic curves, 316, 351, 352, 357,
        360–362
geodesic deviation, 360
geometric algebras, 199–206
geometric optics, 133
Gibson, C.G., 164, 173, 183
Gibson–Hunt classification,
        165–183
    1-systems, 166
    2-systems, 166–175
    3-systems, 166, 175–183
    4-systems, 166
    5-systems, 166
    A systems, 165, 247, 252
    B systems, 165, 310
    C systems, 165, 247, 253
    D systems, 247
    I systems, 165
    II systems, 165, 247, 252, 310
Gilmore, R., 83, 245
global coordinate system, 16
global dimension, 355
Goldstein, H., 305, 306, 316, 330,
        365
Gostiaux, B., 9
Gramm–Schmidt process, 39, 172,
        176, 178, 197, 289, 331
graph theory, 321
Grassmann algebra, 194, 198, 205
Grassmann algebras
    products, 205

Grassmann–Cayley algebra, 194,
        198, 226
Grassmannians, 135–138, 147, 164,
        167, 169, 175, 241, 250,
        257
    dimensions, 138
    oriented, 138
gravity, 293, 294, 296, 299, 301,
        304, 309, 316, 360
Greenberg, M., 259
Griffiths, P., 7, 252, 257, 259, 311,
        354
gripper, 46, 193, 278–283, see also
        end-effector
Grothendieck, A., 5
group actions, 21–23, 36–37, 252,
        255
    transitive, 36, 37, 137, 243
    trivial, 21, 37, 168
group axioms, 12
group manifold, 12, 15, 31, 198, see
        also under individual
        groups
Grübler mobility formula, 356
Guggenheimer, H.W., 217

Halphen's theorem, 257, 261
Hamilton, W.R., 2
Hamiltonian, 306, 309, 317
Hamiltonian mechanics, 17,
        306–309, 316, 361
harmonic screws, 305, 306
Harris, J., 7, 41, 139, 151, 203,
        250, 252, 257, 259, 311
Hartshorne, R., 7, 355
Hausdorff formula, 73, 75
Hausdorff, F., 73
Hawkins, T., 11, 133
Helgason, S., 352, 358
helical joints, 46, 47, 276, 356
helical motion, 128, 360–363
helicoidal surface, 44, 45, 219, 220
helix, 111, 219, 360

Hermite interpolation, 68, 70
Hermitian
    conjugate, 18
    form, 15, 18
    matrices, 52, 59
Hervé, J.M., 41, 44, 358
Hessian, 304, 305
Hestenes, D., 197
hexapod machine tools, 264
Hilbert, D., 4
Hiller, H., 241
hinge joint, 44
Hodge star, 226
Hodge, W.V.D., 41, 137
Hollerbach, J.M, 297
home position, 46, 47, 71–73, 85,
        87, 235, 241, 297–299,
        319, 327, 352
homeomorphisms, 16, 243, 250,
        252, 258
    local, 62, 123, 355
homogeneous coordinates, 5, 16,
        35, 36, 116, 211, 244, 246,
        255
homogeneous space, 34, 137, 250,
        252, 253
homologous, 257
homology class, 259, 261, 263, 266,
        267
homology theory, 257, 259
homomorphisms, 18–21, 24, 31–35,
        42, 78, 141, 202, 203, 207
Hook's law, 277
Hunt, K.H., viii, 3, 25, 102, 103,
        133, 161, 164, 194, 253,
        356, 357, 360
Husemöller, D., 138
Husty, M.L., 268
hydraulic rams, 264, 343, 345
hyperboloid, 124–126, 218, 313
hyperplane, 5, 6, 130, 204, 247, 260
hypersurface, 5, 117, 122

ideals, 4, 79
idempotents, 63–71, 75–77
identity, 12, 245, 247, 248, 252, 253
ignorable coordinate, 312
immersion, 9
implicit function theorem, 9
impulse, 289
incidence relations, 224–225, 227
inclusion, 19
indecomposable representations,
        143, 144, 147, 153
index of a metric, 33
inertia, 272, 287
inertia matrix, 288, 291–293,
        296–298, 300–302, 305
infinitesimal mobility, 355, 357
inflection circle, 103–104, 108, 110
injective maps, 19, 20, 24, 32, 62,
        140, 261
instantaneous screw, see screw
integral curves, 61, 349
intersection theory, 256–269
intersections
    in a Clifford algebra, 202, 203
    of cones, 284
    of feasible sets, 282
    of flags, 242
    of hyperplanes, 5
    of hypersurfaces, 6
    of planes, 117, 242, 248
    of screw systems, 193, 280
    of subgroups, 42, 44
    of varieties, 7, 91, 101, 104,
        107, 109, 125, 126, 130,
        165, 169, 175, 182, 183,
        244, 255, 256
invariant subspace, 143, 144, 153,
        154, 168, 176
invariants, 4, 15, 139, 147
    of a robot, 180
    of motion, 308
    of screw systems, 183–193
    surfaces, 44, 45
    under congruence, 39

under conjugation, 42, 79
under isometries, 120, 122,
    155, 157, 180, 183, 272
under rotations, 152
inverse function theorem, 8, 355
inverse Jacobian, 133, 353
inverse kinematics, 48, 264, 360
    for 3-R robots, 89–97, 263
    for 3-R wrists, 85–89, 263
    for 6-R robots, 92, 221, 263
inverses, 12, 14, 22, 202
inversion geometry, 2
irreducible representations, 143,
    147, 148, 151, 153
Iserles, A., 73, 75, 309
isometry, 23, 317
isomorphisms, 5, 20, 32, 33, 43, 59,
    78, 123, 143, 154, 155,
    159, 199, 201, 204, 207,
    210, 212, 272, 287
isotropic subspace, 250
isotropy group, 36, 123, 137, 138,
    167–183, 194, 207, 243,
    251–253

Jacobi field, 361
Jacobi identity, 58, 77, 167, 308
Jacobian, 8, 9, 54, 78, 88, 94, 141,
    240, 308, 316
    of a robot, 71–73, 78, 133, 271,
        277, 306, 352–354, 358
Jänich, K., 51
jets, 52
join, 224, 226–228
joint screws, 71, 72, 163, 180, 276,
    292, 293, 298, 302, 310,
    352, 356, 358
joint space, 48, 72, 316, 352, 354,
    357, 360, 362
joints, 2, 44, 72, 126, 292, 298, 362
    angles, 46, 89, 92, 235, 310,
        360
    rates, 72, 166, 295, 360
    stiffness, 277

variables, 46, 72, 302, 303,
    306, 352, 362
Jordan normal form, 38
    real, 38

Kanatani, K., vii
Karger, A., 240
Kawasaki, 3
Kempe, A.B., 2
kernel, 31, 150, 203, 207
Killing field, 279
Killing form, 80, 120, 121, 144,
    157, 213, 230, 352
    for se(3), 80, 161, 175
    for so(3), 80
Killing, W., 11, 51
kinematics, 3, 46–49, 71, 83,
    85–111, 355, 356
    planar, 98–111
kinetic energy, 40, 271, 287, 288,
    301, 305, 317
Klein quadric, 22, 113, 117–120,
    122–126, 129, 130, 136,
    138, 165, 169, 173, 180,
    183, 242, 247, 249, 254,
    257, 283, 290, 311
Klein, F., 164
Knuth, D.E., 334
Kobayashi, S., 9
Koditschek, D., 316
Koenigs, G., 2
Kotelnikov, A.P., 159
Kummer surface, 311
Kutzbach mobility formula, 356

Lagrange interpolation, 65
Lagrange multipliers, 330
Lagrangian, 300, 303, 317
Lagrangian Grassmannians, see
    quadric Grassmannians
Lagrangian mechanics, 300–307,
    317
lead screw, 343
Lee, H.Y., 264
left multiplication, 21

left-invariant vector fields, 61, 349, 352, 354
Legendre transform, 306
Leibnitz rule, 58
Liang, C.G., 264
Lie algebra, 51–83, 101, 271, 308, 349, *see also under individual groups*
   commutative, 168
   dual representation, 143
   representation theory, 141–158
Lie bracket, *see* commutators
Lie derivative, 279
Lie, S., vi, 11, 51, 133
line bundles, 113, 253
line complex, 17, 113, 130–133, 240, 279
   singular, 130, 131, 133
linear algebra, 5, 37, 167, 284
linear equations, 5, 7, 27, 101, 104, 115, 118, 125, 165, 173, 180, 183, 246, 248, 256
linear functionals, 142, 154, 271
linear momentum, 287
linear programming, 285
linear projection, 7, 260
linear span, 163, 166, 193, 281, 356
linearised equations of motion, 304, 305
lined planes, 254
lines at infinity, 117, 122, 123, 165, 242
linkages, *see* mechanisms
links, 2, 71, 126, 163, 166, 276, 277, 293, 295, 298, 299, 309, 356, 357, 360, 362
Liouville, J., 11
Lipkin, H., 363
local coordinates, 7
local dimension, 355
local mobility, 357
logarithms, 62, 79
   for $SE(3)$, 68, 69
   for $SO(3)$, 65

   for $SU(2)$, 63
loop equation, 356
Lorentz group, 34
Lounesto, P., vii, 198, 206, 229

main involution, 202, 204, 207
manifold, 7, 21, 48
Manseur, R., 264
Marsden. J.E., 271
mass, 288, 291
matrix groups, 19, 20, 31, 33
   actions, 22
matrix multiplication, 14, 21, 142, 272
matrix normal forms, 37–41
matrix representation, 140
Maunder, C.R.F., 258
Maurier–Cartan form, 354
maximal torus, 249
McAree, P.R., vii, 161, 321, 360, 364
McCarthy, J.M., 207, 245
McClamroch, N.H., 371
McKerrow, P.J., 276
mechanical engineers, 108
mechanism balancing, 339
mechanisms, 2, 3, 46, 93, 113, 193, 262, 263
   4-bar, 2, 108
   banal, 358
   Bennett, 358
   Bricard, 358
   extraordinary, 358
   overconstrained, 355–358
   paradoxical, 358
   planar, 207, 358
   single loop, 356, 358
   spatial, 3
   spherical, 358
   straight line, 2
meet, 224–226
Mercator series, 62
metric, 318
   curvature, 361

Euclidean, 23, 25, 244, 318
flat, 318
hyperbolic, 250
index, 33
invariant, 350, 352, 354, 361
non-degenerate, 33, 250, 351
semi-index, 39
spaces, 15, 33, 316
Miller, W., 83, 147, 153
Milnor, J., 351, 352, 360, 361
minors, 136, 137
Mishra, B., 280
mobility, 355–358
Möbius band, 20
moduli space, 164, 175
moment
    of a force, 273, 274
    of a line, 115, 119
momentum, 271, 287, 292
Morgan, A.P., 260
motors, 163
multi-homogeneous polynomials,
    259
Murray, R.M., 73, 77

National Science Foundation, 1
Nering, E.D., 284
nets of quadrics, 41
Newton's equations of motion, 293
Newton's second law, 292
Newton, I., 4, 7
Nguyen, V.D., 285
nilpotents, 66–71, 76
Nomizu, K., 9
non-commutative group, 14
normal modes, 305
normal subgroups, 34–36, 41, 42,
    79
Norsett, S.P., 73, 75, 309
nuclear industry, 3
nut and bolt, 44

O'Neill, B., 9, 72, 129, 318, 328,
    370
offset distance, 126

one-parameter subgroups, 62, 71,
    73, 111, 245, 349, 351
Open University, 238
orbit, 36, 37, 44, 45, 122, 164,
    167–183
oriented flags, 242
orthogonal complement, 144
orthogonal Grassmannians, see
    quadric Grassmannians
orthogonal group $O(2)$, 42, 124,
    170, 172, 173, 178, 251
orthogonal group $O(3)$, 118, 178,
    182, 247, 250, 251
orthogonal group $O(4)$, 246, 249,
    250
orthogonal group $O(4,4)$, 251
orthogonal group $O(6)$, 289
orthogonal groups $O(n)$, 15–16, 19,
    24, 31, 33, 38, 52, 137,
    198, 203, 204, 206, 249
    adjoint rep., 206
    dimension, 52, 138
    group manifold, 16
    Lie algebra, 206
    representations, 143
orthogonal groups $O(p,q)$, 34
orthogonal matrices, 23, 182, 246,
    248
Ovseevitch, A.I., 363
Owren, B., 76

parabaloid, 180
parallel axis theorem, 288
Park, F.C., 297
Parkin, I.A., 253
partial flags, 252–254
partitioned matrix, 19, 24, 55, 99,
    120, 158, 247, 248, 272,
    288
Pauli spin matrices, 59
payload, 296, 309, 363
Peaucellier, C.N., 2
Pedoe, D., 41, 137
pencils of conics, 41, 91, 92, 95,
    175, 190, 233

pencils of quadrics, 40, 122, 164
perpendicularity, 228–230
perspective, 5
Pfaffian, 136, 137
phase space, 307, 308
Phillips, A.V., 77
Phillips, J.R., 3, 164
Pieper, D.L., 92, 221, 231
Pin(n), 202, 204, 206–208
pitch
    in a representation, 157, 158
    infinite, 122, 123, 165,
        168–170, 176, 178, 274
    of a screw, 26, 27, 43, 45, 122,
        127, 161, 165, 169, 170,
        173, 176–178, 216, 219,
        290, 291, 293, 352, 360
    of a screw system, 166, 184,
        310
    of a wrench, 274, 290
    of an invariant metric, 352
    zero, 67, 113, 122, 169, 170,
        175, 176, 192, 214, 245,
        311
pitch quadrics, 122, 164, 169, 172,
        175, 176, 178, 182, 189
plane, 44, 45, 124, 131, 357
plane star of lines, 119, 131
plethyism, 147
    $SO(3)$, 151–153
Plücker coordinates, 115–117, 119,
        122, 124, 126, 128, 130,
        132, 136, 188
Plücker embedding, 136
Plücker's conoid, 127
Poincaré duality, 261
Poincaré, H., 257
point groups, 42
point of inflection, 103
pointed directed lines, 253
pointed lines, 242
pointed oriented planes, 253
Poisson brackets, 308
polar plane, 6, 165, 260

Porteous, I.R., 18, 101, 106, 107,
        138, 203, 207, 250
poses, see postures
positive definite matrices, 33, 39,
        305
positive grasps, 280–282, 285
positive linear combination, 284
Postnikov, M., 83
postures, 78
    3-R robots, 92, 94
    3-R wrist, 89
    6-R robot, 92, 262–264
potential energy, 40, 301, 304, 317
power, 272, 276, 277
principal axes of curvature, 129
principal axes of inertia, 289–291,
        311
principal screws of inertia, 290–292
principle of transference, 158–161,
        213–215, 217
prismatic joints, 44, 46, 47, 133,
        240, 276
products of groups, 21–23
projection, 252
    linear, 7, 260
    stereographic, 242, 252, 253
projective groups, 25, 37, 41
projective space, 5
    complex, 5, 257, 259
    complex 1-dimensional, 35, 41
    complex 2-dimensional, 107
    complex 3-dimensional, 263
    real, 36, 135, 164, 250
    real 3-dimensional, 16, 20,
        131, 183, 263
    real 5-dimensional, 116, 117,
        120, 122, 125, 126, 128,
        130, 164, 168, 169, 175,
        282
    real 7-dimensional, 211, 244,
        246
    real 9-dimensional, 137
projective variety, 5
PUMA, 3, 46, 93, 134, 238–240

quadratic equations, 104, 122, 126, 137, 211, 267
quadratic form, 39, 40, 120, 122, 146
quadratic line complex, 311
quadric Grassmannians, 250–252
quadrics, 6, 39, 250, 289, *see also* Klein quadric *and* Study quadric
    2-dimensional, 128, 180, 183, 255, 256, 267, 281
    3-dimensional, 15, 282
    4-dimensional, 117, 122
    6-dimensional, 212, 241
    even-dimensional, 249
    linear subspaces, 118, 124, 245–252
    pencils, 40, 122, 164
    singular, 6, 122, 165, 175, 260
quantum physics, 59, 148
quaternions, 2, 13, 17, 200, 204, 211, 213
    conjugate, 13, 201
    modulus, 14
    multiplication, 79, 213
    unit modulus, 19, 20, 59, 78, 200, 204, 211, 212, 244
quotients, 34–35, 42, 77, 79, 137, 150, 211, 263

radius of gyration, 336
Raibert, M.H., 363, 367
rank, 6, 27, 39, 106, 182, 192
rational map, 7
rational normal form, 37
reaction wrenches, 276, 294
reciprocal product, 120, 121, 130, 133, 157, 161, 175, 213, 228, 230, 272, 287, 352
reciprocal screw system, 165, 183, 184, 195, 291, 310, 311
recursion, 297, 299
redundant robots, 46, 356
reflections, 16, 24, 124, 169, 203
reflexive relation, 34

regional manipulator, 89
regular polyhedra, 42
regulus, 124–126, 180, 183, 218, 313
Reid, M., 7
representation theory, 122, 139–161, 274
representations, 21, 23, 24, 37, 52, 54, 56, 66, 198, *see also* *under individual groups*
resolved force matrix, 276
Reuleaux's pairs, 44–45, 278, 280
Reuleaux, F., v, 44
revolute joints, 44, 46–48, 85, 89, 113, 133, 255, 256, 262, 276, 293, 310, 311, 318, 360
Riemann 2-sphere, 35
Riemann curvature tensor, 318, 361
Riemann, B., 7
right multiplication, 21
right-handed coordinate frames, 243
rigid body motions, 164
    in space, 23, 47, 51, 71, 128, 272
    in the plane, 42, 98, 101, 108
Rimon, E., 283
ring of polynomials, 4
Rodrigues formula, 64, 85
roll-pitch-yaw wrist, 87
Rooney, J.J., 44, 159
rooted trees, 321
Rota, G-C., 198
rotation axis, 43, 45
rotations, 15, 16, 24, 27, 42, 119, 122, 123, 128, 155, 168–171, 173, 177, 204, 245, 253, 254, 262, 263, 280
Roth, B., 110
Roth, L., 7, 137, 257
Ruf, A., vii

ruled surfaces, 113, 123–130, 198,
        214–220
    cone, 216, 217
    developable, 124, 216, 217
    distribution parameter, 217,
        219
    non-cylindric, 215, 217
    striction curve, 215, 218, 219
    striction point, 216, 217, 219

Saletan contraction, 245, 253, 254
Salmon, G., 190
Samuel, A.E., 161, 194, 360
Sanz-Sena, J.M., 309
scalar product, 15, 24, 80, 88, 100,
        133, 152, 203, 306
scalar triple product, 55, 89, 96,
        279
SCARA, 3, 238
schemes, 5
Schonflies, A., 2
Schubert, H., 257
Schutz, B.F., 9, 279, 349–351, 360
Schwartz, J.T., 280
scleronomic constraints, 327, 365
screw, 51, 53, 54, 113, 120, 163,
        219, 247, 271, 280–282,
        287, 289, 297, 301, 352,
        361
    axis, 45, 53, 126–128, 133,
        161, 215, 216, 290, 291
screw motion, 26–27, 32, 43, 44,
        51, 53, 126, 352
screw systems, 163–195, 356, 357
    1-systems, 164, 195
    2-systems, 184–188
    3-systems, 189–193, 247, 252,
        310, 311
    4-systems, 188–189
    5-systems, 184, 195, 280
    completion group, 167–183,
        247, 357
    identification, 183–193
    intersection, 280
    operations, 193–195

    that guarantee mobility, 357
    union, 281
screw theory, 2
second fundamental form, 370
sectional curvature, 361
Segre symbols, 38, 40, 41, 175
Selig, J.M., 44, 159, 226, 297, 321,
        363, 364
semi-direct product, 22, 24, 41, 207
semi-index of a metric, 39
semi-simple Lie algebra, 79, 144
Semple, J.G., 7, 137, 257
sensors, 1
Serre, J-P., 204
shaking force, 339
shaking moment, 339
shape operator, 370
Sharir, M., 280
Shigly, J.E., 336, 339
Shilov, G.E., 38
shuffle product, 193, 198, 224–228
similarity, 22, 26, 37, 40, 140–142,
        150, 155, 157
simple Lie algebra, 79
simply connected cover, 78, 141
simply connected space, 77
singular matrices, 14, 27
singularity, 6, 89, 94
    cusp, 97, 105
    node, 106, 107
    of a robot, 72, 78, 240, 358
skew lines, 114, 126
Slotine, J.-J.E., 364
small oscillations, 304
Sobczyk, G., 64, 197
sojourner, 4
solution by radicals, 11, 91
soma, 241–245
soup plate trick, 77
special functions, 147, 153
special linear group $SL(2)$
    group manifold, 15
special linear groups $SL(n)$, 14, 18,
        19

special orthogonal group $SO(2)$,
   20, 42, 44, 98, 170, 173,
   253
   group manifold, 16
special orthogonal group $SO(3)$,
   45, 85, 170, 176, 178, 247,
   249, 252, 263
   adjoint rep., 61, 80
   C-B-H formula, 81
   exponential map, 63, 85, 150
   group manifold, 16, 20
   Lie algebra, 141
   representations, 142, 144,
      148–153
   subgroups, 42
special orthogonal group $SO(3,1)$,
   34
special orthogonal group $SO(3,2)$,
   133
special orthogonal group $SO(4)$,
   243, 244, 248, 250, 252
   group manifold, 243, 250
   Lie algebra, 59
special orthogonal groups $SO(n)$,
   16–17, 19, 31, 52, 138,
   204, 207, 245
   dimension, 52
special orthogonal groups $SO(p,q)$,
   34
special relativity, 34
special unitary group $SU(2)$, 20
   adjoint rep., 141
   exponential map, 63
   group manifold, 18, 20
   Lie algebra, 141
   representations, 142
special unitary groups $SU(n)$,
   18–19, 31, 35, 52
   dimension, 52
sphere, 44, 45
   $(n-1)$-dimensional, 203, 245,
      258
   2-dimensional, 35

3-dimensional, 14, 18, 20, 204,
   243, 252
7-dimensional, 244
sphere geometry, 133
spherical harmonics, 153
spherical indicatrix, 263
Spin(2), 207
Spin(3), 204, 210, 211, 244
Spin(4), 243, 244
Spin(n), 204, 206, 207
Spong, M.W., 316, 318, 363
square matrices, 14
stability, 305, 360, 361
static equillibrium, 273, 276, 277
statically indeterminate sysytem,
   284
statics, 271–285
Sternberg, S., 133, 352
Stewart platform, 193, 264–265,
   267–269, 343–347
stiffness matrix, 277
Stokes, A., 312
strain, 274, 277
stress, 277
striction curve, 215, 218, 219
structure constants, 58, 83, 349
structure equations, 354
Study quadric, 212, 241–269
Study, E., v, 3, 11, 197, 241
subalgebras, 78, 79, 166, 176, 180
subgroups, 31–49
substitutions, 11
surface of revolution, 44, 45
surface of translation, 44, 45
surjective maps, 19, 20, 33, 62
Sylvester's law of inertia, 33, 39,
   125, 172, 178, 199
Sylvester, J.J., 147
symmetric matrices, 6, 23, 33, 38,
   39, 91, 122, 146, 165, 172,
   178, 189, 199, 281, 287,
   301
   pairs, 39, 305
   pencils, 40

symmetric powers, 146, 152, 155
symmetric product, 145
symmetric relation, 34
symmetries, 147
    discrete, 168–170, 173
    of a cylinder, 123
    of a line complex, 132
    of a metric space, 251
    of a sphere, 245
    of a vector space, 140
    of anti-symmetric form, 17
    of bilinear forms, 18
    of Euclidean space, 248
    of line complexes, 113
    of metric spaces, 15
    of projective space, 36
    operations, 21
symplectic geometry, 113
symplectic group $Sp(2, \mathbb{R})$
    group manifold, 20
symplectic group $Sp(4, \mathbb{R})$, 132
symplectic groups $Sp(2n, \mathbb{R})$, 17,
    19, 34
symplectic integration, 309
symplectic matrix, 308

T^3, 234–238
Talman, J.D., 147, 153
tangent space, 6, 8, 51, 52, 54, 106,
    107, 130, 243, 260, 283,
    355
tangent vectors, 8, 51–54, 61, 78,
    218, 243, 307, 351
telechirs, 3
Telequipment, 238
tensor powers, 145
tensor product, 144, 147, 151, 152
Thurston, W.P, 3
tool frame, 47
torque, 274, 276, 278, 280, 284,
    289, 293, 296, 300, 304,
    360, 363
torus, 48, 318
totally geodesic submanifold, 357

trace, 62, 64, 80, 157
traceless matrices, 52, 58, 59, 62
transitive group action, 36, 37, 137
transitive relation, 34
translations, 24, 27, 66, 98, 119,
    122, 123, 128, 155,
    168–171, 177, 208, 246,
    255, 263, 280, 288
triality, 250
trivial group, 31, 32
trivial representation, 140, 147,
    151, 155
twist, 51, see screw
twist angle, 126, 127, 219

Uicker, J.J., 336, 339
umbilic points, 129
undulation, 104
Unimation, 3, 238
union of screw systems, 193
unit ball, 258
unitary group $U(1)$, 20, 35
unitary groups $U(n)$, 18–19, 31,
    35, 52
    dimension, 52

variety, 5, 257, 355
    affine, 5, 123
    non-singular, 9, 259
    projective, 356
    quasi-projective, 5
vector field, 8, 61, 352, 353, 358,
    361
vector product, 53, 59, 68, 88, 152,
    153, 300
vector space, 12, 21–23, 36, 37, 51,
    58, 77, 140–143, 145, 146,
    148, 149, 154, 167, 198,
    200, 202, 205, 271, 284
vector triple product, 68, 114, 185,
    217
velocity
    of a point, 53, 71, 100, 101,
    103, 104, 108, 279

rigid body, 52, 163, 166, 193,
    272, 280, 287, 299, 305,
    360
velocity screw, 72, 74, 76–77, 166,
    271, 273, 288, 292, 293,
    295, 302
vertices, 105
Vidyasagar, M., 363
Vilenkin, N.J., 147, 153

Waldron, J.K., 44
walking robot, 193, 340
Wampler, C.W., vii, 260, 268
Watt, J., 2
Weeks, J.R., 3

Weyl, H., 15, 17
White, N.L., 194, 226
Whittaker, E.T., 303
Woodhouse, N.M.J., 288, 293, 306
work, 273, 276, 289, 294
wrenches, 133, 271–274, 276, 278,
    284, 285, 292, 293,
    297–299
wrist, 85, 92, 239, 274–276

Yaglom, I.M., 243
Yoshikawa, T., 371
Youcef-Toumi, K., 309

zero divisors, 159, 202